T0178610

Universitext

Universitext

Universitext is a series of textbooks that presents material from a wide variety of mathematical disciplines at master's level and beyond. The books, often well class-tested by their author, may have an informal, personal even experimental approach to their subject matter. Some of the most successful and established books in the series have evolved through several editions, always following the evolution of teaching curricula, to very polished texts.

Thus as research topics trickle down into graduate-level teaching, first textbooks written for new, cutting-edge courses may make their way into *Universitext*.

More information about this series at http://www.springer.com/series/223

Jia-An Yan

Introduction to Stochastic Finance

 Science Press
Beijing

 Springer

Jia-An Yan
Academy of Mathematics and System Science
Chineses Academy of Sciences
Beijing, China

ISSN 0172-5939   ISSN 2191-6675   (electronic)
Universitext
ISBN 978-981-13-1656-2   ISBN 978-981-13-1657-9   (eBook)
https://doi.org/10.1007/978-981-13-1657-9

Library of Congress Control Number: 2018952344

Mathematics Subject Classification: 91Gxx, 60Gxx, 60Hxx

This Springer imprint is published by the registered company Springer Nature Singapore Pte Ltd.
The registered company address is: 152 Beach Road, #21-01/04 Gateway East, Singapore 189721,
Singapore

Preface

The history of financial mathematics can be traced back to the French mathematician Louis Bachelier's doctoral dissertation "Théorie de la speculation" in 1900 Bachelier (1900). Bachelier's work, however, was not known to most economists until Paul A. Samuelson mentioned it in an article of 1965. In 1969 and 1971, Robert C. Merton studied the optimal portfolio problem in continuous-time using the stochastic dynamic programming method. In 1973, Fischer Black and Myron S. Scholes used stochastic analysis, in particular, Kiyosi Itô's formula, to derive the famous Black-Scholes formula for option pricing. Almost at the same time, Merton (1973a) improved the Black-Scholes model and developed an idea of using options to evaluate a company's debt in the so-called contingent claims analysis. Harrison and Kreps (1979) proposed a martingale method to characterize a no-arbitrage market and the use of equivalent martingale measure on options pricing and hedging, which had profound influence on the subsequent development of financial mathematics. Since 1970s, in order to study the pricing of interest rate derivatives, many scholars have proposed several interest rate term structure models, including Vasicek, CIR, HJM, and BGM models, which could reflect the market trend of the future spot rate.

For more than half a century, many scholars worked research on theory and applications of financial mathematics (also known as mathematical finance) and published many books in the area. Financial mathematics not only has a direct impact on the innovation of financial instruments and on the efficient functioning of financial markets but also is widely used in investment decisions, valuation of research and development projects, and in risk management.

This book is intended to give a systematic introduction to the basic theory of financial mathematics, with an emphasis on applications of martingale methods in pricing and hedging of contingent claims, interest rate term structure models, and expected utility maximization problems. The book consists of 14 chapters. Chapter 1 introduces the basic theory of probability and discrete time martingales, which is specially designed for readers who do not have much knowledge of probability theory. In Chap. 2, we introduce the theory of discrete time portfolio selection (i.e., Harry M. Markowitz's mean-variance analysis), the capital asset pricing model

(CAPM), the arbitrage pricing theory (APT), and the multistage mean-variance analysis. The basic idea of expected utility theory and the consumption-based asset pricing model are also sketched in this chapter. In Chap. 3, we introduce discrete time financial markets and martingale characterization of arbitrage-free markets and present the martingale method for the expected utility maximization and the risk-neutral pricing principle for European contingent claims. Chapter 4 systematically presents Itô's theory of stochastic analysis (including Itô's integral and Itô's formula, Girsanov's theorem, and the martingale representation theorem) which is the theoretical basis of the martingale method in financial mathematics. This chapter can be used separately as a concise text for postgraduates in probability studying Itô's theory of stochastic analysis. In Chap. 5, for the Black-Scholes model, we introduce the martingale method for pricing and hedging European contingent claims and derive the Black-Scholes formula for pricing European options. The pricing of American options is also briefly discussed. In addition, some examples are given to illustrate applications of the martingale method, and several modifications of the Black-Scholes model are presented. In Chap. 6, we introduce several commonly used exotic options: barrier options, Asian options, lookback options, and reset options. The pricing and hedging of these exotic options are investigated with the martingale method and partial differential equations. Chapter 7 presents Itô process and diffusion process models. The martingale method of contingent claim pricing is presented in detail, including an introduction of time and scale transformation to give some explicit formulas for option pricing. Chapter 8 introduces the bond market and interest rate term structure models, including a variety of single-factor short-term interest rate model, HJM model, and BGM Model, and studies the pricing of interest rate derivatives. Chapter 9 introduces optimal investment portfolios and investment-consumption strategies for diffusion process models. Within L^2-allowable trading strategies, the risk-mean portfolio selection problem, expected utility maximization problem, and the selection of portfolio strategy with consumption are investigated. Chapter 10 introduces the general theory of static risk measures, which includes consistent risk measures, convex risk measures, comonotonically sub-additive risk measures, comonotonically convex risk measure, and a variety of distribution invariant risk measures, as well as their characterizations and representations. In Chap. 11, after a brief overview of semimartingales and stochastic calculus, we introduce some basic concepts and results on markets of semimartingale model and give numeraire-free characterizations of attainable contingent claims. In Chap. 12, we give a survey on convex duality theory for optimal investment and present a numeraire-free and original probability-based framework for financial markets. The expected utility maximization and valuation problems in a general semimartingale setting are studied in Chap. 13. For a market driven by a Lévy process, the optimal portfolio and the related martingale measure are worked out explicitly for some particular types of utility function. Finally, in Chap. 14, we introduce the "optimal growth portfolios" in markets of semimartingale model and work out their expressions in a geometric Lévy process model and a jump-diffusion-like process model.

I have taught from the manuscript of this book in an introductory course of mathematical finance for my former graduate students. Thanks to them for reporting misprints and errors. I am grateful to Prof. Jianming Xia for contributing to the writing of Sect. 9.2 of Chap. 9; to Dr. Yongsheng Song for his PhD thesis, which is the basic material for Chap. 10 of the book; and to Dr. Jun Yan for typographical and grammatical suggestions.

Beijing, China   Jia-An Yan
May, 2018

Contents

Chapter 1
Foundation of Probability Theory and Discrete-Time Martingales

Gambling with dice was very popular in medieval Europe. The study of problems involving probability associated with gambling led to the development of probability theory. However, it was not until the early twentieth century that probability theory was considered as a branch of mathematics. The mathematical foundation of modern probability theory was laid by Andrei N. Kolmogorov in 1933. He adopted Lebesgue's framework of measure theory and created an axiomatic system for probability theory. This chapter introduces some basic concepts and results of modern probability theory, highlights the results related to the conditional mathematical expectation, and then introduces discrete-time martingale theory, including the martingale transform and the Snell envelope. We assume that the reader has basic knowledge of measure theory.

1.1 Basic Concepts of Probability Theory

1.1.1 Events and Probability

Consider a trial or random experiment. Let Ω be the collection of all possible outcomes of the trial, called *sample space*. If the number of outcomes of the trial is finite or countable, we can use combinatorial mathematics to study issues related to probability. However, if the number of outcomes of the trial is uncountably infinite, we may not be able to consider a single outcome of the trial because the probability of its appearance may be zero. Hence, we need to study issues related to probability under the framework of measure theory.

In measure theory, we use Ω to represent a space, i.e., a predefined biggest set for a study. Its elements are denoted by ω; $\omega \in A$ or $\omega \notin A$ stands for the fact that ω belongs to A or does not belong to A, respectively. The set which does not contain any element is called an *empty set* and is denoted by $\emptyset$. We use $A \supset B$ or $B \subset A$ to

© Springer Nature Singapore Pte Ltd. and Science Press 2018
J.-A. Yan, *Introduction to Stochastic Finance*, Universitext,
https://doi.org/10.1007/978-981-13-1657-9_1

express that B is a subset of A and use

$$A \cap B, \ A \cup B, \ A \setminus B, \ A \triangle B$$

to express the intersection, union, difference, and symmetric difference of A and B, respectively. That is:

$$A \cap B = \{\omega : \omega \in A \text{ and } \omega \in B\}, \ A \cup B = \{\omega : \omega \in A \text{ or } \omega \in B\},$$

$$A \setminus B = \{\omega : \omega \in A \text{ and } \omega \notin B\}, \ A \triangle B = (A \setminus B) \cup (B \setminus A).$$

Sometimes we also use AB to represent $A \cap B$. We use A^c to express $\Omega \setminus A$ and call A^c the *complement* of A (in Ω). Consequently, it holds that $A \setminus B = A \cap B^c$. If $A \cap B = \emptyset$, we say that A and B are *disjoint*. Clearly, $A \cap A^c = \emptyset$, $A \cup A^c = \Omega$.

Let $\{A_\lambda, \ \lambda \in \Lambda\}$ be a family of subsets of Ω. We use $\bigcup_{\lambda \in \Lambda} A_\lambda$ and $\bigcap_{\lambda \in \Lambda} A_\lambda$ to express their union and intersection, respectively. Assume that $\{A_n, \ n \geqslant 1\}$ (denoted also as (A_n)) is a finite or countable sequence of subsets of Ω. If (A_n) are mutually disjoint (i.e., $n \neq m \Rightarrow A_n \cap A_m = \emptyset$), we usually use $\sum_n A_n$ to express $\bigcup_n A_n$. If $\sum_n A_n = \Omega$, we say that $\{A_n, \ n \geqslant 1\}$ is a *partition* of Ω.

For any sequence (A_n) of subsets, set

$$\limsup_{n \to \infty} A_n = \bigcap_{n=1}^{\infty} \bigcup_{k=n}^{\infty} A_k, \quad \liminf_{n \to \infty} A_n = \bigcup_{n=1}^{\infty} \bigcap_{k=n}^{\infty} A_k,$$

and call them the *upper limit* and *lower limit* of (A_n). Obviously,

$$\liminf_{n \to \infty} A_n \subset \limsup_{n \to \infty} A_n.$$

If $\liminf_{n \to \infty} A_n = \limsup_{n \to \infty} A_n$, we say that the limit of (A_n) exists, and we denote its same upper and lower limit by $\lim_{n \to \infty} A_n$ and call it the *limit* of (A_n).

A family of subsets of Ω is called a *set class*. A set class $\mathcal{C}$ is called an *algebra*, if $\Omega \in \mathcal{C}$, $\emptyset \in \mathcal{C}$, and $\mathcal{C}$ is closed under finite intersection and complement (thereby, $\mathcal{C}$ is closed under finite union and under difference). We call $\mathcal{C}$ a σ-*algebra*, if $\Omega \in \mathcal{C}, \emptyset \in \mathcal{C}$, and $\mathcal{C}$ are closed under countable intersection and complement (thereby, $\mathcal{C}$ is closed under the countable union and the difference). The smallest σ-algebra containing a set class $\mathcal{C}$ is called a σ-*algebra generated by* $\mathcal{C}$, denoted by $\sigma(\mathcal{C})$.

Let $\mathcal{F}$ be a σ-algebra on Ω. We call ordered couple $(\Omega, \mathcal{F})$ a *measurable space*, and an element of $\mathcal{F}$ an $\mathcal{F}$-*measurable set*. Let μ be a function defined on $\mathcal{F}$ with values in $\overline{\mathbb{R}}_+ = [0, \infty]$. If $\mu(\emptyset) = 0$ and μ is *countably additive* or σ-*additive*, namely,

$$\forall n \geqslant 1, \ A_n \in \mathcal{F}; \ \forall n \neq m, \ A_n \cap A_m = \emptyset$$

$$\Rightarrow \mu\left(\bigcup_{n=1}^{\infty} A_n\right) = \sum_{n=1}^{\infty} \mu(A_n),$$

then we call μ a *measure* on Ω (or $(\Omega, \mathcal{F})$). If $\mu(\Omega) < \infty$, we call μ a *finite measure*. If there is a countable measurable division $(A_i)_{i \geqslant 1}$ of Ω, such that for any A_i, $\mu(A_i) < \infty$, we call μ a σ-*finite measure*. If $\mu(\Omega) = 1$, we call μ a *probability measure* and call the ordered triple $(\Omega, \mathcal{F}, \mu)$ a *probability space*. Usually, we use $\mathbb{P}$ to denote a probability measure.

Let $(\Omega, \mathcal{F}, \mathbb{P})$ be a probability space. If $A \in \mathcal{F}$, and $\mathbb{P}(A) = 0$, we call A a *null set*. If all subsets of any $\mathbb{P}$-null set belong to $\mathcal{F}$, $\mathcal{F}$ is said to be *complete* w.r.t. $\mathbb{P}$, and we call $(\Omega, \mathcal{F}, \mathbb{P})$ a *complete probability space*.

Let $(\Omega, \mathcal{F}, \mathbb{P})$ be a probability space. Set

$$\mathcal{N} = \{N \subset \Omega : \text{there exists } A \in \mathcal{F}, \ \mathbb{P}(A) = 0, \ \text{such that } N \subset A\},$$

$$\overline{\mathcal{F}} = \{B \cup N : B \in \mathcal{F}, N \in \mathcal{N}\},$$

$$\overline{\mathbb{P}}(B \cup N) = \mathbb{P}(B), \quad B \in \mathcal{F}, N \in \mathcal{N}.$$

Then $(\Omega, \overline{\mathcal{F}}, \overline{\mathbb{P}})$ is a complete probability space, which is the smallest complete probability space containing $(\Omega, \mathcal{F}, \mathbb{P})$. We call $(\Omega, \overline{\mathcal{F}}, \overline{\mathbb{P}})$ the *completion* of $(\Omega, \mathcal{F}, \mathbb{P})$ and call $\overline{\mathcal{F}}$ the *completion* of $\mathcal{F}$ w.r.t. $\mathbb{P}$.

1.1.2   Independence, 0-1 Law, and Borel-Cantelli Lemma

Let $(\Omega, \mathcal{F}, \mathbb{P})$ be a probability space. If $A \in \mathcal{F}$, we call A an *event*. The sample space Ω itself is called the *inevitable event*. An event A happens if the outcome ω of the trial is an element of A. Let A and B be two events. If $\mathbb{P}(AB) = \mathbb{P}(A)\mathbb{P}(B)$, events A and B are called *independent*. An event class $(A_t, t \in T)$ is called an *independent class*, if for any finite subset S of T, we have

$$\mathbb{P}\left(\bigcap_{s \in S} A_s\right) = \prod_{s \in S} \mathbb{P}(A_s).$$

In this case we say that the events in this event class are *mutually independent*. This mutual independence is much stronger than pairwise independence.

Two event classes $\mathcal{A}$ and $\mathcal{B}$ are called independent if any event $A \in \mathcal{A}$ and any event $B \in \mathcal{B}$ are independent. More generally, let $(\mathcal{C}_t, t \in T)$ be a family of event classes. If for any event A_t from each event class $\mathcal{C}_t$, event class $(A_t, t \in T)$ is an independent event class, then we call this family an *independent family* and say that the event classes in this family are mutually independent.

It is easy to prove the following result.

Extension theorem of independent class Let $(\mathcal{C}_t, t \in T)$ be an independent class. If each $\mathcal{C}_t$ is a π class (i.e., closed under intersection), then $(\sigma(\mathcal{C}_t), t \in T)$ is also an independent class, where $\sigma(\mathcal{C}_t)$ is the σ-algebra generated by event class $\mathcal{C}_t$.

The following *Kolmogorov 0-1 law* is an important result about event independency.

Kolmogorov 0-1 law Let $(\mathcal{F}_n, n \geqslant 1)$ be a sequence of mutually independent σ-algebras. Set

$$\mathcal{G} = \bigcap_{k=1}^{\infty} \sigma \left(\bigcup_{n=k}^{\infty} \mathcal{F}_n \right).$$

Then for any $A \in \mathcal{G}$, we have $\mathbb{P}(A) = 0$ or 1. We call $\mathcal{G}$ the *tail σ-algebra* of sequence $(\mathcal{F}_n, n \geqslant 1)$.

In fact, since $\sigma \left(\bigcup_{n=k+1}^{\infty} \mathcal{F}_n \right)$ and $\sigma(\mathcal{F}_n, 1 \leqslant n \leqslant k)$ are independent, $\mathcal{G}$ and $\sigma(\mathcal{F}_n, 1 \leqslant n \leqslant k)$ are independent. Set $\mathcal{A} = \bigcup_{k=1}^{\infty} \sigma(\mathcal{F}_n, 1 \leqslant n \leqslant k)$. Then $\mathcal{G}$ and $\mathcal{A}$ are also independent. But $\mathcal{A}$ itself is an algebra (thus a π class), thus $\mathcal{G}$ and $\sigma(\mathcal{A})$ are independent. Then, since $\mathcal{G} \subset \sigma(\mathcal{A})$, this implies that $\mathcal{G}$ is independent with itself. Therefore, for any $A \in \mathcal{G}$, it holds that $\mathbb{P}(A) = \mathbb{P}(AA) = \mathbb{P}(A)^2$, namely, $\mathbb{P}(A) = 0$ or 1.

Let $(A_n)_{n \geqslant 1}$ be a sequence of events. Event $A = \bigcap_{k=1}^{\infty} \bigcup_{n=k}^{\infty} A_n$ (the upper limit of (A_n)) occurs, if and only if an infinite number of A_n occur. We use "A_n i.o." to repress event A. About the probability of this event, we have the following result.

Borel-Cantelli lemma If $\sum_{n=1}^{\infty} \mathbb{P}(A_n) < \infty$, then $\mathbb{P}(A_n$ i.o. $) = 0$. If (A_n) are mutually independent, and $\sum_{n=1}^{\infty} \mathbb{P}(A_n) = \infty$, then $\mathbb{P}(A_n$ i. o.$) = 1$.

In fact, if $\sum_{n=1}^{\infty} \mathbb{P}(A_n) < \infty$, then

$$\mathbb{P}(A_n \text{ i.o.}) = \lim_{k \to \infty} \mathbb{P} \left(\bigcup_{n=k}^{\infty} A_n \right) \leqslant \lim_{k \to \infty} \sum_{n=k}^{\infty} \mathbb{P}(A_n) = 0.$$

If (A_n) are mutually independent, and $\sum_{n=1}^{\infty} \mathbb{P}(A_n) = \infty$, then for $k \geqslant 1$, we have (noting that $1 - x \leqslant e^{-x}$)

$$\mathbb{P} \left(\bigcup_{n=k}^{\infty} A_n \right) = 1 - \mathbb{P} \left(\bigcap_{n=k}^{\infty} A_n^c \right) = 1 - \prod_{n=k}^{\infty} \mathbb{P}(A_n^c)$$

$$\geqslant 1 - \prod_{n=k}^{\infty} e^{-\mathbb{P}(A_n)} = 1 - e^{-\sum_{n=k}^{\infty} \mathbb{P}(A_n)} = 1.$$

Therefore,

$$\mathbb{P}(A_n \text{ i.o.}) = \lim_{k \to \infty} \mathbb{P}\left(\bigcup_{n=k}^{\infty} A_n\right) = 1.$$

1.1.3   Integrals, (Mathematical) Expectations of Random Variables

Let $(\Omega, \mathcal{F})$ and $(E, \mathcal{E})$ be two measurable spaces. Let f be a mapping from Ω to E. If for any $A \in \mathcal{E}$ we have $f^{-1}(A) \in \mathcal{F}$, we call f a *measurable mapping*. If $E = [-\infty, \infty]$ and $\mathcal{E} = \mathcal{B}(\overline{\mathbb{R}})$, where $\mathcal{B}(\overline{\mathbb{R}})$ is the Borel σ-algebra of $\overline{\mathbb{R}}$, such a measurable mapping is called a *measurable function*. A measurable function of the following form is called a *simple measurable function*:

$$f = \sum_{i=1}^{n} a_i I_{A_i},$$

where $a_i \in \mathbb{R}$, $A_i \in \mathcal{F}$, and I_{A_i} are the *indicator function* of A_i. Let $\mathcal{S}^+$ be the collection of all nonnegative simple measurable functions on Ω and $\mathcal{L}$ (respectively, $\overline{\mathcal{L}}$) the collection of all measurable real (respectively, numerical) functions on Ω.

First, we define the integral of a nonnegative simple measurable function w.r.t. μ. Let $f = \sum_{i=1}^{n} a_i I_{A_i} \in \mathcal{S}^+$ be a simple measurable function, and put

$$\int_{\Omega} f d\mu = \sum_{i=1}^{n} a_i \mu(A_i).$$

It is easy to see that $\int_{\Omega} f d\mu$ does not depend on the specific expression of f. We call $\int_{\Omega} f d\mu$ the integral of f w.r.t. μ.

Usually, we simply use $\int f d\mu$ or $\mu(f)$ to denote $\int_{\Omega} f d\mu$.

Let f be a nonnegative measurable function. Take any sequence $f_n \in \mathcal{S}^+$ with $f_n \uparrow f$, and set

$$\mu(f) = \lim_{n \to \infty} \mu(f_n).$$

Then the limit on the above right-side exists and does not depend on the selection of the sequence f_n. We call $\mu(f)$ the integral of f w.r.t. μ.

Assume now that f is a measurable function. Let $f^+ = f \vee 0$, $f^- = (-f) \vee 0$. If $\mu(f^+) < \infty$ or $\mu(f^-) < \infty$, we say that the integral of f w.r.t. μ exists. Put

$$\mu(f) = \mu(f^+) - \mu(f^-).$$

We call $\mu(f)$ the integral of f w.r.t. μ. If $\mu(f^+) < \infty$ and $\mu(f^-) < \infty$ (i.e., $\mu(|f|) < \infty$), we say that f is integrable w.r.t. μ (simply called μ integrable). In measure theory, we do not distinguish almost everywhere equal (abbreviated as a.e. equal) measurable functions.

Let $(\Omega, \mathcal{F}, \mathbb{P})$ be a probability space and X be an $\mathbb{R}^d$-valued function defined on Ω. We use $\sigma(X)$ to express the σ-algebra $\{X^{-1}(A), A \in \mathcal{B}(\mathbb{R}^d)\}$, where $\mathcal{B}(\mathbb{R}^d)$ is the Borel σ-algebra on $\mathbb{R}^d$. We call $\sigma(X)$ the σ-algebra generated by X, and X a *random variable*, which is $\mathcal{F}$-measurable (i.e., $\sigma(X) \subset \mathcal{F}$). Since $\mathcal{B}(\mathbb{R}^d)$ is generated by $\{(-\infty, x] : x \in \mathbb{R}^d\}$, X is a random variable, iff for all $x \in \mathbb{R}^d$, $[X \leqslant x] := \{\omega : X(\omega) \leqslant x\} \in \mathcal{F}$. In probability theory, we do not distinguish almost surely equal (abbreviated as a.s. equal) random variables.

Let X be an $\mathbb{R}^d$-valued random variable. If there exists a finite or countable subset $\{x_1, x_2, \cdots\}$ of $\mathbb{R}^d$ such that for each i $\mathbb{P}(X = x_i) > 0$, and $\sum_i \mathbb{P}(X = x_i) = 1$, X is called a *discrete random variable* with range $\{x_1, x_2, \cdots\}$. If for each $x \in \mathbb{R}^d$, $\mathbb{P}(X = x) = 0$, then X is called a *continuous random variable*.

Let $X = (X_1, \cdots, X_d)^\tau$ be an $\mathbb{R}^d$-valued random variable. Here and henceforth, superscript " τ " represents the transpose of a vector or matrix. Define

$$F(x) = \mathbb{P}(X \leqslant x) = \mathbb{P}\left(\bigcap_{i=1}^d [X_i \leqslant x_i]\right), \quad x = (x_1, \cdots, x_d)^\tau \in \mathbb{R}^d,$$

the *distribution (function)* of X, or the *joint distribution* of the d scalar random variables $\{X_1, \cdots, X_d\}$. For a given $i \in \{1, \cdots, d\}$,

$$F_i(x) = \mathbb{P}(X_i \leqslant x), \quad x \in \mathbb{R},$$

is called the *marginal distribution* of X_i. The distribution function F of an $\mathbb{R}^d$-valued random variable is a right continuous nondecreasing function. If F is absolutely continuous, its derivative f is called the *density function* of F, namely,

$$\frac{\partial^d F}{\partial x_1 \partial x_2 \cdots \partial x_d} = f(x_1, x_2, \cdots, x_d).$$

The integral of a random variable X w.r.t. probability measure $\mathbb{P}$ is called the *(mathematical) expectation* of X w.r.t. $\mathbb{P}$ and denoted by $\mathbb{E}[X]$. If one of $\mathbb{E}[X^+]$ and $\mathbb{E}[X^-]$ is finite, we set $\mathbb{E}[X] = \mathbb{E}[X^+] - \mathbb{E}[X^-]$ and say that the expectation of X w.r.t. $\mathbb{P}$ exists. If both $\mathbb{E}[X^+]$ and $\mathbb{E}[X^-]$ are finite, then $\mathbb{E}[X]$ is finite, and X is called an *integrable random variable*. For $p \geqslant 1$, if $\mathbb{E}[|X|^p]$ is finite, then X is said to be L^p-*integrable*. We use $L^p(\Omega, \mathcal{F}, \mathbb{P})$ to denote the set of all L^p-integrable random variables.

A family $(X_t, t \in T)$ of random variables is called an independent family, if the family of σ-algebras $(\sigma(X_t), t \in T)$ is an independent family. By the extension theorem of independent class, a sequence $X_1, \cdots, X_n$ of random variables are mutually independent iff for any $x_i \in \mathbb{R}$, $i \geqslant 1$,

$$\mathbb{P}(X_1 \leqslant x_1, \cdots, X_n \leqslant x_n) = \prod_{i=1}^{n} \mathbb{P}(X_i \leqslant x_i).$$

If $(X_i, i = 1, \ldots, n)$ are mutually independent integrable random variables, then

$$\mathbb{E}\Big[\prod_i X_i\Big] = \prod_i \mathbb{E}[X_i].$$

1.1.4 Convergence Theorems

Definition 1.1 Let (X_n) be a sequence of random variables and X be a random variable.

(1) If there is a null set N such that $\forall \omega \in N^c$ $\lim_{n \to \infty} X_n(\omega) = X(\omega)$, we say that (X_n) *converges almost surely (a.s.)* to X and denote it by $\lim_{n \to \infty} X_n = X$ a.s., or $X_n \overset{\text{a.s.}}{\to} X$.

(2) If for all $\varepsilon > 0$, $\lim_{n \to \infty} \mathbb{P}([|X_n - X| > \varepsilon]) = 0$, we say that (X_n) *converges in probability* to X and denote it by $X_n \overset{\text{P}}{\to} X$.

(3) Assume that F_n and F are distribution functions of X_n and X, respectively. If at each continuous point t of F we have $F_n(t) \to F(t)$, then we say that (X_n) *converges in destribution* to X and denote it by $X_n \overset{\mathcal{L}}{\to} X$.

(4) If $\mathbb{E}[|X_n|^p] < \infty$, $\mathbb{E}[|X|^p] < \infty$, $\lim_{n \to \infty} \mathbb{E}[|X_n - X|^p] = 0$, we say that (X_n) *converges in L^p-norm* to X, or L^p-*converges* to X and denote it by $X_n \overset{L^p}{\to} X$, or $(L^p) \lim_{n \to \infty} X_n = X$. L^2-convergence is also called *convergence in mean square*.

(5) Let (X_n) be a sequence of integrable random variables, and X be an integrable random variable. If for any bounded random variable η, we have

$$\lim_{n \to \infty} \mathbb{E}[X_n \eta] = \mathbb{E}[X\eta],$$

we say that in L^1, (X_n) *converges weakly* to X.

Theorem 1.2 *A.s. convergence or L^p-convergence implies convergence in probability, and the latter implies convergence in distribution.*

Proof Assume that (X_n) does not converge in probability to X. Then there is $\varepsilon > 0$, such that

$$\limsup_{n \to \infty} \mathbb{P}(|X_n - X| > \varepsilon) > 0.$$

Set $A_n = [|X_n - X| > \varepsilon]$, then

$$\mathbb{P}(A_n \text{ i.o. }) = \lim_{k \to \infty} \mathbb{P}\left(\bigcup_{n=k}^{\infty} A_n \right) \geqslant \limsup_{n \to \infty} \mathbb{P}(A_n) > 0.$$

This implies that X_n does not converge a.s. to X.

Assume X_n L^p-converges to X. Then for any $\varepsilon > 0$, we have

$$\mathbb{P}(|X_n - X| > \varepsilon) \leqslant \varepsilon^{-p} \mathbb{E}[|X_n - X|^p] \to 0, \quad n \to \infty.$$

Thus, (X_n) converges in probability to X.

Finally, assume that (X_n) converges in probability to X. Let t be a continuous point of the distribution of X. Then for any $\eta > 0$, we can chose $\varepsilon > 0$ such that $\mathbb{P}(t - \varepsilon < X \leqslant t + \varepsilon) < \eta$. Since

$$|\mathbb{P}(A) - \mathbb{P}(B)| \leqslant \mathbb{P}(A \cup B) = \mathbb{P}\big((A \setminus B) \cup (B \setminus A)\big),$$

we have

$$
\begin{aligned}
|\mathbb{P}(X_n \leqslant t) - \mathbb{P}(X \leqslant t)| &\leqslant \mathbb{P}([X_n \leqslant t, X > t] \cup [X_n > t, X \leqslant t]) \\
&\leqslant \mathbb{P}(X_n \leqslant t, X > t + \varepsilon) + \mathbb{P}(t - \varepsilon < X \leqslant t + \varepsilon) \\
&\quad + \mathbb{P}(X_n > t, X \leqslant t - \varepsilon) \\
&\leqslant 2\mathbb{P}(|X_n - X| > \varepsilon) + \eta,
\end{aligned}
$$

from which it follows that (X_n) converges in distribution to X. $\qquad\square$

Theorem 1.3 *The following three results are related to the limit theorems on expectations of random variables:*

(1) **Monotone convergence theorem** *Let (X_n) be a sequence of random variables whose expectation exist. If for all $n \geqslant 1$, $X_n \leqslant X_{n+1}$ (respectively, $X_n \geqslant X_{n+1}$) a.s., and $\mathbb{E}[X_1] > -\infty$ (respectively, $\mathbb{E}[X_1] < \infty$), then we have*

$$\mathbb{E}[\lim_{n \to \infty} X_n] = \lim_{n \to \infty} \mathbb{E}[X_n].$$

(2) **Fatou's lemma** *Let (X_n) be a sequence of random variables whose expectation exist. If there is random variable Y with $\mathbb{E}[Y] > -\infty$ (respectively, $\mathbb{E}[Y] < \infty$), such that for all $n \geqslant 1$, we have $X_n \geqslant Y$ (respectively, $X_n \leqslant Y$), then the expectation of $\liminf_{n \to \infty} X_n$ (respectively, $\limsup_{n \to \infty} X_n$) exists, and*

$$\mathbb{E}[\liminf_{n\to\infty} X_n] \leqslant \liminf_{n\to\infty} \mathbb{E}[X_n]$$

(respectively, $\mathbb{E}[\limsup_{n\to\infty} X_n] \geqslant \limsup_{n\to\infty} \mathbb{E}[X_n]$).

(3) ***Dominated convergence theorem*** *Let (X_n) be a sequence of random variables, such that $X_n \overset{a.s.}{\to} X$ or $X_n \overset{P}{\to} X$. If there is an integrable random variable Y such that $|X_n| \leqslant Y$ for all $n \geqslant 1$, then $\mathbb{E}[X_n] \to \mathbb{E}[X]$.*

Proof As (1) is obvious true, its proof is omitted. Set $Y_n = \inf_{k \geqslant n} X_k$ (respectively, $Y_n = \sup_{k \geqslant n} X_k$). Then from (1) we obtain (2). Now we prove (3). For the a.s. convergence case, (3) is a consequence of (2). For the convergence in probability case, since for any given subsequence $(X_{n'})$ of (X_n), there exists a sub-subsequence $(X_{n'_k})$ such that $(X_{n'_k})$ a.s. converges to X, thus $\mathbb{E}[X_{n'_k}] \to \mathbb{E}[X]$, which implies that $\mathbb{E}[X_n] \to \mathbb{E}[X]$. $\qquad\square$

1.2 Conditional Mathematical Expectation

1.2.1 Definition and Basic Properties

Let $(\Omega, \mathcal{F}, \mathbb{P})$ be a probability space, A and B be two events, and $\mathbb{P}(A) > 0$. The probability that B occurs under the condition that A occurs is obviously equal to $\mathbb{P}(AB)/\mathbb{P}(A)$, which is called the *conditional probability* of B given A, and is denoted by $\mathbb{P}(B|A)$.

Let $(B_j)_{1 \leqslant j \leqslant m}$ be a finite partition of Ω, where $B_j \in \mathcal{F}$, $\mathbb{P}(B_j) > 0, 1 \leqslant j \leqslant m$. Let $\mathcal{G}$ be the σ algebra generated by (B_j). For an integrable random variable X, define

$$\mathbb{E}[X|\mathcal{G}] = \sum_{j=1}^{m} \frac{\mathbb{E}[X I_{B_j}]}{\mathbb{P}(B_j)} I_{B_j},$$

the *conditional (mathematical) expectation* of X w.r.t. $\mathcal{G}$. If $(A_i)_{1 \leqslant i \leqslant n}$ is a finite partition of Ω, $A_i \in \mathcal{F}$, for $1 \leqslant i \leqslant n$, and $X = \sum_{i=1}^{n} a_i I_{A_i}$ is a simple random variable, then it is easy to see that

$$\mathbb{E}[X|\mathcal{G}] = \sum_{j=1}^{m} \sum_{i=1}^{n} a_i \mathbb{P}(A_i|B_j) I_{B_j}.$$

Furthermore, $\mathbb{E}(X|\mathcal{G})$ is a $\mathcal{G}$-measurable random variable and satisfies

$$\mathbb{E}[\mathbb{E}[X|\mathcal{G}]I_B] = \mathbb{E}[X I_B], \qquad \forall B \in \mathcal{G}.$$

Below we will use the Radon-Nikodym theorem from measure theory to extend the conditional expectation to general random variables and general σ-algebras. To this end, we first introduce some related results of measure theory.

Let $(\Omega, \mathcal{F})$ be a measurable space. A σ-additive set function ν (not necessarily nonnegative) on $\mathcal{F}$ with $\nu(\emptyset) = 0$ is called a *signed measure*. By the Jordan-Hahn decomposition theorem, any signed measure ν can be expressed as the difference of two measures: $\nu = \mu_1 - \mu_2$, where one of them is a finite measure, and there is a measurable set K such that $\forall A \in \mathcal{F}$, $\mu_1(A) = \nu(A \cap K)$, $\mu_2(A) = -\nu(A \cap K^c)$. We call $\Omega = K \cup K^c$ a *Hahn decomposition* of ν. We set $\nu^+ = \mu_1, \nu^- = \mu_2$ and call $\nu = \nu^+ - \nu^-$ the *Jordan decomposition* of ν. We set $|\nu| = \nu^+ + \nu^-$ and call $|\nu|$ the *variance measure* of ν. Let μ and ν be two signed measures. Measure ν is said to be *absolutely continuous* w.r.t. μ (denoted $\nu \ll \mu$), if $|\mu|(A) = 0 \Rightarrow |\nu|(A) = 0$. Two measures ν and μ are said to be *mutually singular* (denoted $\nu \perp \mu$), if there is a measurable set A, such that $|\mu|(A) = 0$ and $|\nu|(A^c) = 0$.

Let $g \in \mathcal{L}$ such that the integral of g w.r.t. $|\mu|$ exists. We denote $\mu^+(g) - \mu^-(g)$ by $\mu(g)$ and call it the *integral* of g w.r.t. μ. We define a signed measure ν by $\nu(A) = \mu(gI_A)$, $A \in \mathcal{F}$, call it the *indefinite integral* of g w.r.t. μ and denote it by $g.\mu$.

The next theorem shows that: any σ-finite signed measure ν can always be uniquely decomposed into the sum of an absolutely continuous part and a singular part w.r.t. a σ-finite signed measure μ.

Theorem 1.4 *Let μ and ν be two σ-finite signed measures on $(\Omega, \mathcal{F})$. Then ν has the following unique decomposition (called the Lebesgue decomposition):*

$$\nu = \nu_s + \nu_c, \tag{1.1}$$

where ν_s and μ are mutually singular and ν_c is absolutely continuous w.r.t. μ. In addition, ν_s and ν_c are σ-finite, and there is $g \in \mathcal{L}$ such that the integral of g w.r.t. $|\mu|$ exists and $\nu_c = g.\mu$.

Proof First of all, it may be assumed that μ is a measure (otherwise we can replace μ by $|\mu|$), and $\mu(\Omega) > 0$. Then by σ-finiteness of μ, there exists a countable partition $\Omega = \sum_{n=1}^{\infty} A_n$ of Ω, such that $A_n \in \mathcal{F}, 0 < \mu(A_n) < \infty, \forall n \geqslant 1$. Put

$$h = \sum_{n=1}^{\infty} \frac{1}{2^n \mu(A_n)} I_{A_n}.$$

Then h is strictly positive everywhere, and $\mu(h) = 1$. Set $\widetilde{\mu} = h.\mu$. Then $\widetilde{\mu}$ is a measure, and $\widetilde{\mu}(\Omega) = 1$. Since $\widetilde{\mu}$ and μ are equivalent, we can replace μ by $\widetilde{\mu}$ in proving the theorem. Therefore, it may be assumed that μ is a finite measure.

Below we first assume that ν is also a finite measure. Put

$$\mathcal{H} = \left\{ h \in \overline{\mathcal{L}}^+ : \forall A \in \mathcal{F}, \int_A h \, d\mu \leqslant \nu(A) \right\},$$

where $\overline{\mathcal{L}}^+$ is the set of all nonnegative measurable functions on $(\Omega, \mathcal{F})$. Let $h_1, h_2 \in \mathcal{H}, h = h_1 \vee h_2$. Then

$$\int_A h d\mu = \int_{A \cap [h_1 \geq h_2]} h_1 d\mu + \int_{A \cap [h_1 < h_2]} h_2 d\mu$$

$$\leq \nu(A \cap [h_1 \geq h_2]) + \nu(A \cap [h_1 < h_2]) = \nu(A).$$

This shows that $\mathcal{H}$ is closed under the finite maximum operation. Now let $h_n \in \mathcal{H}, h_n \uparrow g$, such that

$$\int_\Omega g d\mu = \sup\left\{ \int_\Omega h d\mu : h \in \mathcal{H} \right\}.$$

Then by the monotone convergence theorem of integration, it is easy to see $g \in \mathcal{H}$. Put

$$\nu_s(A) = \nu(A) - \int_A g d\mu, \ A \in \mathcal{F},$$

then ν_s is a finite measure. Next we prove $\nu_s \perp \mu$. Let $\Omega = D_n + D_n^c$ be the Hahn decomposition of the signed measure $\nu_s - \frac{1}{n}\mu$. Then for all $A \in \mathcal{F}$,

$$\nu_s(A \cap D_n) \geq n^{-1}\mu(A \cap D_n) = n^{-1}\int_A I_{D_n} d\mu.$$

Thus, $\forall A \in \mathcal{F}$,

$$\int_A (g + n^{-1}I_{D_n})d\mu \leq \int_A g d\mu + \nu_s(A \cap D_n) \leq \nu(A).$$

This shows $g + n^{-1}I_{D_n} \in \mathcal{H}$. On the other hand, since $\mu(g) = \sup\{\mu(h) : h \in \mathcal{H}\}$, we must have $\mu(D_n) = 0$. Let $N = \bigcup_n D_n$, then $\mu(N) = 0$. In addition, since $\left(\nu_s - \frac{1}{n}\mu\right)(D_n^c) \leq 0$, we have

$$\nu_s(N^c) \leq \nu_s(D_n^c) \leq n^{-1}\mu(D_n^c) \leq n^{-1}\mu(\Omega) \to 0, \ n \to \infty,$$

which implies $\nu_s \perp \mu$. Let $\nu_c = g.\mu$, then $\nu_c \ll \mu$. Since g is μ-integrable, g can be taken as a real measurable function.

Now let ν be a σ-finite signed measure. In order to complete the proof of the theorem, we may assume that ν is a σ-finite measure (otherwise, we consider ν^+ and ν^-, respectively). We take a countable partition $\Omega = \sum_n A_n$ of Ω, such that $A_n \in \mathcal{F}, \nu(A_n) < \infty, n \geq 1$. Let $\nu^n(A) = \nu(A \cap A_n)$, then each ν^n is a finite measure. Therefore, by the above proof, v^n has the following decomposition:

$$v^n = v_s^n + v_c^n, \ n \geqslant 1,$$

where $v_s^n \perp \mu$, $v_c^n \ll \mu$, and there is a nonnegative real measurable function g_n such that $v_c^n = g_n . \mu$. Obviously, g_n can take zero on A_n^c. Put

$$v_s = \sum_n v_s^n, \ \ v_c = \sum_n v_c^n, \ \ g = \sum_n g_n.$$

Then $v_s \perp \mu$, $v_c \ll \mu$, $v_c = g . \mu$, and (1.1) holds. The uniqueness of the decomposition of v is obvious. $\square$

Assume that μ is a measure on the measurable space $(\Omega, \mathcal{F})$, f is a measurable function whose integral w.r.t. μ exists, and v is the indefinite integral of f w.r.t. μ. Then v is absolutely continuous w.r.t. μ. The next theorem shows that if μ is a σ-finite measure, then the converse proposition holds.

Theorem 1.5 (Radon-Nikodym theorem) *Let $(\Omega, \mathcal{F})$ be a measurable space, μ be a σ-finite measure, v be a signed measure (not necessarily σ-finite). If v is absolutely continuous w.r.t. μ, then there exists a measurable function g whose integral w.r.t. μ exists, such that $v = g . \mu$. In addition, g is uniquely determined in the sense of μ-equivalence (g_1 and g_2 are said μ-equivalent, if $\mu([g_1 \neq g_2]) = 0$). The measurable function g is μ-a.e. finite, if and only if v is σ-finite.*

Proof In order to prove the theorem, it may be assumed that v is a σ-finite measure (otherwise we consider v^+ and v^-, respectively). In addition, by σ-finiteness of μ, we can assume that μ is a finite measure. If v is a finite measure, then by Theorem 1.4, we obtain immediately the conclusion of the theorem. Therefore, below we can assume $\mu(\Omega) < \infty$, $v(\Omega) = \infty$. Put

$$\mathcal{G} = \{C \in \mathcal{F} : v(C) < \infty\}.$$

Obviously, $\mathcal{G}$ is closed under the finite union operation. Thus, there are $C_n \in \mathcal{G}$, $C_n \uparrow C$, such that

$$\mu(C) = \sup\{\mu(G) : G \in \mathcal{G}\}.$$

Let

$$v'(B) = v(B \cap C), \ \ v''(B) = v(B \cap C^c), \ B \in \mathcal{F},$$

then v' is a σ-finite measure, and $v' \ll \mu$. Therefore, there exists a nonnegative real measurable function g', such that $v' = g' . \mu$. On the other hand, from the definition of $\mathcal{G}$, we know that

$$\mu(B \cap C^c) > 0 \Rightarrow v(B \cap C^c) = \infty.$$

Hence, if we let $g'' = (+\infty)I_{C^c}$, $g = g' + g''$, then $\nu'' = g''.\mu$, $\nu = g.\mu$. The remaining claims of the theorem are obvious.   □

Remark We use $\frac{d\nu}{d\mu}$ to represent g in Theorem 1.5 (it is uniquely determined in the sense of μ-equivalence) and call $\frac{d\nu}{d\mu}$ the *Radon-Nikodym derivative* of ν w.r.t. μ.

Let $(\Omega, \mathcal{F}, \mathbb{P})$ be a probability space and $\mathcal{G}$ be a sub-σ-algebra of $\mathcal{F}$. Let X be a random variable, whose mathematical expectation exists, and let $\nu = X.\mathbb{P}$ be the indefinite integral of X w.r.t. $\mathbb{P}$, namely,

$$\nu(A) = \int_A X d\mathbb{P}, \quad \forall A \in \mathcal{F}.$$

Then ν is a signed measure and is absolutely continuous w.r.t. $\mathbb{P}$. If we restrict ν and $\mathbb{P}$ onto $(\Omega, \mathcal{G})$, then we still have $\nu \ll \mathbb{P}$. Let Y be the Radon-Nikodym derivative of ν w.r.t. $\mathbb{P}$ on $(\Omega, \mathcal{G})$, then Y is a $\mathcal{G}$-measurable random variable, and it holds that

$$\mathbb{E}[YI_B] = \mathbb{E}[XI_B], \quad \forall B \in \mathcal{G}. \tag{1.2}$$

We call Y the *conditional (mathematical) expectation* of X w.r.t. $\mathcal{G}$. In the $\mathbb{P}$-equivalent sense, the conditional expectation Y is uniquely determined and denoted by $\mathbb{E}[X|\mathcal{G}]$, which is characterized by (1.2).

Theorem 1.6 *The conditional expectation has the following basic properties:*

(1) $\mathbb{E}[\mathbb{E}[X|\mathcal{G}]] = \mathbb{E}[X]$;

(2) *if X is $\mathcal{G}$-measurable, then $\mathbb{E}[X|\mathcal{G}] = X$ a.s.;*

(3) *if $\mathcal{G} = \{\emptyset, \Omega\}$, then $\mathbb{E}[X|\mathcal{G}] = \mathbb{E}[X]$ a.s.;*

(4) $\mathbb{E}[X|\mathcal{G}] = \mathbb{E}[X^+|\mathcal{G}] - \mathbb{E}[X^-|\mathcal{G}]$ *a.s.;*

(5) $X \geqslant Y$ *a.s.* $\Rightarrow \mathbb{E}[X|\mathcal{G}] \geqslant \mathbb{E}[Y|\mathcal{G}]$ *a.s.;*

(6) *assume that c_1 and c_2 are real numbers, and the expectations of $X, Y, c_1X + c_2Y$ all exist, then*

$$\mathbb{E}[c_1X + c_2Y|\mathcal{G}] = c_1\mathbb{E}[X|\mathcal{G}] + c_2\mathbb{E}[Y|\mathcal{G}] \text{ a.s. },$$

if the summation in the right hand is well defined;

(7) $|\mathbb{E}[X|\mathcal{G}]| \leqslant \mathbb{E}[|X||\mathcal{G}]$ *a.s.;*

(8) *if $0 \leqslant X_n \uparrow X$ a.s., then $\mathbb{E}[X_n|\mathcal{G}] \uparrow \mathbb{E}[X|\mathcal{G}]$ a.s.;*

(9) *assume the expectations of X and XY exist, and Y is $\mathcal{G}$-measurable, then*

$$\mathbb{E}[XY|\mathcal{G}] = Y\mathbb{E}[X|\mathcal{G}] \text{ a.s. }; \tag{1.3}$$

(10) *(smoothness of conditional expectation) if $\mathcal{G}_1$ and $\mathcal{G}_2$ are sub-σ-algebras of $\mathcal{F}$ and $\mathcal{G}_1 \subset \mathcal{G}_2$, then*

$$\mathbb{E}[\mathbb{E}[X|\mathcal{G}_2]|\mathcal{G}_1] = \mathbb{E}[X|\mathcal{G}_1] \text{ a.s. }; \tag{1.4}$$

(11) *if X and $\mathcal{G}$ are mutually independent (i.e., $\sigma(X)$ and $\mathcal{G}$ are mutually indepen-dent), then $\mathbb{E}[X|\mathcal{G}] = \mathbb{E}[X]$ a.s.*

Proof (1)–(7) can be seen directly from the definition of conditional expectation.

(8) By (5), $\mathbb{E}[X_n|\mathcal{G}] \uparrow Y$ a.s., Y is a $\mathcal{G}$-measurable random variable. Thus, for any $B \in \mathcal{G}$, we have

$$\int_B Y d\mathbb{P} = \lim_{n\to\infty} \int_B \mathbb{E}[X_n|\mathcal{G}]d\mathbb{P} = \lim_{n\to\infty} \int_B X_n d\mathbb{P} = \int_B X d\mathbb{P}.$$

Consequently, $Y = \mathbb{E}[X|\mathcal{G}]$ a.s.

(9) It may be assumed that X and Y are nonnegative random variables. First we assume $Y = I_A$, $A \in \mathcal{G}$, then $\mathbb{E}[X|\mathcal{G}]$ is $\mathcal{G}$-measurable, and for all $B \in \mathcal{G}$, we have

$$\int_B Y\mathbb{E}[X|\mathcal{G}]d\mathbb{P} = \int_{A\cap B} \mathbb{E}[X|\mathcal{G}]d\mathbb{P} = \int_{A\cap B} X d\mathbb{P}$$
$$= \int_B X I_A d\mathbb{P} = \int_B Y X d\mathbb{P}.$$

Therefore, (1.3) holds. Then we can use (8) to make a transition from simple random variables to general nonnegative random variables.

(10) Assume $B \in \mathcal{G}_1$, then

$$\int_B \mathbb{E}[\mathbb{E}[X|\mathcal{G}_2]|\mathcal{G}_1]d\mathbb{P} = \int_B \mathbb{E}[X|\mathcal{G}_2]d\mathbb{P} = \int_B X d\mathbb{P},$$

from which (1.4) follows.

(11) It may be assumed that X is a nonnegative random variable. Let $A \in \mathcal{G}$. Since I_A and X are independent, we have

$$\int_A \mathbb{E}[X]d\mathbb{P} = \mathbb{E}[X]\mathbb{P}(A) = \mathbb{E}[X I_A] = \int_A X d\mathbb{P},$$

from which we get $\mathbb{E}[X] = \mathbb{E}[X|\mathcal{G}]$ a.s.   □

1.2.2   Convergence Theorems

For the conditional expectations, we have also monotone convergence theorem, Fatou's Lemma, and dominated convergence theorem. Their proofs are similar to that for the case of integrals. Therefore, we describe only the results and omit their proofs. Note that in the probability space case a.s. convergence always implies convergence in probability.

In the next theorems, $(\Omega, \mathcal{F}, \mathbb{P})$ is a probability space, $\mathcal{G}$ is a sub-σ-algebra of $\mathcal{F}$.

Theorem 1.7 (Monotone convergence theorem) *Let (X_n) be a sequence of random variables whose conditional expectations w.r.t. $\mathcal{G}$ exist. If $X_n \uparrow X$ a.s. and $\mathbb{E}[X_1^- | \mathcal{G}] < \infty$ a.s., then the conditional expectation of X w.r.t. $\mathcal{G}$ exists and $\mathbb{E}[X_n | \mathcal{G}] \uparrow \mathbb{E}[X | \mathcal{G}]$ a.s.*

Theorem 1.8 (Fatou's lemma) *Let (X_n) be a sequence of random variables whose conditional expectations w.r.t. $\mathcal{G}$ exist.*

(1) *If there exists a random variable Y such that $\mathbb{E}[Y^- | \mathcal{G}] < \infty$ a.s., and for each $n \geqslant 1$, $X_n \geqslant Y$ a.s., then the conditional expectation of $\liminf\limits_{n \to \infty} X_n$ w.r.t. $\mathcal{G}$ exists, and we have:*

$$\mathbb{E}[\liminf_{n \to \infty} X_n | \mathcal{G}] \leqslant \liminf_{n \to \infty} \mathbb{E}[X_n | \mathcal{G}];$$

(2) *If there exists a random variable Y, such that $\mathbb{E}[Y^+ | \mathcal{G}] < \infty$ a.s., and for each $n \geqslant 1$, $X_n \leqslant Y$ a.s., then the conditional expectation of $\limsup\limits_{n \to \infty} X_n$ w.r.t. $\mathcal{G}$ exists, and we have*

$$\mathbb{E}[\limsup_{n \to \infty} X_n | \mathcal{G}] \geqslant \limsup_{n \to \infty} \mathbb{E}[X_n | \mathcal{G}].$$

Theorem 1.9 (Dominated convergence theorem) *Assume $X_n \xrightarrow{a.s.} X$ (respectively, $X_n \xrightarrow{p} X$). If there exists a nonnegative random variable Y, such that $|X_n| \leqslant Y$ a.s., then X is integrable, and we have $\lim\limits_{n \to \infty} \mathbb{E}[X_n | \mathcal{G}] = \mathbb{E}[X | \mathcal{G}]$ a.s. (respectively, $\mathbb{E}[X_n | \mathcal{G}] \xrightarrow{p} \mathbb{E}[X | \mathcal{G}]$).*

1.2.3 Two Theorems About Conditional Expectation

In this book we will often utilize the following two theorems about conditional expectations.

Theorem 1.10 *Let $(\Omega, \mathcal{F}, \mathbb{P})$ be a probability space, $\mathcal{G}$ be a sub-σ-algebra of $\mathcal{F}$, X be a $\mathbb{R}^m$-valued $\mathcal{G}$-measurable random variable, and Y be a $\mathbb{R}^n$-valued random variable. If Y and $\mathcal{G}$ are independent (i.e., $\sigma(Y)$ and $\mathcal{G}$ are independent), then for any nonnegative Borel function $g(x, y)$ on $\mathbb{R}^m \times \mathbb{R}^n$, it holds*

$$\mathbb{E}[g(X, Y) | \mathcal{G}] = \mathbb{E}[g(x, Y)]|_{x=X}. \tag{1.5}$$

Proof In order to prove (1.5), we need only to prove that for any nonnegative $\mathcal{G}$-measurable random variable Z,

$$\mathbb{E}[g(X, Y)Z] = \mathbb{E}[f(X)Z],$$

where $f(x) = \mathbb{E}[g(x, Y)]$. To this end, set

$$\mu_Y(A) = \mathbb{P}(Y^{-1}(A)), \quad A \in \mathcal{B}(\mathbb{R}^n),$$

$$\mu_{X,Z}(E) = \mathbb{P}((X, Z)^{-1}(E)), \quad E \in \mathcal{B}(R^m \times R).$$

Then we have

$$f(x) = \int g(x, y)\mu_Y(dy).$$

Since Y and (X, Z) are independent,

$$\mathbb{E}[g(X, Y)Z] = \int zg(x, y)\mu_Y(dy)\mu_{X,Z}(dx, dz)$$

$$= \int zf(x)\mu_{X,Z}(dx, dz) = \mathbb{E}[Zf(X)].$$

The theorem is proved.   □

The following theorem is the *Bayes' rule* of conditional expectations.

Theorem 1.11 *Assume that $\mathbb{Q}$ is a probability measure which is absolutely contin-uous w.r.t. $\mathbb{P}$ and that $\mathcal{G}$ is a sub-σ-algebra of $\mathcal{F}$. Set*

$$\xi = \frac{d\mathbb{Q}}{d\mathbb{P}}, \qquad \eta = \mathbb{E}[\xi \mid \mathcal{G}].$$

Then $\eta > 0$, $\mathbb{Q}$-a.s. If X is a $\mathbb{Q}$-integrable random variable, then

$$\mathbb{E}_{\mathbb{Q}}[X \mid \mathcal{G}] = \eta^{-1}\mathbb{E}[X\xi \mid \mathcal{G}], \quad \mathbb{Q}\text{-a.s.} . \tag{1.6}$$

Proof First, since $[\eta > 0] \in \mathcal{G}$, we have

$$\mathbb{Q}([\eta > 0]) = \mathbb{E}[\xi I_{[\eta>0]}] = \mathbb{E}[\eta I_{[\eta>0]}] = \mathbb{E}[\eta] = \mathbb{E}[\xi] = 1.$$

Let X be a $\mathbb{Q}$-integrable random variable, then

$$\mathbb{E}[X\xi I_A] = \mathbb{E}_{\mathbb{Q}}[XI_A] = \mathbb{E}_{\mathbb{Q}}[\mathbb{E}_{\mathbb{Q}}[X \mid \mathcal{G}]I_A]$$

$$= \mathbb{E}[\mathbb{E}_{\mathbb{Q}}[X \mid \mathcal{G}]\xi I_A] = \mathbb{E}[\mathbb{E}_{\mathbb{Q}}[X \mid \mathcal{G}]\eta I_A], \quad \forall A \in \mathcal{G}.$$

This shows that

$$\mathbb{E}[X\xi \mid \mathcal{G}] = \mathbb{E}_{\mathbb{Q}}[X \mid \mathcal{G}]\eta, \quad \mathbb{P}\text{-a.s.} .$$

Hence, the above equality holds $\mathbb{Q}$-a.s., whence it follows (1.6).   □

1.3   Duals of Spaces $L^\infty(\Omega, \mathcal{F})$ and $L^\infty(\Omega, \mathcal{F}, m)$

Let $(\Omega, \mathcal{F})$ be a measurable space. Let $L^\infty(\Omega, \mathcal{F})$ denote the set of all bounded measurable functions on $(\Omega, \mathcal{F})$. For any $f \in L^\infty(\Omega, \mathcal{F})$, put

$$\|f\| = \sup_{\omega \in \Omega} |f(\omega)|.$$

Then under this norm $L^\infty(\Omega, \mathcal{F})$ is a Banach space.

Let μ be a finitely additive set function on $\mathcal{F}$. Set

$$\|\mu\|_{\mathrm{var}} = \sup \left\{ \sum_{i=1}^{n} |\mu(A_i)| \, \big| \, A_i \in \mathcal{F}, i = 1, \cdots, n, \{A_i\} \text{is a finite partition of } \Omega \right\}.$$

We call $\|\mu\|_{\mathrm{var}}$ the *total variation* of μ. We use $ba(\Omega, \mathcal{F})$ to denote the set of all finitely additive set functions with finite total variation. Besides, let $\mu \in ba(\Omega, \mathcal{F})$, and $f = \sum_{i=1}^{n} a_i I_{A_i}$ be a simple measurable function, where $a_i \in \mathbb{R}$, $A_i \in \mathcal{F}$. Put

$$\int_\Omega f d\mu = \sum_{i=1}^{n} a_i \mu(A_i).$$

It is easy to prove that $\int_\Omega f d\mu$ does not depend of the specific expression of f, and it holds

$$\left| \int_\Omega f d\mu \right| \leqslant \|f\| \|\mu\|_{\mathrm{var}}. \tag{1.7}$$

Since the set of all simple measurable functions is dense in $L^\infty(\Omega, \mathcal{F})$ under the norm, inequality (1.7) allows us to expend the above definition to be a continuous linear functional on $L^\infty(\Omega, \mathcal{F})$ and keep the inequality. We call the value at f of this functional the integral of f with respect to μ and denote it by $\int_\Omega f d\mu$ or simply by $\mu(f)$.

Let $(\Omega, \mathcal{F}, m)$ be a measure space. Assume $\mu \in ba(\Omega, \mathcal{F})$, and μ is absolutely continuous w.r.t. m. Put

$$ba(\Omega, \mathcal{F}, m) = \{\mu \in ba(\Omega, \mathcal{F}) \, | \, \mu \ll m\}.$$

Assume $\mu \in ba(\Omega, \mathcal{F}, m)$, $f \in L^\infty(\Omega, \mathcal{F}, m)$. Obviously, we can choose an element $\tilde{f}$ of $L^\infty(\Omega, \mathcal{F})$ as a representative of f and define $\mu(\tilde{f})$ as the integral of f w.r.t μ, still by $\int_\Omega f d\mu$, or simply by $\mu(f)$. Then we have:

$$\left| \int_\Omega f d\mu \right| \leqslant \|f\|_\infty \|\mu\|_{\mathrm{var}}.$$

The following theorem shows that $ba(\Omega, \mathcal{F})$ and $ba(\Omega, \mathcal{F}, m)$ can be regarded as the dual spaces of $L^\infty(\Omega, \mathcal{F})$ and $L^\infty(\Omega, \mathcal{F}, m)$, respectively.

Theorem 1.12

(1) *Assume* $\mu \in ba(\Omega, \mathcal{F})$. *Set*

$$T_\mu(f) = \mu(f), \ f \in L^\infty(\Omega, \mathcal{F}).$$

Then T_μ *is a norm-preserved linear isomorphic mapping from the dual space* $L^\infty(\Omega, \mathcal{F})^*$ *of* $(\Omega, \mathcal{F})$ *onto* $ba(\Omega, \mathcal{F})$.
(2) *Assume* $\mu \in ba(\Omega, \mathcal{F}, m)$. *Set:*

$$T_\mu(f) = \mu(f), \ f \in L^\infty(\Omega, \mathcal{F}, m).$$

Then T_μ *is a norm-preserved linear isomorphic mapping from the dual space* $L^\infty(\Omega, \mathcal{F}, m)^*$ *of* $(\Omega, \mathcal{F}, m)$ *onto* $ba(\Omega, \mathcal{F}, m)$.

Proof By (1.7), we have $T_\mu \in L^\infty(\Omega, \mathcal{F})^*$, and $\|T_\mu\| \leqslant \|\mu\|_{\mathrm{var}}$. Conversely, assume $l \in L^\infty(\Omega, \mathcal{F})^*$. Put

$$\mu(A) = l(I_A), \ \ A \in \mathcal{F}.$$

Then μ is a finitely additive set function on $\mathcal{F}$. Clearly, we have $\|\mu\|_{\mathrm{var}} \leqslant \|l\|$, and consequently, $\mu \in ba(\Omega, \mathcal{F})$, and $T_\mu = l$. Thus, finally, we have $\|T_\mu\| = \|\mu\|_{\mathrm{var}}$.
The proof of (2) is similar and is left for the reader to complete. $\qquad\square$

1.4   Family of Uniformly Integrable Random Variables

Definition 1.13 Let $(\Omega, \mathcal{F}, \mathbb{P})$ be a probability space and $\mathcal{H}$ be a family of integrable random variables. $\mathcal{H}$ is said to be *uniformly integrable*, if as $C \to \infty$, the integral

$$\int_{[|\xi| \geqslant C]} |\xi| d\mathbb{P}, \ \xi \in \mathcal{H}$$

converges uniformly to 0.

The following theorem gives a criterion of uniform integrability.

Theorem 1.14 *Let* $\mathcal{H} \subset L^1(\Omega, \mathcal{F}, \mathbb{P})$. *Then* $\mathcal{H}$ *is a uniformly integrable family if and only if it satisfies the following conditions:*

(1) $a = \sup\{\mathbb{E}|\xi|, \xi \in \mathcal{H}\} < +\infty$;
(2) *for any* $\varepsilon > 0$, *there exists* $\delta > 0$, *such that for any* $A \in \mathcal{F}$ *satisfying* $\mathbb{P}(A) \leqslant \delta$, *we have*

$$\sup_{\xi \in \mathcal{H}} \int_A |\xi| d\mathbb{P} \leqslant \varepsilon.$$

Proof Necessity. Assume that $\mathcal{H}$ is a uniformly integrable family. For a given $\varepsilon > 0$, we take a constant C large enough such that

$$\sup_{\xi \in \mathcal{H}} \int_{[|\xi| \geqslant C]} |\xi| d\mathbb{P} \leqslant \frac{\varepsilon}{2}.$$

On the other hand, we have

$$\int_A |\xi| d\mathbb{P} \leqslant C\mathbb{P}(A) + \int_{[|\xi| \geqslant C]} |\xi| d\mathbb{P}. \tag{1.8}$$

Taking $A = \Omega$ in (1.8) yields condition (1); taking $\delta = \varepsilon/2C$ yields condition (2).

Sufficiency. Assume that conditions (1) and (2) hold. For any $\varepsilon > 0$, we take $\delta > 0$ such that (2) holds. Let $C \geqslant a/\delta$. Then

$$\mathbb{P}([|\xi| \geqslant C]) \leqslant \frac{1}{C}\mathbb{E}[|\xi|] \leqslant \frac{a}{C} \leqslant \delta, \ \xi \in \mathcal{H}.$$

From condition (2) it follows that

$$\int_{[|\xi| \geqslant C]} |\xi| d\mathbb{P} \leqslant \varepsilon, \ \xi \in \mathcal{H}.$$

This shows that $\mathcal{H}$ is a uniformly integrable family. $\qquad\square$

The following theorem gives a criterion of L^1 convergence.

Theorem 1.15 *Let (ξ_n) be a sequence of integrable random variables, ξ be a real random variable. Then the following conditions are equivalent:*

(1) $\xi_n \xrightarrow{L^1} \xi$;

(2) $\xi_n \xrightarrow{p} \xi$, *and (ξ_n) is uniformly integrable.*

Proof (1)$\Rightarrow$(2). Assume $\xi_n \xrightarrow{L^1} \xi$. Let $A \in \mathcal{F}$, we have

$$\int_A |\xi_n| d\mathbb{P} \leqslant \int_A |\xi| d\mathbb{P} + E[|\xi_n - \xi|]. \tag{1.9}$$

For any given $\varepsilon > 0$, we take a positive number N, such that for $n > N$, it holds $\mathbb{E}[|\xi_n - \xi|] \leqslant \varepsilon/2$. Then we take $\delta > 0$ such that for any $A \in \mathcal{F}$ satisfying $\mathbb{P}(A) \leqslant \delta$,

$$\int_A |\xi| d\mathbb{P} \leqslant \frac{\varepsilon}{2}, \ \int_A |\xi_n| d\mathbb{P} \leqslant \frac{\varepsilon}{2}, \quad n = 1, 2, \cdots, N. \tag{1.10}$$

Thus, by (1.9) and (1.10), for any $A \in \mathcal{F}$ satisfying $\mathbb{P}(A) \leqslant \delta$, we have $\sup_n \int_A |\xi_n| d\mathbb{P} \leqslant \varepsilon$. Besides, we have $\sup_n \mathbb{E}[|\xi_n|] < \infty$. Therefore, by Theorem 1.14, (ξ_n) is a uniformly integrable family. Finally, it is clear that $\xi_n \xrightarrow{P} \xi$.

(2)$\Rightarrow$(1). Assume that (ξ_n) is uniformly integrable and $\xi_n \xrightarrow{P} \xi$. By Fatou's lemma,

$$\mathbb{E}[|\xi|] \leqslant \sup_n \mathbb{E}[|\xi_n|] < +\infty,$$

and ξ is integrable. Hence, $(\xi_n - \xi)$ is uniformly integrable. For any $\varepsilon > 0$, by Theorem 1.14, there exists a $\delta > 0$, such that for any $A \in \mathcal{F}$ satisfying $\mathbb{P}(A) < \delta$, it holds

$$\sup_n \int_A |\xi_n - \xi| d\mathbb{P} \leqslant \varepsilon.$$

We take N large enough, such that for $n \geqslant N$, it holds $\mathbb{P}([|\xi_n - \xi| \geqslant \varepsilon]) < \delta$. Thus, when $n \geqslant N$, we have

$$\mathbb{E}[|\xi_n - \xi|] = \int_{[|\xi_n - \xi| < \varepsilon]} |\xi_n - \xi| d\mathbb{P} + \int_{[|\xi_n - \xi| \geqslant \varepsilon]} |\xi_n - \xi| d\mathbb{P} \leqslant 2\varepsilon.$$

This shows $\xi_n \xrightarrow{L^1} \xi$.   □

Theorem 1.16 Let $(\Omega, \mathcal{F}, \mathbb{P})$ be a probability space, ξ be an integrable random variable, and $(\mathcal{G}_i)_{i \in I}$ be a family of sub-σ algebras of $\mathcal{F}$. Set $\eta_i = \mathbb{E}[\xi | \mathcal{G}_i]$. Then $(\eta_i, i \in I)$ is a uniformly integrable family.

Proof For any $C > 0$, we have

$$\mathbb{P}([|\eta_i| \geqslant C]) \leqslant \frac{1}{C} \mathbb{E}[|\eta_i|] \leqslant \frac{1}{C} \mathbb{E}[|\xi|], \quad i \in I.$$

Thus (noting $|\eta_i| \geqslant C \in \mathcal{G}_i$), we have

$$\int_{[|\eta_i| \geqslant C]} |\eta_i| d\mathbb{P} \leqslant \int_{[|\eta_i| \geqslant C]} |\xi| d\mathbb{P} \leqslant \delta \mathbb{P}([|\eta_i| \geqslant C]) + \int_{[|\xi| \geqslant \delta]} |\xi| d\mathbb{P}$$

$$\leqslant \frac{\delta}{C} \mathbb{E}[|\xi|] + \int_{[|\xi| \geqslant \delta]} |\xi| d\mathbb{P}.$$

For any $\varepsilon > 0$, we take $\delta > 0$, such that $\int_{[|\xi| \geqslant \delta]} |\xi| d\mathbb{P} \leqslant \varepsilon/2$. Then for any $C \geqslant (2\delta/\varepsilon)\mathbb{E}[|\xi|]$, we have

$$\int_{[|\eta_i| \geqslant C]} |\eta_i| d\mathbb{P} \leqslant \varepsilon, \quad \forall i \in I.$$

This shows that $(\eta_i, i \in I)$ is a uniformly integrable family.   □

The following theorem is called the *Vitali-Hahn-Saks theorem*. For its proof see, e.g., Yan (2009).

Theorem 1.17 *Let* $(\Omega, \mathcal{F}, \mathbb{P})$ *be a probability space,* (μ_n) *be a sequence of finite signed measures on* $(\Omega, \mathcal{F})$ *which are absolutely continuous w.r.t.* $\mathbb{P}$. *If for any* $A \in \mathcal{F}$, *the limit* $\mu(A) = \lim_{n\to\infty} \mu_n(A)$ *exists and is finite, then*

(1) μ *is a signed measure, and* $\sup_n \|\mu_n\| < \infty$, *where* $\|\mu_n\|$ *represents the total variation of* μ_n;
(2) *for any* $\varepsilon > 0$, *there exists* $\eta > 0$, *such that*

$$A \in \mathcal{F}, \ \mathbb{P}(A) \leqslant \eta \Rightarrow \sup_n |\mu_n|(A) \leqslant \varepsilon.$$

Lemma 1.18 *Assume that* (ξ_n) *is a sequence of integrable random variables on* $(\Omega, \mathcal{F}, \mathbb{P})$. *Then in order that* (ξ_n) *is weakly convergent to an integrable random variable* ξ *in* L^1, *it is necessary and sufficient that for any* $A \in \mathcal{F}$, *the limit of* $\mathbb{E}[\xi_n I_A]$ *exists and is finite.*

Proof The necessity is obvious. We will prove the sufficiency. Assume that the conditions of the lemma are satisfied. Let μ_n be the indefinite integral of ξ_n w.r.t. $\mathbb{P}$. By the Vitali-Hahn-Saks theorem, $\sup_n \|\mu_n\| = \sup_n \mathbb{E}[|\xi_n|] < \infty$. Besides, there is a finite measure μ on $\mathcal{F}$, such that for all $A \in \mathcal{F}$, it holds $\mu(A) = \lim_{n\to\infty} \mu_n(A)$, and $\mu \ll \mathbb{P}$. Set $\xi = d\mu/d\mathbb{P}$. It is easy to see that ξ_n converges weakly to ξ. (Here the fact that $\sup_n E[|\xi_n|] < \infty$ is used.) □

The following theorem is a part of the famous Dunford-Pettis *weak compactness criterion*, which is the most useful part for probability theory.

Theorem 1.19 *Let* $\mathcal{H} \subset L^1(\Omega, \mathcal{F}, \mathbb{P})$. *Then the following conditions are equivalent:*

(1) $\mathcal{H}$ *is a uniformly integrable family;*
(2) *For any sequence* (ξ_n) *in* $\mathcal{H}$, *there exists a subsequence* (ξ_{n_k}), *such that it is weakly convergent in* L^1.

Proof (1)$\Rightarrow$(2). Assume that $\mathcal{H}$ is a uniformly integrable family. Let (ξ_n) be a sequence in $\mathcal{H}$, $\mathcal{G} = \sigma(\xi_1, \xi_2, \cdots)$. Then $\mathcal{G}$ is a separable σ-algebra. Thus, there is a countable algebra $\mathcal{A} = \{A_1, A_2, \cdots\}$, such that $\sigma(\mathcal{A}) = \mathcal{G}$. By the diagonal rule, we can choose a subsequence (ξ_{n_k}) of (ξ_n) such that for all $j \geqslant 1$, $\lim_{k\to\infty} \mathbb{E}[\xi_{n_k} I_{A_j}]$ exists and finite. Put

$$\mathcal{H} = \{A \in \mathcal{G} \mid \lim_{k\to\infty} E[\xi_{n_k} I_A] \text{ exists and finite}\}.$$

From the uniform integrability of (ξ_{n_k}) we observe easily that $\mathcal{H}$ is a monotone class. Since $\mathcal{A} \subset \mathcal{H}$, by the monotone class theorem, $\mathcal{H} = \mathcal{G}$. Then by Lemma 1.18, (ξ_{n_k}) converges weakly in $L^1(\Omega, \mathcal{G}, \mathbb{P})$. Thus, for all bounded $\mathcal{G}$-measurable random variable η, $\lim_{k \to \infty} \mathbb{E}[\xi_{n_k} \eta]$ exists and finite. Now assume $A \in \mathcal{F}$, and let $\eta = \mathbb{E}[I_A | \mathcal{G}]$. Then we have $\mathbb{E}[\xi_{n_k} I_A] = \mathbb{E}[\mathbb{E}[\xi_{n_k} I_A | \mathcal{G}]] = E[\xi_{n_k} \eta]$, and consequently, $\lim_{k \to \infty} \mathbb{E}[\xi_{n_k} I_A]$ exists and finite. Once again by Lemma 1.18, (ξ_{n_k}) converges weakly in $L^1(\Omega, \mathcal{F}, \mathbb{P})$.

(2)$\Rightarrow$(1). We prove it by contradiction. Assuming that (1) does not hold. Then there exists a sequence (ξ_n) in $\mathcal{H}$, such that $\lim_{n \to \infty} \mathbb{E}[|\xi_n|] = \infty$, or there is $\varepsilon > 0$ and a sequence $(A_n, n \geqslant 1)$ of sets in $\mathcal{F}$, satisfying

$$\lim_{n \to \infty} \mathbb{P}(A_n) = 0, \quad \inf_n \int_{A_n} |\xi_n| d\mathbb{P} \geqslant \varepsilon.$$

By the Vitali-Hahn-Saks theorem, the sequence (ξ_n) cannot have a weakly convergent subsequence. $\qquad \square$

1.5   Discrete Time Martingales

The term martingale comes from a gambling policy. It was first introduced into probability theory by J. Ville in 1939. He borrowed its French word meaning "betting strategy " (i.e., after losing a bet, the gambler doubles his bet). This chapter briefly describes some results on discrete-time martingale, such as Doob's optional stopping theorem, Doob's inequalities, martingale convergence theorems, martingale transforms, and Snell's envelope, among others.

1.5.1   Basic Definitions

Suppose that a gambler participates in a gamble and at each game the probabilities of his winning or losing are equally $1/2$. If he loses, then he keeps doubling his bet until he wins a game and then exits. With this so-called *doubling bet policy*, he will finally win the same amount of his initial bet, but this conclusion is based on an unrealistic premise that he has infinite funds. We say that a kind of gambling is fair if no one can increase his expected wealth by limited times of gambling. To see this point, let $X_1, X_2, \cdots$ be a sequence of independent and identically distributed random variables, with $\mathbb{P}(X_i = 1) = \mathbb{P}(X_i = -1) = 1/2$, where $[X_i = 1]$ and $[X_i = -1]$ represent the events corresponding to the gambler's winning and losing at the i-th gambling, respectively. Let $\mathcal{F}_k$ be the σ-algebra generated by $\{X_1, \cdots, X_k\}$. If W_k represents the k-th gambling bet, then W_k must be $\mathcal{F}_{k-1}$-measurable, because the gambler can determine the k-th gambling bet only according to the results of previous $k - 1$ gambles. The gambler's wealth at time n is

$$Y_n = Y_0 + \sum_{i=1}^{n} W_i X_i = Y_{n-1} + W_n X_n.$$

Since W_n is $\mathcal{F}_{n-1}$-measurable and X_n is independent of $\mathcal{F}_{n-1}$, we have

$$\mathbb{E}[Y_n|\mathcal{F}_{n-1}] = Y_{n-1} + W_n\mathbb{E}[X_n|\mathcal{F}_{n-1}]$$
$$= Y_{n-1} + W_n\mathbb{E}[X_n] = Y_{n-1}.$$

This property of the sequence (Y_n) is called the *martingale property*. In particular, if each W_k is integrable, then the martingale property implies that $\mathbb{E}[Y_n] = Y_0$ for all n.

Now we give a general definition of martingale and study its basic properties. Let $(\Omega, \mathcal{F}, \mathbb{P})$ be a probability space and $(\mathcal{F}_n, n \geqslant 0)$ be a monotone increasing sequence of sub-σ-algebras of $\mathcal{F}$. Put $\mathcal{F}_\infty \hat{=} \sigma(\bigcup_n \mathcal{F}_n)$, $\mathcal{F}_{-1} = \mathcal{F}_0$. A sequence of random variables $(X_n, n \geqslant 0)$ is said to be $(\mathcal{F}_n)$-*adapted* (respectively, *predictable*), if each X_n is $\mathcal{F}_n$-(respectively, $\mathcal{F}_{n-1}$-)measurable.

We denote $\overline{\mathbb{N}}_0 = \{0, 1, 2, \cdots\}$. Let τ be an $\overline{\mathbb{N}}_0$-valued random variable. If $\forall n \in \overline{\mathbb{N}}_0, [\tau = n] \in \mathcal{F}_n$, then we call τ an $(\mathcal{F}_n)$-*stopping time*. For a stopping time τ , set

$$\mathcal{F}_\tau = \{A \in \mathcal{F}_\infty : A \cap [\tau = n] \in \mathcal{F}_n, \forall n \geqslant 0\}. \tag{1.11}$$

Then $\mathcal{F}_\tau$ is a σ-algebra, called *σ-algebra of events before τ*.

Theorem 1.20 *Let $(X_n, n \geqslant 0)$ be an $(\mathcal{F}_n)$-adapted sequence of random variables, T be a stopping time. Put*

$$X_n^T(\omega) = X_{T \wedge n}(\omega) = X_{T(\omega) \wedge n}(\omega).$$

Then $(X_n^T, n \geqslant 0)$ is an $(\mathcal{F}_n)$-adapted sequence of random variables, and $X_T I_{[T < \infty]}$ is $\mathcal{F}_T$-measurable.

Definition 1.21 Let $(X_n, n \geqslant 0)$ be an $(\mathcal{F}_n)$-adapted sequence of random variables. It is called a *martingale (respectively, supermartingale and submartingale)* w.r.t. $(\mathcal{F}_n)$, if each X_n is integrable and

$$\mathbb{E}[X_{n+1} | \mathcal{F}_n] = X_n \text{ (respectively, } \leqslant X_n, \geqslant X_n) \text{ a.s..} \tag{1.12}$$

It is called a *local martingale* if there is a monotone increasing sequence of stopping times (T_n), satisfying $\lim_n T_n = \infty$, such that for each k, $(X_{n \wedge T_k} I_{[T_k > 0]}, n \geqslant 0)$ is a martingale, or equivalently, for each k, $(X_{n \wedge T_k} - X_0, n \geqslant 0)$ is a martingale.

Let $(X_n, n \geqslant 0)$ be a sequence of random variables and $\mathcal{F}_n$ be the σ-algebra generated by $\{X_1, \cdots, X_n\}$. We call $(\mathcal{F}_n)$ the *natural σ-filtration* of the sequence (X_n). If (X_n) is a martingale (supermartingale, submartingale) with respect to its natural σ-filtration $(\mathcal{F}_n)$, we simply call it a martingale (supermartingale, submartingale).

Henceforth, if a σ-filtration $(\mathcal{F}_n)$ is pre-specified, or unambiguous from the context, then when we state that a sequence is a martingale (supermartingale, submartingale), we may omit to mention " w.r.t. $(\mathcal{F}_n)$ ". In this case, it is not necessary that $(\mathcal{F}_n)$ is the natural σ-filtration of the sequence.

The following conclusions are evident:

(1) If (X_n) and (Y_n) are (super-)martingales, then $(X_n + Y_n)$ is also a (super-)martingale, and $(X_n \wedge Y_n)$ ia a supermartingale.
(2) If (X_n) is a (sub-)martingale and $f : \mathbb{R} \longrightarrow \mathbb{R}$ is a (non-decreasing) convex function on $\mathbb{R}$, such that each $f(X_n)$ in integrable, then by Jensen's inequality, $\big(f(X_n)\big)$ is a submartingale.
(3) Let (X_n) be a supermartingale. Put

$$A_n = X_0 + \sum_{j=1}^{n}(X_{j-1} - \mathbb{E}[X_j|\mathcal{F}_{j-1}]); \quad M_n = X_n + A_n, \quad n\geqslant 1, \qquad (1.13)$$

and $A_0 = 0$, $M_0 = X_0$. Then (A_n) is a non-decreasing predictable process, and (M_n) is a martingale. We call expression $X = M - A$ *Doob's decomposition* of supermartingale X.

1.5.2 Basic Theorems

Theorem 1.22 (Doob's optional stopping theorem)   *Let (X_n) be a martingale (submartingale). If S and T are bounded stopping times, then X_S and X_T are integrable, and*

$$\mathbb{E}[X_T \mid \mathcal{F}_S] = X_{T\wedge S} \ (\geqslant X_{T\wedge S}) \ a.s. \ . \qquad (1.14)$$

In particular, for any stopping time T, $(X_n^T, n\geqslant 0)$ is a martingale (submartingale).

Proof We only consider the submartingale case. Let (X_n) be a submartingale. Assume that stopping times S and T are less than a certain natural number m. First assume $S\leqslant T$. Since $|X_T|\leqslant \sum_{j=0}^{m} |X_j|, |X_S|\leqslant \sum_{j=0}^{m} |X_j|$, X_T and X_S are integrable. Let $j\geqslant 0$, $A \in \mathcal{F}_S$. We have

$$A_j := A \cap [S = j] \cap [T > j] \in \mathcal{F}_j.$$

If $T - S\leqslant 1$, since (X_n) is a submartingale, we have

$$\int_A (X_S - X_T)d\mathbb{P} = \sum_{j=0}^{m} \int_{A_j} (X_j - X_{j+1})d\mathbb{P}\leqslant 0.$$

For the general case, let $R_j = T \wedge (S + j)$, $1 \leq j \leq m$. Then each R_j is a stopping time, and

$$S \leq R_1 \leq \cdots \mathbb{R}_m; \quad R_1 - S \leq 1, R_{j+1} - R_j \leq 1, \quad 1 \leq j \leq m - 1.$$

Thus, we have:

$$\int_A X_S d\mathbb{P} \leq \int_A X_{R_1} d\mathbb{P} \leq \cdots \leq \int_A X_T d\mathbb{P},$$

which proves (1.14) for the case of $S \leq T$.

For the case of two bounded stopping times, we can apply the above result to stopping times S and $T \vee S$ and obtain

$$\mathbb{E}[X_T \,|\, \mathcal{F}_S] = \mathbb{E}[X_{T \vee S} I_{[T \geq S]} + X_{T \wedge S} I_{[T < S]} \,|\, \mathcal{F}_S]$$

$$\geq X_S I_{[T \geq S]} + X_{T \wedge S} I_{[T < S]} = X_{T \wedge S}.$$

The proof is completed.   □

Remark Theorem 1.22 asserts that in a fair gamble, one cannot use a bounded stopping time strategy to increase his expected wealth.

Theorem 1.23 *Let* $(X_n)_{n \leq N}$ *be a nonnegative submartingale. Put* $X_N^* = \sup_{n \leq N} X_n$. *Then* $\forall \lambda > 0$, *we have*

$$\lambda \mathbb{P}(X_N^* \geq \lambda) \leq \int_{[X_N^* \geq \lambda]} X_N d\mathbb{P}; \tag{1.15}$$

for $p > 1$, *we have*

$$(\mathbb{E}[(X_N^*)^p])^{1/p} \leq \frac{p}{p-1} (\mathbb{E}[X_N^p])^{1/p} . \tag{1.16}$$

(1.15) *and* (1.16) *are called   Doob's maximum inequality and   Doob's inequality, respectively.*

Proof For $\lambda > 0$, set $T = \inf\{n : X_n \geq \lambda\} \wedge N$. Then on $[X_N^* \geq \lambda]$, it holds $X_T \geq \lambda$, and on $[X_N^* < \lambda]$ we have $T = N$. Since (X_n) is a submartingale, by Theorem 1.22, we obtain:

$$\mathbb{E}[X_N] \geq \mathbb{E}[X_T] = \int_{[X_N^* \geq \lambda]} X_T d\mathbb{P} + \int_{[X_N^* < \lambda]} X_N d\mathbb{P}$$

$$\geq \lambda \mathbb{P}(X_N^* \geq \lambda) + \int_{[X_N^* < \lambda]} X_N d\mathbb{P},$$

which implies (1.15).

Next we prove (1.16). By Fubini theorem and from (1.15),

$$
\mathbb{E}[(X_N^*)^p] = \mathbb{E} \int_0^\infty p\lambda^{p-1} I_{[X_N^* \geq \lambda]} d\lambda
$$

$$
= \int_0^\infty p\lambda^{p-1} \mathbb{P}(X_N^* \geq \lambda) d\lambda
$$

$$
\leq \int_0^\infty p\lambda^{p-2} \int_{[X_N^* \geq \lambda]} X_N d\mathbb{P} d\lambda
$$

$$
= \frac{p}{p-1} \mathbb{E}[X_N (X_N^*)^{p-1}].
$$

By Hölder's inequality for mathematical expectation $|\mathbb{E}[\xi\eta]| \leq (\mathbb{E}[|\xi|^p])^{\frac{1}{p}}$ $(\mathbb{E}[|\eta|^{\frac{p}{p-1}}])^{\frac{p-1}{p}}$, we further obtain

$$
\mathbb{E}[(X_N^*)^p] \leq \frac{p}{p-1} (\mathbb{E}[X_N^p])^{\frac{1}{p}} (\mathbb{E}[(X_N^*)^p])^{\frac{p-1}{p}},
$$

which implies (1.16).   □

We give below the supermartingale convergence theorem and omit its proof.

Theorem 1.24 (Supermartingale convergence theorem) *Let (X_n) be a super-martingale. If $\sup_n E[X_n^-] < \infty$ (or equivalently, $\sup_n \mathbb{E}[|X_n|] < \infty$, because $E[|X_n|] = E[X_n] + 2E[X_n^-]$), then when $n \to \infty$, X_n a.s. converges to an integrable random variable X_∞. If (X_n) is a nonnegative supermartingale, then for each $n \geq 0$*

$$
E[X_\infty \mid \mathcal{F}_n] \leq X_n \quad a.s.
$$

In particular, if ξ is an integrable random variable, then we have

$$
\lim_{n \to \infty} \mathbb{E}[\xi|\mathcal{F}_n] = \mathbb{E}[\xi| \vee_n \mathcal{F}_n].
$$

Now we consider index set $-\mathbb{N}_0 = \{\cdots, -2, -1, 0\}$. Let $(\mathcal{F}_n)_{n \in -\mathbb{N}_0}$ be a sequence of sub-σ-algebra of $\mathcal{F}$ such that for all $n \in -\mathbb{N}_0$, $\mathcal{F}_{n-1} \subset \mathcal{F}_n$. We call an $(\mathcal{F}_n)$-adapted random sequence $(X_n)_{n \in -\mathbb{N}_0}$ a *reverse martingale* (respectively, *reverse supermartingale*), if for each $n \in -\mathbb{N}_0$, X_n is integrable, and

$$
\mathbb{E}[X_n \mid \mathcal{F}_{n-1}] = X_{n-1} \text{ (respectively, } \leq X_{n-1}) \quad a.s. .
$$

Theorem 1.25 (Reverse supermartingale convergence theorem) *Let $(X_n)_{n \in -\mathbb{N}_0}$ be a reverse supermartingale, then $\lim_{n \to -\infty} X_n$ a.s. exists. If $\lim_{n \to -\infty} \mathbb{E}[X_n] < +\infty$ a.s., then (X_n) is uniformly integrable, and X_n a.s. L^1-converges to $X_{-\infty}$. In* *particular, if ξ is an integrable random variable, then we have*

$$\lim_{n \to -\infty} \mathbb{E}[\xi | \mathcal{F}_n] = \mathbb{E}[\xi | \cap_n \mathcal{F}_n].$$

1.5.3 Martingale Transforms

Definition 1.26 Let $(M_n, n \geqslant 0)$ be an adapted sequence of random variables and (H_n) be a predictable sequence. We define $\Delta M_n = M_n - M_{n-1}$,

$$X_0 = H_0 M_0, \ X_n = H_0 M_0 + \sum_{i=1}^{n} H_i \Delta M_i, \ n \geqslant 1,$$

and denote X by $H.M$. If M is a martingale, we call $H.M$ the *martingale transform* of M via H.

In order to study the martingale transforms, we need to introduce the notion of σ-integrable random variable.

Definition 1.27 Let $(\Omega, \mathcal{F}, \mathbb{P})$ be a probability space and $\mathcal{G}$ be a sub-σ-algebra of $\mathcal{F}$. A random variable ξ is said to be *σ-integrable* w.r.t. $\mathcal{G}$ if there exists $\Omega_n \in \mathcal{G}$, $\Omega_n \uparrow \Omega$, such that ξI_{Ω_n} is integrable for each n.

Remark It is easy to prove that a random variable ξ is σ-integrable w.r.t. $\mathcal{G}$, if and only if there is a $\mathcal{G}$-measurable real random variable $\eta > 0$, such that $\xi \eta$ is integrable.

Theorem 1.28 *Let ξ be a random variable which is σ-integrable w.r.t. $\mathcal{G}$. Put*

$$\mathcal{C} = \{A \in \mathcal{G} : \mathbb{E}[|\xi| I_A] < +\infty\}.$$

Then there is an a.s. unique $\mathcal{G}$-measurable real random variable η such that for all $A \in \mathcal{C}$, we have

$$\mathbb{E}[\xi I_A] = \mathbb{E}[\eta I_A]. \tag{1.17}$$

We call η the conditional expectation of ξ w.r.t. $\mathcal{G}$ and denote it by $\mathbb{E}[\xi | \mathcal{G}]$.

Proof It may be assumed that ξ is nonnegative. We take $\Omega_n \in \mathcal{G}$, $\Omega_n \uparrow \Omega$, such that ξI_{Ω_n} is integrable. Let $\eta_n = \mathbb{E}[\xi I_{\Omega_n} | \mathcal{G}]$. Then $\eta_{n+1} I_{\Omega_n} = \eta_n$ a.s., $\eta_n \uparrow \eta$ a.s., where η is a $\mathcal{G}$-measurable real random variable. For any $A \in \mathcal{C}$,

$$\mathbb{E}[\xi I_A] = \lim_n \mathbb{E}[\xi I_A I_{\Omega_n}] = \lim_n \mathbb{E}[\eta_n I_A] = \mathbb{E}[\eta I_A],$$

which is (1.17). By (1.17), ηI_{Ω_n} is the conditional expectation of ξI_{Ω_n} w.r.t. $\mathcal{G}$, and hence η is a.s. uniquely determined. $\square$

It is easy to show that the above extended conditional expectation preserves the properties of the usual conditional expectation such as linearity and the monotone convergence theorem.

The following theorem is from Meyer (1972, pp. 47–48).

Theorem 1.29 *Let $X = (X_n, n \geqslant 0)$ be an adapted sequence. Then the following propositions are equivalent:*

(1) *X is a local martingale;*
(2) *For each $n \geqslant 0$, X_{n+1} is σ-integrable w.r.t. $\mathcal{F}_n$, and $\mathbb{E}[X_{n+1} \mid \mathcal{F}_n] = X_n$ a.s.;*
(3) *X is a martingale transform.*

Proof (1)$\Rightarrow$(2). Let $T_n \uparrow \infty$ be a sequence of stopping times such that for each k, the sequence $Z_n = X_{n \wedge T_k} I_{[T_k > 0]}$, $n \geqslant 0$, is a martingale. Then for any $A \in \mathcal{F}_n$, since $A \cap [T_k > n] \in \mathcal{F}_n$, we have

$$\mathbb{E}[Z_{n+1} I_{A \cap [T_k > n]}] = \mathbb{E}[Z_n I_{A \cap [T_k > n]}],$$

namely, $\mathbb{E}[X_{n+1} I_{A \cap [T_k > n]}] = \mathbb{E}[X_n I_{A \cap [T_k > n]}]$. Because $[T_k > n] \uparrow \Omega$, when $k \to \infty$, we obtain (2).

(2)$\Rightarrow$(3). Put $H_0 = X_0$, $H_n = \mathbb{E}[|X_n - X_{n-1}| \mid \mathcal{F}_{n-1}]$, $n \geqslant 1$; $V_0 = 0$, $V_n = \frac{1}{H_n} I_{[H_n > 0]}$, $n \geqslant 1$. Then the sequences (H_n) and (V_n) are predictable. Let $M = 1 + V.X$, then it is easy to see that M is a martingale and $X = H.M$.

(3)$\Rightarrow$(1). Assume that $X = H.M$ is a martingale transform, where M is a martingale and H is a predictable sequence. Put

$$T_k = \inf\{n : |H_{n+1}| \geqslant k\}.$$

Then each T_k is a stopping time, $T_k \uparrow \infty$, and on $[T_k > 0]$, $|H_{n \wedge T_k}| \leqslant k$. Hence, each $X_{n \wedge T_k} I_{[T_k > 0]}$ is integrable, and we have

$$\mathbb{E}[(X_{n+1 \wedge T_k} - X_{n \wedge T_k}) I_{[T_k > 0]} \mid \mathcal{F}_n] = H_{n+1} I_{[T_k > 0]} \mathbb{E}[(M_{n+1 \wedge T_k} - M_{n \wedge T_k}) \mid \mathcal{F}_n] = 0.$$

This shows that $(X_{n \wedge T_k} I_{[T_k > 0]}, n \geqslant 0)$ is a martingale. Therefore, X is a local martingale. $\square$

Corollary 1.30

(1) *Let M be a local martingale. If each M_n is integrable, then M is a martingale. In particular, if M is a nonnegative local martingale and M_0 is integrable, then M is a martingale.*
(2) *Let L be a local martingale, and (H_n) is a predictable sequence, then $X = H.L$ is a martingale transform.*

Proof (1) is a direct consequence of (1)$\Rightarrow$ (2) in Theorem 1.29. To prove (2), by (1)$\Rightarrow$ (3) in Theorem 1.29, $L = K.M$, where M is a martingale and K a predictable sequence. Thus, $X = H.L = (HK).M$. So X is a martingale transform. $\qquad \square$

Theorem 1.31 *If $(M_n, 0 \leqslant n \leqslant N)$ is an adapted sequence of integrable r.v.'s, such that for any bounded predictable sequence $(H_n, 1 \leqslant n \leqslant N)$, we have $\mathbb{E}[\sum_{j=1}^{N} H_j \Delta M_j] = 0$, then (M_n) is a martingale.*

Proof For any $1 \leqslant j \leqslant N$, and $A \in \mathcal{F}_{j-1}$, we put $H_n = 0, n \neq j, H_j = I_A$. Then (H_n) is a bounded predictable sequence and by assumption $\mathbb{E}[I_A(M_j - M_{j-1})] = 0$. This means that $\mathbb{E}[M_j | \mathcal{F}_{j-1}] = M_{j-1}$. Thus, (M_n) is a martingale. $\qquad \square$

Lemma 1.32 *Let ξ be a random variable which is σ-integrable w.r.t. $\mathcal{G}$ and $\mathbb{E}[\xi \mid \mathcal{G}] = 0$. If there is an integrable random variable η such that $\xi \geqslant \eta$ or $\xi \leqslant \eta$, then ξ is also integrable. In particular, ξ is integrable if and only if ξ^+ or ξ^- is integrable.*

Proof Assume $\xi \geqslant \eta$. Let $X = \xi - \eta$. Then X is a nonnegative random variable. We have

$$\mathbb{E}[X] = \mathbb{E}[\mathbb{E}[\xi - \eta \mid \mathcal{G}]] = \mathbb{E}[\mathbb{E}[-\eta \mid \mathcal{G}]] = \mathbb{E}[-\eta] < \infty.$$

Thus, X is integrable, which implies that ξ is integrable. For the case where $\xi \leqslant \eta$, let $X = \eta - \xi$. Then a similar proof implies that ξ is integrable. $\qquad \square$

The following theorem gives an important example of the martingale transform.

Theorem 1.33 (Exponential martingale) *Let L be a local martingale with $L_0 = 0$. Put*

$$Z_0 = 1, \quad Z_n = \prod_{k=1}^{n}(1 + \Delta L_k), n \geqslant 1. \tag{1.18}$$

Then Z is a local martingale. We call Z the exponential martingale of L. If furthermore $\Delta L_n \geqslant -1$, then Z is a nonnegative martingale, and L itself is a martingale. Besides, every nonnegative martingale Z with $Z_0 = 1$ is of the form (1.18), where L is a martingale with $\Delta L_n \geqslant -1$.

Proof From (1.18) we obtain

$$Z_n - Z_{n-1} = Z_{n-1} \Delta L_n, \quad n \geqslant 1.$$

By Corollary 1.30, $Z - 1 = H.L$ is a martingale transform, where $H_n = Z_{n-1}$. Thus, by Theorem 1.29, Z is a local martingale.

If $\Delta L_n \geqslant -1$, then by Corollary 1.30 (1), Z is a nonnegative martingale. In this case, by Lemma 1.32, ΔL_n is integrable, so that L itself is a martingale.

Now assume that Z is a nonnegative martingale and $Z_0 = 1$. Put

$$L_0 = 0, \quad L_n = L_{n-1} + (Z_n/Z_{n-1} - 1)I_{[Z_{n-1}>0]}, \quad n \geqslant 1.$$

Then by Theorem 1.29 L is a local martingale. Due to $\Delta L_n \geqslant -1$, by Lemma 1.32 ΔL_n is integrable, and consequently, L is a martingale. Finally, it is easy to verify that (1.18) holds. □

1.5.4 Snell Envelop

The following theorem solves an optimal stopping problem.

Theorem 1.34 (Snell envelope)   *Let $(Z_n)_{0 \leqslant n \leqslant N}$ be an adapted sequence of integrable r.v.'s. Define by backward induction a sequence (U_n):*

$$U_N = Z_N, \quad U_n = \text{Max}(Z_n, \mathbb{E}[U_{n+1}|\mathcal{F}_n]), \quad n \leqslant N - 1.$$

(1) *(U_n) is the smallest supermartingale dominating (Z_n) (i.e., $U_n \geqslant Z_n$ for all n). (U_n) is called the Snell envelope of (Z_n).*

(2) *Let $\mathcal{T}_{j,N}$ be the set of stopping times taking values in $\{j, \cdots, N\}$ and $T_j = \inf\{l \geqslant j : U_l = Z_l\}$, where $\inf \emptyset := N$. Then each T_j is a stopping time, $(U_n^{T_j})$ is a martingale, and for all $j \leqslant N$,*

$$U_j = \mathbb{E}[Z_{T_j} \mid \mathcal{F}_j] = \text{ess sup}\{\mathbb{E}[Z_T \mid \mathcal{F}_j] : T \in \mathcal{T}_{j,N}\}.$$

In particular, the supremum of expected values $\mathbb{E}[Z_T]$ on $\mathcal{T}_{j,N}$ is attained at T_j, and the optimal value is equal to $\mathbb{E}[U_j]$, namely,

$$\mathbb{E}[U_j] = \mathbb{E}[Z_{T_j}] = \sup\{\mathbb{E}[Z_T] : T \in \mathcal{T}_{j,N}\}.$$

Proof

(1) Since $U_n \geqslant \mathbb{E}[U_{n+1} \mid \mathcal{F}_n]$, and $U_n \geqslant Z_n$, (U_n) is a supermartingale dominating (Z_n). Let (V_n) be any other supermartingale dominating (Z_n). By backward induction, it is easy to see that (V_n) dominates (U_n). So (U_n) is the smallest supermartingale dominating (Z_n).

(2) The claim that T_n's are stopping times is easy to verify. Since $U_n^{T_j} = U_{n \wedge T_j}$, we have for $n \leqslant N - 1$,

$$U_{n+1}^{T_j} - U_n^{T_j} = I[T_j \geqslant n + 1](U_{n+1} - U_n).$$

On the other hand, by the definitions of T_j and U_n,

$$U_n = \mathbb{E}[U_{n+1} \mid \mathcal{F}_n] \text{ on} [T_j \geqslant n + 1].$$

So we have

$$U_{n+1}^{T_j} - U_n^{T_j} = I[T_j \geqslant n + 1](U_{n+1} - \mathbb{E}[U_{n+1} \mid \mathcal{F}_n]).$$

Noting that $[T_j \geqslant n + 1] = [T_j \leqslant n]^c \in \mathcal{F}_n$, the above equation implies that

$$\mathbb{E}[U_{n+1}^{T_j} - U_n^{T_j} \mid \mathcal{F}_n] = 0.$$

Thus, for each j, $(U_n^{T_j})$ is a martingale. Since $U_{T_j} = Z_{T_j}$, we have

$$U_j = U_j^{T_j} = \mathbb{E}[U_N^{T_j} \mid \mathcal{F}_j] = \mathbb{E}[Z_{T_j} \mid \mathcal{F}_j] = \mathbb{E}[Z_{T_j} \mid \mathcal{F}_j].$$

Now for any $T \in \mathcal{T}_{j,N}$, since $U_T \geqslant Z_T$ and (U_n) is a supermartingale, we have

$$\mathbb{E}[Z_T \mid \mathcal{F}_j] \leqslant \mathbb{E}[U_T \mid \mathcal{F}_j] \leqslant U_j.$$

Claim (2) is proved.   □

1.6 Markov Sequences

Let $(\Omega, \mathcal{F}, \mathbb{P})$ be a probability space. Let $\mathcal{G}, \mathcal{G}_1$ and $\mathcal{G}_2$ be sub-σ-algebras of $\mathcal{F}$. If for any $B_1 \in \mathcal{G}_1$, $B_2 \in \mathcal{G}_2$ we have

$$\mathbb{P}[B_1 B_2 | \mathcal{G}] = \mathbb{P}[B_1 | \mathcal{G}] \mathbb{P}[B_2 | \mathcal{G}] \text{ a.s.},$$

then we say that $\mathcal{G}_1$ and $\mathcal{G}_2$ are *conditionally independent* given σ-algebra $\mathcal{G}$. It is easy to prove that $\mathcal{G}_1$ and $\mathcal{G}_2$ are conditionally independent given σ-algebra $\mathcal{G}$, if and only if for any $B_2 \in \mathcal{G}_2$,

$$\mathbb{P}[B_2 | \mathcal{G}_1 \vee \mathcal{G}] = \mathbb{P}[B_2 | \mathcal{G}] \text{ a.s.}, \qquad (1.19)$$

where $\mathcal{G}_1 \vee \mathcal{G}$ is the smallest σ-algebra containing $\mathcal{G}_1$ and $\mathcal{G}$.

Let X and Y be two $\mathbb{R}^d$-valued random variables. If $\sigma(X)$ and $\sigma(Y)$ are conditionally independent given $\mathcal{G}$, then we say that X and Y are *conditionally independent* given $\mathcal{G}$. By (1.19), it means that for any $A \in \mathcal{B}(\mathbb{R}^d)$,

$$\mathbb{P}[X \in A | \sigma(Y) \vee \mathcal{G}] = \mathbb{P}[X \in A | \mathcal{G}], \qquad (1.20)$$

because $\sigma(X) = \{[X \in A] : A \in \mathcal{B}(\mathbb{R}^d)\}$. If X and Y are conditionally independent nonnegative (or integrable) random variables given $\mathcal{G}$, then we have:

$$\mathbb{E}[XY|\mathcal{G}] = \mathbb{E}[X|\mathcal{G}]\mathbb{E}[Y|\mathcal{G}] \text{ a.s..} \qquad (1.21)$$

Let $(\Omega, \mathcal{F}, \mathbb{P})$ be a probability space. A sequence of $\mathbb{R}^d$-valued random variables $\{X_0, X_1, \cdots\}$ is called a *Markov sequence*, if for all $n \geqslant 1$ and $A \in \mathcal{B}(\mathbb{R}^d)$,

$$\mathbb{P}[X_{n+1} \in A|X_0, X_1, \cdots, X_n] = \mathbb{P}[X_{n+1} \in A|X_n]. \qquad (1.22)$$

By (1.19), this means that the sequence (X_n) has the following *Markov property*: the past evolution $\{X_0, \cdots, X_{n-1}\}$ and the future state X_{n+1} are conditionally independent, given the present state X_n.

Let $\mathcal{S}$ be a non-empty set which consists of at most countable elements. If $\{X_0, X_1, \cdots\}$ are discrete random variables with range $\mathcal{S}$, then the Markov property reduces to the requirement that

$$\mathbb{P}(X_{n+1} = x_{n+1}|X_0 = x_0, \cdots, X_n = x_n) = \mathbb{P}(X_{n+1} = x_{n+1}|X_n = x_n),$$

where $x_i \in \mathcal{S}, i = 0, 1, \cdots, n + 1$. In this case the sequence $\{X_0, X_1, \cdots\}$ is called a *Markov chain* with *state space* $\mathcal{S}$, and conditional probabilities $\mathbb{P}(X_{n+1} = x_{n+1}|X_n = x_n)$ are called one-step *transition probabilities* of the chain. If they do not depend on n, the chain is called *stationary*. In this case we put $P(x, y) = \mathbb{P}(X_1 = y|X_0 = x), x, y \in \mathcal{S}$ and call $P(x, y)$ the *transition function* of the chain.

A simple example of stationary Markov chain is the *birth and death chain*, which has state space $\mathcal{S} = \{0, 1, 2, \cdots\}$ and transition function

$$P(x, x-1) = q_x, \text{ if } x \geqslant 1; \quad P(x, x+1) = p_x; \quad P(x, x) = r_x; \quad P(x, y) = 0, \text{ elsewhere.}$$

Here a transition from state x to $x + 1$ (resp. $x - 1$) corresponds to a "birth" (resp. "death").

Chapter 2
Portfolio Selection Theory in Discrete-Time

When a broker (economic agent) invests in a stock market, he must select an appropriate portfolio. As a saying goes: do not put all your eggs in one basket. A core issue of portfolio selection is the trade-off between profit and risk. To tackle this issue, Harry M. Markowitz (1952) developed a mean-variance analysis, which was the first quantitative treatment of portfolio selection under uncertainty. This approach was further developed by James Tobin (1958). In the mid-1960s, based on Markowitz's mean-variance analysis and Tobin's two-fund separation theorem, William F. Sharpe (1964), John Lintner (1965), and Jan Mossin (1966) independently observed that there exists a linear relation among expected rates of return on portfolios; in a competitive equilibrium market, the coefficients of the linear relation are related to the so-called betas of rates of return on portfolios with respect to the rate of return on the market portfolio. This relation is the so-called *capital asset pricing model* (CAPM). CAPM implicitly assumes that the returns on risky assets depend on a single market factor. Ross (1976) proposed a multifactor model for the behavior of asset prices in capital markets, which is called the *arbitrage pricing theory* (APT).

Markowitz's theory is known as "Wall Street's first revolution." Markowitz and Sharpe shared the 1990 Nobel Prize in economics with Merton H. Miller, who was awarded on the basis of the theory of corporate finance contribution.

In this chapter we will present the main results on the mean-variance analysis, CAPM and APT. Markowitz's original paper in 1952 and monograph in 1958 studied the mean-variance analysis, assuming that shot-selling is not allowed. In that case the problem becomes a quadratic programming problem with both equality and inequality constrains and has no explicit solution. So we only introduce the theory of mean-variance analysis for the case where shot-selling is allowed. We follow closely Huang and Litzenberger (1988) and Wang (2006).

© Springer Nature Singapore Pte Ltd. and Science Press 2018
J.-A. Yan, *Introduction to Stochastic Finance*, Universitext,
https://doi.org/10.1007/978-981-13-1657-9_2

2.1 Mean-Variance Analysis

Assume that an economic agent makes a portfolio selection in a security market only once at the initial date, time 0, and that at time 1, he gets the return. We suppose that the security market consists of one risk-free asset, indexed by 0, and d risky assets, indexed by $1, \cdots, d$. We use $X_i(0)$ and $X_i(1)$ to denote the asset i's prices at times 0 and 1, respectively. The *rate of return* and *total return* of asset i are defined by $r_i = X_i(1)/X_i(0) - 1$ and $R_i = 1 + r_i$, respectively. The rate of return of a risky asset is a random variable, while the rate of return of the risk-free asset is a positive constant, denoted by r_f. We set $e_i = \mathbb{E}[r_i]$, the expectation of r_i, and put

$$r = (r_1, \cdots, r_d)^\tau, \quad e = (e_1, \cdots, e_d)^\tau,$$

where τ denotes the transpose of a vector. We denote by V the covariance matrix of random vector r; i.e., $V = \mathbb{E}[(r - e)(r - e)^\tau]$. Henceforth, we always assume that V is of full rank (and thus is positive definite), namely, $r_1, \cdots, r_d$ are linearly independent. In this case the market has no redundant asset in the sense that the rate of return of any asset cannot be expressed as a linear combination of rates of return of other assets.

A portfolio selection consists in determining the weight of investment on each risky asset. Let w_i be the weight on asset i, and set $w = (w_1, \cdots, w_d)^\tau$. Then $1 - w^\tau \mathbf{1}$ is the weight of the risk-free asset, where and henceforth $\mathbf{1}$ denotes the d-dimensional vector of ones. A $w_i < 0$ means a short position on asset i; $1 - \omega^\tau \mathbf{1} < 0$ means that we borrow money with an interest rate r_f. For convenience we call a weight vector w a *portfolio*.

Let w be a portfolio. The rate of return on w, denoted by $r(w)$, and the variance of $r(w)$, denoted by $\sigma^2(w)$, are

$$r(w) = w^\tau \rangle + (1 - w^\tau \mathbf{1})r_f, \qquad \sigma^2(w) = w^\tau V w, \tag{2.1}$$

respectively. Let $\mu(w) = \mathbb{E}[r(w)]$ be the *expected rate of return* of portfolio w. We call $r(w) - r_f$ and $\mu(w) - r_f$ the *excess return* and *risk premium* of portfolio w, respectively. The ratio $(\mu(w) - r_f)/\sigma(w)$ is called the *Sharpe ratio* of w, which represents the risk premium brought by a unit risk (as measured by the standard deviation of $r(\omega)$).

For a portfolio we regard its expected rate of return as the profit and its variance as the risk. For a given $\mu > 0$, a portfolio having the minimum variance among all portfolios with the same expected rate of return μ is called the *mean-variance frontier portfolio* with expected rate of return μ. Our task is to work out all mean-variance frontier portfolios.

2.1.1 *Mean-Variance Frontier Portfolios Without Risk-Free Asset*

First we consider those portfolios without the risk-free asset. In this case, searching for a mean-variance frontier portfolio consists in solving the following quadratic programing problem:

$$\begin{cases} w(\mu) = \arg \min_{w} w^\tau V w, \\ w^\tau e = \mu, \qquad w^\tau 1 = 1. \end{cases} \tag{2.2}$$

Using the method of Lagrangian multipliers, we solve $\min_{(w,\lambda,\gamma)} L(w, \lambda, \gamma)$, where

$$L(w, \lambda, \gamma) = \frac{1}{2} w^\tau V w + \lambda(\mu - w^\tau e) + \gamma(1 - w^\tau 1). \tag{2.3}$$

The first order necessary and sufficient conditions for $w(\mu)$ to be the solution of $\min_{(w,\lambda,\gamma)} L(w, \lambda, \gamma)$ are

$$\frac{\partial L}{\partial w} = V w - \lambda e - \gamma 1 = 0, \tag{2.4a}$$

$$\frac{\partial L}{\partial \lambda} = \mu - w^\tau e = 0, \tag{2.4b}$$

$$\frac{\partial L}{\partial \gamma} = 1 - w^\tau 1 = 0, \tag{2.4c}$$

where 0 denotes the d-dimensional vector of zeros. Solving (2.4a) gives

$$w(\mu) = \lambda(V^{-1} e) + \gamma(V^{-1} 1). \tag{2.4d}$$

From (2.4b), (2.4c) and (2.4d) we get two equations on λ and γ:

$$\begin{cases} \lambda(e^\tau V^{-1} e) + \gamma(e^\tau V^{-1} 1) = \mu, \\ \lambda(1^\tau V^{-1} e) + \gamma(1^\tau V^{-1} 1) = 1. \end{cases}$$

Solving the equations gives

$$\lambda = \frac{C\mu - A}{D}, \quad \gamma = \frac{B - A\mu}{D}, \tag{2.5}$$

where

$$A = 1^\tau V^{-1} e = e^\tau V^{-1} 1, \quad B = e^\tau V^{-1} e,$$

$$C = 1^\tau V^{-1} 1, \quad D = BC - A^2.$$

It is easy to see that $B > 0$ and $C > 0$. Since

$$(Ae - B1)^\tau V^{-1}(Ae - B1) = B(BC - A^2) = BD,$$

we have also $D > 0$. However, A can be either positive, negative, or zero. Finally, we obtain the following theorem.

Theorem 2.1 *The mean-variance frontier portfolio with expected rate of return* μ, *denoted by* $w(\mu)$, *is given by:*

$$w(\mu) = g + \mu h, \tag{2.6}$$

the variance of whose rate of return is

$$w(\mu)^\tau V w(\mu) = \frac{C}{D}\left(\mu - \frac{A}{C}\right)^2 + \frac{1}{C} := \sigma(\mu)^2, \tag{2.7}$$

where

$$g = \frac{1}{D}(BV^{-1}1 - AV^{-1}e), \quad h = \frac{1}{D}(CV^{-1}e - AV^{-1}1). \tag{2.8}$$

Proof (2.6) can be deduced from (2.4d) and (2.5). In addition, it is easy to show

$$g^\tau V g = \frac{B}{D}, \quad g^\tau V h = -\frac{A}{D}, \quad h^\tau V h = \frac{C}{D},$$

from which it follows (2.7).   □

From (2.6) we get immediately the following *two-fund separation theorem*, due to Tobin (1958), see also Tobin (1965).

Theorem 2.2 *Assume that* $d \geqslant 2$, $\mu_1 > 0$, $\mu_2 > 0$, *and* $\mu_1 \neq \mu_2$. *Let* $w(\mu_1)$ *and* $w(\mu_2)$ *be the mean-variance frontier portfolios with expected rates of return* μ_1 *and* μ_2, *respectively. Then* $\forall \mu > 0$, *the mean-variance frontier portfolios with expected rate of return* μ *is an affine combination of* $w(\mu_1)$ *and* $w(\mu_2)$:

$$w(\mu) = \alpha w(\mu_1) + (1 - \alpha)w(\mu_2),$$

where $\alpha = (\mu - \mu_2)/(\mu_1 - \mu_2)$ *solves the following equation:*

$$\mu = \alpha\mu_1 + (1 - \alpha)\mu_2.$$

Conversely, for any real number α *satisfying* $\mu = \alpha\mu_1 + (1 - \alpha)\mu_2 > 0$, $\alpha w(\mu_1) + (1 - \alpha)w(\mu_2)$ *is a mean-variance frontier portfolio.*

Remark The two-fund separation theorem tells us: under the mean-variance analysis criteria, an investor may just invest two trusted funds (i.e., mean-variance frontier portfolios) with different expected rates of return.

Fig. 2.1 Portfolio frontier

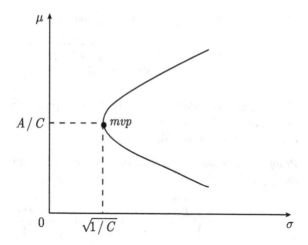

Let p and q be two mean-variance frontier portfolios. By (2.6) we have

$$p = g + \mathbb{E}[r(p)]h, \quad q = g + \mathbb{E}[r(q)]h,$$

where g and h are given by (2.8). Thus, the covariance of the rates of return on portfolios p and q is given by

$$\text{cov}(r(p), r(q)) = p^\tau V q = \frac{C}{D}\left(\mathbb{E}[r(p)] - \frac{A}{C}\right)\left(\mathbb{E}[r(q)] - \frac{A}{C}\right) + \frac{1}{C}. \tag{2.9}$$

In particular, if $p = q$, we obtain

$$\frac{\sigma^2(r(p))}{\frac{1}{C}} - \frac{\left(\mathbb{E}[r(p)] - \frac{A}{C}\right)^2}{\frac{D}{C^2}} = 1. \tag{2.10}$$

This is just (2.7). If we denote by $\sigma^2(\mu)$, the variance of the rate of return on the portfolio $w(\mu)$ (i.e., $\sigma^2(\mu) := \sigma^2(w(\mu))$), from (2.10) we see that the graph of $(\sigma(\mu), \mu)$ is a hyperbola in the $\sigma(r(p))$-$\mathbb{E}[r(p)]$ plane with center $(0, A/C)$ and asymptotes of slopes $\pm\sqrt{D/C}$. We call this hyperbola the *portfolio frontier* (Fig. 2.1). The point $(\sqrt{1/C}, A/C)$ corresponds to the mean-variance frontier portfolio which has the minimum variance of all portfolios. We call this portfolio the *minimum variance portfolio* (MVP). Those mean-variance frontier portfolios with expected rates of return higher than or equal to A/C are called *efficient portfolios*. They correspond to the upside of the portfolio frontier. It is easy to see that any convex combination of efficient portfolios will be an efficient portfolio. For investment purpose one only needs to consider efficient portfolios.

In the mean-variance model, we can also consider the following problem: for a given variance of rate of return, how to select a portfolio which has the biggest expected rate of return? The next theorem gives an answer to the problem.

Theorem 2.3 *For any given* $\sigma \geqslant \sqrt{1/C}$, *put*

$$\mu(\sigma) = \sqrt{\frac{D\sigma^2}{C} - \frac{D}{C^2} + \frac{A}{C}}.$$

Then among all portfolios whose standard deviation of the return rate equals σ, *the highest expected rate of return is* $\mu(\sigma)$, *and the corresponding portfolio* $\tilde{w}(\sigma)$ *is an efficient portfolio* $w(\mu(\sigma))$.

Proof If the standard deviation of return rate of an efficient portfolio $w(\mu)$ is σ, then by (2.7)

$$\mu = \sqrt{\frac{D\sigma^2}{C} - \frac{D}{C^2} + \frac{A}{C}} \text{ or } \mu = -\sqrt{\frac{D\sigma^2}{C} - \frac{D}{C^2} + \frac{A}{C}}.$$

Hence, among portfolios whose standard deviation of return rate is σ, the highest expected rate of return is $\mu(\sigma)$, and the corresponding portfolio $\tilde{w}(\sigma)$ is the efficient portfolio $w(\mu(\sigma))$ whose expected rate of return is $\mu(\sigma)$.   □

Remark 1 It is easy to see that $w(\mu(\sigma))$ is the optimal solution to the following problem:

$$\begin{cases} \tilde{w}(\sigma) = \arg\max_{w} w^\tau e, \\ w^\tau V w = \sigma^2, \qquad w^\tau \mathbf{1} = 1. \end{cases}$$

Remark 2 This theorem shows that in the mean-variance model, "given the variance of rate of return and select a portfolio which has the biggest expected rate of return" and "given expected rate of return and select a portfolio which has the smallest variance of rate of return" are usually not equivalent. In general, the former is better than the latter, namely, $\tilde{\mu}(\sigma) \geqslant \mu$, where $\sigma = \sigma(\mu)$.

2.1.2 Revised Formulations of Mean-Variance Analysis Without Risk-Free Asset

In this subsection, based on Cui et al. (2015), we revisit the mean-variance portfolio selection problem. We continue considering only portfolios without risk-free asset. Suppose that an agent (or investor) has an initial wealth $x_0 > 0$ and at time 1 his wealth is x_1. The difference $x_1 - x_0$ is called *capital gain* (in abbreviation, *gain*). The question is among portfolios whose expected gains are given as μ, how do we look for a portfolio so that it reaches the minimum variance of gain. Let $\hat{\mu} = \mu/x_0$. Then $\hat{\mu}$ is the expected rate of return and the optimal portfolio is just $w(\hat{\mu})$. It is the solution to problem (2.2) corresponding to $\hat{\mu}$. At this time, the amounts of money invested in various risky assets constitute a money vector $\hat{w}(x_0, \mu) = x_0 w(\hat{\mu})$.

Therefore, if we consider the money vector (also known as portfolio) instead of the weight vector, then $\hat{w}(x_0, \mu)$ is the solution to the following quadratic programming problem:

$$\begin{cases} \hat{w}(x_0, \mu) = \arg\min_{w} w^\tau V w, \\ w^\tau e = \mu, \qquad w^\tau \mathbf{1} = x_0, \end{cases} \tag{2.2*}$$

where μ is the expected gain at time 1 and $w^\tau V w$ is the variance of the gain at time 1. The variance of the gain of portfolio $\hat{w}(x_0, \mu)$ at time 1 is

$$\hat{w}(x_0, \mu)^\tau V \hat{w}(x_0, \mu) = \frac{C}{D}\left(\mu - \frac{A}{C}x_0\right)^2 + \frac{x_0^2}{C} =: \sigma(\mu, x_0)^2. \tag{2.7*}$$

Given x_0, if we let $\mu = Ax_0/C$, then $\sigma(\mu, x_0)$ attains the minimum $|x_0|/\sqrt{C}$. The set of pairs (mean, variance) with variance larger than the minimum variance

$$\left\{ (\mu, \sigma^2) : \sigma^2 = \frac{C}{D}\left(\mu - \frac{A}{C}x_0\right)^2 + \frac{x_0^2}{C}, \ \mu \geqslant \frac{A}{C}x_0 \right\}$$

forms the so-called *mean-variance efficient frontier*.

Theorem 2.4 *We have*

$$\sigma^2(\mu, x_0) = \frac{1}{BD}(A\mu - Bx_0)^2 + \frac{\mu^2}{B}. \tag{2.11}$$

Given μ, if we let $x_0 = A\mu/BD$, then $\sigma(\mu, x_0)$ attains the minimum $|\mu|/\sqrt{B}$. The set of pairs (initial wealth, variance) with variance larger than the minimum variance

$$\left\{ (x_0, \sigma^2) : \sigma^2 = \frac{1}{BD}(A\mu - Bx_0)^2 + \frac{\mu^2}{B}, \ x_0 \leqslant \frac{A}{B}\mu \right\}$$

forms the so-called initial wealth-variance efficient frontier*.*

Proof Noting that $BC - D = A^2$, from (2.7*) we deduce

$$D(B\sigma^2(\mu, x_0) - \mu^2) = BC\left(\mu - \frac{A}{C}x_0\right)^2 + \frac{BD}{C}x_0^2 - D\mu^2$$

$$= (BC - D)\mu^2 - \frac{2AB\mu x_0}{C} + \frac{B}{C}(A^2 + D)x_0^2$$

$$= A^2\mu^2 - 2AB\mu x_0 + B^2 x_0^2$$

$$= (A\mu - Bx_0)^2.$$

Thus, (2.11) is proved. The rest conclusion is a direct consequence of (2.11). $\qquad\square$

Remark 1 If $A\mu < Bx_0$, $x_0^* = A\mu/B$, then portfolio $\hat{w}(x_0^*, \mu)$ achieves the same expected gain as portfolio $\hat{w}(x_0, \mu)$, but with less initial investment x_0^*, and achieves a smaller standard deviation of gain $|\mu|/\sqrt{B}$.

Remark 2 If $\mu < Ax_0/C$, $\mu^* = Ax_0/C$, then $\hat{w}(x_0, \mu^*)$ achieves a higher expected gain and a smaller standard deviation of the gain than portfolio $\hat{w}(x_0, \mu)$, but with the initial investment remaining the same as x_0.

Both remarks above tell us that if we use the mean and variance to measure the benefit and risk, at a given level of risk, a higher investment amount does not necessarily gives higher expected return, and a higher expected return does not necessarily involves greater risk.

In the following, we always assume $x_0 > 0$ and $\mu > 0$. According to the above two observations, when we consider investment gain, a more reasonable expression of the mean-variance analysis should be

$$\begin{cases} \hat{w}(x_0, \mu) = \arg\min\limits_{w} w^\tau V w, \\ w^\tau e \geqslant \mu, \qquad w^\tau \mathbf{1} \leqslant x_0, \end{cases} \tag{2.12}$$

or,

$$\begin{cases} \tilde{w}(x_0, \mu) = \arg\min\limits_{w} w^\tau V w, \\ w^\tau e + (x_0 - w^\tau \mathbf{1}) \geqslant \mu, \qquad w^\tau \mathbf{1} \leqslant x_0. \end{cases} \tag{2.13}$$

In order to solve problem (2.12), for any given $v \geqslant \mu$, we first solve the following problem

$$\begin{cases} \hat{w}(x_0, v) = \arg\min\limits_{w} w^\tau V w, \\ w^\tau e = v, \qquad w^\tau \mathbf{1} \leqslant x_0, \end{cases} \tag{2.12*}$$

to find the optimal investment level $\hat{x}_0(v)$:

$$\hat{x}_0(v) = \arg\min\limits_{x \leqslant x_0} \frac{C}{D}\left(v - \frac{A}{C}x\right)^2 + \frac{x^2}{C}. \tag{2.14}$$

The solution is

$$\hat{x}_0(v) = \begin{cases} x_0, & Av \geqslant Bx_0, \\ \frac{A}{B}v, & Av < Bx_0. \end{cases} \tag{2.15}$$

If $Av < Bx_0$, then

$$\sigma(\hat{x}_0(v), v)^2 = \frac{C}{D}\left(v - \hat{x}_0(v)\frac{A}{C}\right)^2 + \frac{\hat{x}_0(v)^2}{C} = \frac{1}{B}v^2.$$

Consequently, we know $\hat{w}(\hat{x}_0(\mu), \mu)$ is the solution to problem (2.12). This shows that if $A\mu < Bx_0$, the optimization problem (2.12) is in fact equivalent to

$$\begin{cases} \hat{w}(x_0, \mu) = \arg\min_w w^\tau V w, \\ w^\tau e = \mu, \qquad w^\tau \mathbf{1} \leqslant x_0. \end{cases} \tag{2.16}$$

On the other hand, to solve problem (2.12), for any $x \leqslant x_0$, we can also first solve the following problem

$$\begin{cases} \hat{w}(x, v) = \arg\min_w w^\tau V w, \\ w^\tau e \geqslant \mu, \qquad w^\tau \mathbf{1} = x \end{cases} \tag{2.12'}$$

to find the optimal investment level $\hat{\mu}(x)$:

$$\hat{\mu}(x) = \arg\min_{v \geqslant \mu} \frac{C}{D}\left(v - \frac{A}{C}x\right)^2 + \frac{x^2}{C}. \tag{2.14'}$$

The solution is

$$\hat{\mu}(x) = \begin{cases} \mu, & C\mu > Ax, \\ \frac{A}{C}x, & C\mu \leqslant Ax. \end{cases} \tag{2.15'}$$

If $C\mu \leqslant Ax_0$ and $A > 0$, let $x_0^* = \frac{C}{A}\mu$, then $\hat{w}(x_0^*, \mu)$ is the solution to problem (2.12'). If $C\mu > Ax_0$, then $\hat{w}(x_0, \mu)$ is the solution to problem (2.12').

This shows that if $C\mu \leqslant Ax_0$ and $A > 0$, the optimization problem (2.12) is in fact equivalent to

$$\begin{cases} \hat{w}(x_0, \mu) = \arg\min_w w^\tau V w, \\ w^\tau e = \mu, \qquad w^\tau \mathbf{1} \leqslant x_0. \end{cases} \tag{2.17}$$

In order to solve problem (2.13), for any given $v \geqslant \mu$, we first solve the following problem

$$\begin{cases} \tilde{w}(x_0, v) = \arg\min_w w^\tau V w, \\ w^\tau e + (x_0 - w^\tau \mathbf{1}) = v, \qquad w^\tau \mathbf{1} \leqslant x_0, \end{cases} \tag{2.13*}$$

to find the optimal investment level $\hat{x}_0(v)$:

$$\tilde{x}_0(v) = \arg\min_{x \leqslant x_0} \frac{C}{D}\left(v - x_0 + x - \frac{A}{C}x\right)^2 + \frac{x^2}{C}.$$

The solution is

$$\tilde{x}_0(v) = \begin{cases} x_0, & (A - C)v \geq (B - A)x_0, \\ \frac{A-C}{B+C-2A}(v - x_0), & (A - C)v < (B - A)x_0. \end{cases} \tag{2.18}$$

If $(A - C)v < (B - A)x_0$, then

$$\sigma^2(\tilde{x}_0(v), v) = \frac{C}{D}\left(v - x_0 + \tilde{x}_0(v) - \frac{A}{C}\tilde{x}_0(v)\right)^2 + \frac{\tilde{x}_0^2(v)}{C}$$

$$= \frac{A^2 - 4AC + 2C^2 + CB}{C(B + C - 2A)^2}(v - x_0)^2.$$

To find the optimal v, we need to solve the following problem:

$$v^* = \arg\min_{v \geq \mu}(v - x_0)^2.$$

Noting that

$$B + C - 2A = (1 - e)^\tau V^{-1}(1 - e) > 0,$$

we have $(A - C) < (B - A)$.

Consider the following two cases:

(1) $A > C$. If $(A - C)\mu < (B - A)x_0$, then $\mu < x_0$, $v^* = x_0$, and $\tilde{w}(\tilde{x}_0(x_0), x_0)$ is the solution to (2.13); if $(A - C)\mu \geq (B - A)x_0$, then $v^* = \mu$, $\tilde{x}_0(\mu) = x_0$, and problem (2.13) reduces to problem (2.2*).
(2) $A \leq C$, then we have $(A - C)\mu < (B - A)x_0$. If $\mu \leq x_0$, then $v^* = x_0$, and $\tilde{w}(\tilde{x}_0(x_0), x_0)$ is the solution to (2.13); if $\mu > x_0$, then $v^* = \mu$, and $\tilde{w}(\tilde{x}_0(\mu), \mu)$ is the solution to (2.13).

The above result shows that if $\mu \leq x_0$, $(A - C)\mu < (B - A)x_0$, optimization problem (2.13) is in fact equivalent to

$$\begin{cases} \tilde{w}(x_0, \mu) = \arg\min_{w} w^\tau V w, \\ w^\tau e = w^\tau 1, \quad w^\tau 1 \leq x_0. \end{cases} \tag{2.19}$$

Remark Since $B + C - 2A > 0$, if $A \leq C$, and $\mu \geq x_0 > 0$, then the optimal investment level $\tilde{x}_0(v)$ $(= \frac{A-C}{B+C-2A}(\mu - x_0))$ is negative. In particular, if $A \leq 0$ and $\mu > x_0 > 0$, then the optimal investment level $\tilde{x}_0(v)$ is strictly negative.

For other issues, see Cui et al. (2015).

2.1.3 Mean-Variance Frontier Portfolios with Risk-Free Asset

Now we assume that in the market, a risk-free asset with rate of return r_f is available. In this case, for any $\mu \geqslant r_f$, the mean-variance frontier portfolio with expected return μ is the solution to the following quadratic program problem:

$$\begin{cases} w(\mu) = \arg\min_{w} w^\tau V w, \\ w^\tau e + (1 - w^\tau 1)r_f = \mu. \end{cases} \tag{2.20}$$

If $\mu = r_f$, the solution is trivial: $w(\mu) = \mathbf{0}$. Now we assume $\mu \neq r_f$. Using the method of Lagrangian multipliers, we solve $\min_{(w,\lambda)} L(w, \lambda)$, where

$$L(w, \lambda) = \frac{1}{2} w^\tau V w + \lambda(\mu - w^\tau e - (1 - w^\tau 1)r_f).$$

The first order necessary and sufficient conditions are

$$\frac{\partial L}{\partial w} = V w - \lambda(e - r_f 1) = 0 \tag{2.21}$$

and

$$\frac{\partial L}{\partial \lambda} = \mu - w^\tau e - (1 - w^\tau 1)r_f = 0. \tag{2.22}$$

Solving Eqs. (2.21) and (2.22) gives

$$w(\mu) = \frac{(\mu - r_f)}{H} V^{-1}(e - r_f 1), \tag{2.23}$$

where

$$H = (e - r_f 1)^\tau V^{-1}(e - r_f 1) = B - 2r_f A + r_f^2 C. \tag{2.24}$$

Since $A^2 < BC$, we have $H > 0$. From (2.23) we get

$$\sigma^2(\mu) = w(\mu)^\tau V w(\mu) = \frac{(\mu - r_f)^2}{H}, \tag{2.25}$$

where $\sigma^2(\mu)$ represents the variance of the rate of return $r(w(\mu))$ of $w(\mu)$. From (2.25) we get

$$\sigma(\mu) = \frac{1}{\sqrt{H}} |\mu - r_f|. \tag{2.26}$$

Fig. 2.2 Portfolio frontier
and capital market line

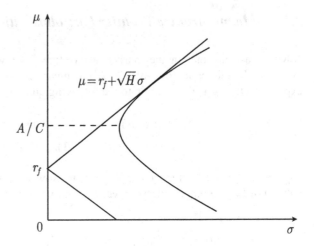

Therefore, as shown in Fig. 2.2., the graph $(\sigma(\mu), \mu)$ forms two half-lines emanating from $(0, r_f)$ in the $\sigma(r(w))$-$\mathbb{E}[r(w)]$ plane with slopes $\sqrt{H}$ and $-\sqrt{H}$, respectively. Those mean-variance frontier portfolios with expected rates of return higher than or equal to r_f are called *efficient portfolios*. They correspond to the half-line with slope $\sqrt{H}$, called the *capital market line* (see Fig. 2.2). From (2.26) we know that the slope $\sqrt{H}$ of the capital market line is the common Sharpe ratio of all efficient portfolios (i.e., $\forall \mu > r_f$, $\frac{\mu - r_f}{\sigma(\mu)} = \sqrt{H}$). It is called the *market price of risk*. In comparison, the rate of return on the risk-free asset can be considered as the *price of time*.

It is easy to see that for any portfolio p, $(\sigma(r(p)), \mathbb{E}[r(p)])$ is in the interior of two half-lines. Therefore, when there is a risk-free asset, Theorem 2.3 in the last section still holds.

In the last section, we have proved that the expected return rate of risky asset portfolio with the minimum variance is A/C. Therefore, the following three cases may occur: $A/C > r_f$, $A/C < r_f$, and $A/C = r_f$. When $A/C = r_f$, the capital market line is one of the asymptotes of the portfolio frontier of risky assets and has no intersection with the efficient portfolio frontier of risky assets; when $A/C < r_f$, the slope of the capital market line $\sqrt{H}$ is steeper than the slope of the asymptote of the efficient portfolio frontiers of risk assets $\sqrt{D/C}$, and hence in this case the capital market line has also no intersection with the efficient portfolio frontier of risky assets.

Now we only consider the case of $A/C > r_f$.

Theorem 2.5 *Assume $A/C > r_f$. Then when and only when $\mu^* = r_f + H/(A - r_f C)$, we have $\mathbf{1}^\tau w(\mu^*) = 1$, namely, the efficient frontier portfolio $w(\mu^*)$ does not contain the risk-free asset, and the standard deviation of its return rate is $\sigma(\mu^*) = \sqrt{H}/(A - r_f C)$. Besides, the capital market line and the efficient portfolio frontier of risky assets are tangent at $(\sigma(\mu^*), \mu^*)$. The portfolio corresponding to the tangent point is called the tangent portfolio.*

Proof Since

$$1^\tau V^{-1}(e - r_f 1) = A - r_f C,$$

from (2.23) and (2.26), we get immediately the first conclusion. On the other hand, since the efficient frontier portfolio $w(\mu^*)$ does not contain the risk-free asset, $w(\mu^*)$ must be the efficient frontier portfolio in the last section where the risk-free asset was not considered, and consequently, point $(\sigma(\mu^*), \mu^*)$ lies in the portfolio frontier of risky assets. Therefore, the capital market line and the efficient portfolio frontiers of risk assets are tangent at $(\sigma(\mu^*), \mu^*)$. $\qquad\square$

Remark 1 We use w^* to express the tangent portfolio, i.e., the abovementioned $w(\mu^*)$. Then by (2.13) we have

$$w^* = \frac{1}{A - r_f C} V^{-1}(e - r_f 1). \tag{2.27}$$

For efficient frontier portfolio w corresponding to the points on the capital market line which are above and below the tangent point, we have $1^\tau w > 1$ and $1^\tau w < 1$, respectively. The former represents that the investor borrows some money at the risk-free rate and invests all the funds into risky assets in accordance with the weights of the tangent portfolio; the latter represents that the investor uses part of funds to buy certain risk-free asset and then invests the remaining funds into risky assets in accordance with the weights of the tangent portfolio.

Remark 2 From (2.23) we can see that the portfolio $w(\mu)$ is linear in μ. Therefore, when a risk-free asset is available, we also have the two-fund separation theorem. In particular, any efficient portfolio of an investor can be a linear combination of the risk-free asset and the tangent portfolio. Therefore, an investor invests part of the funds into risky assets in accordance with the weights of tangent portfolio.

From the definition of the tangent portfolio, we deduce easily the following result.

Theorem 2.6 *Assume* $A/C > r_f$. *Then the Sharpe ratio of the tangent portfolio attains the maximum among all portfolios of risky assets.*

2.1.4 Mean-Variance Utility Functions

How to choose portfolios in a security market for an economic agent? From the viewpoint of financial economics, it depends on the individual's preference or attitude toward risk. The latter can be represented by a so-called *utility function* U. The objective of an agent is to maximize his expected utility $\mathbb{E}[U(W)]$ of the total wealth W of portfolios at time 1. If the agent is risk averse, then his utility function is an increasing and strictly concave function. Assume that his utility function is a quadratic function or the joint distribution function of the rates of return on risky assets is Gaussian, then it is easy to show that the expected utility of the total wealth of a portfolio is reduced to a function of the form $u(\sigma, \mu)$, known as a *mean-variance utility function*, where μ is the expected rate of return on the portfolio and σ is the standard deviation of the rate of return on the portfolio. More generally, Chamberlain (1983) proved the following fact: in a market with risk-free asset, the distribution of the return rate of a portfolio relies only on its mean and variance if and only if the joint distribution of risky assets in the portfolio obeys an *elliptical distribution*. In this case, the expected utility is reduced to a mean-variance utility function. A random vector $X = (X_1, \cdots, X_n)^\tau$ is said to obey an ellipsoidal distribution if its characteristic function has the following form:

$$\varphi_X(t) = \exp\{it^\tau \mu\}\psi(t^\tau \Sigma t),$$

where μ is an n-dimensional vector, Σ is an $n \times n$ positive definite symmetric matrix, and ψ is a real function.

The monotonicity and strictly concavity of the utility function U imply that the agent prefers a higher expected rate of return and a lower standard deviation. Consequently, in general, we should have

$$\partial u/\partial \sigma < 0, \quad \partial u/\partial \mu > 0. \tag{2.28}$$

Thus, for each constant c, equation $u(\sigma, \mu) = c$ determines a positively sloped curve on the $\sigma - \mu$ plane, called the *indifference curve* at utility level c.

It is easy to show that there exists a unique indifference curve which is tangent to the portfolio frontier when no risk-free asset is available or to the capital market line when a risk-free asset is available. The tangent point (σ, μ) is the unique solution to the following equation:

$$-\frac{\partial u/\partial \sigma}{\partial u/\partial \mu} = \sqrt{H} = \frac{\mu - r_f}{\sigma}. \tag{2.29}$$

For such a μ, from (2.23), we can find the wanted efficient portfolio. If an agent has a mean-variance utility function as his/her preference, this efficient portfolio is the optimal one in the sense that it has the highest utility level among all efficient portfolios (see Fig. 2.3).

Formula (2.29) shows that for an optimal portfolio, the marginal rate of substitution between the risk and the expected rate of return must equal the market price of risk. As an example, if the agent's mean-variance utility function is

$$u(\sigma, \mu) = \mu - \frac{1}{\tau}\sigma^2, \tag{2.30}$$

where $\tau > 0$ is called *risk tolerance*, then the solution to Eq. (2.29) is

$$\mu = r_f + \frac{\tau}{2}H, \qquad \sigma = \frac{\tau}{2}\sqrt{H}. \tag{2.31}$$

Fig. 2.3 Indifference curves and the optimal portfolio

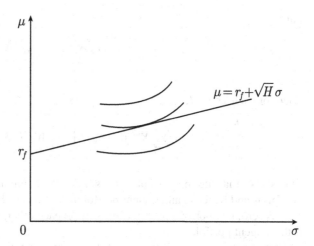

2.2 Capital Asset Pricing Model (CAPM)

2.2.1 Market Competitive Equilibrium and Market Portfolio

We assume that the market is *frictionless*, in the sense that there are no transaction costs, no bid-ask spread, no short sale restriction, no taxes, borrowing and lending are at the same interest rate, and asset shares are arbitrarily divisible. Furthermore, we assume that all agents in the market have their own mean-variance utility functions and hold the corresponding optimal portfolios. Thus, all agents hold efficient portfolios. If *market-clearing* is realized, namely, the agents' ownership of securities is equal to the sum of the individual supplies of securities, we say that the market attains a *competitive equilibrium*, in abbreviation, *equilibrium*.

Assume that the total value of risky asset i is W_i and the total value of the risk-free asset is W_f. Put

$$(w_M)_i = \frac{W_i}{\left(\sum_{j=1}^{d} W_j + W_f\right)}, \tag{2.32}$$

and call portfolio $w_M = ((w_M)_1, \cdots, (w_M)_d)^\tau$ the *market portfolio*. In practice, the market portfolio can be approximated by an index fund.

We will show that in a competitive equilibrium market, the market portfolio is an efficient portfolio. In fact, assume that there are K agents in the market. Let $w^{(k)}$ be the optimal portfolio of the k-th agent at time 0 and $X^{(k)}(0)$ be the total value of his ownership of securities at time 0. Then we have

$$\sum_{j=1}^{d} W_j + W_f = \sum_{k=1}^{K} X^{(k)}(0), \quad W_i = \sum_{k=1}^{K} X^{(k)}(0) w_i^{(k)}.$$

Put

$$\alpha_k = X^{(k)}(0) \Big/ \Big(\sum_{j=1}^{d} W_j + W_f \Big).$$

Then for each $1 \leqslant i \leqslant d$,

$$\sum_{k=1}^{K} \alpha_k w_i^{(k)} = \sum_{k=1}^{K} X^{(k)}(0) w_i^{(k)} \Big/ \Big(\sum_{j=1}^{d} W_j + W_f \Big) = (w_M)_i.$$

This shows that the market portfolio is a convex combination of certain efficient portfolios and hence is an efficient portfolio. Besides, if the net supply of risk-free asset is zero, i.e., borrowing and lending offset each other, then the market portfolio is the tangent portfolio.

Henceforth, we will call the risk of the market portfolio the *market risk* and call the risk premium of the market portfolio the *market risk premium*. Since the market portfolio is an efficient portfolio, the Sharpe ratio of the market portfolio is just the market price of risk $\sqrt{H}$ defined in Sect. 2.1.3.

According to the two-fund separation theorem, any efficient portfolio can be expressed as a linear combination of the risk-free asset and the market portfolio. For example, assume that there are K agents in the market and each agent's mean-variance utility function is of the form (2.30) with risk tolerances $\tau_1, \cdots, \tau_K$. Then by (2.23) and (2.31), the k-th agent's optimal portfolio is

$$w^{(k)} = \frac{\tau_k}{2} V^{-1}(e - r\mathbf{1}).$$

On the other hand, in (2.23) letting $w(\mu) = w_M$ gives

$$V^{-1}(e - r\mathbf{1}) = \frac{H}{\mathbb{E}[r(w_M)] - r} w_M.$$

Hence, we have

$$w^{(k)} = \frac{\tau_k H}{2(\mathbb{E}[r(w_M)] - r)} w_M. \tag{2.33}$$

Since $\sum_{k=1}^{K} \alpha_k w^{(k)} = w_M$, from (2.33) we get

$$w_M = \frac{\tau H}{2(\mathbb{E}[r(w_M)] - r)} w_M, \tag{2.34}$$

where

$$\tau = \sum_{k=1}^{K} \alpha_k \tau_k. \tag{2.35}$$

Comparing (2.34) with (2.33), we see that the market portfolio can be regarded as the optimal portfolio corresponding to the mean-variance utility function of the form (2.30) with risk tolerance τ, where τ is the weighted average of the risk tolerances of all agents, the weight α_k being the proportion of the k-th agent's wealth on risky assets at time 0.

2.2.2 CAPM with Risk-Free Asset

In this subsection we assume that the market attains competitive equilibrium and the risk-free asset with interest rate r_f is available. The next theorem gives a relationship between the expected rates of return of two portfolios, one of which is a mean-variance frontier portfolio.

Theorem 2.7 *Let p be a mean-variance frontier portfolio. Then for any portfolio q (not necessarily a mean-variance frontier portfolio), we have*

$$\mathbb{E}[r(q)] = r_f + \beta_{q,p}(\mathbb{E}[r(p)] - r_f), \tag{2.36}$$

where

$$\beta_{q,p} = \frac{\mathrm{cov}(r(q), r(p))}{\sigma^2(r(p))}. \tag{2.37}$$

In particular, let p be market portfolio w_M. Then

$$\mathbb{E}[r(q)] = r_f + \beta_{q,w_M}(\mu(w_M) - r_f). \tag{2.38}$$

We call β_{q,w_M} the β coefficient of portfolio q.

Proof Since p is a mean-variance frontier portfolio, by (2.21) we have

$$Vp = \lambda(e - r_f \mathbf{1}),$$

where $\lambda > 0$ is a constant depending on p. Let q be a portfolio. Since $q^\tau e = \mathbb{E}[r(q)] - (1 - q^\tau \mathbf{1})r_f$ and $q^\tau Vp = \mathrm{Cov}(r(q), r(p))$, we have

$$\mathrm{Cov}(r(q), r(p)) = \lambda(\mathbb{E}[r(q)] - r_f). \tag{2.39}$$

In particular, if $q = p$, then

$$\sigma(r(p))^2 = \lambda(\mathbb{E}[r(p)] - r_f).$$

Substituting it into (2.39) gives (2.36). $\qquad\square$

Remark 1 From (2.36) and (2.37), we get

$$r(q) - r_f = \beta_{q,p}(r(p) - r_f) + \xi, \tag{2.40}$$

where $\mathbb{E}[\xi] = 0$, $\text{cov}(\xi, r(p)) = 0$. Besides, from (2.40) and (2.37), it is easy to see:

$$\beta_{q,p} = \arg\min_{\alpha} \text{Var}[(r(q) - r_f) - \alpha(r(p) - r_f)].$$

Therefore, using the least square method to make a linear regression for the observed data of $r(q) - r_f$ and $r(p) - r_f$, the regression coefficient is just the optimal estimate of $\beta_{q,p}$.

Remark 2 From (2.40) we get

$$\sigma^2(r(q)) = \beta_{q,w_M}^2 \sigma^2(r(w_M)) + \sigma^2(\xi).$$

Let $\rho_{r(q),r(w_M)}$ represent the correlation coefficient of $r(q)$ and $r(w_M)$, i.e.,

$$\rho_{r(q),r(w_M)} = \frac{\text{cov}(r(q), r(w_M))}{\sigma(r(q))\sigma(r(w_M))}.$$

Then

$$\beta_{q,w_M}\sigma(r(w_M)) = \rho_{r(q),r(w_M)}\sigma(r(q)). \tag{2.41}$$

Thus, we have

$$\sigma^2(r(q)) = \rho_{r(q),r(w_M)}^2 \sigma^2(r(q)) + \sigma^2(\xi),$$

$$\sigma^2(\xi) = (1 - \rho_{r(q),r(w_M)}^2)\sigma^2(r(q)).$$

We call $\sigma^2(r(q))$ and $\rho_{r(q),r(w_M)}^2 \sigma^2(r(q))$ the *total risk* and *systematic risk* of portfolio q, respectively. The latter is caused by the market environment. We call $(1 - \rho_{r(q),r(w_M)}^2)\sigma^2(r(q))$ the *non-systematic risk* of portfolio q, which is independent of the market environment.

It is easy to prove that the non-systematic risk of portfolio q is zero if and only if portfolio q is a mean-variance frontier portfolio.

Since $(\mu(w_M) - r_f)/\sigma(r(w_M)) = \sqrt{H}$, by (2.36) and (2.41), Eq. (2.38) can be rewritten as

$$\mathbb{E}[r(q)] = r + \rho_{r(q),r(w_M)}\sigma(r(q))\sqrt{H}, \tag{2.38'}$$

Fig. 2.4 Security market line

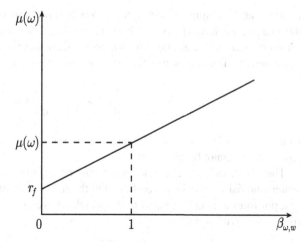

where H is given by (2.34). Equation (2.38) or (2.38′) is called the *capital asset pricing model* (CAPM). It was established independently by Sharpe (1964), Lintner (1965), and Mossin (1966).

CAPM expressed by (2.38′) shows that the risk premium of a portfolio is proportional to its systematic risk, and its scaling factor is the market price of risk (i.e., $\sqrt{H}$). CAPM expressed by (2.38) shows that the risk premium of a portfolio is equal to its β-coefficient multiplied by the market risk premium.

From (2.41) we see that the β-coefficient of a portfolio q is also equal to the ratio of the systematic risk of q and the market risk, called the *relative systematic risk*. From (2.38) we see that the risk premium $\mathbb{E}[r(q)] - r_f$ of a portfolio q depends only on its systematic risk and does not depend on its non-systematic risk.

On the $\beta - \mu$ plane, the line passing through $(0, r_f)$ with slope $\sigma(w_M)\sqrt{H}$ (i.e., $\mu(w_M) - r_f$) is called the *security market line* (see Fig. 2.4).

The security market line provides another expression for CAPM. It vividly demonstrates the following fact: the risk premium of a portfolio is proportional to the relative systematic risk of the portfolio. Its scale factor is the market risk premium. In particular, $\beta_i = \mathrm{Cov}(r_i, r(w_M))/\sigma^2(r(w_M))$ is called the β-coefficient of the asset i. In practice, the β-coefficient of a risky asset can be statistically estimated using observed data from the companies and the market, and the market risk premium $\mu(w_M) - r_f$ can be approximated by the risk premium of the index fund.

From (2.38),

$$e_i = r_f + \beta_i\big(\mathbb{E}[r(w_M)] - r_f\big). \tag{2.42}$$

The systematic risk $\beta_i\sigma(r(w_M))$ of the i-th asset, determined by the market, cannot be eliminated through a diversified investment. Non-systematic risk $\sqrt{\sigma^2(r_i) - \beta_i^2\sigma^2(w)}$ of the i-th asset, however, caused by company's internal factors, can be eliminated through a diversified investment. In fact, assume that in the market, we have d assets whose return rates have the same variance σ^2 and same covariance Cov among any two assets. Consider the simplest portfolio p with equal weight. The variance of the return rate of this portfolio is

$$\sigma^2(p) = \frac{1}{d}\sigma^2 + \left(1 - \frac{1}{d}\right)\text{Cov}.$$

Hence, the variances of return rates of individual assets are dispersed, but the covariances cannot be dispersed.

The β-coefficient of a risky asset can be regarded as a measure of relative systematic risk. Since the β-coefficient of the risk-free asset is zero, the β-coefficient of a portfolio w is in fact equal to the weighted average of the β-coefficients of the risky assets with weight w, i.e.,

$$\beta_{w,w_M} = \sum_{i=1}^{d} w_i \beta_i.$$

Because the systematic risk of a portfolio is proportional to its β-coefficient, selecting those assets with small β-coefficient can reduce the risk of the portfolio, although these assets have lower expected return rates.

2.2.3 CAPM Without Risk-Free Asset

Assume now the risk-free asset is not available (or restricted to use) in the market. Let p be a mean-variance frontier portfolio which is not the minimum variance portfolio, i.e., $\mathbb{E}[r(p)] \neq A/C$. We will show that there exists a unique mean-variance frontier portfolio, denoted by $zc(p)$, which has a zero covariance with p. In order to find this portfolio $zc(p)$ with $\text{Cov}(r(p), r(zc(p))) = 0$, by (2.9), we solve the following equation:

$$\frac{C}{D}\left(\mathbb{E}[r(p)] - \frac{A}{C}\right)\left(\mathbb{E}[r(zc(p))] - \frac{A}{C}\right) + \frac{1}{C} = 0$$

to get

$$\mathbb{E}[r(zc(p))] = \frac{A}{C} - \frac{\frac{D}{C^2}}{\mathbb{E}[r(p)] - \frac{A}{C}}.$$

Then the mean-variance frontier portfolio with the expected rate of return given by the right side of the above formula is the portfolio $zc(p)$ wanted.

The following theorem is due to Black (1972).

Theorem 2.8 *Let p be a mean-variance frontier portfolio which is not the minimum variance portfolio. Then for any portfolio q (not necessarily a mean-variance frontier portfolio), we have*

$$\mathbb{E}[r(q)] = \mathbb{E}[r(zc(p))] + \beta_{q,p}(\mathbb{E}[r(p)] - \mathbb{E}[r(zc(p))]). \tag{2.43}$$

Proof Since p is a mean-variance frontier portfolio, by (2.4a) we have

$$Vp = \lambda e + \gamma \mathbf{1},$$

where $\lambda > 0$ and $\gamma > 0$ are constants depending on p. Let q be a portfolio. Since $q^\tau e = \mathbb{E}[r(q)], q^\tau \mathbf{1} = 1$, and $q^\tau Vp = \text{Cov}(r(p), r(q))$, we have

$$\text{Cov}(r(p), r(q)) = \lambda \mathbb{E}[r(q)] + \gamma. \tag{2.43'}$$

In particular, letting $q = p$ and $q = zc(p)$ gives

$$\sigma^2(r(p)) = \lambda \mathbb{E}[r(p)] + \gamma, \quad 0 = \lambda \mathbb{E}[r(zc(p))] + \gamma,$$

from which we get

$$\lambda^{-1} = \frac{\mathbb{E}[r(p)] - \mathbb{E}[r(zc(p))]}{\sigma^2(r(p))}, \quad \lambda^{-1}\gamma = -\mathbb{E}[r(zc(p))].$$

Thus, we deduce (2.43) from (2.43').   □

Remark If in (2.43) we let p be market portfolio w_M, then we get

$$\mathbb{E}[r(q)] = \mathbb{E}[r(zc(w_M))] + \beta_{q,w_M}(\mathbb{E}[r(w_M)] - \mathbb{E}[r(zc(w_M))]). \tag{2.44}$$

Since the β-coefficient of $zc(w_M)$ is zero, Eq. (2.44) is called the *zero-β CAPM*. Comparing (2.44) to (2.38), $zc(w_M)$ in the zero-β CAPM is a counterpart of the risk-free asset in CAPM.

2.2.4   Equilibrium Pricing Using CAPM

The main application of CAPM is the pricing of contingent claims. In this subsection we assume that the risk-free asset is available and all agents have their own, mean-variance utility functions and hold the corresponding optimal portfolios. Furthermore, assume that the market is in competitive equilibrium. Let X be an uncertain payoff at time 1, called a *contingent claim*. Assume that its statistic characters (the mean and variance) and correlation coefficient with the return rate of the market portfolio are known. We want to calculate its price at time 0, denoted by $e(X)$. By (2.38'), its rate of return $r_X = X/e(X) - 1$ should satisfy the following equation:

$$\mathbb{E}[r_X] = r + \rho_{r_X, r(w_M)} \sigma(r_X) \sqrt{H}. \tag{2.45}$$

Since

$$\mathbb{E}[X] = (1 + \mathbb{E}[r_X])e(X), \quad \rho_{r_X, r(w_M)} = \rho_{X, r(w_M)}, \quad \sigma(X) = e(X)\sigma(r_X),$$

from (2.45) we get

$$e(X) = \frac{\mathbb{E}[X] - \rho_{X, r(w_M)} \sigma(X) \sqrt{H}}{1 + r}. \tag{2.46}$$

We call $e(X)$ the *equilibrium price* of contingent claim X.

Because $\rho_{X, r(w_M)} \sigma(X) = Cov(X, r(w_M))/\sigma(r(w_M))$, from (2.46) we see that equilibrium pricing is a linear operation w.r.t. contingent claims. This linear property is important for a company in making the capital budget. A company should invest in those projects whose present equilibrium values of the future uncertain returns are higher than the costs.

Another application of CAPM is that one can compare the equilibrium prices to the actual prices of risky assets to find those under-valued or over-valued assets and make profit through selling high and buying low. However, one should take care of the non-systematic risks of individual assets and do not invest only in certain under-valued assets.

2.3 Arbitrage Pricing Theory (APT)

CAPM is a single factor model. It implicitly assumes that the factor influencing the returns on risky assets is only the market factor, which is not observable. In reality, there are many common factors, such as gross domestic product (GDP), rate of full employment, interest rate, and inflation index, which influence the rates of return of most assets. This shows that the application of CAPM is quite restrictive. A more serious drawback of CAPM is that it is based on the assumption that all agents are risk averse and use mean-variance utility functions to determine their optimal portfolios and on the assumption that the market is in competitive equilibrium. These assumptions are obviously unreasonable. Without making those assumptions, Ross (1976) proposed a multifactor model for the rates of return on assets:

$$r_i = e_i + \sum_{j=1}^{M} b_{ij} f_j + \varepsilon_i, \quad 1 \leqslant i \leqslant d, \tag{2.47}$$

where r_i and e_i are the rates of return and expected one on asset i, respectively, $f_1, \cdots, f_M$ are mean zero random variables representing common factors influencing the rates of return on assets, and b_{ij} is the so-called *sensitive coefficient* of asset i w.r.t. the factor j, $\varepsilon_1, \cdots, \varepsilon_d$ are mutually unrelated mean zero random variables representing model errors, which are unrelated with factors $(f_1, \cdots, f_M)$.

The so-called Ross' *arbitrage pricing theory* (APT) can be roughly described as follows. Assume that the number d of risky assets are much larger than the number M of common factors and the market is *approximately arbitrage-free* in the sense that when d tends to infinity arbitrage opportunities will disappear. Then there exist M portfolios $w_1, \cdots, w_M$, such that the sensitive coefficient of j-th portfolio w.r.t. the factor j is 1 and that w.r.t. other factors are 0, and there exists approximately a linear relationship among the expected rates of return on most of assets and the sensitivities of assets w.r.t. common factors:

$$e_i = r_f + \sum_{j=1}^{M} b_{ij}(\lambda_j - r_f), \qquad (2.48)$$

where λ_j is the expected rate of return on w_j. Like CAPM, APT can be used for pricing of risky assets.

A rigorous formulation of APT needs an assumption that there are infinitely many of assets, the rate of return on each asset satisfies (2.47), and $\mathbb{E}\epsilon_i^2$ are uniformly bounded. If the market is approximately arbitrage-free, then APT affirms that there are $N + 1$ constants $\mu_0, \cdots \mu_M$, independent of n, such that

$$\lim_{n \to \infty} \frac{1}{n} \sum_{i=1}^{n} (e_i - \mu_0 - \sum_{j=1}^{M} b_{ij}\mu_j)^2 = 0. \qquad (2.49)$$

If there is a risk-free asset in the market, then μ_0 is the rate of return on the risk-free asset.

As Huberman (1983) pointed out, the main advantage of Ross' APT is that its empirical tests do not need to know actually unobserved market portfolio. Unfortunately, Ross (1976) did not provide a clear definition of asymptotically free arbitrage, and his proof is not rigorous. Here we give a clear statement of the APT and a rigorous proof which is essentially based on Huberman (1983). To this end, we require a sequence of markets with growing number of risky assets. At the n-th market, there are n assets whose rates of return are given by the multifactor model of the form (2.47):

$$r_i^n = e_i^n + \sum_{j=1}^{M} b_{ij}^n f_j^n + \varepsilon_i^n, \quad i = 1, 2 \cdots, n, \qquad (2.50)$$

where $\varepsilon_1^n, \cdots, \varepsilon_n^n$ are uncorrelated random variables with mean zero, $\mathbb{E}[(\varepsilon_i^n)^2]$ are uniformly bounded on n and i. In terms of vectors and matrices, (2.50) can be rewritten as

$$r^n = e^n + B^n f^n + \varepsilon^n, \qquad (2.51)$$

where $r^n = (r_1, \cdots, r_n)^\tau$, $B^n = (B_{i,j}^n)_{n \times M}$ and $f^n = (f_1^n, \cdots, f_M^n)^\tau$.

Previously, we always take into consideration those investments with a total value that is non-zero and define a weight vector w as a portfolio. Now we define a value vector W as a portfolio, where W_i is the value invested in asset i. Then we can consider portfolios with total value of zero. According to Huberman (1983), such a portfolio is known as *"arbitrage portfolio."* Note that the so-called arbitrage portfolio is not really able to generate an arbitrage. A market is said to have *asymptotic arbitrage*, if there is a sequence of "arbitrage portfolios" $W^n \in \mathbb{R}^n$ such that

$$\lim_{n \to \infty} (W^n)^\tau e^n = \infty, \quad \lim_{n \to \infty} (W^n)^\tau V^n W^n = 0.$$

Here $(W^n)^\tau e^n$ and $(W^n)^\tau V^n W^n$ are the mathematical expectation and variance of the rate of return of portfolio W^n, where $V^n = \mathbb{E}[\varepsilon^n (\varepsilon^n)^\tau]$.

The following theorem is from Huberman (1983).

Theorem 2.9 *Consider a sequence of markets. The rates of return of risky assets are given by (2.40) and satisfy the abovementioned conditions. If these markets have no asymptotic arbitrage, then for $n = 1, 2, \cdots$, there are constants $\rho^n, \mu_1^n, \cdots, \mu_M^n$ and A, such that for $n = 1, 2, \cdots$,*

$$L_n := \sum_{i=1}^{n} (e_i^n - \rho^n - \sum_{j=1}^{M} b_{ij}^n \mu_j^n)^2 \leqslant A. \tag{2.52}$$

Proof The following proof is adapted from Huberman (1983). Projecting e^n into the linear space spanned by $\mathbf{1}^n$ and the column vectors of B^n, we get

$$e^n = \rho^n \mathbf{1}^n + B^n \mu^n + c^n, \tag{2.53}$$

where

$$\mu^n \in \mathbb{R}^M, \quad (c^n)^\tau \mathbf{1}^n = 0, \quad (B^n)^\tau c^n = 0.$$

We denote $\|c^n\|^2 = \sum_{i=1}^{n} (c_i^n)^2 = L_n$. Suppose that the statement of the theorem does not hold. Then there is a subsequence (n') such that $L_{n'} \to \infty$. We fix a $p \in (-1, -1/2)$ and consider "arbitrage portfolios" $W^{n'} = L_{n'}^p c^{n'}$. By (2.53), the expected return of $W^{n'}$ is $L_{n'}^{1+p}$, and it tends to ∞ when $n' \to \infty$. On the other hand, by assumption, for certain $\sigma > 0$, it holds $\sup_n \text{Var}(\varepsilon_i^n) \leqslant \sigma^2$. We have the following estimate for the variance of return of $W^{n'}$:

$$(W^{n'})^\tau V^{n'} W^{n'} \leqslant \sigma^2 L_{n'}^{2p} \|c^{n'}\|^2 = \sigma^2 L_{n'}^{1+2p}.$$

When $n' \to \infty$ it tends to zero. This contradicts the assumption that these markets have no asymptotic arbitrage. The theorem is proved. $\qquad\square$

Ross (1976) considered a stationary model, which contains infinitely many assets, and in (2.40) $e_i^n = e_i$, $b_{i,j}^n = b_{i,j}$, for all i, j and n. In other words, (2.51) is replaced by

$$r^n = e + Bf^n + \varepsilon^n. \tag{2.54}$$

Ross has proved the following result. For its proof see Huberman (1983).

Theorem 2.10 *Consider a market where the return rates of risky assets are given by (2.54) and satisfy the abovementioned conditions. If the market has no asymptotic arbitrage, then for $n = 1, 2, \cdots$, there are constants $\rho, \mu_1, \cdots, \mu_M$, such that*

$$\sum_{i=1}^{\infty} (e_i - \rho - \sum_{j=1}^{M} b_{ij}\mu_j)^2 < \infty. \tag{2.55}$$

If there is a risk-free asset in the market, then in Theorem 2.8 (correspondingly, Theorem 2.9), ρ^n (correspondingly, ρ) is the return rate of the risk-free asset.

2.4 Mean-Semivariance Model

Regarding the variance of the return rate, the risk of a portfolio can simplify the calculation. From the perspective of risk management, however, variance is not a satisfactory risk measure, because it "punishes" gains and losses at the same time. Markowitz (1959) suggested to use the semivariance as a risk measure and gave a procedure to calculate one-period mean-semivariance. However, when the restricted set is unbounded, the existence of efficient set was not established. Since then a lot of work focused on the numerical calculation of semivariance efficient frontier, and the existence of efficient frontier for the case where the restricted set is a closed subset (usually unbounded) of $\mathbb{R}^d$ was established by Jin et al. (2006). We will give below, in our framework, a simple proof of the main result in their paper. For simplicity, we only consider a market without risk-free asset.

Let $\mu > 0$ be given. The *mean-semivariance model* is the following optimization problem:

$$\begin{cases} w(\mu) = \arg\min_{w} \mathbb{E}[(w^\tau r - \mu)_-^2], \\ w^\tau e = \mu, \quad w^\tau \mathbf{1} = 1, \end{cases} \tag{2.56}$$

where $x_- = \max(-x, 0)$. Our task is to prove the existence of the solution to (2.56).

Put $\mathcal{D} = \{w \in \mathbb{R}^d : w^\tau e = \mu, w^\tau \mathbf{1} = 1\}$, and $f(w) = \mathbb{E}[(w^\tau r - \mu)_-^2]$, $w \in \mathcal{D}$. Then $\mathcal{D}$ is a closed subset of $\mathbb{R}^d$. Take a sequence $\{w_n, n \geqslant 1\}$ from $\mathcal{D}$ such that $\lim_{n \to \infty} f(w_n) = \inf_{w \in \mathcal{D}} f(w)$. We will prove that $\{w_n, n \geqslant 1\}$ is bounded, and consequently, any limiting point of the sequence is a solution to (2.56). We prove by contradiction that $\{w_n, n \geqslant 1\}$ is bounded. In fact, if otherwise, we can take a subsequence $\{w_{n(k)}, k \geqslant 1\}$ such that $\lim_{k \to \infty} \|w_{n(k)}\| = \infty$, and the limit

$\lim_{k \to \infty} w_{n(k)}/\|w_{n(k)}\|$ exists, denoted by y. Then $\|y\| = 1$, and we have

$$0 = \lim_{k \to \infty} \frac{f(w_{n(k)})}{\|w_{n(k)}\|^2} = \mathbb{E}[\left(y^\tau (r - e)\right)^2_-],$$

which implies $\left(y^\tau(r-e)\right)_- = 0$ a.s. However, since $\mathbb{E}[y^\tau(r-e)] = y^\tau \mathbb{E}[r-e] = 0$,

we must also have $\left(y^\tau(r-e)\right)_+ = 0$ a.s. Hence, we obtain $y^\tau(r-e) = 0$ a.s., which contradicts the original assumption that $r_1 - e_1, \cdots, r_d - e_d$ are linearly independent (i.e., V is of full rank).

More generally, for a given target $b > 0$ of the return rate of portfolio, the following optimization problem is the *target-semivariance model*:

$$\begin{cases} w(b) = \arg \min_w \mathbb{E}[(w^\tau r - b)^2_-], \\ w^\tau \mathbf{1} = 1. \end{cases} \tag{2.57}$$

For the existence of the solution to this problem, see Jin et al. (2006).

2.5   Multistage Mean-Variance Model

So far, we have assumed that agents only make investment decisions at the current time (time 0). How can this result be extended to investment decisions at time 0, time 1, $\cdots$ time $T - 1$? Here T is a positive integer. This problem was not solved until 2000 by Li and Ng (2000). Below we will introduce the main result of this article without giving the detailed proof.

Suppose that there are $d + 1$ assets in the market, represented by indices 0 and $1, \cdots, d$, respectively. The asset indexed by 0 generally represents a risk-free asset, but it may also be a risky asset; other assets are risky assets. The rate of return of the asset i from time $t - 1$ to time t is represented by $r_i(t)$ which is a random variable. Let $e_i(t) = E[r_i(t)]$ denote the expectation of $r_i(t)$, and set

$$r(t) = (r_1(t), \cdots, r_d(t))^\tau, \qquad e(t) = (e_1(t), \cdots, e_d(t))^\tau,$$

where τ represents the transpose. We use $V(t)$ to represent the covariance matrix of random vector $r(t)$, i.e.,

$$V(t) = E[(r(t) - e(t))(r(t) - e(t))^\tau].$$

Assume that the means and covariance matrix of the return rates are known and the covariance matrix is of full rank. Let $\mathcal{F}_t$ be the σ-algebra generated by $\{r(s), s \leqslant t\}$. An investment strategy is $(\mathcal{F}_t)$ adapted vector valued stochastic sequence $u(t) = (u_1(t), u_2(t), \cdots, u_d(t))^\tau$, $t = 0, 1, 2, \cdots, T - 1$, where $u_i(t)$ is the value of asset i owned by an agent at time t after the transaction. Then it is $\mathcal{F}_t$ measurable.

Assume that the initial wealth of an agent is $x(0)$. Let $x^u(t) = \sum_{i=0}^{d} u_i(t)$ represent the total value at time t of assets owned by the agent using the portfolio $u(t)$. We call an investment strategy *self-financing* if it satisfies the following "self-financing condition":

$$x^u(t+1) = \sum_{i=1}^{d} u_i(t) r_i(t) + (x^u(t) - \sum_{i=1}^{d} u_i(t)) r_0(t)$$

$$= r_0(t) x^u(t) + P(t)^\tau u(t), \quad t = 0, 1, \cdots, T-1,$$

where

$$P(t) = ((r_1(t) - r_0(t)), (r_2(t) - r_0(t)), \cdots, (r_d(t) - r_0(t)))^\tau.$$

Under the criterion of mean-variance, an investor looks for the optimal self-financing investment strategy u in the sense that at time T, the variance of the wealth is not more than a pre-specified level, making the biggest expectation of the final wealth; or at time T, the expectation of the final wealth is not less than a pre-specified level, making the minimum variance of the wealth at time T. Their mathematical descriptions are as follows:

$(P1(\sigma))$ $\qquad\qquad$ $\text{argmax}_{u \in \mathcal{U}} \mathbb{E}(x^u(T))$

$\qquad\qquad\qquad$ s. t. $\text{Var}(x^u(T)) \leqslant \sigma,$

$(P2(\mu))$ $\qquad\qquad$ $\text{argmin}_{u \in \mathcal{U}} \text{Var}(x^u(T))$

$\qquad\qquad\qquad$ s. t. $\mathbb{E}(x^u(T)) \geqslant \mu,$

where $\mathcal{U}$ is the set of all self-financing strategies with initial wealth $x(0)$.

The above two different expressions can be unified as follows: the investor makes a trade-off between increasing the expected wealth at time T and reducing the variance of the wealth at time T. If we pre-specify a trade-off coefficient $w \in [0, \infty)$, then $(P1(\sigma))$ and $(P2(\mu))$ can be uniformly written as the following equivalent form $(E(w))$:

$(E(w))$ $\qquad\qquad$ $\text{argmax}_{u \in \mathcal{U}} \mathbb{E}[x^u(T)] - w \text{Var}(x^u(T)).$

It is easy to see that if u^* is a solution to problem $(E(w))$ and the wealth process is $(x^*(t))$, then u^* is a solution to problem $(P1(\sigma))$, where $\sigma = \text{Var}(x^*(T))$; u^* is also a solution to problem $(P2(\mu))$, where $\mu = \mathbb{E}[x^*(T)]$.

We will solve the problem $(E(w))$ in three steps.

First step, we introduce the auxiliary problem $(A(\lambda, w))$, $\lambda \geqslant 0$, $w \geqslant 0$:

$(A(\lambda, w))$ $\qquad\qquad$ $\text{argmax}_{u \in \mathcal{U}} \mathbb{E}\{\lambda x^u(T) - w x^u(T)^2\}.$

It is easy to prove that the solution set of problem $(E(w))$ is a subset of the solution set of problem $(A(\lambda, w))$. More exactly, if u^* is a solution to problem $(E(w))$, then it is also a solution to problem $(A(\lambda^*, w))$, where $\lambda^* = 1 + 2w\mathbb{E}[x^{u^*}(T)]$.

Second step, we use the dynamic programming method to solve the auxiliary problem $(A(\lambda, w))$. Since this problem actually depends only on parameter $\gamma = \frac{\lambda}{w}$, we use u_γ to represent the solution to problem $(A(\lambda, w))$; the corresponding wealth process is denoted by $x_\gamma(t)$. Then we have

$$u_\gamma(t) = -K(t)x_\gamma(t) + v_\gamma(t), \quad t = 0, 1, \cdots, T - 1,$$

where

$$K(t) = P(t)^\tau \mathbb{E}(P(t)P(t)^\tau)^{-1}\mathbb{E}(r_0(t)P(t)),$$

$$v_\gamma(t) = \frac{\gamma}{2}\left(\prod_{j=t+1}^{T-1}\frac{A_j}{B_j}\right)\mathbb{E}(P(t)P(t)^\tau)^{-1}\mathbb{E}(P(t)),$$

$$A_t = \mathbb{E}(r_0(t)) - \mathbb{E}(P(t)^\tau)\mathbb{E}(P(t)P(t)^\tau)^{-1}\mathbb{E}(r_0(t)P(t)),$$

$$B_t = \mathbb{E}(r_0(t)^2) - \mathbb{E}(r_0(t)P(t)^\tau)\mathbb{E}(P(t)P(t)^\tau)^{-1}\mathbb{E}(r_0(t)P(t)).$$

Since we assume that the covariance matrix of the return rates is of full rank, i.e., $\mathbb{E}(P(t)P(t)^\tau)$ is strictly positive definite, we have $B_t > 0, t = 0, 1, \cdots, T - 1$.

Third step, we fix a suitable γ^*. We substitute optimal solution u_γ of auxiliary problem $(A(\lambda, w))$ into the wealth process satisfying the self-financing condition, and let it be squared, then we obtain

$$x_\gamma(t + 1) = \left(r_0(t) - P(t)^\tau K(t)\right)x_\gamma(t) + P(t)^\tau v_\gamma(t),$$

$$x_\gamma(t + 1)^2 = \left[r_0(t)^2 - 2r_0(t)P(t)^\tau K(t) + K(t)^\tau P(t)P(t)^\tau K(t)\right]x_\gamma(t)^2$$

$$+ 2\left(r_0(t) - P(t)^\tau K(t)\right)x_\gamma(t)P(t)^\tau v_\gamma(t) + v_\gamma(t)^\tau P(t)P(t)^\tau v_\gamma(t),$$

Taking the expectation on both sides of the equation gives

$$\mathbb{E}(x_\gamma(t + 1)) = A_t\mathbb{E}(x_\gamma(t)) + \frac{\gamma}{2}\left(\prod_{j=t+1}^{T-1}\frac{A_j}{B_j}\right)C_t,$$

$$\mathbb{E}(x_\gamma(t + 1)^2) = B_t\mathbb{E}\left(x_\gamma(t)^2\right) + \frac{\gamma^2}{4}\left(\prod_{j=t+1}^{T-1}\frac{A_j}{B_j}\right)^2 C_t,$$

where

$$C_t = \mathbb{E}(P(t)^\tau)\mathbb{E}(P(t)P(t)^\tau)^{-1}\mathbb{E}(P(t)).$$

We further solve these two recursive equations to get

$$\mathbb{E}(x_\gamma(T)) = \mu x(0) + \nu\gamma,$$

$$\mathbb{E}(x_\gamma(T)^2) = \tau x(0)^2 + \frac{\nu}{2}\gamma^2,$$

$$\text{Var}(x_\gamma(T)) = a(\gamma - bx(0))^2 + cx(0)^2,$$

where

$$\mu = \prod_{k=0}^{T-1} A_t, \quad \tau = \prod_{t=0}^{T-1} B_t,$$

$$\nu = \sum_{k=0}^{T-1} C_k \frac{\left(\prod\limits_{j=k+1}^{T-1} A_j\right)^2}{2\prod\limits_{j=k+1}^{T-1} B_j},$$

$$a = \frac{\nu}{2} - \nu^2, \quad b = \frac{\mu\nu}{a}, \quad c = \tau - \mu^2 - ab^2.$$

Set

$$U(\gamma) = \mathbb{E}[\gamma x_\gamma(T) - x_\gamma(T)^2].$$

Substituting the above mean and variance on the wealth at time T into the target function of original problem ($E(w)$) gives

$$U(\gamma) = \mu x(0) + \nu\gamma - w[a(\gamma - bx(0))^2 + cx(0)^2].$$

Obviously, U is a concave function of γ. Differentiating w.r.t γ, we obtain

$$\frac{dU}{d\gamma} = \nu - 2wa(\gamma - bx(0)).$$

The optimal γ must satisfy the condition $dU/d\gamma = 0$, namely,

$$\gamma^* = bx(0) + \frac{\nu}{2wa}.$$

Finally, by substituting γ^* into $u_\gamma(t)$, we get the optimal strategy u^* of problem ($E(w)$):

$$u^*(t) = -K(t)x^*(t) + \frac{1}{2}\left(bx(0) + \frac{\nu}{2wa}\right)\left(\prod_{k=t+1}^{T-1} \frac{A_k}{B_k}\right)(\mathbb{E}(P(t)P(t)^\tau)^{-1}\mathbb{E}(P(t)),$$

$$\forall t = 1, \cdots, t = T - 2,$$

$$u^*(T-1) = -K(T-1)x^*(T-1) + \frac{1}{2}\left(bx(0) + \frac{v}{2wa}\right)$$
$$(\mathbb{E}(P(T-1)P(T-1)^\tau)^{-1}\mathbb{E}(P(T-1)).$$

The mean and variance of the corresponding final wealth are

$$\mathbb{E}(x^*(T)) = (\mu + bv)x(0) + \frac{v^2}{2wa},$$

$$\mathrm{Var}(x^*(T)) = \frac{v^2}{4w^2a} + cx(0)^2.$$

2.6   Expected Utility Theory

In the traditional microeconomic theory, a basic assumption is that the decision-making of market participants is based on the satisfaction levels for commodities or a combination of commodities; the order can be represented by ordinal numbers (i.e., preferences): $A \succeq B$(or $B \not\succeq A$) stands for that A is better than B. This theory is called the theory of ordinal utility. Assume that the collection of alternative commodities is $\mathcal{X}$. About the *preference relation* $\succeq$ on $\mathcal{X}$, we have the following three basic assumptions:

(1) *Completeness*: for any two commodities A and B in $\mathcal{X}$, either $A \succeq B$ or $B \succeq A$, or both hold;
(2) *Reflexivity*: for any commodity A in $\mathcal{X}$, one has $A \succeq A$;
(3) *Transitivity*: if $A \succeq B$ and $B \succeq C$, then $A \succeq C$.

If $A \succeq B$ and $B \succeq A$, then we say that A and B have no difference and denote it by $A \sim B$.

From the preference relation $\succeq$ on $\mathcal{X}$, we can introduce another preference relation $\succ$ on $\mathcal{X}$: $B \succ A \Leftrightarrow A \not\succeq B$. Here $B \succ A$ represents that B is "strictly better than" A. The preference relation $\succ$ satisfies:

(1) *Antisymmetry*: If $A \succ B$, then $B \not\succ A$;
(2) *Negative transitivity*: If $A \succ B$, $C \in \mathcal{X}$, then either $A \succ C$, or $C \succ B$, or both hold.

In 1944, von Neumann and Morgenstern in their book *Game Theory and Economic Behavior* studied decision-making under uncertainty, where the preference is not only defined on the set of alternative commodities but also on their uncertain choices. The book proved the following result: if an ordinal utility theory-based decision-maker faces the uncertainty described with the probability, and the preference satisfies the so-called continuity axiom and independence axiom, then there is a utility function on the set of alternative commodities such that there exists a one-to-one correspondence between the numerical size (called *cardinal utility*) defined by the *expected utility* of selection of the commodity and the order of the merits described by the preference relation. Detailed derivation can be found in Föllmer and Schied (2004). More general results are in Fishburn (1970). Thus, one can replace cardinal utility by easy-to-handle ordinal utility. In this way, when a preference-based decision-maker selects commodities or a combination of commodities, he/she will maximize the expected utility. This important expected utility theory was established by von Neumann and Morgenstern (1944). It is the mathematical framework of decision analysis under uncertain conditions and is the cornerstone of modern microeconomics.

2.6.1 Utility Functions

When we study financial economics with expected utility theory, we usually consider uncertain returns as an alternative collection. For convenience, below we assume that the utility functions of market participants are twice continuously differentiable. Since for market participants, the more wealth, the greater the sense of satisfaction, so we always assume that the utility function u is strictly increasing (i.e., the first derivative u' is greater than zero). Economics divides the risk attitudes of market participants into three categories: risk averse, risk-loving (or risk seeking), and risk-neutral; this means that when one makes a choice between a certain return and an uncertain return with the same expectation, a risk averse one chooses the former, and a risk lover chooses the latter, while for a risk-neutral one, both options are equivalent. Assuming that the utility function of market participants is u. The above classification corresponds, respectively, to the case $\mathbb{E}[u(W)] < u(\mathbb{E}[W])$, $\mathbb{E}[u(W)] > u(\mathbb{E}[W])$, and $\mathbb{E}[u(W)] = u(\mathbb{E}[W])$, where W is any nonconstant return with finite expectation. Below we will prove that this classification can be characterized by second derivative u'' of utility function u.

Let $(\Omega, \mathcal{F}, \mathbb{P})$ be a probability space. A set $A \in \mathcal{F}$ is an *atom* w.r.t. $\mathbb{P}$, if $\mathbb{P}(A) > 0$, and $\forall B \subset A$, $B \in \mathcal{F}$, we have $\mathbb{P}(B) = \mathbb{P}(A)$ or $\mathbb{P}(B) = 0$. It is easy to prove that a probability space $(\Omega, \mathcal{F}, \mathbb{P})$ is *atom-free*, if and only if there is a random variable which is uniformly distributed on [0,1].

Theorem 2.11 *Assume that the utility function of a market participant is strictly increasing concave (convex) function, namely, its second derivative is less than zero (greater than zero), then the market participant is risk averse (risk-loving). Conversely, if the space probability is atom-free, then the utility function of a risk averter (risk lover) must be strictly increasing concave (convex) function. In addition, a market participant is risk-neutral, if and only if his utility function is a linear function.*

Proof First, by Jensen's inequality we deduce immediately the first conclusion of the theorem. Now we assume that the probability space is atom-free, a market participant is risk averse, and his utility function is u, that is, for any real number x and any zero-mean random variable X, it holds $\mathbb{E}[u(x + X)] < u(x)$, where $\mathbb{P}(X \neq$

$0) > 0$. Let $x_2 > x_1$, $p \in (0, 1)$. Set $x = px_1 + (1 - p)x_2$, $X = x_1 I_A + x_2 I_{A^c} - x$, where A is chosen such that $\mathbb{P}(A) = p$; this is guaranteed by the assumption of atom-free probability space. Then we have $\mathbb{E}[X] = 0$, and

$$u(px_1 + (1 - p)x_2) = u(x) > \mathbb{E}[u(x + X)] = pu(x_1) + (1 - p)u(x_2).$$

This indicates that u is a strictly increasing concave function. The proof for the risk-loving case is similar. The statement about the risk-neutral case is obvious. $\square$

Remark Henceforth, under any circumstances (i.e., even without the assumption that the probability space is atom-free), a market participant having a strictly concave utility function (resp. strictly convex function) is referred to as a risk averse one (resp. risk lover). The marginal utility of the corresponding utility function (i.e., the first derivative of the utility function) is decreasing and increasing, respectively.

2.6.2 Arrow-Pratt's Risk Aversion Functions

Let x be the initial wealth of a market participant. His performance of investment in the market is like a game, and after the game, his wealth $W = x + X$ is an uncertain return. If the expected value of W is x, the game is called fair. In a fair game, if the utility function of a risk averse investor is u, then there exists a nonnegative real number $\pi(W)$ such that $\mathbb{E}[u(W)] = u(x - \pi(W))$. The value $\pi(W)$ is called the *risk premium* of W; it is the wealth value that a risk averse investor is willing to give up in order to avoid the risk. The value $x - \pi(W)$ is called the *certainty equivalent* of W; it is the wealth level which a risk averse investor requires to ensure in order to achieve the level of expected utility of the uncertain return W.

Assume that the utility functions of two risk averse investors A and B are u and v, respectively. We say that A is more risk averse than B, if for any uncertain return W, we have

$$\pi_A(W) \geqslant \pi_B(W),$$

where $\pi_A(W)$ and $\pi_B(W)$ represent the risk premiums of W corresponding to the utility functions u and v, respectively.

Let $W = x + X$ be a uncertain return, whose expectation is x. Then we have

$$u(W) = u(x + X) = u(x) + u'(x)X + \frac{1}{2}u''(x)X^2 + o(X^2).$$

Assume the variance of W is very small. Since $\mathbb{E}[u(W)] = u(x - \pi(W))$, we have approximately the following equation:

$$u(x) + \frac{1}{2}u''(x)\text{Var}(W) + o(\text{Var}(W)) = u(x) - u'(x)\pi(W) + o(\pi(W)).$$

So it holds approximately

$$\pi(W) = \frac{1}{2}\left[-\frac{u''(x)}{u'(x)}\right]\text{Var}(W).$$

This shows that if the variance of W is very small, then the risk premium is proportional to the size of the risk. Thus, Pratt (1964) and Arrow (1965) proposed to let $-u''(x)/u'(x)$ be as a measure of absolute risk aversion and called it the *absolute risk aversion function*, or the Arrow-Pratt *measure of risk aversion*, denoted by

$$A(x) = -\frac{u''(x)}{u'(x)}.$$

It gives a quantitative indicator of the degree of risk aversion (see Theorem 2.12 below).

We call the reciprocal of $A(x)$ the *risk tolerance function*, denoted by

$$T(x) = -\frac{u'(x)}{u''(x)}. \tag{2.58}$$

Let $W = x + X$ be a uncertain return, whose expectation is x. When a risk aversion utility function u is concerned, there is a nonnegative real number $\pi_R(W)$, such that $\mathbb{E}[u(W)] = u(x(1 - \pi_R(W)))$. We call $\pi_R(W)$ the *relative risk premium*. Similar to the above reasoning, if the variance of W is very small, we have approximately

$$\pi_R(W) = \frac{1}{2}\left[-\frac{xu''(x)}{u'(x)}\right]\text{Var}(X).$$

We call function $-xu''(x)/u'(x)$ the *relative risk aversion function*, denoted by

$$R(x) = -\frac{xu''(x)}{u'(x)}. \tag{2.59}$$

In financial economics the most commonly used utility functions belong to the family of *hyperbolic absolute risk aversion* (HARA), namely,

$$u(x) = \frac{1-\gamma}{\gamma}\left[\frac{ax}{1-\gamma} + b\right]^{\gamma}, \quad b \geqslant 0, \tag{2.60}$$

In order to ensure that a utility function is strictly increasing and concave, we need to properly determine its domain. For example, when $\gamma < 1$, the domain of u is $(-(1-\gamma)b/a, \infty)$; when $\gamma > 1$, the domain of u is $(-\infty, (\gamma - 1)b/a)$. The risk tolerance function of a HARA utility function is

$$T(x) = A(x)^{-1} = \frac{b}{a} + \frac{x}{1-\gamma}. \tag{2.61}$$

In (2.60), if we let $b = 0$ and $a = 1$, we obtain a power utility function $u(x) = x^\gamma/\gamma$; if we let $b = 0, a = 1$ and $\gamma \to 0$, we obtain a logarithm utility function $u(x) = \ln x$ with domain $(0, \infty)$; if we let $b = 1$ and $\gamma \to -\infty$, we obtain an exponential utility function $u(x) = -e^{-ax}$.

2.6.3 Comparison of Risk Aversion Functions

The next theorem due to Pratt (1964) gives several equivalent characterizations for comparing degrees of risk aversion.

Theorem 2.12 *Let u_1 and u_2 be two utility functions, A_1 and A_2 be the corresponding absolute risk aversion functions. Then the following conditions are equivalent:*

(1) $A_1(x) \geqslant A_2(x)$, $\forall x$;
(2) $u_1(u_2^{-1})$ *is a strictly increasing twice continuously differentiable concave function;*
(3) *there is a strictly increasing twice continuously differentiable concave function f, such that $u_1(x) = f(u_2(x))$;*
(4) *for any uncertain return W of a fair game, it holds $\pi_1(W) \geqslant \pi_2(W)$.*

Proof (1) $\Rightarrow$ (2). Let $f(x) = u_1(u_2^{-1}(x))$. Function f is strictly increasing and twice continuously differentiable. It is easily verified that

$$f''(x) = -(A_1(y) - A_2(y))\frac{u_1'(y)}{(u_2'(y)^2)} \leqslant 0,$$

where $y = u_2^{-1}(x)$. Consequently, f is a concave function.

(2) $\Rightarrow$ (3) can be deduced directly.

(3) $\Rightarrow$ (4). Let $W = x + X$ be an uncertain return of a fair game. Then by Jensen's inequality,

$$u_1(x - \pi_1(W)) = \mathbb{E}[u_1(W)] = \mathbb{E}[f(u_2(W))] \leqslant f(\mathbb{E}[u_2(W)]) = f(u_2(x - \pi_2(W)),$$

so that $\pi_1(W) \geqslant \pi_2(W)$.

(4) $\Rightarrow$ (1) is obvious. $\square$

2.6.4 Preference Defined by Stochastic Orders

Let X and Y be two random variables. We say that X dominates Y in the sense of *first-order stochastic dominance*, if for any $x \in \mathbb{R}$, we have $F_X(x) \leqslant F_Y(x)$, where F_X is the distribution function of X. We say that X dominates Y in the sense of *second-order stochastic dominance*, if for any $t \in \mathbb{R}$, we have

$$\int_{-\infty}^{t} F_X(s)ds \leqslant \int_{-\infty}^{t} F_Y(s)ds.$$

Obviously, first-order stochastic dominance implies second-order stochastic dominance. The introduction of the concept of stochastic dominance defines a preference in the set of uncertain returns. In some literature the preference defined by the second-order stochastic dominance is called the *uniform preference*.

The following theorem gives a characterization of the second-order stochastic dominance:

Theorem 2.13 *Let X and Y be two random variables. Then X dominates Y in second-order stochastic dominance if and only if for any $c \in \mathbb{R}$, $E[(c - X)_+] \leqslant E[(c - Y)_+]$.*

Proof By Fubini's theorem,

$$\int_{-\infty}^{c} F_X(y)dy = \int_{-\infty}^{c} \int_{(-\infty,y]} dF_X(y)dy$$

$$= \int I_{[z \leqslant y \leqslant c]} dy dF_X(z)$$

$$= \int (c - z)^+ dF_X(z)$$

$$= \mathbb{E}[(c - X)^+].$$

$\square$

The next theorem gives a characterization of stochastic orders from the perspective of expected utility.

Theorem 2.14 *Let X and Y be two random variables.*

(1) *X dominates Y in first-order stochastic dominance if and only if for all increasing functions u, $\mathbb{E}[u(X)] \geqslant \mathbb{E}[u(Y)]$.*
(2) *X dominates Y in second-order stochastic dominance if and only if for all concave increasing functions u, $\mathbb{E}[u(X)] \geqslant \mathbb{E}[u(Y)]$.*

Proof (1) Assume that u is an increasing function on $\mathbb{R}$ with inverse function u^{-1}. If $u(X)$ is integrable, then

$$\mathbb{E}[u(X)] = \int_{-\infty}^{0} [\mathbb{P}(u(X) \geqslant x) - 1]dx + \int_{0}^{\infty} \mathbb{P}(u(X) \geqslant x)dx$$

$$= -\int_{-\infty}^{0} F_X(u^{-1}(x))dx + \int_{0}^{\infty} [1 - F_X(u^{-1}(x))]dx.$$

From this we know that (1) holds.

(2) Sufficiency. By Fubini's theorem,

$$\mathbb{E}[(t - X)^+] = \mathbb{E}[\int_{-\infty}^t I_{[X \leqslant x]} dx] = \int_{-\infty}^t F_X(x) dx. \tag{2.62}$$

Let $u(x) = -(t - x)^+$. Since u is a concave increasing function, the sufficiency is proved.

Necessity. Assume that u is a concave increasing function on $\mathbb{R}$. Let u' be the right derivative of u. Then there is a measure μ on $\mathbb{R}$ such that for $x \leqslant y$ we have $u'(x) = u'(y) + \mu((x, y])$, and then by Fubini's theorem

$$u(x) = u(y) - u'(y)(y - x) - \int_x^y \int_{(t, y]} \mu(dz) dt$$

$$= u(y) - u'(y)(y - x) - \int_{(-\infty, y)} (z - x)^+ \mu(dz), \quad x \leqslant y.$$

Assume that X dominates Y in second-order stochastic dominance. Then for any $y \in \mathbb{R}$, from the above equation and (2.62), we get

$$\mathbb{E}[u(X)I_{[X \leqslant y]}] = u(y)\mathbb{P}(X \leqslant y) - u'(y)\mathbb{E}[(y - X)^+] - \int_{(-\infty, y)} \mathbb{E}[(z - X)^+] \mu(dz)$$

$$\geqslant u(y)\mathbb{P}(Y \leqslant y) - u'(y)\mathbb{E}[(y - Y)^+] - \int_{(-\infty, y)} \mathbb{E}[(z - Y)^+] \mu(dz)$$

$$= \mathbb{E}[u(Y)I_{[Y \leqslant y]}].$$

Let $y \to \infty$, finally we get $\mathbb{E}[u(X)] \geqslant \mathbb{E}[u(Y)]$.   □

The next proposition shows that if the rates of return of risky assets are normally distributed, Markowitz' mean-variance analysis is a reasonable choice for a risk averse person.

Proposition 2.15 *Assume that random variables X and Y follow distributions $N(m, \sigma^2)$ and $N(\widetilde{m}, \widetilde{\sigma}^2)$, respectively. Then X dominates Y in second-order stochastic dominance if and only if $m \geqslant \widetilde{m}$ and $\sigma^2 \leqslant \widetilde{\sigma}^2$.*

Proof Necessity. From Theorem 2.14, if X dominates Y in second-order stochastic dominance, then $\forall \alpha > 0$,

$$\exp\left\{-\alpha m + \frac{1}{2}\alpha^2\sigma^2\right\} = \mathbb{E}[e^{-\alpha X}] \leqslant \mathbb{E}[e^{-\alpha Y}] = \exp\left\{-\alpha\widetilde{m} + \frac{1}{2}\alpha^2\widetilde{\sigma}^2\right\}.$$

Namely, we have

$$m - \frac{1}{2}\alpha\sigma^2 \geqslant \widetilde{m} - \frac{1}{2}\alpha\widetilde{\sigma}^2, \quad \forall \alpha > 0,$$

from which it follows $m \geqslant \widetilde{m}$ and $\sigma^2 \leqslant \widetilde{\sigma}^2$.

Sufficiency. Assume $m \geqslant \widetilde{m}$ and $\sigma^2 \leqslant \widetilde{\sigma}^2$. Then we can directly verify that for any $t \in \mathbb{R}$,

$$\int_{-\infty}^{t} F_X(s)ds \leqslant \int_{-\infty}^{t} F_Y(s)ds.$$

Therefore, X dominates Y in second-order stochastic dominance. □

The next proposition comes from Föllmer and Schied (2004). Here we make some modifications to the original proof.

Proposition 2.16 *Assume that random variables X and Y follow distributions $N(m, \sigma^2)$ and $N(\widetilde{m}, \widetilde{\sigma}^2)$, respectively. Then e^X dominates e^Y in second-order stochastic dominance if and only if*

$$\sigma^2 \leqslant \widetilde{\sigma}^2, \quad m + \frac{1}{2}\sigma^2 \geqslant \widetilde{m} + \frac{1}{2}\widetilde{\sigma}^2.$$

In particular, the fact that e^X dominates e^Y in second-order stochastic dominance implies that X dominates Y in second-order stochastic dominance.

Proof Necessity. Assume that e^X dominates e^Y in second-order stochastic dominance. Put $u_\alpha(x) = x^\alpha$, $\forall 0 < \alpha < 1$. Then u_α is a strictly increasing concave function on $(0, \infty)$. Thus, by Theorem 2.14, we have

$$\exp\left\{\alpha m + \frac{1}{2}\alpha^2\sigma^2\right\} = \mathbb{E}[u_\alpha(e^X)] \geqslant \mathbb{E}[u_\alpha(e^Y)] = \exp\left\{\alpha\widetilde{m} + \frac{1}{2}\alpha^2\widetilde{\sigma}^2\right\},$$

namely,

$$m + \frac{1}{2}\alpha\sigma^2 \geqslant \widetilde{m} + \frac{1}{2}\alpha\widetilde{\sigma}^2, \quad \forall 0 < \alpha < 1.$$

In the above formula, letting $\alpha \uparrow 1$ gives $m + \frac{1}{2}\sigma^2 \geqslant \widetilde{m} + \frac{1}{2}\widetilde{\sigma}^2$. On the other hand, $\forall \varepsilon > 0$, let $f_\varepsilon(x) = \ln(\varepsilon + x)$. Then f_ε is a strictly increasing concave function on $[0, \infty)$, and for any increasing concave function u on $\mathbb{R}$, $u \circ f_\varepsilon$ is an increasing concave function on $[0, \infty)$. Thus, by Theorem 2.14, we have

$$\mathbb{E}[u(X)] = \lim_{\varepsilon \to 0} \mathbb{E}[u \circ f_\varepsilon(e^X)] \geqslant \lim_{\varepsilon \to 0} \mathbb{E}[u \circ f_\varepsilon(e^Y)] = \mathbb{E}[u(Y)],$$

which implies that X dominates Y in second-order stochastic dominance, and, consequently, $\sigma^2 \leqslant \widetilde{\sigma}^2$.

For the proof of the sufficiency, one can see Föllmer and Schied (2004) or directly verify that for any $t \in \mathbb{R}$,

$$\int_{-\infty}^{t} F_{e^X}(s)ds \leqslant \int_{-\infty}^{t} F_{e^Y}(s)ds,$$

namely,

$$\int_0^{\ln t} s^{-1} F_X(s) ds \leqslant \int_0^{\ln t} s^{-1} F_Y(s) ds.$$

$\square$

2.6.5   Maximization of Expected Utility and Initial Price of Risky Asset

Below we study in a single period market the problem about the maximization of expected utility and the initial prices of risky assets. For simplicity, we assume that the yield of the risk-free asset is a constant $r > 0$ and there is another risky asset whose initial price is S_0 and price at time 1 is S_1. Assume that a certain market participant has initial wealth w and his utility function u is a strictly concave and continuously differentiable increasing function. The question is how to invest his wealth so that he maximizes expected utility of the wealth in the end. Let $0 \leqslant \lambda \leqslant 1$. Assume that he invests λw of funds in the risk-free asset and $(1 - \lambda)w$ of funds in the risky asset. Then at time 1, his wealth becomes

$$X_\lambda = \frac{(1-\lambda)w}{S_0} S_1 + \lambda w (1 + r).$$

The question is to find $\lambda^* \in [0, 1]$, such that

$$\lambda^* = \arg \max_{\lambda \in [0,1]} \mathbb{E}[u(X_\lambda)].$$

The following theorem is from Föllmer and Schied (2004).

Theorem 2.17 *Assume that S_1 is a bounded random variable. Let $R = \frac{S_1 - S_0}{S_0}$, we have the following conclusions:*

(1) *if $r \geqslant \mathbb{E}[R]$, then $\lambda^* = 1$; if $u((1 + r)w) > \mathbb{E}[u((1 + R)w)]$, then $\lambda^* > 0$.*
(2) *if u is differentiable, then $\lambda^* = 1$, if and only if $r \geqslant \mathbb{E}[R]$; $\lambda^* = 0$, if and only if*

$$r \leqslant \frac{\mathbb{E}[Ru'((1 + R)w)]}{\mathbb{E}[u'((1 + R)w)]}.$$

Proof

(1) Set $c = w(1 + r), X = w(1 + R), f(\lambda) = \mathbb{E}[u(X_\lambda)]$, then by Jensen's inequality,

$$f(\lambda) \leqslant u(\mathbb{E}[X_\lambda]) = u((1 - \lambda)\mathbb{E}[X] + \lambda c),$$

equality holds, if and only if $\lambda = 1$. If $r \geqslant \mathbb{E}[R]$, then the right side of the above formula is a strictly increasing function of λ, and hence $\lambda^* = 1$. Moreover, the strict concavity of u implies

$$f(\lambda) \geqslant \mathbb{E}[(1 - \lambda)u(X) + \lambda u(c)] = (1 - \lambda)u(c(X)) + \lambda u(c),$$

where $c(X)$ is the certainty equivalent of X. If R is not identically equal to r, then the above equality holds, if and only if $\lambda = 0$ or 1. If $u((1 + r)w) > \mathbb{E}[u((1 + R)w)]$, i.e., $c > c(X)$, then the right side of the above formula is strictly increasing, and hence $\lambda^* > 0$.

(2) If u is differentiable, then

$$f'_+(0) = \mathbb{E}[u'(X)(c - X)], \quad f'_-(1) = u'(c)(c - \mathbb{E}[X]).$$

Hence, $\lambda^* = 1$ if and only if $f'_-(1) \geqslant 0$, i.e., $c \geqslant \mathbb{E}[X]$, or equivalently, $r \geqslant \mathbb{E}[R]$; $\lambda^* = 0$, if and only if $f'_+(0) \leqslant 0$, i.e.,

$$c \leqslant \frac{E[Xu'(X)]}{\mathbb{E}[u'(X)]},$$

or equivalently,

$$r \leqslant \frac{\mathbb{E}[Ru'((1 + R)w)]}{\mathbb{E}[u'((1 + R)w)]}.$$

The theorem is proved. $\qquad\square$

In the following we discuss the initial price of the risky asset from the point of view of utility function. From Theorem 2.17 we know that for an investor whose preference is determined by utility function u, only when the expected rate of return of risky asset $\mathbb{E}[R]$ is strictly greater than the rate of return r of the risk-free asset, he will consider making some investments in the risky asset; only when

$$\frac{\mathbb{E}[Ru'((1 + R)w)]}{\mathbb{E}[u'((1 + R)w)]} \geqslant r,$$

he will choose not to invest in the risk-free asset. Therefore, the initial price of the risky asset such that the above equality holds is an acceptable and reasonable price for investors whose preference is determined by utility function u. It is given by

$$S_0 = (1 + r)^{-1} \frac{\mathbb{E}[S_1 u'((1 + R)w)]}{\mathbb{E}[u'((1 + R)w)]},$$

namely,

$$S_0 = (1 + r)^{-1} \mathbb{E}^*[S_1],$$

where

$$\frac{d\mathbb{P}^*}{d\mathbb{P}} = \frac{u'((1+R)w)}{\mathbb{E}[u'((1+R)w)]}.$$

Here $\mathbb{P}^*$ is the probability measure such that $\mathbb{E}^*[R] = r$, which is known as *risk-neutral measure*.

2.7 Consumption-Based Asset Pricing Models

Assume that in the market, there are d risky assets and the current price and the (unknown) price of the next time of the j-th asset are S_0^j and S_1^j, respectively. Put

$$S_0 = (S_0^1, \cdots, S_0^d)^\tau, \quad S_1 = (S_1^1, \cdots, S_1^d)^\tau.$$

Assume that there are many investors with the same preference or, equivalently, there is a *representative agent*. Suppose that his utility function is u and the initial wealth is w_0. He wants to use part of the money for initial consumption and invest the remaining money in risky assets to prepare for the next time consumption. His goal is to determine the amount c_0 of initial consumption and shares $\theta = (\theta_1, \cdots, \theta_d)^\tau$ of risky assets, such that the following expected utility of consumptions attaints the maximum:

$$U(c_0, \theta) = u(c_0) + \beta\mathbb{E}[u(c_1)], \tag{2.63}$$

where

$$c_0 = w_0 - \theta^\tau S_0 \text{ and } c_1 = w_1 + \theta^\tau S_1 \tag{2.64}$$

are the amount of initial consumption as well as the summation of the certainty income w_1 and the value of risky assets at the next time, respectively. Here $0 < \beta \leqslant 1$ is the (subjective) discount factor reflecting investors' discount about the cash utility in the next time period. Quantity $\beta^{-1} - 1$ can also be seen as the rate of return of the risk-free asset in this period. A necessary condition for a solution to the optimization problem is that the first-order derivatives of $U(c_0, \theta)$ w.r.t. c_0 and each θ_j are all zero. Noting that $c_1 = w_1 + \theta^\tau(S_1 - S_0) + w_0 - c_0$, optimal strategy (c_0^*, θ^*) satisfies

$$u'(c_0^*) - \beta\mathbb{E}[u'(c_1^*)] = 0, \tag{2.65}$$

$$S_0^j u'(c_0^*) = \beta\mathbb{E}[u'(c_1^*)S_1^j], \quad j = 1, \cdots, d. \tag{2.66}$$

From another perspective, we can regard (c_0, Θ) satisfying (2.64) as a viable *consumption-investment plan* of the investor. If this plan is the best one, then the current price S_0^j of asset j should be given by the following formula:

$$S_0^j = \beta \mathbb{E}\left[\frac{u'(c_1^*)}{u'(c_0^*)} S_1^j\right]. \tag{2.67}$$

This formula is called *consumption-based asset pricing*.

Theoretically, in a competitive equilibrium market which is achieved by participants who pursue their expected utility maximizations, the equilibrium price of an asset may be calculated by the above consumption-based asset pricing formula using a weighted utility function of a representative agent. This theory, however, has a serious mismatch with the actual data. Mehra and Prescott (1985) pointed out that the consumption-based asset pricing formula cannot explain the phenomenon that in the US stock market the average return rate of stocks is 4–6 percentage point higher than that on government bonds. This phenomenon is known as the *equity premium puzzle* and has not yet received a reasonable explanation.

Chapter 3
Financial Markets in Discrete Time

In this chapter, after introducing the basic concepts of financial markets, we first use the binomial tree model to illustrate the risk-neutral valuation principle. Then we study the general discrete-time model and give the martingale characterization of arbitrage-free market and European contingent claims pricing. In addition, we discuss the investment strategy via expected utility maximization and utility function-based contingent claims pricing and market equilibrium pricing. In the end, we study the pricing of American contingent claims.

3.1 Basic Concepts of Financial Markets

A *contingent claim* is a kind of financial products, which is a claim that can be realized before or at some time point in the future. If the value of a contingent claim is derived from the prices of some basic assets, called *underlying assets*, such as stocks, rates of foreign exchanges, and commodities, then this contingent claim is also called a *derivative (asset)*. Typical examples of derivatives are *options*. An option is a contract that entitles its owner with the right but no obligation to buy or sell a specified quantity of an underlying asset, such as stock, currency, and commodity, at a specified price, known as the *exercise price* or *strike price*, at or before a certain date, known as the *expiration date* or *maturity*. There are two basic types of option: call and put. Here "call" refers to the right to buy and "put" refers to the right to sell. If an option can be executed at any time before maturity, it is called an *American option*; if it can only be exercised at maturity, it is called a *European option*.

The call or put option described above is called a *vanilla option*. An option that is not a vanilla put or call is called an *exotic option*. Most exotic options are *path-dependent* in the sense that their payoff depends on the current and past values of underlying assets. An option that has no expiry date is called a *perpetual option*. A *Russian option* is a perpetual American-style option which, at any time chosen by the owner, pays the maximum realized asset price up to that date.

© Springer Nature Singapore Pte Ltd. and Science Press 2018
J.-A. Yan, *Introduction to Stochastic Finance*, Universitext,
https://doi.org/10.1007/978-981-13-1657-9_3

3.1.1   Numeraire

In some cases, for example in option pricing, one often needs to choose a strictly positive process to denominate the asset price processes. Such a process is called a *numeraire*. The resulting relative price is called the *deflated price*. In many situations, one takes the price process of a particular asset as a numeraire. Such an asset is called a *numeraire asset*. If one takes the bank account as a numeraire asset, the related deflated prices are often called *discounted prices*. In this case, an important concept concerning interest rate is the *present value*. If the bank account earns a constant interest rate r, then in continuous-time setting, the present value at time t of a value ξ at a future time T is defined as $e^{-r(T-t)}\xi$. If we deposit this amount of money in a bank at time t, we get exactly the value ξ at time T.

3.1.2   Pricing and Hedging

An option provides a nonnegative payoff which is not identically null. It must have a value (called the *price*) at any time before maturity. The initial price of an option is usually called the *premium*. There are two fundamental problems related to an option: pricing and hedging. Pricing an option is to determine its market value at any time t before maturity. The minimal value with which one can build a trading strategy to generate (resp. cover) the payoff of an option is called the *fair price* (resp. *upper price* or *selling price*) of the option. Such a trading strategy is called *hedging strategy* or *replicating strategy* (resp. *super-hedging strategy* or *super-replicating strategy*). Note that the term hedging has also other meanings in different contexts. For example, taking opposite positions in different financial instruments in order to reduce (but not necessary eliminate) a risk is also called hedging. To distinguish these two types of hedge, we sometimes call a hedge with replicating strategy a *perfect hedge*.

   In option pricing theory, one of the basic assumptions on the market is the absence of arbitrage opportunity. It means that there is no riskless and profitable opportunity in the market. This assumption implies the so-called law of one price, which states that two financial packages having identical payoffs must sell at the same price. Another basic assumption is the existence of a riskless investment that gives a guaranteed return with no chance of default. A good approximation to such an investment is a government bond or a deposit in a sound bank. The latter is called a *bank account* (or *money market account, savings account*). A bank account earns a return at an *interest rate*.

3.1.3   Put-Call Parity

From two basic assumptions on the market we can deduce an intrinsic relationship between the values of a European vanilla call and a vanilla put written on a same asset with same maturity T and same strike price K. We consider a portfolio consisting of a long position in a share of stock and in one put option and a short position in one call option. Let S_t denote the price of the stock at time t, and let C_t and P_t denote the values at time t of such a call and a put, respectively. The value of this portfolio at time T is given by

$$S_T + P_T - C_T = S_T + (K - S_T)^+ - (S_T - K)^+ = K,$$

where a^+ stands for $\max(a, 0)$ for a real number a. Thus, this portfolio is riskless. Assume that the riskless interest rate of the bank account is a constant r. According to the law of one price, the wealth of this portfolio at time $t < T$ should be $Ke^{-r(T-t)}$, because if at time t we invest such money in the bank account, we get also money K at time T. Thus, we obtain the following equality:

$$S_t + P_t - C_t = Ke^{-r(T-t)},$$

which is called the *put-call parity*.

3.1.4 Intrinsic Value and Time Value

At any time t before maturity an option can be referred to as *in the money, at the money,* or *out of the money.* An in-the-money call (resp. put) option is one whose strike price is less (resp. greater) than the current price of the underlying asset. An at-the-money call or put option is one whose strike price is equal to the current price of the underlying asset. An out-the-money call (resp. put) option is one whose strike price is greater (resp. less) than the current price of the underlying asset.

Let S_t be the price of an asset at time t and K the strike price of a call (resp. put) written on that asset. We call $(S_t - K)^+$ (resp. $(K - S_t)^+$) the *intrinsic value* of the call (resp. put) option. An in-the-money American option must be worth at least as much as its intrinsic value, since the holder can realize a positive intrinsic value by exercising the option immediately. In this case the option is said to have a *time value* which is equal to the difference between the market value and the intrinsic value of the option.

3.1.5 Bid-Ask Spread

One of the important characteristics for a financial market is the *liquidity*, the ability to buy or sell significant quantities of a security easily. Market makers stand ready to buy and sell, with different prices for purchases and sells. They buy low at the *bid price* and sell high at the *ask price*. The investor pays the ask price to buy and sell for the bid price. The difference between these two prices is called the *bid-ask spread*. The price quoted in the newspapers is usually the average of the bid and ask prices.

3.1.6   Efficient Market Hypothesis

A financial market is called *efficient*, if participants can quickly obtain all information relevant to trading, the market is liquid, and there are low trading costs (including transaction costs, taxes, bid-ask spreads). The *efficient market hypothesis* (EMH) states that in an efficient market the present prices of assets contain all information about the market development. In mathematical language, EMH means that the price process of assets is a Markov process.

3.2   Binomial Tree Model

It is a common knowledge that an asset price moves randomly and one cannot predict future values of the prices. However, by a statistical analysis of historical data, one can establish some models for asset price movements. There are two types of models: the discrete-time model and continuous-time model. The simplest discrete-time model is the so-called *binomial tree model*, which was introduced by Cox et al. (1979) as a technical tool for pricing a contingent claim. This model is far from being realistic. We shall only use this model to illustrate two important methods for pricing options, *arbitrage pricing* and *risk-neutral valuation*, which are essentially equivalent.

Suppose that there are two assets in the market. One is a risk-free asset with interest rate r per unit time; another one is a no-dividend-paying stock whose current price (i.e., the price at time zero) is S_0. Assume that at time $n + 1$ the stock price S_{n+1} will be either $u S_n$ or $d S_n$, where d and u are constants and $d < u$ (see Fig. 3.1). Obviously, the absence of arbitrage opportunity is equivalent to the condition "$d < 1+r < u$." We are interested in valuing a European contingent claim ξ with maturity N. We assume that ξ depends only on stock price S_N at time N.

3.2.1   The One-Period Case

In this subsection we consider the one-period case (i.e., $N = 1$). The multi-period case will be studied in the next subsection. In the one-period case, at time 1 the contingent claim takes one of two values, say, ξ_u or ξ_d, which corresponds, respectively, to stock prices $u S_0$ or $d S_0$ at time 1. Consider a portfolio consisting of a long position in α_0 shares of stock and a short position in contingent claim. The wealth of the portfolio at time 1 is equal to $\alpha_0 S_1 - \xi$. In order to find a value α_0 that makes the portfolio risk-free (i.e., the wealth of the portfolio does not depend on the up or down movements of the stock price), we solve the equation:

$$\alpha_0 u S_0 - \xi_u = \alpha_0 d S_0 - \xi_d,$$

and obtain $\alpha_0 = \frac{\xi_u - \xi_d}{(u-d)S_0}$. The wealth of the portfolio at time 1 is then $X_1 = \frac{d\xi_u - u\xi_d}{u-d}$. Thus, by the law of one price, the wealth X_0 of the portfolio at time zero is equal to $X_1/1 + r$. Consequently, the price of contingent claim ξ at time 0 is given by

$$C_0 = \alpha_0 S_0 - X_0 = \frac{(1+r-d)\xi_u + (u-(1+r))\xi_d}{(1+r)(u-d)}. \tag{3.1}$$

This method of pricing by no-arbitrage argument is called the *arbitrage pricing*.

A careful reader may notice that the contingent claim pricing formula (3.1) does not involve the probabilities of the stock price moving up or down. An explanation of this somewhat surprising fact is that the probabilities of future up or down movements are already embedded into the current price of the stock. Another explanation is that if the stock is expected to go up with a high probability and the price of a call option raises, the investor may invest his money on the stock directly instead of buying the call option.

If we put $q = \frac{1+r-d}{u-d}$, then (3.1) can be rewritten as

$$C_0 = (1+r)^{-1}[q\xi_u + (1-q)\xi_d]. \tag{3.2}$$

If we imagine $\{q, 1-q\}$ as the probability distribution of the stock price moving up and down, then (3.2) states that today's price of the contingent claim is the expectation of its discounted future value under this new probability measure. Since $qu + (1-q)d = 1+r$, it is readily seen that this new probability measure is the unique measure under which the world is *risk-neutral* in the sense that the expected return on the stock, i.e., $\mathbb{E}[S_1]/S_0 - 1$, is equal to the riskless rate r. This new probability measure is called the *risk-neutral probability measure*, and pricing by (3.2) is called the *risk-neutral valuation*.

3.2.2 The Multistage Case

Now we turn to the multistage binomial tree model (see Fig. 3.1). Our objective is to determine the value at any time n of a European contingent claim ξ with maturity N. Let Ω denote the set of all stock price paths running from time zero up to time N. It represents the uncertainty of the stock price movements. Clearly, Ω contains 2^N elements. Each element is a possible realization of the movements of the stock price.

At time n, we have $n+1$ nodes. We number these nodes from top to bottom. For each $\omega \in \Omega$, we denote by $\omega(n)$ the serial number of the node passed through by path ω at time n. Put

$$\Omega_{n,j} = \{\omega : \omega(n) = j\}, \quad 1 \leqslant j \leqslant n+1.$$

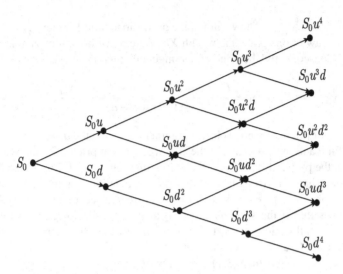

Fig. 3.1 Binomial tree model

Then $\Omega_{n,j} \subset \Omega_{n+1,j} \cup \Omega_{n+1,j+1}$, and

$$\omega, \omega' \in \Omega_{n,j} \Longleftrightarrow S_n(\omega) = S_n(\omega'). \tag{3.3}$$

If we regard each node as an origin point and consider one-period movement of the stock price from this node, we come to the situation of one-period binomial tree model. So by the "backward induction" and by a repeated application of Eq. (3.2), we can give the value of the claim at any time $n = 0, 1, \cdots, N-1$. More precisely, let us define a probability measure on Ω by

$$\mathbb{P}^*(\omega) = q^{\sum_{i=0}^{N-1} \alpha_i(\omega)} (1-q)^{N - \sum_{i=0}^{N-1} \alpha_i(\omega)}, \quad \omega \in \Omega, \tag{3.4}$$

where $\alpha_i(\omega) = 1$ or 0, when the stock price goes up or down at the i-th step of path ω. We denote by C_n the value at time n of the contingent claim ξ. From (3.1), (3.2) and (3.3) we see that

$$\omega, \omega' \in \Omega_{n,j} \Longleftrightarrow C_n(\omega) = C_n(\omega'), \tag{3.5}$$

and

$$C_{n,j} = (1+r)^{-1}[qC_{n+1,j} + (1-q)C_{n+1,j+1}], \tag{3.6}$$

where $C_{n,j}(\omega) := C_n(\omega), \forall \omega \in \Omega_{n,j}$. Since by (3.3) the σ-algebra $\sigma(S_n)$ is generated by the family $\{\Omega_{n,j}, 1 \leqslant j \leqslant n+1\}$ and

$$\mathbb{P}^*(\Omega_{n+1,j}|\Omega_{n,j}) = q, \quad \mathbb{P}^*(\Omega_{n+1,j+1}|\Omega_{n,j}) = 1 - q,$$

we can rewrite (3.6) as

$$C_n = (1+r)^{-1}\mathbb{E}^*[C_{n+1}|S_n]. \tag{3.7}$$

Thus, if we denote by $\mathcal{F}_n$ the σ-algebra generated by the sequence $(S_j, 0 \leqslant j \leqslant n)$, then by Markov property of (C_n), from (3.7), we get

$$C_n = (1+r)^{-1}\mathbb{E}^*[C_{n+1}|\mathcal{F}_n]. \tag{3.8}$$

In other words, the sequence of discounted values of the contingent claim $\{(1 + r)^{-n}C_n, 0 \leqslant n \leqslant N\}$ forms a $\mathbb{P}^*$-martingale. Here $(1 + r)^{-n}$ is called the *discount factor* at time n. In particular, from the last statement of the previous section, we know that $\mathbb{P}^*$ is the unique probability measure on Ω under which the sequence $\{(1 + r)^{-n}S_n, 0 \leqslant n \leqslant N\}$ of the discounted stock prices forms a martingale. We call a probability measure with such property a *risk-neutral probability measure* or *martingale measure*.

From (3.8), we get a formula for pricing contingent claim ξ:

$$C_n = (1+r)^{-(N-n)}\mathbb{E}^*[\xi|\mathcal{F}_n]. \tag{3.9}$$

This formula is an example of an important general principle in contingent claim pricing, known as the *risk-neutral valuation principle*, which states that any contingent claim can be valuated under the risk-neutral world assumption.

Now assume that $\xi = f(S_N)$ with f being a positive function. We are going to deduce an explicit expression for C_n. Put $T_n = S_n/S_{n-1}$, for $n = 1, \cdots, N$. It is easy to verify that random variables $T_1, \cdots, T_n$ are independent identically distributed (i.i.d.) and their distribution is $\mathbb{P}^*(T_1 = u) = q = 1 - \mathbb{P}^*(T_1 = d)$. In particular, for each $i \geqslant n + 1$, T_i is independent of $\mathcal{F}_n$. Consequently, since $S_N = S_n \prod_{i=n+1}^{N} T_i$, from (3.9), we can apply Theorem 1.10 to get

$$C_n = (1+n)^{-(N-n)}\mathbb{E}^*\left[f(S_n \prod_{i=n+1}^{N} T_i) \Big| \mathcal{F}_n\right]$$

$$= (1+n)^{-(N-n)}\mathbb{E}^*\left[f(x \prod_{i=n+1}^{N} T_i)\right]\Big|_{x=S_n}$$

$$= (1+r)^{-(N-n)} \sum_{j=0}^{N-n} \binom{N-n}{j} q^j(1-q)^{N-n-j} f(S_n u^j d^{N-n-j}).$$

3.2.3   The Approximately Continuous Trading Case

Let $\xi = f(S_T)$ be a contingent claim, where T is a calendar time. If the trading takes place at times $\{\frac{T}{N}, \frac{2T}{N}, \cdots\}$ with an extremely large N, what can we say about the price of ξ at time 0? In order to answer this question, we adjust N-dependent parameters u, d, and p in such a way that a meaningful limit exists in the expression C_0 when N tends to ∞. Since C_0 does not depend on p, we take $p = \frac{1}{2}$ for every N. Assume that the interest rate is a constant r per unit time. Then in each period, the interest rate should be $\frac{rT}{N}$. Let μ and σ be two positive constants. If we choose

$$u = \exp\left\{\frac{\mu T}{N} + \sigma\sqrt{\frac{T}{N}}\right\}, \quad d = \exp\left\{\frac{\mu T}{N} - \sigma\sqrt{\frac{T}{N}}\right\},$$

then the stock price at time T is

$$S_T^{(N)} = S_0 \exp\left\{\left(\frac{\mu T}{N} + \sigma\sqrt{\frac{T}{N}}\right) X_N + \left(\frac{\mu T}{N} - \sigma\sqrt{\frac{T}{N}}\right)(N - X_N)\right\}$$

$$= S_0 \left\{\mu T + \sigma\sqrt{T}\frac{2X_N - N}{\sqrt{N}}\right\},$$

with

$$X_N = \sum_{i=0}^{N-1} \alpha_i,$$

where $\alpha_i(\omega) = 1$ or 0, when the stock price goes up or down at the i-th step of path ω. Under the objective probability, X_N has the binomial distribution with parameter $(N, \frac{1}{2})$. So by the central limit theorem, the distribution of random variable $\frac{2X_N - N}{\sqrt{N}}$ converges to the standard normal distribution. Put

$$q = \frac{1 + rT/N - d}{u - d}.$$

It is easy to see that q is approximately equal to

$$q = \frac{1}{2}\left(1 - \sqrt{\frac{T}{N}}\frac{\mu + \frac{1}{2}\sigma^2 - r}{\sigma}\right).$$

Thus, we have $0 < q < 1$ when N is sufficiently large. For such an N, the market has no arbitrage, and we can define a risk-neutral probability $\mathbb{P}^*$ using q as in (3.4). Under risk-neutral probability $\mathbb{P}^*$, X_N has the binomial distribution with parameter (N, q). According to the central limit theorem, the distribution of random variable $\ln S_t^{(N)}$ under $\mathbb{P}^*$ is approximately the normal distribution with mean $\ln S_0 + (r -$

$\frac{1}{2}\sigma^2)t$ and variance $\sigma^2 t$. Consequently, the price of ξ at time 0 is approximately equal to

$$C_0 = e^{-rT}\mathbb{E}^*[f(S_T)] = e^{-rT}\int_{-\infty}^{\infty} f\left(S_0 e^{(r-\sigma^2/2)T+\sigma y\sqrt{T}}\right)\frac{e^{-y^2/2}}{\sqrt{2\pi}}dy.$$

We shall see in Chap. 5 that this equation gives the famous Black-Scholes formula for pricing call (resp. put) options, when $f(x) = (x-K)^+$ (resp. $f(x) = (K-x)^+$).

3.3 The General Discrete-Time Model

Now we turn to the general discrete-time model. This model was introduced by Michael Harrison and Stanley R. Pliska (1981) and further discussed by Taqqu and Willinger (1987) for the case of finite states. We shall show that the existence of an equivalent probability measure under which the discounted price processes of assets are martingales is equivalent to the absence of arbitrage, and that the uniqueness of such a martingale measure is equivalent to the completeness of the market. This unique martingale measure enables us to price uniquely contingent claims. We follow closely the first chapter of Lamberton and Lapeyre (1996).

3.3.1 The Basic Framework

We assume that the trading activity takes place at discrete times $\{0 = t_0 < t_1 < \cdots < t_{N-1}\}$, where 0 is the initial time. We let $t_N > t_{N-1}$ be a specified time horizon. At time t_N there is no trading. For notational simplicity, we usually denote these times by $\{0, 1, \cdots, N-1, N\}$. Uncertainty up to time N is represented by a probability space $(\Omega, \mathcal{F}, \mathbb{P})$, where Ω represents the set of all possible states. Let $\mathcal{F}_n$ be a sub-σ-algebra of $\mathcal{F}$, which represents the information available up to time n. Then, $\{\mathcal{F}_n, 0 \leqslant n \leqslant N\}$ constitutes a filtration, i.e., an increasing sequence of sub-σ-algebras of $\mathcal{F}$. For notational convenience, we put $\mathcal{F}_{-1} = \mathcal{F}_0$.

Assume that the market consists of $d+1$ assets whose prices constitute an adapted sequence of $\mathbb{R}^{d+1}$-valued strictly positive random variables $(S_n^0, \cdots, S_n^d)$. We take asset indexed by 0 as the numeraire, and assume $S_0^0 = 1$. Let γ_n be deflator $(S_n^0)^{-1}$ at time n. We denote by S_n the vector of stock prices $(S_n^1, \cdots, S_n^d)$ at time n, and denote by $\widetilde{S}_n$ deflated prices $\gamma_n S_n$.

A *trading strategy* of an investor is a predictable $\mathbb{R}^{d+1}$-valued stochastic sequence:

$$\phi_n = (\phi_n^0, \cdots, \phi_n^d), \ 0 \leqslant n \leqslant N,$$

where for $n \geqslant 1$, ϕ_n^i represents the number of shares of asset i held from time $n-1$ to n before trading takes place at time n. So $\phi_n = (\phi_n^0, \cdots, \phi_n^d)$ is the *portfolio* readjusted at the trading time $n - 1$. This is the reason that ϕ_n is $\mathcal{F}_{n-1}$ measurable. If $\phi_n^i < 0$, we say that we are *selling in short* a number ϕ_n^i of asset i. Short selling is allowed. We denote by φ_n the vector $(\phi_n^1, \cdots, \phi_n^d)$ and by φ the process $(\varphi_n)_{0 \leqslant n \leqslant N}$. In this way, we can write $\phi_n = (\phi_n^0, \varphi_n)$ and $\phi = (\phi^0, \varphi)$.

The wealth at time n of a trading strategy ϕ_n is

$$V_n(\phi) = \phi_n^0 S_n^0 + \varphi_n \cdot S_n = \sum_{i=0}^{d} \phi_n^i S_n^i,$$

where "$\cdot$" denotes the inner product in $\mathbb{R}^d$. The discounted wealth $\tilde{V}_n(\phi) := \gamma_n V_n(\phi)$ is given by

$$\tilde{V}_n(\phi) = \phi_n^0 + \varphi_n \cdot \tilde{S}_n.$$

For a trading strategy $(\phi_n) = (\phi_n^0, \varphi_n)$, we define

$$G_n(\phi) := \sum_{i=1}^{n} \phi_i^0 (S_i^0 - S_{i-1}^0) + \sum_{i=1}^{n} \varphi_i \cdot \Delta S_i, \quad \forall 1 \leqslant n \leqslant N, \tag{3.10}$$

where $\Delta S_i = S_i - S_{i-1}$. We call (G_n) the *cumulative gain process* associated with strategy ϕ.

A trading strategy $(\phi_n) = (\phi_n^0, \varphi_n)$ is called *self-financing* if

$$\phi_n^0 S_n^0 + \varphi_n \cdot S_n = \phi_{n+1}^0 S_n^0 + \varphi_{n+1} \cdot S_n, \quad \forall 0 \leqslant n \leqslant N - 1. \tag{3.11}$$

It means that at time n, once the price vector S_n is quoted, the investor readjusts his/her positions from ϕ_n to ϕ_{n+1} without bringing or withdrawing any wealth. It is easy to prove that (3.11) is equivalent to

$$V_n(\phi) = V_0(\phi) + G_n(\phi), \quad \forall 1 \leqslant n \leqslant N, \tag{3.12}$$

or

$$\tilde{V}_n(\phi) = V_0(\phi) + \sum_{i=1}^{n} \varphi_i \cdot \Delta \tilde{S}_i, \quad \forall 1 \leqslant n \leqslant N. \tag{3.13}$$

In fact, by (3.11) ϕ is self-financing if and only if

$$\tilde{V}_n(\phi) = \phi_{n+1}^0 + \varphi_{n+1} \cdot \tilde{S}_n,$$

or equivalently

$$\tilde{V}_{n+1}(\phi) - \tilde{V}_n(\phi) = \varphi_{n+1} \cdot \Delta \tilde{S}_{n+1},$$

which means (3.13). Similarly, we can prove that (3.11) is equivalent to (3.12).

3.3.2 Arbitrage, Admissible, and Allowable Strategies

Let (ϕ_n) be a self-financing strategy. It is called *admissible* if its wealth process $V_n(\phi)$ is nonnegative. It is called *tame* if its discounted wealth process is bounded from below by some real constant c. The above two notions are standard in the literature. However, they have at least two defects. One is that a strategy with a short selling of assets is in general not an admissible or tame strategy. Another is that the notion of tame strategy is not invariant under the change of numeraires. So we propose to define a new class of trading strategies. A self-financing strategy (ϕ_n) is called *allowable* if there exists a positive constant c such that $V_n(\phi) \geqslant - c \sum_{i=0}^{d} S_n^i$, $0 \leqslant n \leqslant N$.

An *arbitrage strategy* is defined as a self-financing strategy with initial wealth zero and nonnegative and non-zero final wealth at time N. Here and in the sequel, a random variable ξ is called non-zero, if $\mathbb{P}(\xi \neq 0) > 0$. If we say that a market has no arbitrage or is arbitrage-free, we will indicate clearly that for which class of trading strategies the market excludes arbitrage opportunities.

Lemma 3.1 *Let* $\varphi_n = (\phi_n^1, \cdots, \phi_n^d)$, $0 \leqslant n \leqslant N$, *be an* $\mathbb{R}^d$-*valued predictable process. Put*

$$W_n(\varphi) = \sum_{i=1}^{n} \varphi_i \cdot \Delta \widetilde{S}_i, \quad \forall 1 \leqslant n \leqslant N. \tag{3.14}$$

Then there exists a predictable process (ϕ_n^0) *such that* $\phi = (\phi_n^0, \varphi_n)$ *is a self-financing strategy with initial wealth zero and* $\widetilde{V}_n(\phi) = W_n(\varphi)$. *If the market has no arbitrage for admissible strategies, then* $W_N(\varphi) \notin L_{++}^0$, *where* L_{++}^0 *denotes the convex cone of nonnegative and non-zero random variables. In particular, the market has no arbitrage for admissible self-financing strategies imply that it has no arbitrage for all self-financing strategies.*

Proof We put

$$\phi_0^0 = -\varphi_0 \cdot S_0, \qquad \phi_n^0 = W_{n-1}(\varphi) - \varphi_n \cdot \widetilde{S}_{n-1}, \quad n \geqslant 1. \tag{3.15}$$

Then (ϕ_n^0) is a predictable process. Let $\phi_n = (\phi_n^0, \varphi_n)$. Then by (3.15),

$$\begin{aligned}
\widetilde{V}_n(\phi) &= \phi_n^0 + \varphi_n \cdot \widetilde{S}_n \\
&= \phi_n^0 + \varphi_n \cdot \widetilde{\Delta} S_n + \varphi_n \cdot \widetilde{S}_{n-1} \\
&= W_{n-1}(\varphi) + \varphi_n \cdot \widetilde{\Delta} S_n \\
&= \sum_{i=1}^{n} \varphi_i \cdot \Delta \widetilde{S}_i, \quad \forall 1 \leqslant n \leqslant N.
\end{aligned}$$

Since $V_0(\phi) = 0$, by the equivalence of (3.13) and (3.11) we see that $\phi = (\phi^0, \cdots, \phi^d)$ is a self-financing strategy and $\widetilde{V}_n(\phi) = W_n(\varphi)$.

Now we assume that the market has no arbitrage for admissible strategies. Suppose that $W_N(\varphi) \in L_{++}^0$. Then above constructed self-financing strategy ϕ cannot be an admissible strategy. So if we put

$$m = \sup\{k : \mathbb{P}(\tilde{V}_k(\phi) < 0) > 0\}, \quad A = [\tilde{V}_m(\phi) < 0],$$

then $1 \leqslant m \leqslant N - 1$ and $\mathbb{P}(A) > 0$. For $1 \leqslant j \leqslant N$, we put

$$\psi_j(\omega) = \begin{cases} 0 & \text{if } j \leqslant m, \\ I_A(\omega)\phi_j(\omega) & \text{if } j > m. \end{cases}$$

Then $\psi = (\psi_1, \cdots, \psi_N)$ is an $\mathbb{R}^d$-valued predictable process and

$$W_j(\psi) = \begin{cases} 0 & \text{if } j \leqslant m, \\ I_A(\tilde{V}_j(\phi) - \tilde{V}_m(\phi)) & \text{if } j > m. \end{cases}$$

Thus, $W_j(\psi) \geqslant 0$ for all $j \in \{0, \cdots, N\}$ and $W_N(\psi) > 0$ on A. This contradicts the assumption, because $(W_n(\psi))$ is the discounted wealth process of an admissible self-financing strategy with initial wealth zero. Therefore, $W_N(\phi) \notin L_{++}^0$.   □

3.4 Martingale Characterization of No-Arbitrage Markets

Characterizing those stochastic processes which can be transformed into martingales by an equivalent change of measure is of particular interest in financial economics. Harrison and Kreps (1979), Harrison and Pliska (1981), and Kreps (1981) studied the relationship between the existence of equivalent martingale measures for discounted price processes of securities and the economic concept of "no arbitrage." Under appropriate assumptions on the price processes, such as integrability, they obtained some fundamental results. The most important fact in option pricing theory, as pointed out in Harrison and Kreps (1979), is that the absence of arbitrage follows from the existence of an equivalent martingale measure for (discounted) price processes of securities. Fortunately, the proof of this fact is quite easy. The converse that the absence of arbitrage implies the existence of an equivalent measure is, however, rather difficult. In discrete-time case, the proof of this fact for an arbitrary probability space was given by Dalang et al. (1990).

3.4.1 The Finite Market Case

We start with the assumption that Ω is a finite set, $\mathcal{F}$ is the family of all subsets of Ω, and $\mathbb{P}(\{\omega\}) > 0$ for all $\omega \in \Omega$. Moreover, we assume that $\mathcal{F}_0 = \{\emptyset, \Omega\}$ and $\mathcal{F}_N = \mathcal{F}$. Under these assumptions every real-valued random variable takes only a finite number of values and is thus bounded.

The following theorem gives a complete characterization of no arbitrage.

Theorem 3.2 *There exists no admissible arbitrage strategy if and only if there exists a probability measure* $\mathbb{P}^*$ *equivalent to* $\mathbb{P}$ *(in the present case, this means that* $\mathbb{P}^*(\{\omega\}) > 0$ *for all* $\omega \in \Omega$*) such that the* $\mathbb{R}^d$*-valued process of discounted prices* $(\tilde{S}_n)_{0 \leqslant n \leqslant N}$ *of assets is a* $\mathbb{P}^*$*-martingale.*

Proof Sufficiency. Assume that there exists a probability $\mathbb{P}^*$ equivalent to $\mathbb{P}$ such that $(\tilde{S}_n)$ is a martingale. By Theorem 1.7 and (3.13), for any admissible self-financing strategy (ϕ_n), the discounted wealth process $\tilde{V}_n(\phi)$ is a $\mathbb{P}^*$-martingale. In particular, if $V_0(\phi) = 0$, then $\mathbb{E}[\tilde{V}_N(\phi)] = 0$. Since $V_N(\phi) \geqslant 0$ and $\mathbb{P}(\{\omega\}) > 0, \forall \omega \in \Omega$, we must have $V_N(\phi) = 0$. It means that there exists no-arbitrage strategy.

Necessity. Assume that there exists no-arbitrage strategy. Let $\mathcal{V}$ be the set of all random variables $W_N(\varphi)$ with $\varphi = (\phi_n^1, \cdots, \phi_n^d)$ being ($\mathbb{R}^d$-valued) predictable process. By Lemma 3.1, $\mathcal{V} \cap L_{++}^0 = \emptyset$. In particular, $\mathcal{V}$ does not intersect convex compact set $K = \{x \in L_{++}^0 : \sum_\omega x(\omega) = 1\}$. Put $K - \mathcal{V} = \{x - y : x \in K, y \in \mathcal{V}\}$. Since $\Omega = \{\omega_1, \cdots, \omega_m\}$ is a finite set, we can regard a random variable defined on Ω as a vector in $\mathbb{R}^m$. Thus, $K - \mathcal{V}$ is a closed convex subset of $\mathbb{R}^m$ which does not contain the origin. By the convex sets separation theorem (see Dudley (1989), p. 152, or Lamberton and Lapeyre (1996), p.178) there exists a linear functional f, defined on $\mathbb{R}^m$, such that $f(x) \geqslant \alpha, \forall x \in K - \mathcal{V}$ for an $\alpha > 0$. Since $\mathcal{V}$ is a subspace of $\mathbb{R}^m$, we must have $f(x) = 0, \forall x \in \mathcal{V}$ and $f(x) \geqslant \alpha > 0, \forall x \in K$. It means that there exists $(\lambda(\omega))_{\omega \in \Omega}$ such that (1) $\forall x \in K, \sum_\omega \lambda(\omega) x(\omega) > 0$; and (2) for any $\mathbb{R}^d$-valued predictable process $\varphi, \sum_\omega \lambda(\omega) W_N(\varphi) = 0$. From (1) we get $\lambda(\omega) > 0$ for all $\omega \in \Omega$. Put

$$\mathbb{P}^*(\{\omega\}) = \frac{\lambda(\omega)}{\sum_{\omega' \in \Omega} \lambda(\omega')}.$$

Then $\mathbb{P}^*$ is equivalent to $\mathbb{P}$, and by (2), for any $\mathbb{R}^d$-valued predictable process φ,

$$\mathbb{E}^* \sum_{j=1}^N \varphi_j \cdot \Delta \tilde{S}_j = 0.$$

Therefore, according to Theorem 1.31, $(\tilde{S}_n)_{0 \leqslant n \leqslant N}$ is a $\mathbb{P}^*$-martingale. $\qquad \square$

3.4.2 The General Case: Dalang-Morton-Willinger Theorem

Now we turn to an arbitrary probability space case. In this case we have still a martingale characterization for no arbitrage. The result, due to Dalang et al. (1990), is usually referred to as the *fundamental theorem of asset pricing*. Note that the integrability of the price process is not assumed in the theorem. Its original proof is mainly based on a measurable selection theorem. Alternative proofs can be found in

Schachermayer (1992), Kabanov and Kramkov (1994), and Rogers (1994). Based on the Kreps-Yan separation theorem, Kabanov and Stricker (2001) gave a rather simple proof.

Theorem 3.3 *The market has no arbitrage for admissible strategies if and only if there exists a probability measure $\mathbb{P}^*$ equivalent to $\mathbb{P}$ such that the discounted price process $(\widetilde{S}_n)_{0 \leqslant n \leqslant N}$ of assets is a $\mathbb{P}^*$-martingale. If it is the case, $\mathbb{P}^*$ may be chosen such that Radon-Nikodym derivative $d\mathbb{P}^*/d\mathbb{P}$ is bounded.*

The proof of the "if" part is similar to that of Theorem 3.2 and will be omitted. Now we follow Kabanov and Stricker (2001) to give a proof of the "only if" part. First of all we prepare several lemmas.

Lemma 3.4 *Let (η_n) be a sequence of $\mathbb{R}^d$-valued random variables such that $\eta := \liminf |\eta_n| < \infty$. Then there exists a sequence of $\mathbb{R}^d$-valued random variables (ξ_n) such that for all ω the sequence $(\xi_n(\omega))$ is convergent and is a subsequence of $(\eta_n(\omega))$.*

Proof Instead of working on one component after another component, we can assume that $d = 1$. Let $\tau(0) = 0$ and $\tau(k) = \inf\{n > \tau(k-1) : |\,|\eta_n| - \eta| \leqslant \frac{1}{k}\}$. Put $\zeta_k = \eta_{\tau(k)}$. Then each ζ_k is a real-valued random variable, and $\sup_k |\zeta_k| < \infty$. Let $\xi = \liminf \zeta_n$. Let $\alpha(0) = 0$ and $\alpha(k) = \inf\{n > \alpha(k-1) : |\zeta_n - \xi| \leqslant \frac{1}{k}\}$. Put $\xi_n = \zeta_{\alpha(n)}$. Then the sequence (ξ_n) meets the requirements. □

The following lemma is a result due independently to Yan (1980a) (for the $p = 1$ case) and Kreps (1981) (for the $p = \infty$ case). In the literature this result is named as the Kreps-Yan *separation theorem*. Ansel and Stricker (1990) have remarked that the proof given in Yan (1980a) is valid also for $1 < p < \infty$ case. Stricker (1990) has applied this lemma to mathematical finance for the first time.

Lemma 3.5 *Let $1 \leqslant p, q \leqslant \infty$ with $\frac{1}{p} + \frac{1}{q} = 1$. Let K be a convex cone in $L^p := L^p(\Omega, \mathcal{F}, \mathbb{P})$ containing $-L^p_+$, which is closed in L^p if $1 \leqslant p < \infty$; and closed in L^∞ under the weak* topology if $p = \infty$. Then the following two conditions are equivalent:*

(1) $K \cap L^p_+ = \{0\}$;
(2) *There exists a probability measure $\mathbb{Q}$ equivalent to $\mathbb{P}$ with $d\mathbb{Q}/d\mathbb{P} \in L^q$ such that $\mathbb{E}_\mathbb{Q}[\xi] \leqslant 0$ for all $\xi \in K$.*

Proof (1)$\Rightarrow$ (2). Assume that (1) holds. Since L^q (resp. L^1) is the dual space of L^p (resp. L^∞ with weak* topology) if $1 \leqslant p < \infty$ (resp. if $p = \infty$), by the Hahn-Banach separation theorem, for any $x \in L^p_+, x \neq 0$ there is $z_x \in L^q$ such that

$$\mathbb{E}[z_x \xi] < \mathbb{E}[z_x x], \quad \forall \xi \in K.$$

Since $0 \in K$ and K is a cone, we must have $\mathbb{E}[z_x x] > 0$ and $\sup_{\xi \in K} \mathbb{E}[z_x \xi] \leqslant 0$. On the other hand, since $\xi_\alpha := -\alpha I_{[z_x < 0]} \in K$ for any $\alpha > 0$, we must have $z_x \geqslant 0$ a.s., because otherwise we get a contradiction:

$$\lim_{\alpha \to \infty} \mathbb{E}[z_x \xi_\alpha] = \alpha \lim_{\alpha \to \infty} \mathbb{E}[z_x^-] = \infty.$$

Let $\Theta = \{\theta : \theta \in L_+^q, \sup_{\xi \in K} \mathbb{E}[\theta \xi] \leqslant 0\}$. Then Θ is non-empty. We let $b = \sup\{\mathbb{P}(\theta > 0) : \theta \in \Theta\}$, and let $\theta_n \in \Theta$ be such that $\mathbb{P}(\theta_n > 0) \to b$. Put

$$\theta = \sum_{n=1}^{\infty} 2^{-n} \frac{\theta_n}{(1 + ||\theta_n||_q)^n}.$$

Then $\theta \in L_+^q$ and $\mathbb{P}(\theta > 0) = b$. Now we show that $b = 1$. To this end, suppose that $b < 1$. We take $\theta \in \Theta$ such that $\mathbb{P}(\theta > 0) = b$. Let $A = [\theta = 0]$. Since $\mathbb{P}(A) > 0$ and $I_A \in L_+^p$, we can find a $\theta' \in \Theta$, associated to I_A (i.e., $\theta' = z_{I_A}$), such that $\mathbb{E}[I_A \theta'] > 0$. In particular, we have $\mathbb{P}([\theta' > 0] \cap A) > 0$. Put $\eta = \theta + \theta'$, we have $\eta \in \Theta$ and $\mathbb{P}(\eta > 0) > \mathbb{P}(\theta > 0) = b$. This contradicts the definition of b. Finally, if we take a $\theta \in \Theta$ with $\mathbb{P}(\theta > 0) = 1$ and define a probability measure $\mathbb{Q}$ with $\mathbb{Q} = c\theta.\mathbb{P}$, where $c^{-1} = \mathbb{E}[\theta]$, then $\mathbb{Q}$ meets the requirements of (2).

(2)$\Rightarrow$ (1). Suppose that (1) does not hold. Then there exists a $\xi \in K$ such that $\xi \geqslant 0$ a.s. and $\mathbb{P}(\xi > 0) > 0$. Thus, for any probability measure $\mathbb{Q}$ equivalent to $\mathbb{P}$, we must have $\mathbb{E}_\mathbb{Q}[\xi] > 0$. This shows that (2) does not hold. $\qquad\square$

Lemma 3.6 *Let (ξ_n) be a sequence of nonnegative real random variables defined on a probability space $(\Omega, \mathcal{F}, \mathbb{P})$. Then there exists a probability measure $\mathbb{P}'$ equivalent to $\mathbb{P}$ with bounded $d\mathbb{P}'/d\mathbb{P}$ such that each ξ_n is $\mathbb{P}'$-integrable.*

Proof We take a sequence (a_n) of positive numbers such that $\sum_{n=1}^{\infty} \mathbb{P}(\xi_n > a_n) < \infty$. Let $A_n = [\xi_n > a_n]$. By Borel-Cantelli's lemma, we have $\mathbb{P}(A_n, \text{i.o.}) = 0$. This means that for almost all ω, there exists an integer $N(\omega)$ such that for all $n \geqslant N(\omega)$, we have $\xi_n(\omega) \leqslant a_n$. Thus, if we let $c_n = (2^n a_n)^{-1}$ and $X = \sum_{n=1}^{\infty} c_n \xi_n$, then $X < \infty$ a.s. Now let $Y = e^{-X}$ and $\mathbb{P}' = cY.P$ with $c^{-1} = \mathbb{E}[Y]$, then $\mathbb{P}'$ meets the requirements. $\qquad\square$

In the following we denote by $\Phi(x)$ the set of all self-financing strategies with initial wealth x.

Proof of Theorem 3.3 According to Lemma 3.6, instead of making a change of measure, we may assume that all $\widetilde{S}_n$ are $\mathbb{P}$-integrable $\mathbb{R}^d$-valued random variables. Put

$$K_1 = \{\widetilde{G}_N(\phi) : \phi \in \Phi(0)\},$$

and let $K = (K_1 - L_+^0) \cap L^1$, where L^0 denotes the set of all $\mathcal{F}_N$-measurable real random variables. Note that for $\phi \in \Phi(0)$, we have $\widetilde{G}_N(\phi) = W_n(\varphi)$. Thus, by the definition of K_1 and Lemma 3.1, the no-arbitrage assumption implies that $K_1 \cap L_+^1 = \{0\}$. Consequently, $K \cap L_+^1 = \{0\}$. Obviously, K contains $-L_+^\infty$. If we can show that K is closed in L^1, then by Lemma 3.5, there exists a probability measure $\mathbb{Q}$, equivalent to $\mathbb{P}$ and with bounded $d\mathbb{Q}/d\mathbb{P}$, such that $\mathbb{E}_\mathbb{Q}[\xi] \leqslant 0$ for all $\xi \in K$. Since $K_1 \cap L^1$ is a linear subspace of L^1, we must have $\mathbb{E}_\mathbb{Q}[\xi] = 0$ for all $\xi \in K_1 \cap L^1$. In particular, for any bounded predictable process $\varphi_n = (\phi_n^1, \cdots, \phi_n^d)$, $0 \leqslant n \leqslant N$, we have $\mathbb{E}_\mathbb{Q}[\sum_{i=1}^N \varphi_i \cdot \Delta \widetilde{S}_i] = 0$. Thus, $(\widetilde{S}_n)$ is a $\mathbb{Q}$-martingale.

Therefore, we only need to show that $K_1 - L_+^0$ is closed in probability, because this will imply that K is closed in L^1. To this end we proceed by induction. Let $N = 1$. Suppose that when $n \to \infty$, $\phi_1^n \Delta \tilde{S}_1 - r^n \to \zeta$ a.s., where ϕ_1^n is $\mathcal{F}_0$-measurable and $r^n \in L_+^0$. It is sufficient to find a sequence of $\mathcal{F}_0$-measurable random variables $\tilde{\phi}_1^k$ which is convergent a.s. and $\tilde{r}^k \in L_+^0$, such that when $k \to \infty$, $\tilde{\phi}_1^k \Delta \tilde{S}_1 - \tilde{r}^k \to \zeta$ a.s.

Let $\Omega_i \in \mathcal{F}_0$ form a finite partition of Ω. Obviously, we may argue on each Ω_i separately as on an autonomous measure space (considering the restrictions of random variables and traces of σ-algebras).

Let $\underline{\phi}_1 := \liminf |\phi_1^n|$. On the set $\Omega_1 := \{\underline{\phi}_1 < \infty\}$ we can take, using Lemma 3.4, a sequence of $\mathcal{F}_0$-measurable random variables $\tilde{\phi}_1^k$ such that $\tilde{\phi}_1^k(\omega)$ is a convergent subsequence of $\phi_1^n(\omega)$ for every $\omega \in \Omega_1$; $\tilde{r}^k$ are defined correspondingly. Thus, if Ω_1 is of full measure, the goal is achieved.

On $\Omega_2 := \{\underline{\phi}_1 = \infty\}$ we put $g_1^n := \phi_1^n/|\phi_1^n|$ and $h_1^n := r_1^n/|\phi_1^n|$ and observe that $g_1^n \Delta \tilde{S}_1 - h_1^n \to 0$ a.s. when $n \to \infty$. By Lemma 3.4 we can find a sequence of $\mathcal{F}_0$-measurable random variables $\tilde{g}_1^k$ and consequently $\tilde{h}_1^k$ such that $\tilde{g}_1^k(\omega)$ (resp. $\tilde{h}_1^k$) is a convergent subsequence of $g_1^n(\omega)$ (resp. h_1^n) for every $\omega \in \Omega_2$. Denoting the limit by $\tilde{g}_1$ (resp. $\tilde{h}_1$), we obtain that $\tilde{g}_1 \Delta \tilde{S}_1 = \tilde{h}_1$ on Ω_2, where $\tilde{h}_1$ is nonnegative. Hence, by no-arbitrage assumption, $\tilde{g}_1 \Delta S_1 = \tilde{h}_1 = 0$ on Ω_2.

As $\tilde{g}_1 = (\tilde{g}_1(1), \cdots, \tilde{g}_1(d)) \neq 0$, there exists a partition of Ω_2 into d disjoint subsets $\Omega_2^i \in \mathcal{F}_0$ (some subsets can be empty sets or with probability 0) such that $\tilde{g}_1(i) \neq 0$ on Ω_2^i. Define $\bar{\phi}_1^n := \phi_1^n - \beta^n \tilde{g}_1$ where $\beta^n := \phi_1^n(i)/\tilde{g}_1(i)$ on Ω_2^i. Then $\bar{\phi}_1^n \Delta \tilde{S}_1 = \phi_1^n \Delta \tilde{S}_1$ on Ω_2. We repeat the procedure on each Ω_2^i with the sequence $\bar{\phi}_1^n$ (noting that $\bar{\phi}_1^n(i) = 0$ on Ω_2^i for all n). Apparently, after a finite number of steps we can construct a desired sequence.

Suppose that the claim be true for $N - 1$ and let $\sum_{i=1}^{N} \phi_i^n \Delta \tilde{S}_i - r^n \to \zeta$ a.s., where ϕ_i^n are $\mathcal{F}_{i-1}$-measurable and $r^n \in L_+^0$. By the same arguments based on the elimination of non-zero components of the sequence ϕ_1^n and using the induction hypothesis, we replace ϕ_i^n and r^n by $\tilde{\phi}_i^k$ and $\tilde{r}^k$ such that $\tilde{\phi}_1^k$ converges a.s. This means that the problem is reduced to the one with $N - 1$ steps.   □

Remark 1 For an extension of Theorem 3.3 to the infinite-horizon case see Schachermayer (1994).

3.5   Pricing of European Contingent Claims

In the following we assume that the market has no arbitrage, i.e., there exists equivalent martingale measures. Let $\mathcal{M}$ denote the set of all equivalent martingale measures. A *European contingent claim* with maturity time N is an $\mathcal{F}_N$-measurable random variable. It expresses a claim which can be realized at a future time N. Usually we assume that contingent claims are nonnegative.

A contingent claim ξ is called *replicatable*, if there exists an admissible strategy such that its terminal wealth at time N is ξ. Then discounted wealth process $(\widetilde{V}_n)$ of an admissible strategy replicating ξ is a martingale transform under any equivalent martingale measure $\mathbb{P}^*$ and hence a local martingale. Since it is nonnegative, it is a martingale (see Corollary 1.30). In this case, for any equivalent martingale measure $\mathbb{P}^*$,

$$\widetilde{V}_n = \mathbb{E}^*[\widetilde{\xi}|\mathcal{F}_n],$$

where $\mathbb{E}^*$ denotes the expectation operator corresponding to $\mathbb{P}^*$. In particular, we have

$$V_0 = \mathbb{E}^*[\widetilde{\xi}]. \tag{3.16}$$

Thus, if we add this contingent claim (whose price process is defined as (V_n)) to the market as a new risky asset, then $\mathbb{P}^*$ is still an equivalent martingale measure in the expanded market, so the market is arbitrage-free. The initial wealth of the admissible strategy replicating ξ is equal to V_0 given by formula (3.16), which is an arbitrage-free price of contingent claim ξ, and does not depend on the choice of equivalent martingale measure $\mathbb{P}^*$.

Let ξ be a contingent claim. If there is an adapted nonnegative sequence (V_n), such that $V_T = \xi$, and we add ξ to the original market as a new risky asset with price process (V_n), the market is still arbitrage-free, then we call V_0 an *arbitrage-free price* of ξ. In this case, there exists a $\mathbb{P}^* \in \mathcal{M}$ such that V_0 is given by formula (3.16).

Let ξ be a contingent claim. We use $\Pi(\xi)$ to denote the set of all arbitrage-free prices of ξ. Put

$$\mathcal{M}(\xi) = \{\mathbb{P}^* : \mathbb{P}^* \in \mathcal{M}, \ \mathbb{E}^*[\widetilde{\xi}] < \infty\}.$$

Then

$$\Pi(\xi) = \{\mathbb{E}^*[\widetilde{\xi}] : \mathbb{P}^* \in \mathcal{M}(\xi)\},$$

which is obviously an interval on $\mathbb{R}$, because $\mathcal{M}(\xi)$ is a convex set.

We put

$$\pi_{\sup}(\xi) = \sup_{Q \in \mathcal{M}(\xi)} \mathbb{E}_Q[\widetilde{\xi}], \quad \pi_{\inf}(\xi) = \inf_{Q \in \mathcal{M}(\xi)} \mathbb{E}_Q[\widetilde{\xi}]. \tag{3.17}$$

The following theorem is from Föllmer and Schied (2004).

Theorem 3.7 *Let ξ be a contingent claim. If ξ is replicatable, then the arbitrage-free price of ξ is uniquely determined, and is equal to $\mathbb{E}^*[\widetilde{\xi}]$, where $\mathbb{P}^*$ is any equivalent martingale measure. If ξ is not replicatable, then we have $\pi_{\inf}(\xi) < \pi_{\sup}(\xi)$, and*

$$\Pi(\xi) = (\pi_{\inf}(\xi), \pi_{\sup}(\xi)).$$

Proof The first conclusion has been explained in the preceding discussion. In order to prove the second conclusion, we need only to prove that for any $\pi = \mathbb{E}^*[\tilde{\xi}] \in \Pi(\xi)$ there exists $\pi_1, \pi_2 \in \Pi(\xi)$, such that $\pi_1 < \pi < \pi_2$. For details see Föllmer and Schied (2004). $\qquad\square$

An arbitrary-free market is called *complete* if each contingent claim ξ is replicatable.

The next theorem gives a characterization of market completeness. It was first established by Willinger and Taqqu (1988).

Theorem 3.8 *An arbitrage-free market is complete if and only if the equivalent martingale measure is unique. In this case, the probability space $(\Omega, \mathcal{F}, \mathbb{P})$ must be purely atomic and the number of atoms is finite. In particular, each contingent claim in a complete market is bounded.*

Proof Suppose that the equivalent martingale measure $\mathbb{P}^*$ is unique. Theorem 3.7 shows that the market is complete. Furthermore, by Theorem 3.3, all real-valued random variables are integrable with respect to $\mathbb{P}^*$, so that $(\Omega, \mathcal{F}, \mathbb{P})$ must be purely atomic, and the number of atoms is finite. Conversely, if the equivalent martingale measure is not unique, namely, there are $\mathbb{Q}_1, \mathbb{Q}_2 \in \mathcal{M}$ and $A \in \mathcal{F}$, such that $Q_1(A) \neq Q_2(A)$, then by Theorem 3.7, the contingent claim $S_N^0 I_A$ is not replicatable, because its arbitrage-free price is not unique. $\qquad\square$

3.6 Maximization of Expected Utility and Option Pricing

In this section we study the problem of maximizing the expected utility in discrete-time case. We will use a martingale approach which was introduced by Karatzas et al. (1991). As an application, we investigate the European contingent claim pricing based on utility functions and market equilibrium pricing. The content of this section is from Li et al. (2001).

Assume that in the market there are $d+1$ assets whose price process is an adapted sequence $(S_n^0, \cdots, S_n^d)^\tau$ of $\mathbb{R}^{d+1}$-valued nonnegative random vectors. We take asset 0 as the numeraire asset, and always assume $S_0^0 = 1$. The discount factor $(S_n^0)^{-1}$ at time n is denoted by β_n.

3.6.1 General Utility Function Case

We assume that a market participant has an initial wealth w. His utility function u is a strictly concave continuously differentiable increasing function: $u : (D_u, \infty) \longrightarrow \mathbb{R}$, $-\infty \leqslant D_u < \infty$, satisfying

$$u'(D_u) \hat{=} \lim_{x \downarrow D_u} u'(x) = \infty, \qquad u'(\infty) \hat{=} \lim_{x \to \infty} u'(x) = 0.$$

The inverse function of u' is denoted by $I : (0, \infty) \longrightarrow (D_u, \infty)$, which is a strictly decreasing continuous function. By the concavity of u, we have the following inequality:

$$u(I(y)) \geqslant u(x) + y[I(y) - x], \quad \forall x > D_u, \quad y > 0. \tag{3.18}$$

We denote Ψ_s^z as the set of all admissible strategies with initial wealth $z > 0$. Let $(V_n(\psi))$ represent the wealth process of an admissible strategy ψ. Then for any equivalent martingale measure $\mathbb{Q}$, $(\widetilde{V}_n(\psi))$ is a $\mathbb{Q}$-local martingale and consequently a $\mathbb{Q}$-martingale (see Corollary 1.30). Our concern is how to choose an admissible strategy which maximizes the expected utility of the wealth at end time N.

We discuss the cases $D_u > -\infty$ and $D_u = -\infty$, respectively.

First consider the case $D_u > -\infty$. For a given initial wealth $z > 0$, we let

$$\Psi_s^z(D_u) \hat{=} \{\psi : \psi \in \Psi_s^z, \quad V_N(\psi) > D_u\}.$$

We consider the problem of maximizing the expected utility of the wealth at end time N:

$$\arg \max_{\psi \in \Psi_s^z(D_u)} \mathbb{E}[u(V_N(\psi))].$$

Let $\mathcal{P}$ denote the set of all equivalent martingale measures. For $\mathbb{Q} \in \mathcal{P}$, we put

$$Z_n^{\mathbb{Q}} \hat{=} \mathbb{E}\left[\frac{d\mathbb{Q}}{d\mathbb{P}} \middle| \mathcal{F}_n\right].$$

Then $(Z_n^{\mathbb{Q}})$ is a strictly positive martingale. Put

$$\mathcal{P}_n \hat{=} \{\mathbb{Q} \in \mathcal{P} : |\mathbb{E}[\beta_n Z_n^{\mathbb{Q}} I(y\beta_n Z_n^{\mathbb{Q}})]| < \infty, \forall y \in (0, \infty)\}, \quad n = 0, 1, \cdots, N.$$

Suppose that each $\mathcal{P}_n$ is non-empty. For $n = 0, 1, \cdots, N$ and $\mathbb{Q} \in \mathcal{P}_n$, let

$$\mathcal{X}_n^{\mathbb{Q}}(y) \hat{=} \mathbb{E}\left[\beta_n Z_n^{\mathbb{Q}} I(y\beta_n Z_n^{\mathbb{Q}})\right], \quad 0 < y < \infty.$$

It is easy to see that $\mathcal{X}_n^{\mathbb{Q}}$ is a strictly decreasing continuous function from $(0, \infty)$ to $(D_u \mathbb{E}_{\mathbb{Q}}[\beta_n], \infty)$, and its inverse function $\mathcal{Y}_n^{\mathbb{Q}}$ is a strictly decreasing continuous function from $(D_u \mathbb{E}_{\mathbb{Q}}[\beta_n], \infty)$ to $(0, \infty)$. We define

$$\xi_n^{\mathbb{Q}}(x) \hat{=} I(\mathcal{Y}_n^{\mathbb{Q}}(x)\beta_n Z_n^{\mathbb{Q}}), \quad 0 \leqslant n \leqslant N, \quad \mathbb{Q} \in \mathcal{P}_n, \quad x \in (D_u \mathbb{E}_{\mathbb{Q}}[\beta_n], \infty). \tag{3.19}$$

Then for $\psi \in \Psi_s^z(D_u)$ and $\mathbb{Q} \in \mathcal{P}$, $(\beta_n V_n(\psi))$ is a $\mathbb{Q}$-martingale. From (3.18) and (3.19) we get

$$\mathbb{E}\left[u(\xi_N^{\mathbb{Q}}(z))\right] \geqslant \mathbb{E}[u(V_N(\psi))], \quad \forall \mathbb{Q} \in \mathcal{P}_N, \quad \psi \in \Psi_s^z(D_u). \tag{3.20}$$

If there is a probability measure $\mathbb{Q}^* \in \mathcal{P}_N$ and an admissible strategy $\widehat{\psi} \in \Psi_s^z(D_u)$ such that $\xi_N^{\mathbb{Q}^*}(x) = V_N(\widehat{\psi})$, then from (3.20), we know that $\widehat{\psi}$ is an optimal admissible strategy. Since $Z_N^{\mathbb{Q}}$ is uniquely determined by $\xi_N^{\mathbb{Q}}(x)$, such a $\mathbb{Q}^*$ is unique. In addition, by (3.20), $\mathbb{Q}^*$ satisfies

$$\mathbb{E}\left[u(\xi_N^{\mathbb{Q}^*}(x))\right] \geqslant \mathbb{E}\left[u(\xi_N^{\mathbb{Q}}(x))\right], \quad \forall \mathbb{Q} \in \mathcal{P}_N. \tag{3.21}$$

Now we consider case $D_u = -\infty$. Let

$$\mathcal{P}' \widehat{=} \{\mathbb{Q} \in \mathcal{P} : \frac{d\mathbb{Q}}{d\mathbb{P}} \in L^2(\Omega, \mathcal{F}_N, \mathbb{P})\},$$

$$\widehat{\Psi}_s^z \widehat{=} \{\psi : \psi \in \Psi_s^z, \ \beta_n V_n(\psi) \in L^2(\Omega, \mathcal{F}_n, \mathbb{P}), \ 0 \leqslant n \leqslant N\},$$

and assume that $\mathcal{P}'$ is non-empty. Then for $\psi \in \widehat{\Psi}_s^z$ and $\mathbb{Q} \in \mathcal{P}'$, it is easy to prove that $(\beta_n V_n(\psi))$ is a $\mathbb{Q}$-martingale. Now we consider martingale measures in $\mathcal{P}'$ (but not in $\mathcal{P}$) and corresponding $\mathcal{P}'_n$ and $\widehat{\Psi}_s^z$. Similarly to case $D_u > -\infty$, we can see that if there is a probability measure $\mathbb{Q}^* \in \mathcal{P}'_N$ and an admissible strategy $\widehat{\psi} \in \widehat{\Psi}_s^z$ such that $\xi_N^{\mathbb{Q}^*}(z) = V_N(\widehat{\psi})$, then $\widehat{\psi}$ is an optimal admissible strategy in $\widehat{\Psi}_s^z$ and $\mathbb{Q}^*$ satisfies

$$\mathbb{E}\left[u(\xi_N^{\mathbb{Q}^*}(x))\right] \geqslant \mathbb{E}\left[u(\xi_N^{\mathbb{Q}}(x))\right], \quad \forall \mathbb{Q} \in \mathcal{P}'_N. \tag{3.22}$$

3.6.2 HARA Utility Functions and Their Duality Case

Below we assume that (S_n^0) is a deterministic function, and hence so is (β_n). Consider a HARA utility function:

$$U_\gamma(x) = \begin{cases} \frac{1}{\gamma}(x^\gamma - 1), & \gamma < 0, \\ \ln x, & \gamma = 0. \end{cases}$$

For $u = U_\gamma, (\gamma \leqslant 0)$, we have $D_u = 0, I(x) = x^{\frac{1}{\gamma-1}}$, and $\mathcal{P}_n = \mathcal{P}, \ n = 0, 1, \cdots, N$. Put $\delta \widehat{=} \frac{\gamma}{\gamma-1} \in [0, 1)$. Then for $\gamma < 0$, it holds $\frac{1}{\delta} + \frac{1}{\gamma} = 1$. Thus, we have

$$\xi_n^{\mathbb{Q}}(x) = \frac{x(Z_n^{\mathbb{Q}})^{\frac{1}{\gamma-1}}}{\beta_n \mathbb{E}[(Z_n^{\mathbb{Q}})^\delta]}, \quad n = 0, 1, \cdots, N.$$

For HARA utility function $u = U_\gamma$, if $\gamma < 0$ (or $\gamma = 0$), then the inequality corresponding to (3.22) is following (3.23) (correspondingly, (3.24)):

$$\mathbb{E}\left[(Z_N^{Q^*})^\delta\right] \geqslant \mathbb{E}\left[(Z_N^{Q})^\delta\right], \quad \forall \mathbb{Q} \in \mathcal{P}, \tag{3.23}$$

$$\mathbb{E}[\ln Z_N^{Q^*}] \geqslant \mathbb{E}[\ln Z_N^{Q}], \quad \forall \mathbb{Q} \in \mathcal{P}. \tag{3.24}$$

Definition 3.9 Assume that the probability measure $\mathbb{Q}$ is absolutely continuous w.r.t. $\mathbb{P}$.

(1) Put

$$I_\mathbb{Q}(\mathbb{P}) = \mathbb{E}_\mathbb{Q}\left[\frac{d\mathbb{P}}{d\mathbb{Q}}\ln\frac{d\mathbb{P}}{d\mathbb{Q}}\right].$$

$I_\mathbb{Q}(\mathbb{P})$ is called the *relative entropy* of $\mathbb{Q}$ w.r.t. $\mathbb{P}$;
(2) For $\delta \in (0, 1)$, put

$$H_\delta(\mathbb{Q}, \mathbb{P}) \hat{=} \mathbb{E}_\mathbb{P}\left[\left(\frac{d\mathbb{Q}}{d\mathbb{P}}\right)^\delta\right] = \mathbb{E}_\mathbb{P}\left[(Z_N^{Q})^\delta\right], \quad d_\delta(\mathbb{Q}, \mathbb{P}) \hat{=} 2(1 - H_\delta(\mathbb{Q}, \mathbb{P})).$$

$H_\delta(\mathbb{Q}, \mathbb{P})$ and $d_\delta(\mathbb{Q}, \mathbb{P})$ are called δ order Hellinger *integral* and Hellinger-Kakutani *distance* of $\mathbb{Q}$ w.r.t. $\mathbb{P}$, respectively.

We consider a HARA utility function $u = U_\gamma$, $\gamma \leqslant 0$. If there is a probability measure $\mathbb{Q}^* \in \mathcal{P}$ and an admissible strategy $\hat{\psi} \in \Psi_s^z(D_u)$ such that $\xi_N^{Q^*}(z) = V_N(\hat{\psi})$, then from results of the previous section, we know that $\hat{\psi}$ is optimal, and we have the following conclusions:

(i) For $u = U_\gamma$, $(\gamma < 0)$, the δ-order Hellinger-Kakutani distance of $\mathbb{Q}^*$ w.r.t. $\mathbb{P}$ attains the minimum on $\mathcal{P}$, namely, $d_\delta(\mathbb{Q}^*, \mathbb{P}) = \min_{\mathbb{Q}\in\mathcal{P}} d_\delta(\mathbb{Q}, \mathbb{P})$, where δ satisfies $\frac{1}{\delta} + \frac{1}{\gamma} = 1$;

(ii) For $u(x) = \ln x$, the relative entropy of $\mathbb{Q}^*$ w.r.t. $\mathbb{P}$ attains the minimum on $\mathcal{P}$, namely, $I_{\mathbb{Q}^*}(\mathbb{P}) = \min_{\mathbb{Q}\in\mathcal{P}} I_\mathbb{Q}(\mathbb{P})$.

Now consider a utility function of the following form:

$$W_\gamma(x) = \begin{cases} -(1 - \gamma x)^{\frac{1}{\gamma}}, & \gamma < 0, \\ -e^{-x}, & \gamma = 0. \end{cases}$$

Since $U_\gamma(-W_\gamma(x)) = -x$, $\gamma \leqslant 0$, we call W_γ the *dual utility function* of U_γ. For W_γ, we have

$$D_u = \begin{cases} \frac{1}{\gamma}, & \gamma < 0 \\ -\infty, & \gamma = 0 \end{cases}, \quad I(x) = \begin{cases} \frac{1-x^{\frac{\gamma}{1-\gamma}}}{\gamma}, & \gamma < 0 \\ -\ln x, & \gamma = 0. \end{cases}$$

If $\gamma < 0$, then $\mathcal{P}_n = \mathcal{P}$, $n = 0, 1, \cdots, N$; if $\gamma = 0$, then $\mathcal{P}_N = \{\mathbb{Q} \in \mathcal{P} : |I_\mathbb{P}(\mathbb{Q})| < \infty\}$. Put $\delta \hat{=} \frac{\gamma}{\gamma-1}$, then $\frac{1}{\delta} + \frac{1}{\gamma} = 1$, $\delta \in [0, 1)$.

If $\gamma < 0$ (or $\gamma = 0$), we use $\zeta_n^Q(x)$ (correspondingly, $\eta_n^Q(x)$) to replace $\xi_n^Q(x)$. From (3.19) we obtain

$$\zeta_n^Q(x) = \frac{1}{\gamma}\left\{1 - \frac{\beta_n - \gamma x}{\beta_n}\frac{(Z_n^Q)^{\frac{\gamma}{1-\gamma}}}{\mathbb{E}\left[(Z_n^Q)^{\frac{1}{1-\gamma}}\right]}\right\},$$

$$\eta_n^Q(x) = \frac{x}{\beta_n} + \mathbb{E}[Z_n^Q \ln Z_n^Q] - \ln Z_n^Q.$$

For utility function $W_\gamma(x)$, if $\gamma < 0$ (or $\gamma = 0$), then the inequality corresponding to (3.22) is following (3.25) (correspondingly, (3.26)):

$$\mathbb{E}\left[(Z_N^{Q^*})^{\frac{1}{1-\gamma}}\right] \leqslant \mathbb{E}\left[(Z_N^Q)^{\frac{1}{1-\gamma}}\right], \quad \forall Q \in \mathcal{P}, \tag{3.25}$$

$$\mathbb{E}[Z_N^{Q^*} \ln Z_N^{Q^*}] \leqslant \mathbb{E}[Z_N^Q \ln Z_N^Q], \quad \forall Q \in \mathcal{P}_N. \tag{3.26}$$

Therefore, if there is a probability measure $Q_1^* \in \mathcal{P}$ and an admissible strategy $\widehat{\psi}_1 \in \Psi_s^z(D_u)$, such that $\xi_N^{Q_1^*}(z) = V_N(\widehat{\psi})$, then $\widehat{\psi}_1$ is optimal, and the δ-order Hellinger-Kakutani distance of Q_1^* w.r.t. $\mathbb{P}$ attains the minimum on $\mathcal{P}$, namely,

$$d_\delta(\mathbb{P}, Q_1^*) = \min_{Q \in \mathcal{P}} d_\delta(\mathbb{P}, Q),$$

where δ satisfies $\frac{1}{\delta} + \frac{1}{\gamma} = 1$; if there is a probability measure $Q_2^* \in \mathcal{P}'$ and an admissible strategy $\widehat{\psi}_2 \in \widehat{\Psi}_s^z$, such that $\xi_N^{Q_2^*}(z) = V_N(\widehat{\psi}_2)$, then $\widehat{\psi}_2$ is optimal, and the relative entropy of Q_2^* w.r.t. $\mathbb{P}$ attains the minimum on $\mathcal{P}'_N$, namely,

$$I_\mathbb{P}(Q_2^*) = \min_{Q \in \mathcal{P}} I_\mathbb{P}(Q).$$

3.6.3 Utility Function-Based Pricing

In Chap. 2, for a single risky asset and a single-period model, we start from the investor's utility function give the initial price of the risky asset that the investor can accept, and then construct a risk-neutral measure. In the following, we will further give some general results.

For simplicity, we still consider a single-period model. Assume that there are a risk-free asset and d risky assets in the market. Their prices at time 0 are denoted by π^0 and $\pi = (\pi^1, \ldots, \pi^d)$, respectively, where $\pi^0 = 1$. We put $\overline{\pi} = (\pi^0, \pi) \in \mathbb{R}_+^{d+1}$. In addition, their prices at time 1 are denoted by S^0 and $S = (S^1, \ldots, S^d)$, respectively, where $S^0 = 1 + r$. We put $\overline{S} = (S^0, S)$. It is a $d + 1$-dimensional nonnegative random vector.

A portfolio held by a market participant at time 0 is denoted by $\overline{\xi} = (\xi^0, \xi) = (\xi^0, \xi^1, \ldots, \xi^d)$, where ξ^i represents the holding share of the asset i. Therefore, the initial investment amount of this investor is $\overline{\pi} \cdot \overline{\xi}$; the wealth at time 1 is $\overline{\xi} \cdot \overline{S}$. The market is called *nonredundant* if

$$\overline{\xi} \cdot \overline{S} = 0, \ \mathbb{P}\text{–a.s.} \implies \overline{\xi} = 0.$$

Assume that a market participant's preference is determined by a strictly concave and strictly increasing utility function $\widetilde{u}$, and the initial wealth is w. Then his optimal investment strategy is such that the expected utility of the wealth at the end of period reaches the maximum:

$$\max_{\overline{\xi}} \mathbb{E}[\widetilde{u}(\overline{\xi} \cdot \overline{S})]$$

subject to $\overline{\pi} \cdot \overline{\xi} \leqslant w$.

Noting that $\widetilde{u}$ is strictly increasing, the above optimal solution must satisfy $\overline{\pi} \cdot \overline{\xi} = w$. If we let $u(y) = \widetilde{u}((1 + r)(y + w))$, then u is a strictly concave and strictly increasing utility function. We further let

$$Y = S/(1 + r) - \pi, \quad \mathcal{S}(D) = \{\xi \in \mathbb{R}^d | \xi \cdot Y \in D, \mathbb{P}\text{–a.s.}\},$$

where D is the definition domain of u. Then the original expectation maximization problem is transformed into the following unconstrained optimization problem:

$$\arg \max_{\xi \in \mathcal{S}(D)} \mathbb{E}[u(\xi \cdot Y)]. \tag{3.27}$$

If the market is nonredundant, then this optimization problem has at most one solution.

The next theorem is from Föllmer and Schied (2004). Since the proof of the theorem is rather complex, we omit it here.

Theorem 3.10 *We assume that the market is nonredundant. If the utility function u satisfies one of the following two conditions:*

(1) $D = \mathbb{R}$, *and function u is bounded from above;*
(2) $D = [a, \infty), a < 0$, *and* $\mathbb{E}[u(\xi \cdot Y)] < \infty$, $\forall \xi \in \mathcal{S}(D)$,
   then optimization problem (3.27) *has solutions if and only if the market is arbitrage-free.*

When u is continuously differentiable, the next theorem gives the first-order necessary condition for the optimal solutions and thus determines an equivalent risk-neutral measure.

Theorem 3.11 *Assume that u is continuously differentiable and for any $\xi \in S(D)u(\xi \cdot Y)$ is integrable. If ξ^* is an optimal solution to problem (3.27), and satisfies one of the following two conditions:*

(1) $D = \mathbb{R}$, *and the function u is bounded from above;*
(2) $D = [a, \infty)$, $a < 0$; ξ^* *is an interior point of $S(D)$,*

then $u'(\xi^* \cdot Y)|Y|$ in integrable and ξ^* satisfies the following first-order condition:

$$\mathbb{E}[u'(\xi^* \cdot Y)Y] = 0.$$

In this case, $u'(\xi^* \cdot Y)$ is also integrable and probability measure $\mathbb{P}^*$, defined by

$$\frac{d\mathbb{P}^*}{d\mathbb{P}} = \frac{u'(\xi^* \cdot Y)}{\mathbb{E}[u'(\xi^* \cdot Y)]},$$

is a risk-neutral measure equivalent to $\mathbb{P}$.

Proof For $\xi \in S(D)$, $\varepsilon \in (0, 1]$, we let $\xi_\varepsilon = \varepsilon\xi + (1 - \varepsilon)\xi^*$ and define

$$\Delta_\varepsilon = \frac{u(\xi_\varepsilon \cdot Y) - u(\xi^* \cdot Y)}{\varepsilon}.$$

By the concavity of u, if $\varepsilon \leqslant \delta$, then $\Delta_\varepsilon \geqslant \Delta_\delta$. Thus, when $\varepsilon \downarrow 0$,

$$\Delta_\varepsilon \uparrow u'(\xi^* \cdot Y)(\xi - \xi^*) \cdot Y.$$

By the assumption, we know $\Delta_1 \in L^1(\mathbb{P})$. Consequently, by the monotone convergence theorem, we obtain

$$0 \geqslant E[\Delta_\varepsilon] \uparrow E[u'(\xi^* \cdot Y)(\xi - \xi^*) \cdot Y].$$

In particular, the expectation of the right side of the above equation is finite. By the assumption we know that ξ^* is an interior point of $S(D)$. Thus, by letting $\eta = \xi - \xi^*$ in the above equation, we deduce that for any η taking values in a small ball of $\mathbb{R}^d$ centered at the origin, we have

$$E[u'(\xi^* \cdot Y)\eta \cdot Y] \leqslant 0.$$

Replacing η by $-\eta$ gives $\mathbb{E}[u'(\xi^* \cdot Y)Y] = 0$.

Due to the fact that $\mathbb{P}^*$ is a risk-neutral measure equivalent to $\mathbb{E}^*[Y] = 0$, the last conclusion in the theorem is obvious. $\square$

As a result, in an arbitrage-free market, we can construct an equivalent martingale measure $\mathbb{P}^*$ directly by using the utility function of the investor, such that $d\mathbb{P}^*/d\mathbb{P}$ is bounded. Now assume that a market participant's preference is determined by exponential utility function $u(x) = 1 - e^{-\alpha x}$, $\alpha > 0$. Without loss of generality, we can assume that for any $\xi \in \mathbb{R}^d$ $u(\xi \cdot Y)$ is integrable. Otherwise, by Lemma 3.6 and approximation by rational valued strategy, we can introduce an equivalent probability measures $\widetilde{\mathbb{P}}$ such that $d\widetilde{\mathbb{P}}/d\mathbb{P}$ is bounded and $\widetilde{\mathbb{E}}[u(\xi \cdot Y)] < \infty$. From Theorem 3.11 we can construct an equivalent martingale measure $\mathbb{P}^*$ as follows:

$$\frac{dP^*}{dP} = \frac{u'(\xi^* \cdot Y)}{\mathbb{E}[u'(\xi^* \cdot Y)]} = \frac{e^{-\alpha \xi^* \cdot Y}}{E[e^{-\alpha \xi^* \cdot Y}]}.$$

This completes the pricing of various types of asset in the market.

Probability measure $\mathbb{P}^*$ defined by the above formula occupies a special position in equivalent martingale measure family $\mathcal{P}$. It is called the *Esscher transform* of original probability measure $\mathbb{P}$. It minimizes the relative entropy w.r.t. $\mathbb{P}$ (see Definition 3.9), i.e., $\mathbb{P}^*$ is the only solution to the following optimization problem:

$$\arg \min_{\hat{\mathbb{P}} \in \mathcal{P}} H_{\hat{\mathbb{P}}}(\mathbb{P}).$$

3.6.4 Market Equilibrium Pricing

Examining the financial market from a microeconomic point of view, we will find that an exogenously given pricing rule is not appropriate. The market prices of various assets and contingent claims should be determined endogenously by the market. Specifically, the prices of assets in a financial market are determined by market characteristics (such as the aggregate market supply level) as well as market participants' personal preferences and subjective estimates of the market and other factors. Thus, we need to proceed to look for an "endogenous" pricing rule such that, under this pricing rule, not only is the expected utility of each market participant's individual maximized, but also the total market supply of the same period equals the total demand, so as to achieve the market equilibrium.

The following single-period model is an example of studying the equilibrium pricing problem.

Let $(\Omega, \mathcal{F}, \mathbb{P})$ be a probability space used to describe the market uncertainty at time 1. Let $L^0(\Omega, \mathcal{F}, \mathbb{P})$ denote the set of all real random variables, which is used to describe all possible contingent claims at time 1. Consider a finite set $\mathcal{A}$ of market participants and a convex set $\mathcal{X}$ of admissible contingent claims. Assume that market participant $a \in \mathcal{A}$ holds an *endowment* at time 0 and can get a discounted wealth W_a at time 1. Then the total market supply at time 1 is $W = \Sigma_{a \in \mathcal{A}} W_a$. Each market participant can change the initial allocation, based on his own utility function u_a, with an expectation to, at the future time 1, obtain a contingent claim $X_a \in \mathcal{X}$, which maximizes the expected utility. This gives rise to the total market demand $\Sigma_{a \in \mathcal{A}} X_a$ at time 1. If an allocation $(X_a)_{a \in \mathcal{A}} \subset \mathcal{X}$ satisfies the *market clearing* condition (i.e., the total market supply equals the total market demand), we call allocation $(X_a)_{a \in \mathcal{A}}$ a *viable allocation*.

In our market model, we call any strictly positive random variable φ with expectation 1 a *pricing density* or *pricing kernel*, because we can use it to define a *pricing measure* $\mathbb{P}^* = \varphi.\mathbb{P}$ which determines a pricing rule: for a contingent claim X, define its price at time 0 by $\mathbb{E}^*[\tilde{X}] = \mathbb{E}[\varphi\tilde{X}]$, where $\tilde{X}$ is the discounted contingent claim X. Then the *budget set* of market participant a is given by

$$\mathcal{B}_a(\varphi^*) = \{X \in \mathcal{X} \cap L^1(\Omega, \mathcal{F}, \mathbb{P}) : \mathbb{E}[\varphi^* X] \leqslant \mathbb{E}[\varphi^* W_a]\},$$

where the inequality $\mathbb{E}[\varphi^* X] \leqslant \mathbb{E}[\varphi^* W_a]$ is called the *budget constraint* condition.

Definition 3.12 A pricing density φ^* together with a viable allocation $(X_a)_{a \in \mathcal{A}}$ is said to construct an *Arrow-Debreu equilibrium*, if for each $a \in \mathcal{A}$, X_a is a solution to the following optimization problem:

$$\arg \max_{X_a \in \mathcal{B}_a(\varphi^*)} \mathbb{E}[u_a(X_a)].$$

Remark Assume that X_a is a solution to the above optimization problem and $\mathbb{E}[u(X_a)] < \infty$. Since $\mathcal{B}_a(\varphi^*)$ is a convex set and u_a is strictly concave, X_a is the unique solution. If we further assume $\mathcal{X} = L^0(\Omega, \mathcal{F}, \mathbb{P})$ or $\mathcal{X} = L^0_+(\Omega, \mathcal{F}, \mathbb{P})$, then $\mathbb{E}[\varphi^* X_a] = \mathbb{E}[\varphi^* W_a]$.

Definition 3.13 Put

$$\Lambda = \{\lambda \in [0, 1]^{|\mathcal{A}|} : \Sigma_{a \in \mathcal{A}} \lambda_a = 1\},$$

where $|\mathcal{A}|$ represents the number of elements in $\mathcal{A}$, then Λ is convex compact set. For any given $\lambda \in \Lambda$, consider the following weighted average optimization problem:

$$\arg\max U^\lambda(X) = \Sigma_{a \in \mathcal{A}} \lambda_a E[u_a(X_a)],$$

$$\Sigma_{a \in \mathcal{A}} X_a = W.$$

If a viable allocation $(X_a)_{a \in \mathcal{A}}$ is a solution to the above optimization problem, then we call it λ-*effective*.

The next theorem shows that under certain assumptions there is an Arrow-Debreu equilibrium in the market.

Theorem 3.14 *Assume that each market participant $a \in \mathcal{A}$ satisfies*

$$\limsup_{x \downarrow 0} x u_a'(x) < \infty, \quad \mathbb{E}[u_a'(W/|\mathcal{A}|)] < \infty,$$

and $\mathbb{E}[W] < \infty$, then there is an Arrow-Debreu equilibrium in the market.

In the proof we will use the following lemma from Föllmer and Schied (2004).

Lemma 3.15 *Under the condition of Theorem 3.14, we have the following conclusions:*

(1) *For any $\lambda \in \Lambda$, there is a unique λ-effective allocation $(X_a^\lambda)_{a \in \mathcal{A}}$.*
(2) *A viable allocation $(X_a)_{a \in \mathcal{A}}$ is λ-effective if and only if it satisfies the following first-order condition w.r.t. certain pricing density φ:*

$$\lambda_a u_a'(X_a) \leqslant \varphi, \text{ and the equality holds on } [X_a > 0], \ \forall a \in \mathcal{A}.$$

In this case, $(X_a)_{a \in \mathcal{A}}$ is just a λ-effective allocation $(X_a^\lambda)_{a \in \mathcal{A}}$ and φ can be taken as

$$\varphi^\lambda = \max_{a \in \mathcal{A}} \lambda_a u_a'(X_a^\lambda).$$

(3) *For each $a \in \mathcal{A}$, X_a^λ is a solution to the following optimization problem:*

$$\max \mathbb{E}[u_a(X)], \text{ subject to } \mathbb{E}[\phi^\lambda X] \leqslant \mathbb{E}[\phi^\lambda W_a].$$

Proof of Theorem 3.14 From 3.15(1) we know that if there is λ, such that the corresponding pricing density φ^λ and λ-effective allocation $(X_a^\lambda)_{a \in \mathcal{A}}$ satisfies

$$\mathbb{E}[\phi^\lambda X_a^\lambda] = \mathbb{E}[\phi^\lambda W_a], \quad \forall a \in \mathcal{A}, \tag{3.28}$$

then φ^λ together with $(X_a^\lambda)_{a \in \mathcal{A}}$ constructs an Arrow-Debreu equilibrium. If otherwise, we consider the following mapping $g(\lambda) = (g_a(\lambda))_{a \in \mathcal{A}}$:

$$g_a(\lambda) = \lambda_a + \frac{1}{E[V]} \cdot E[\varphi^\lambda(W_a - X_a^\lambda)],$$

where

$$V = \kappa(1 + W), \ \kappa = \max_{a \in \mathcal{A}} \sup_{0 < x \leqslant 1} x u_a'(x) < \infty,$$

and $u_a'(X_a)X_a \leqslant V \in L^1(\mathbb{P})$. Lemma 3.13(2) insures $g_a(\lambda) \geqslant 0$, and thus $g(\lambda) \in \Lambda$. By the definition of g, any fixed point of g satisfies condition (3.28), and thus an Arrow-Debreu equilibrium is constructed.

Therefore, in order to prove the theorem, we only need to verify that mapping g has fixed points. According to the Brouwer fixed-point theorem in functional analysis, we only need to prove that g is continuous. Now we take a sequence $(\lambda_n) \subset \Lambda$, $\lambda_n \to \lambda \in \Lambda$ and let $X_n = X^{\lambda_n}$, $\varphi_n = \varphi^{\lambda_n}$. We define a random variable

$$F = \max_{a \in \mathcal{A}} u_a'(W/|\mathcal{A}|) \in L^1(\mathbb{P}).$$

Then $\forall \lambda \in \Lambda$ we have $\varphi^\lambda \leqslant F$, so that

$$W_a \varphi_n \leqslant W \varphi_n \leqslant W F, \quad X_n \varphi_n \leqslant W \varphi_n \leqslant W F.$$

On the other hand,

$$W F \leqslant |\mathcal{A}| F I_{[W \leqslant \mathcal{A}]} + \max_{a \in \mathcal{A}} u'_a(1) \cdot W \in L^1(\mathbb{P}).$$

Therefore, if $\varphi_n \overset{p}{\to} \varphi^\lambda$ and $X_n \overset{p}{\to} X^\lambda$, then by the dominated convergence theorem, we get the continuity of g.

Now we only prove $\varphi_n \overset{p}{\to} \varphi^\lambda$, the proof of $X_n \overset{p}{\to} X^\lambda$ is similar. Let I_a^+ denote the inverse function of u'_a. Consider the following mapping $f : \Lambda \times [0, +\infty] \to [0, +\infty]$:

$$f(\lambda, y) = \sum_{a \in \mathcal{A}} I_a^+(\lambda_a^{-1} y).$$

For any given λ, function $f(\lambda, \cdot)$ is continuous on $[0, \infty]$ and is strictly decreasing on $(a(\lambda), b(\lambda))$, where

$$a(\lambda) = \max_{a \in A} \lim_{x \uparrow \infty} \lambda_a u'_a(x) \geqslant 0, \quad b(\lambda) = \max_{a \in A} \lambda_a u'_a(0+) \leqslant +\infty.$$

In addition, if $y \leqslant a(\lambda)$, then $f(\lambda, y) = \infty$; if $y \geqslant b(\lambda)$, then $f(\lambda, y) = 0$. Thus, for any initial wealth ω, there is a unique $y^\lambda \in (a(\lambda), b(\lambda))$ such that

$$f(\lambda, y^\lambda) = \omega.$$

Since $[0, +\infty]$ is a compact set, there exists a subsequence (λ_{n_k}) of (λ_n) such that the solution $y_k := y^{\lambda_{n_k}}$ to equation $f(\lambda_{n_k}, y) = \omega$ converges to a limit $y_\infty \in [a(\lambda), b(\lambda)]$ when k tends to ∞. By the continuity of f,

$$f(\lambda, y_\infty) = \lim_{k \uparrow +\infty} f(\lambda_{n_k}, y_k) = \omega,$$

and consequently, $y^\lambda = y_\infty$.

Simultaneously, since X_a^λ is a solution to the expected utility maximization problem, it has the following form:

$$X_a^\lambda = I_a^+(\lambda_a^{-1} \varphi^\lambda).$$

Thus, $W = f(\lambda, \varphi^\lambda)$, and $\varphi^{\lambda_{n_k}}$ a.e. converges to φ^λ when k tends to ∞. The theorem is proved. $\qquad\square$

As a result, among the markets satisfying the conditions of Theorem 3.13, we can use Arrow-Debreu equilibrium pricing density φ^* to give reasonable prices for nonnegative contingent claims. In particular, assume that in the market there are a risk-free asset and d risky assets, and their prices at time $t = 1$ are denoted by S^0 and S^i, $1 \leqslant i \leqslant d$, respectively. If the price vector $(\mathbb{E}[\varphi^* S^0], \mathbb{E}[\varphi^* S^1], \cdots, \mathbb{E}[\varphi^* S^d])$ given by φ^* is just equal to the constant vector $\pi = (\pi^0, \pi^1, \cdots, \pi^d)$ given in advance, then the probability measure $\mathbb{P}^*$ becomes an equivalent risk-neutral measure, where $d\mathbb{P}^*/d\mathbb{P} = \varphi^*$. In this case, the prices of a contingent claim given by the equilibrium pricing and arbitrage pricing coincide. In addition, from an economic perspective the risk-free interest rate in a financial market should also be endogenously determined by the market. Similar to the above equilibrium pricing idea, we can get an endogenous interest rate in an equilibrium sense. Compared with the arbitrage pricing, the advantage of the equilibrium pricing is no longer limited by the completeness of the market. Thus, the scope of its application is wider.

3.7 American Contingent Claims Pricing

In the following we discuss the pricing of American contingent claims. Unlike European contingent claims, American contingent claims can be executed at any time prior to the expiry of the contract. Generally speaking, an American contingent claim with maturity N can be described by an $(\mathcal{F}_n)$-adapted nonnegative sequence (Z_n), where Z_n represents the payoff of the contract seller to the buyer if the contract is executed at time n. For example, for an American call (resp. put) option on a stock, $Z_n = (S_n - K)^+$ (resp. $Z_n = (K - S_n)^+$), where S_n is the price of the stock at time n, K is the agreed or executed price of the option contract. This section studies hedging and pricing of American contingent claims.

3.7.1 Super-Hedging Strategies in Complete Markets

Now assume that the market is complete and $\mathbb{P}^*$ is the unique equivalent martingale measure. Let U_n denote the seller's price of an American contingent claim at time n; then $U_N = Z_N$. If a contract seller wants to ensure that he can cover payoff Z_{N-1} at time $N - 1$ and payoff Z_N at time N, then he should let

$$U_{N-1} = \max\left(Z_{N-1}, \beta_{N-1}\mathbb{E}^*[\tilde{Z}_N|\mathcal{F}_{N-1}]\right).$$

By induction, for $n = 0, \cdots, N - 1$, he should let

$$U_n = \max\left(Z_n, \beta_n\mathbb{E}^*[\gamma_{n+1}U_{n+1}|\mathcal{F}_n]\right). \tag{3.29}$$

From Theorem 1.34 (Snell's envelop) we obtain

Theorem 3.16 *The random sequence* $(\widetilde{U}_n)_{0\leqslant n\leqslant N}$ *defined by (3.29) is a super-martingale under* $\mathbb{P}^*$. *It is the smallest supermartingale dominating sequence* $(\widetilde{Z}_n)_{0\leqslant n\leqslant N}$ *from the upper.*

A *consumption-investment strategy* (ϕ_n, c_n) is a trading strategy (ϕ_n) together with an adapted nonnegative sequence (c_n) with $c_0 = 0$ such that

$$\phi_{n+1} \cdot S_n = \phi_n \cdot S_n - c_n, \quad 0\leqslant n\leqslant N - 1, \tag{3.30}$$

where c_n represents the amount of wealth taken out at time n for consumption. We still denote the wealth at time n of a consumption-investment strategy (ϕ_n, c_n) by $V_n(\phi) = \phi_n \cdot S_n$. Then its discounted wealth process $(\widetilde{V}_n(\psi))$ is a $\mathbb{P}^*$-supermartingale. It is easy to see that (3.30) is equivalent to

$$V_n(\phi) = V_0(\phi) + G_n(\phi) - \sum_{j=1}^{n-1} c_j, \quad 1\leqslant n\leqslant N. \tag{3.31}$$

From Doob's decomposition of supermartingale $\widetilde{U}$, it is easy to see that there is a consumption-investment strategy (ϕ_n, c_n) such that for all $0\leqslant n\leqslant N$, $V_n(\phi) = U_n$. Obviously, this strategy *super-hedges* American contingent claim (Z_n), i.e., for all $0\leqslant n\leqslant N$, $V_n(\phi)\geqslant Z_n$. On the other hand, if a consumption-investment strategy (ϕ_n, c_n) super-hedges American contingent claim (Z_n), then obviously, for all $0\leqslant n\leqslant N$, $V_n(\phi)\geqslant U_n$. This shows that random sequence (U_n) defined by (3.29) is the smallest one among all wealth processes of super-hedging strategies. Clearly, U_0 is the price at time 0 which the seller of American contingent claim (Z_n) can accept. It is given by

$$U_0 = \sup_{\tau\in\mathcal{T}} \mathbb{E}^*[\widetilde{Z}_\tau], \tag{3.32}$$

where $\mathcal{T}$ is the set of all stopping times taking values in $\{0, 1, \cdots, N\}$.

3.7.2 Arbitrage-Free Pricing in Complete Markets

Now we continue to assume that the market is complete and $\mathbb{P}^*$ is the unique equivalent martingale measure. The payoff Z_τ obtained by the buyer in executing American contingent claim (Z_n) at a stopping time τ before time N can be regarded as a European contingent claim at stopping time τ. We can also convert it to a usual European contingent claim with expiry date N as follows: we use the wealth Z_τ to buy risk-free asset 0 at time τ; the wealth at time N becomes $\xi = \frac{Z_\tau}{S_\tau^0} S_N^0$. Then ξ is an European contingent claim, and its arbitrary-free price is $\mathbb{E}^*[\widetilde{\xi}] = \mathbb{E}^*[\widetilde{Z}_\tau]$. By

Theorem 1.34 (Snell's envelop), in order to maximize the expected value of Z_τ, an optimal executing time τ can be taken as:

$$\tau = \inf\{j \geqslant 0 : U_j = Z_j\}, \tag{3.33}$$

where (U_n) is defined by (3.29). Then the process $(\widetilde{U}_{\tau \wedge n}, 0 \leqslant n \leqslant N)$ is a martingale. In fact, any stopping time σ such that the process $(\widetilde{U}_{\sigma \wedge n}, 0 \leqslant n \leqslant N)$ is a martingale is an optimal executing time, but only the above defined τ is the smallest stopping time among all optimal executing times. As a result, the arbitrage-free price $\mathbb{E}^*[\widetilde{Z}_\tau]$ of Z_τ, obtained by the buyer at executing time τ, is just the seller's acceptable price U_0. Therefore, in a complete market, U_0, defined by (3.32), is a unique arbitrage-fee price of American contingent claim (Z_n).

3.7.3 Arbitrage-Free Pricing in Non-complete Markets

Now we assume the market is not necessarily complete. Let (Z_n) be an American contingent claim. We assume that under any equivalent martingale measure $\mathbb{Q}$, each $\widetilde{Z}_n$ is integrable. If $\tau \in \mathcal{T}$, we let $\Pi(Z_\tau)$ denote the set of all arbitrage-free prices of European contingent claim Z_τ at time τ.

A real number π is called an *arbitrage-free price* of the American contingent claim (Z_n), if it satisfies the following two conditions:

(1) there is a stopping time $\tau \in \mathcal{T}$ and a $\pi' \in \Pi(Z_\tau)$ such that $\pi \leqslant \pi'$;
(2) there is no stopping time $\tau' \in \mathcal{T}$ such that for all $\pi' \in \Pi(Z_{\tau'})$ it holds $\pi < \pi'$.

The set of all arbitrary-free prices of American contingent claim (Z_n) is denoted by $\Pi^a(Z)$. Put

$$U_0^{\mathbb{Q}} = \sup_{\tau \in \mathcal{T}} \mathbb{E}^{\mathbb{Q}}[\widetilde{Z}_\tau],$$

$$\pi_{\sup}^a(Z) = \sup_{\mathbb{Q} \in \mathcal{M}} U_0^{\mathbb{Q}}, \quad \pi_{\inf}^a(Z) = \inf_{\mathbb{Q} \in \mathcal{M}} U_0^{\mathbb{Q}}, \tag{3.34}$$

Föllmer and Schied (2004) proved the following deep result: if $\pi_{\inf}^a(Z) < \pi_{\sup}^a(Z)$, then $\Pi^a(Z)$ is a real interval with endpoints $\pi_{\inf}^a(Z)$ and $\pi_{\sup}^a(Z)$ which does not contain upper endpoint $\pi_{\sup}^a(Z)$ but may contain lower endpoint $\pi_{\inf}^a(Z)$; $\Pi^a(Z)$ can also reduce to a single point.

An American contingent claim (Z_n) is called *attainable*, if there is a stopping time $\tau \in \mathcal{T}$ and a self-financing strategy such that its wealth process (V_n) satisfies $V_n \geqslant Z_n, \forall n$ and $V_\tau = Z_\tau$. In this case, trading strategy φ is called a *hedging strategy* of (Z_n). One can prove that (see Föllmer and Schied (2004)) (Z_n) is attainable if and only if $\Pi^a(Z)$ is a single point set, or equivalently, $\pi_{\sup}^a(Z) \in \Pi^a(Z)$.

Chapter 4
Martingale Theory and Itô Stochastic Analysis

In this chapter, we will briefly introduce the martingale theory and Itô's stochastic analysis. First we introduce some basic concepts of continuous time stochastic processes and the definitions of four basic types of process: Markov process, martingale, Poisson process, and Brownian motion, as well as their basic properties, including Doob-Meyer's decompositions of continuous local submartingales and quadratic variation processes of continuous local martingales and continuous semimartingales. Then we introduce the stochastic integrals of measurable adapted processes w.r.t. the Brownian motion. Finally, we introduce some useful tools for Itô calculus, such as Itô's formula, Girsanov's theorem, and the martingale representation theorem. Itô stochastic differential equations and Feynman-Kac formula are also presented. This chapter is self-contained. For a small number of results, we omit their proofs and refer to Karatzas and Shreve (1991) or Revuz and Yor (1999).

4.1 Continuous Time Stochastic Processes

4.1.1 Basic Concepts of Stochastic Processes

A *stochastic process* is simply a collection of random variables $\{X_t, t \in \Lambda\}$, defined on a common probability space $(\Omega, \mathcal{F}, \mathbb{P})$, where Λ is a time parameter set. If Λ is an interval of $\mathbb{R} = (-\infty, \infty)$, (X_t) is called a continuous time process. For a fixed $\omega \in \Omega$, the function $t \mapsto X_t(\omega)$ defined on Λ is called a *sample path* of the process (X_t). A process having continuous paths is called a continuous process.

Assume that (X_t) and (X_t') are two stochastic processes. If for all t, $X_t = X_t'$ a.s., we call X a *version* of X'. If their sample paths are almost surely the same, we call these two processes *indistinguishable*. A process is called *right (left) continuous*, if its sample paths are almost surely right (left) continuous. A process is called *right*

© Springer Nature Singapore Pte Ltd. and Science Press 2018
J.-A. Yan, *Introduction to Stochastic Finance*, Universitext,
https://doi.org/10.1007/978-981-13-1657-9_4

continuous with left limits, if its sample paths are almost surely right continuous with left limits; it is also called *cadlag*, where "cadlag" is an acronym from the French "continu à droit, limité à gauche." Similarly, a "cadlag" process is left continuous with right limits.

Let (X_t) be a process. For any finite sequence $(t_i)_{1 \leqslant i \leqslant n}$ with $0 \leqslant t_1 < t_2 < \cdots < t_n < \infty$, the joint distribution of n random variables $\{X_{t_1}, \cdots, X_{t_n}\}$ is called an n-dimensional distribution of process X. If all finite dimensional distributions of a process (X_t) are normal distributions, we call (X_t) a *Gaussian process*.

An $\mathbb{R}^d$-valued process (X_t) is called a *process with independent increments*, if for all $0 \leqslant t_0 < t_1 < \cdots < t_n$, $X_{t_0}, X_{t_1} - X_{t_0}, \cdots, X_{t_n} - X_{t_{n-1}}$ are independent. It is easy to see that an $\mathbb{R}^d$-valued process (X_t) is a process with independent increments if and only if for any $0 \leqslant s < t$, $X_t - X_s$ is independent of $\mathcal{F}_s^X$, where $\mathcal{F}_s^X = \sigma(X_u, u \leqslant s)$. A process with independent increments is called *homogeneous* if for any $s > 0$, $X_{t+s} - X_t$ has the same distribution as $X_s - X_0$, for all $t > 0$.

Assume that the time parameter set Λ is $\mathbb{R}_+$ or $[0, T]$. We consider a given complete probability space $(\Omega, \mathcal{F}, \mathbb{P})$. We call an increasing family $(\mathcal{F}_t)$ of sub-σ-algebras of $\mathcal{F}$ a *filtration*. If for each t, $\mathcal{F}_t$ contains the null sets of $\mathcal{F}$, and $(\mathcal{F}_t)$ is right continuous (i.e., $\cap_{s>t}\mathcal{F}_s = \mathcal{F}_t$ for all t), we say that the filtration satisfies the *usual condition*. A complete probability space $(\Omega, \mathcal{F}, \mathbb{P})$ together with a usual filtration $(\mathcal{F}_t)$ is called a *filtered probability space* or *stochastic basis* and denoted by $(\Omega, \mathcal{F}, (\mathcal{F}_t), \mathbb{P})$.

Let (X_t) be a process defined on a stochastic basis $(\Omega, \mathcal{F}, (\mathcal{F}_t), \mathbb{P})$. If for each t, X_t is $\mathcal{F}_t$-measurable, we call (X_t) an $(\mathcal{F}_t)$-*adapted process*. If $X_t(\omega)$, as a function of (ω, t), is $\mathcal{F} \times \mathcal{B}(\mathbb{R}_+)$-measurable, X is called *measurable*; if for each $t \in \mathbb{R}_+$, the restriction of X on $\Omega \times [0, t]$ is $\mathcal{F}_t \times \mathcal{B}([0, t])$-measurable, X is called *progressively measurable* (or simply, *progressive*). A progressively measurable process is an adapted process. It is easy to prove that a right (left) continuous adapted process is a progressively measurable process.

Let (X_t) be a process defined on a complete probability space $(\Omega, \mathcal{F}, \mathbb{P})$. For each t, let $\mathcal{F}_t$ denote the σ-algebra generated by $(X_s, s \leqslant t)$ and the null sets of $\mathcal{F}$; we call the family $(\mathcal{F}_t)$ the *natural filtration* of process (X_t).

4.1.2 Poisson and Compound Poisson Processes

We are going to model the occurrences of successive events that take place at discrete instants with random waiting times. Arrivals of customers in a queue, calls coming into a telephone exchange, and traffic accidents in a city are typical examples of such events. We consider the arrivals of customers in a queue. Let $\xi_i, i = 1, 2, \cdots$ be a sequence of waiting times which are assumed to be independent identically distributed (i.i.d.) random variables having exponential distribution $E(\lambda)$, i.e., $\mathbb{P}(\xi_i \leqslant x) = 1 - e^{-\lambda x}$, $x \geqslant 0$. Put

$$S_0 = 0, \quad S_n = \sum_{i=1}^{n} \xi_i, \; n \geqslant 1,$$

$$N_0 = 0, \quad N_t = \sum_{n=1}^{\infty} I_{(0,t]}(S_n), \; t > 0.$$

Then N_t is the number of arrivals in time interval $(0, t]$. For each t, N_t is a discrete random variable taking values in the set of nonnegative integers. The process (N_t) is a right-continuous increasing process.

Since S_n has density function $\lambda^n t^{n-1} e^{-\lambda t}/(n-1)!$, $t > 0$, we have

$$\mathbb{P}(N_t = n) = \mathbb{P}(S_n \leqslant t < S_{n+1}) = \mathbb{P}(S_n \leqslant t) - \mathbb{P}(S_{n+1} \leqslant t)$$

$$= \frac{\lambda^n}{(n-1)!} \int_0^t u^{n-1} e^{-\lambda u} du - \frac{\lambda^{n+1}}{n!} \int_0^t u^n e^{-\lambda u} du$$

$$= e^{-\lambda t} \sum_{k=n}^{\infty} \frac{(\lambda t)^k}{k!} - e^{-\lambda t} \sum_{k=n+1}^{\infty} \frac{(\lambda t)^k}{k!}$$

$$= e^{-\lambda t} \frac{(\lambda t)^n}{n!}$$

Thus, N_t follows a Poisson distribution with parameter λt.

One can prove that process (N_t) constructed above has the following properties:

(1) N is a homogeneous process with independent increments and $N_0 = 0$;
(2) $N_t - N_s$ follows a Poisson distribution with parameter $\lambda(t - s)$.

A process $(N_t)_{t \geqslant 0}$ having the above two properties is called a *Poisson process* with *intensity* λ. The term intensity comes from the fact that the expected number of arrivals in time arrival $(0, t]$ ($\mathbb{E}[N_t] = \lambda t$) is proportional to the length of the interval, and λ is the constant of proportionality.

Let (N_t) be a Poisson process with intensity λ. Let $(U_j)_{j \geqslant 1}$ be a sequence of i.i.d. integrable random variables and independent of (N_t). We denote the common distribution of U_j's by F. Put

$$C_t = \sum_{j=1}^{N_t} U_j.$$

We call (C_t) a *compound Poisson process* with arrival rate λ and jump law F. The characteristic function of C_t is

$$\phi_t(u) = \mathbb{E}[e^{iuC_t}] = \sum_{k=0}^{\infty} \mathbb{E}\Big[e^{iu \sum_{j=1}^{k} U_j}\Big|N_t = k\Big]\mathbb{P}(N_t = k)$$

$$= \sum_{k=0}^{\infty} \Big(\int_{\mathbb{R}} e^{iux} F(dx)\Big)^k \frac{e^{-\lambda t}(\lambda t)^k}{k!}$$

$$= \exp\Big\{\lambda t \int_{\mathbb{R}} (e^{iux} - 1) F(dx)\Big\}, \quad u \in \mathbb{R}.$$

Let S_j denote the jth jump time of the Poisson process (N_t). We can rewrite C_t as

$$C_t = \sum_{j=1}^{\infty} U_j I_{[S_j \leqslant t]}.$$

In this expression U_j is the jth jump size. If $U_j = 1$ for all j, the compound Poisson process reduces to the Poisson process.

4.1.3 Markov Processes

Let $(S, \mathcal{S})$ be a measurable space. An S-valued $(\mathcal{F}_t)$-adapted process (X_t) is called a (continuous time) *Markov process* with state space S, if for all $s < t$ and $A \in \mathcal{S}$,

$$\mathbb{P}[X_t \in A \mid \mathcal{F}_s] = \mathbb{P}[X_t \in A \mid X_s] \text{ a.s.}$$

If $(\mathcal{F}_t)$ is the natural filtration of (X_t), it is easy to prove that this property is equivalent to the following one: for $t_1 \leqslant t \leqslant t_2$, $A_1 \in \sigma(X_s, s \leqslant t_1)$ and $A_2 \in \sigma(X_s, s \geqslant t_2)$,

$$\mathbb{P}(A_1 A_2 \mid X_t) = \mathbb{P}(A_1 \mid X_t)\mathbb{P}(A_2 \mid X_t).$$

The latter is just a mathematical formulation of the so-called *Markov property*: the past and future are statistically independent when the present is known. If the state space S consists of countably many states and $\mathcal{S}$ is the class of all subsets of S, a Markov process is called a *Markov chain* (in continuous time).

4.1.3.1 Transition Probabilities Function

Now we only consider Markov processes with state space $\mathbb{R}^d$. A function $P(s, x; t, B)$ defined for $0 \leqslant s \leqslant t, x \in \mathbb{R}^d, B \in \mathcal{B}(\mathbb{R}^d)$, is called a *transition probability function* if it has the following properties:

(1) $P(s, x; t, \cdot)$ is a probability on $\mathcal{B}(\mathbb{R}^d)$ for given s, t and x;
(2) $P(s, \cdot; t, B)$ is $\mathcal{B}(\mathbb{R}^d)$-measurable for given s, t and B;
(3) $P(s, x, s, B) = I_B(x)$;
(4)   for $s \leqslant u \leqslant t$ and $B \in \mathcal{B}(\mathbb{R}^d)$, we have the so-called *Chapman-Kolmogorov equation*:

$$P(s, x; t, B) = \int_{\mathbb{R}^d} P(u, y; t, B) P(s, x; u, dy).$$

Let (X_t) be a Markov process and $P(s, x; t, \cdot)$ be a transition probability function. If

$$P(s, X_s; t, B) = \mathbb{P}(X_t \in B | X_s) \text{ a.s.,}$$

or, more symbolically,

$$P(s, x; t, B) = \mathbb{P}(X_t \in B | X_s = x),$$

we call $P(s, x; t, \cdot)$ the *transition probability function* of (X_t). If furthermore $P(s, x; t, \cdot)$ has a density $p(s, x; t, y)$ w.r.t. Lebesgue measure and

$$p(s, x; t, z) = \int_{\mathbb{R}^d} p(u, y; t, z) p(s, x; u, y) dy,$$

we call $p(s, x; t, y)$ the *transition density function* of (X_t).

Let (X_t) be a Markov process with transition probability function $P(s, x; t, \cdot)$. It is called *time-homogeneous* if for any $s < t$ and $h > 0$ we have

$$P(s, x; t, B) = P(s + h, x; t + h, B).$$

In this case we can put

$$P(t, x, B) = P(0, x; t, B),$$

and call also $P(t, x, \cdot)$ the *transition probability function* of (X_t).

It is easy to see that any process with homogeneous independent increments is a time-homogeneous Markov process.

4.1.3.2   Diffusion Processes

Let (X_t) be a Markov process with transition probability function $P(s, x; t, \cdot)$. It is called a *diffusion process* if for any $\epsilon > 0$, $P(s, x; t, \cdot)$ satisfies the following three conditions:

(1) $\lim_{t \downarrow s} \frac{1}{t-s} \int_{|x-y|>\epsilon} P(s, x; t, dy) = 0$;

(2)   there exists an $\mathbb{R}^d$-valued function $b(s, x)$ such that

$$\lim_{t \downarrow s} \frac{1}{t-s} \int_{|x-y| \leqslant \epsilon} (y-x) P(s, x; t, dy) = b(s, x);$$

(3)   there exists a $d \times d$ matrix-valued function $a(s, x)$ such that

$$\lim_{t \downarrow s} \frac{1}{t-s} \int_{|x-y| \leqslant \epsilon} (y-x)(y-x)^\tau P(s, x; t, dy) = a(s, x).$$

The functions b and a are called the *coefficients* of the diffusion process (X_t). Vector b is called the *drift vector*, and a is called the *diffusion matrix*. Matrix a is symmetric and nonnegative definite. If (X_t) is time-homogeneous, then b and a are independent of t.

For every $t \geqslant 0$, each diffusion process (X_t) is associated with a second-order differential operator

$$\mathcal{A}_t = \frac{1}{2} \sum_{i,k=1}^{d} a_{ik}(t, x) \frac{\partial^2}{\partial x_i \partial x_k} + \sum_{i=1}^{d} b_i(t, x) \frac{\partial}{\partial x_i}.$$

If (X_t) is time-homogeneous, the operator $\mathcal{A}_t, t \geqslant 0$, is independent of t and reduces to

$$\mathcal{A} = \frac{1}{2} \sum_{i,k=1}^{d} a_{ik}(x) \frac{\partial^2}{\partial x_i \partial x_k} + \sum_{i=1}^{d} b_i(x) \frac{\partial}{\partial x_i}.$$

In the latter case, $\mathcal{A}$ is called the *generator* of diffusion (X_t).

Let (X_t) be a diffusion process with continuous coefficients. Assume that its transition density function $p(s, x; t, y)$ exists.

(1)   If the derivatives $\partial p/\partial x_i$ and $\partial p/\partial x_i \partial x_j$ exist and are continuous w.r.t. s, then p is a fundamental solution to *Kolmogorov's backward equation*

$$\frac{\partial p}{\partial s} + \mathcal{A}_s p = 0.$$

(2)   If the limits in the definition of coefficients of diffusion process hold uniformly in s and x, derivatives $\partial p/\partial t$, $\partial(b_i p)/\partial y_i$ and $\partial(a_{ij} p)/\partial y_i \partial y_j$ exist and are continuous, then p is a fundamental solution to *Kolmogorov's forward equation* or the *Fokker-Planck equation*

$$\frac{\partial p}{\partial t} + \sum_{i=1}^{d} \frac{\partial}{\partial y_i}(b_i(t, y)p) - \frac{1}{2} \sum_{i,j=1}^{d} \frac{\partial^2}{\partial y_i \partial y_j}(a_{ij}(t, y)p) = 0.$$

4.1.4 Brownian Motion

It was Robert Brown, a botanist, who first investigated in 1827 a phenomenon that small pollen grains suspended in water are in a state of constant irregular motion. This phenomenon is named as *Brownian motion*. This irregular motion is caused by the random bombardment of moving mater molecules. Based on the assumption that the motions of these molecules are statistically independent, Albert Einstein (1905) derived a statistical description of the Brownian motion. In 1923 Robert Wiener constructed a continuous process which meets Einstein's description of the Brownian motion. So in the literature, the Brownian motion is also called the *Wiener process*. We will see below that the Wiener process is in fact more general than the Brownian motion.

A continuous process $(B_t)_{t \geqslant 0}$ defined on a probability space $(\Omega, \mathcal{F}, \mathbb{P})$ is called a Brownian motion with parameter $\sigma^2 > 0$, if it satisfies the following conditions:

(1) $B_0 = 0$;
(2) for $t_1 < t_2 < \cdots < t_n$, $B_{t_2} - B_{t_1}, B_{t_3} - B_{t_2}, \cdots, B_{t_n} - B_{t_{n-1}}$ are independent;
(3) $B_t - B_s$ has a normal distribution with mean 0 and variance $\sigma^2(t - s)$ for $s < t$.

If $\sigma = 1$, the process is called a *standard Brownian motion*.

If $B_t^1, \cdots, B_t^d$ are independent Brownian motions, we call $\mathbb{R}^d$-valued process $(B_t) = (B_t^1, \cdots, B_t^d)^{\tau}$ a d-dimensional Brownian motion. Here and henceforth, $(x^1, \cdots, x^d)^{\tau}$ represses the transpose of row vector $(x^1, \cdots, x^d)$. Let $\mathcal{F}_t^B$ denote the natural filtration of Brownian motion (B_t). It is a well-known result that $(\mathcal{F}_t^B)_{t \geqslant 0}$ satisfies the usual condition.

A d-dimensional Brownian motion is a time-homogeneous Markov process. Its transition density function is

$$p(t, x, y) = \frac{1}{(2\pi t)^{d/2}} \exp \left\{ -\frac{|x - y|^2}{2t} \right\},$$

and p is the fundamental solution to the following heat equation:

$$\frac{\partial u}{\partial t} = \frac{1}{2} \sum_{i=1}^{d} \frac{\partial^2 u}{\partial x_i}.$$

More generally, an $\mathbb{R}^d$-valued $(\mathcal{F}_t)$-adapted continuous process (B_t) defined on a stochastic basis $(\Omega, \mathcal{F}, (\mathcal{F}_t), \mathbb{P})$ is called an $(\mathcal{F}_t)$-*Brownian motion* or *Wiener process* with parameter σ^2, if $B_0 = 0$, $B_t - B_s$ is independent of the σ-algebra $\mathcal{F}_s$ and is normally distributed with mean vector 0 and covariance matrix $\sigma^2(t - s)I$ for $s < t$. We often omit the prefix $(\mathcal{F}_t)$. If $\sigma = 1$, the process (B_t) is called a standard Brownian motion or Wiener process. Here $(\mathcal{F}_t)$ is not necessarily the natural filtration of (B_t). For example, let $(B_t) = (B_t^1, \cdots, B_t^d)^{\tau}$ be a d-dimensional Brownian motion. Then each component (B_t^j) is an $(\mathcal{F}_t^B)$-Brownian motion.

4.1.5 Stopping Times, Martingales, Local Martingales

Let $(\Omega, \mathcal{F}, (\mathcal{F}_t), \mathbb{P})$ be a stochastic basis. We denote $\overline{\mathbb{R}}_+ = \mathbb{R}_+ \cup \{+\infty\}$. A $\overline{\mathbb{R}}_+$-valued random variable T is called a $(\mathcal{F}_t)$-stopping time, if for each $t \geqslant 0$, it holds $[T \leqslant t] \in \mathcal{F}_t$. Let $\mathcal{F}_\infty = \mathcal{F}$, for any $(\mathcal{F}_t)$-stopping time T, define

$$\mathcal{F}_T = \{A \in \mathcal{F}_\infty : \forall t \in \mathbb{R}_+, A \cap [T \leqslant t] \in \mathcal{F}_t\}.$$

Then $\mathcal{F}_T$ is a σ-algebra.

A real-valued $(\mathcal{F}_t)$-adapted process (X_t) is called a *martingale (respectively, supermartingale, submartingale)*, if each X_t is integrable and for all $s < t$ we have

$$\mathbb{E}[X_t \mid \mathcal{F}_s] = X_s \text{ (respectively, } \leqslant X_s, \geqslant X_s) \text{ a.s.}$$

Let M be a cadlag adapted process. If there exists a sequence of stopping times $T_n \uparrow +\infty$, such that each $M^{T_n} - M_0$ is a uniformly integrable martingale (super-martingale, submartingale), M is called a *local martingale (local supermartingale, local submartingale)*. Assume that M is defined only on a finite interval $[0, T]$. If there exists a sequence of stopping times $T_n \uparrow T$, and $\bigcup_{n=1}^{\infty}[T_n = T] = \Omega$, such that each $M^{T_n} - M_0$ is a uniformly integrable martingale (supermartingale, submartingale), M is called a *local martingale (local supermartingale, local submartingale)* on $[0, T]$.

It is easy to prove that uniformly integrable local martingales are martingales; nonnegative local martingales with initial value as an integrable random variables are supermartingales.

We present below three fundamental results in martingale theory. They are counterparts of the corresponding results in discrete-time case. For the first two results, the proof is immediate by taking a limit procedure.

Doob's optional stopping theorem Let (M_t) be a right-continuous martingale (resp., supermartingale) w.r.t. the filtration $(\mathcal{F}_t)$. If τ_1 and τ_2 are two bounded stopping times, then M_{τ_2} is integrable and

$$\mathbb{E}[M_{\tau_2} | \mathcal{F}_{\tau_1}] = M_{\tau_1 \wedge \tau_2} \text{ (resp. } \leqslant M_{\tau_1 \wedge \tau_2}), \quad \mathbb{P}-\text{a.s.}.$$

Doob's inequality Let $T > 0$ and $(X_t)_{0 \leqslant t \leqslant T}$ be a right-continuous nonnegative submartingale. Put $X_T^* = \sup_{0 \leqslant t \leqslant T} X_t$. Then for $\lambda > 0$ and $p > 1$, we have

$$\lambda \mathbb{P}(X_T^* \geqslant \lambda) \leqslant \mathbb{E}[X_T]$$

and

$$(\mathbb{E}[(X_T^*)^p])^{1/p} \leqslant \frac{p}{p-1} (\mathbb{E}[X_T^p])^{1/p} .$$

4.1.5.1 Regularity of Supermartingale Paths

The next theorem is a basic result on the regularity of continuous time supermartingale paths. We will use it in the study of Itô integrals. Its proof can be found in any book on stochastic analysis (e.g., He et al. (1992) or Yan et al. (1997)).

Theorem 4.1 *Let (X_t) be a supermartingale. If its almost all sample paths are right continuous, then its almost all sample paths have left limits on $(0, \infty)$. If $\mathbb{F} = (\mathcal{F}_t)$ is right continuous, then (X_t) has a right-continuous version if and only if $t \mapsto \mathbb{E}[X_t]$ is a right-continuous function on $\mathbb{R}_+$. In particular, if $\mathbb{F} = (\mathcal{F}_t)$ is right continuous, then any $\mathbb{F}$-martingale has a cadlag version.*

4.1.6 Finite Variation Processes

A process is called an *increasing process* if it is a cadlag process with nonnegative initial values and its almost all paths are increasing functions. A process is called a *finite variation (FV) process* or *process of finite variation*, if it is a difference of two increasing processes, namely, a cadlag whose almost all paths are of finite variation on each compact interval of $\mathbb{R}_+$.

Let $A = (A_t)_{t \geqslant 0}$ be a finite variation process. For each $\omega \in \Omega$, the finite variation function $A_\cdot(\omega)$ on $\mathbb{R}_+$ can be uniquely decomposed as $A_\cdot(\omega) = A_\cdot^c(\omega) + A_\cdot^d(\omega)$, where $A_\cdot^c(\omega)$ is continuous finite variation function and $A_\cdot^d(\omega)$ is a purely discontinuous finite variation function:

$$A_t^d(\omega) = \sum_{0 < s \leqslant t} \Delta A_s(\omega).$$

We call process A^c the *continuous part* of A and process A^d the *purely discontinuous part* (or *jump part*) of A.

Let A be a finite variation process. If $A^c = 0$, we call A *purely discontinuous*.

Let $A = (A_t)_{t \geqslant 0}$ be an adapted finite variation process. Then the variation process $B_t = |A_0| + \int_0^t |dA_s|$ is an adapted increasing process, and A can be represented as a difference of two adapted increasing processes.

The following result shows that a continuous local martingale with finite variation is equal to its initial value. This result is often used later.

Theorem 4.2 *If (X_t) is a continuous local martingale with finite variation, then $\forall t \geqslant 0$, $X_t = X_0$ a.s.*

Proof We may assume $X_0 = 0$. Let (V_t) be the variation process of X. Put $T_n = \inf\{t > 0 : |X_t| \geqslant n, \text{ or } V_t \geqslant n\}$, then $(X_t^{T_n})$ is a martingale and $|X_t^{T_n}| \leqslant n$, $V_\infty^{T_n} \leqslant n$. Thus, we can further assume that (X_t) is a bounded martingale and its total variation is bounded: $|X_t| \leqslant K$, $V_\infty \leqslant K$. Let $\Pi_n : 0 = t_0^n < t_1^n < \cdots < t_{k_n}^n = t$ be a sequence of finite partitions of $[0, t]$ such that when $n \to \infty$ the step sizes $\delta(\Pi_n)$ of the partitions tend to 0. Then we have

$$
\mathbb{E}[X_t^2] = \mathbb{E}\left[\sum_{i=1}^{k_n}\left(X_{t_i^n}^2 - X_{t_{i-1}^n}^2\right)\right] = \mathbb{E}\left[\sum_{i=1}^{k_n}(X_{t_i^n} - X_{t_{i-1}^n})^2\right]
$$

$$
\leqslant \mathbb{E}\left[V_t \max_{1\leqslant i\leqslant k_n}|X_{t_i^n} - X_{t_{i-1}^n}|\right] \leqslant K\mathbb{E}\left[\max_{1\leqslant i\leqslant k_n}|X_{t_i^n} - X_{t_{i-1}^n}|\right].
$$

By the dominated convergence theorem, the right-hand side of the above inequality tends to 0 when $n \to \infty$. Thus, $\forall t \geqslant 0$, $X_t = X_0$ a.s.   □

4.1.7 Doob-Meyer Decomposition of Local Submartingales

Henceforth, if a stochastic basis $(\Omega, \mathcal{F}, (\mathcal{F}_t), \mathbb{P})$ is given, we often omit the prefixes $(\mathcal{F}_t)$ in describing Brownian motion, martingales, or adapted processes.

A progressive measurable process X is of class (D) if the family of r.v.'s $\{X_\tau I_{\tau<\infty}, \ \tau \in \mathcal{T}\}$ is uniformly integrable, $\mathcal{T}$ being the set of all stopping times. It is of local class (D) if there is a sequence of stopping times $T_n \uparrow +\infty$ such that each $X^{T_n} - X_0$ is of class (D).

The *Doob-Meyer* decomposition of submartingales of class (D) is a cornerstone of modern martingale theory and stochastic analysis. It was first proved by Meyer (1962, 1963). Later Rao (1969) gave an elementary proof. To avoid using some advanced results from the general theory of stochastic processes, we present below an elementary proof given by Bass (1995, 1996) for the case of continuous submartingales. To this end, we first prove a lemma.

Lemma 4.3 *Let* $(A_n^{(1)}, 0\leqslant n\leqslant\infty)$ *and* $(A_n^{(2)}, 0\leqslant n\leqslant\infty)$ *be increasing predictable sequences of random variables w.r.t.* $(\mathcal{F}_n)$, $A_\infty^{(1)}$ *and* $A_\infty^{(2)}$ *be bounded random variables, and* $A_0^{(1)} = 0$, $A_0^{(2)} = 0$. *Put* $B_k = A_k^{(1)} - A_k^{(2)}$. *If there exist a constant* $C > 0$ *such that*

$$
\mathbb{E}[A_\infty^{(i)} - A_k^{(i)}|\mathcal{F}_k]\leqslant C, \quad i = 1, 2; \quad k = 0, 1, 2, \cdots,
$$

as well as a random variable $W\geqslant 0$ *with* $\mathbb{E}[W^2] < \infty$ *such that*

$$
\mathbb{E}[|B_\infty - B_k||\mathcal{F}_k]\leqslant\mathbb{E}[W|\mathcal{F}_k], \quad k = 0, 1, 2, \cdots,
$$

then we have

$$
\mathbb{E}[\sup_k B_k^2]\leqslant 8\mathbb{E}[W^2] + 32\sqrt{2}C(\mathbb{E}[W^2])^{1/2}. \tag{4.1}
$$

Proof First by assumption (letting $k = 0$), we have $\mathbb{E}[A_\infty^{(i)}]\leqslant C$. Let $a_k^{(i)} = A_{k+1}^{(i)} - A_k^{(i)}$, by the "summation by parts formula,"

$$A_\infty^{(i)2} = 2 \sum_{k=0}^\infty (A_\infty^{(i)} - A_k^{(i)}) a_k^{(i)} - \sum_{k=0}^\infty a_k^{(i)2},$$

we get (noting that $a_k^{(i)}$ is $\mathcal{F}_k$-measurable, and $a_k^{(i)} \geqslant 0$)

$$\mathbb{E}[A_\infty^{(i)2}] = 2\mathbb{E}\Big[\sum_{k=0}^\infty \mathbb{E}[A_\infty^{(i)} - A_k^{(i)}|\mathcal{F}_k] a_k^{(i)}\Big] - \mathbb{E}\Big[\sum_{k=0}^\infty a_k^{(i)2}\Big]$$

$$\leqslant 2C\mathbb{E}\Big[\sum_{k=0}^\infty a_k^{(i)}\Big] = 2C\mathbb{E}[A_\infty^{(i)}] \leqslant 2C^2.$$

Now let $b_k = B_{k+1} - B_k$. Once again by the "summation by parts formula," we obtain

$$\mathbb{E}[B_\infty^2] = 2\mathbb{E}\Big[\sum_{k=0}^\infty \mathbb{E}[B_\infty - B_k|\mathcal{F}_k] b_k\Big] - \mathbb{E}\Big[\sum_{k=0}^\infty b_k^2\Big]$$

$$\leqslant 2\mathbb{E}\Big[\sum_{k=0}^\infty \mathbb{E}[W|\mathcal{F}_k](a_k^{(1)} + a_k^{(2)})\Big]$$

$$\leqslant 2\mathbb{E}[W(A_\infty^{(1)} + A_\infty^{(2)})].$$

Since

$$\mathbb{E}[(A_\infty^{(1)} + A_\infty^{(2)})^2] \leqslant 2\mathbb{E}[A_\infty^{(1)2} + A_\infty^{(2)2}] \leqslant 8C^2,$$

by Schwarz inequality, we get $\mathbb{E}[B_\infty^2] \leqslant 4\sqrt{2}C(\mathbb{E}[W^2])^{1/2}$.

On the other hand, let $M_k = \mathbb{E}[B_\infty|\mathcal{F}_k]$, $X_k = M_k - B_k$. Then we have

$$|X_k| = |\mathbb{E}[B_\infty - B_k|\mathcal{F}_k]| \leqslant \mathbb{E}[W|\mathcal{F}_k] =: N_k.$$

Thus, by Doob's inequality we obtain

$$\mathbb{E}[\sup_k X_k^2] \leqslant \mathbb{E}[\sup_k N_k^2] \leqslant 4\mathbb{E}[N_\infty^2] = 4\mathbb{E}[W^2],$$

$$\mathbb{E}[\sup_k M_k^2] \leqslant 4\mathbb{E}[M_\infty^2] = 4\mathbb{E}[B_\infty^2] \leqslant 16\sqrt{2}C(\mathbb{E}[W^2])^{1/2}.$$

Due to $\sup_k B_k^2 \leqslant 2[\sup_k X_k^2 + \sup_k M_k^2]$, (4.1) is proved. $\qquad\square$

The next theorem is Doob-Meyer decomposition theorem for continuous local submartingales.

Theorem 4.4 *If X is a continuous local submartingale, then X can be uniquely decomposed as $X = M + A$, where M is a continuous local martingale and A is a continuous increasing process with initial value zero.*

Proof The uniqueness of the decomposition is an immediate consequence of Theorem 4.2. Now we prove the existence of the decomposition. We may assume that X is a submartingale and $X_0 = 0$. Let $T_n = \inf\{t > 0 : |X_t| > n\} \wedge n$. Then $|X_t^{T_n}| \leqslant n$ and X^{T_n} is a bounded submartingale. By the uniqueness of the decomposition, we can assume that X itself is a bounded submartingale. Furthermore, we only need to consider the following bounded submartingale: there is a real number $T > 0$, such that $X_t = X_T, \forall t \geqslant T$. In this case, almost all paths of the process X are uniformly continuous.

Put

$$W(\delta) = \sup_s \sup_{s \leqslant t \leqslant s+\delta} |X_t - X_s|.$$

Then when $\delta \to 0$, we have $W(\delta) \to 0$ a.s., so that $\lim_{\delta \to 0} \mathbb{E}[W(\delta)^2] = 0$.

Now let $t_j^n = j/2^n$, $\Delta^n = \{t_j^n, j = 0, 1, \cdots\}$, $A_0^n = 0$,

$$A_t^n = \sum_{j < 2^n t} \mathbb{E}[X_{t_{j+1}^n} - X_{t_j^n} | \mathcal{F}_{t_j^n}], \quad t > 0, \; n \in \mathbb{N}.$$

For $n \leqslant m \in \mathbb{N}$, $k \in \mathbb{N}$, there exists a unique $k(n, m) \in \mathbb{N}$ such that $k(n, m)/2^n < k/2^m \leqslant (k(n, m) + 1)/2^n$. By the definition of A_t^n and the smoothing property of the conditional expectation, we get

$$
\mathbb{E}[A_\infty^n - A_{k/2^m}^n | \mathcal{F}_{k/2^m}] = \mathbb{E}\Big[\mathbb{E}\big[\sum_{j \geqslant k(n,m)+1} \mathbb{E}[X_{t_{j+1}^n} - X_{t_j^n} | \mathcal{F}_{t_j^n}] | \mathcal{F}_{(k(n,m)+1)/2^n}\big] \big| \mathcal{F}_{k/2^m}\Big]
$$

$$
= \mathbb{E}\big[\mathbb{E}[X_\infty - X_{(k(n,m)+1)/2^n} | \mathcal{F}_{(k(n,m)+1)/2^n}] | \mathcal{F}_{k/2^m}\big]
$$

$$
= \mathbb{E}[X_\infty - X_{(k(n,m)+1)/2^n} | \mathcal{F}_{k/2^m}].
$$

On the other hand, by the Doob decomposition theorem for discrete-time submartingales,

$$\mathbb{E}[A_\infty^m - A_{k/2^m}^m | \mathcal{F}_{k/2^m}] = \mathbb{E}[X_\infty - X_{k/2^m} | \mathcal{F}_{k/2^m}], \quad \forall k = 0, 1, 2, \cdots.$$

Consequently, for $n \leqslant m \in \mathbb{N}$, $k \in \mathbb{N}$,

$$
|\mathbb{E}[A_\infty^m - A_{k/2^m}^m | \mathcal{F}_{k/2^m}] - \mathbb{E}[A_\infty^n - A_{k/2^m}^n | \mathcal{F}_{k/2^m}]|
$$

$$
\leqslant \mathbb{E}\big[|X_{(k(n,m)+1)/2^n} - X_{k/2^m}| \big| \mathcal{F}_{k/2^m}\big]
$$

$$
\leqslant \mathbb{E}[W(2^{-n}) | \mathcal{F}_{k/2^m}].
$$

Since $\sup_{t\geqslant 0}|A_t^m - A_t^n| = \sup_{t\in\Delta^m}|A_t^m - A_t^n|$, by Lemma 4.3, we get

$$\lim_{m,n\to\infty} \mathbb{E}[\sup_t |A_t^m - A_t^n|^2] = 0.$$

Thus, A_t^n converges to a limit in L^2, denoted by A_t. Obviously, process (A_t) is nonnegative, and its almost all paths are nondecreasing.

Now we prove that almost all paths of process (A_t) are continuous. Each process A_t^n has jumps only on Δ^n, and jump sizes ΔA_t^n are

$$\Delta A_t^n = \mathbb{E}[X_{k/2^n} - X_{(k-1)/2^n}|\mathcal{F}_{(k-1)/2^n}]\leqslant\mathbb{E}[W(2^{-n})|\mathcal{F}_{(k-1)/2^n}], \quad t = (k-1)/2^n.$$

Noting that $(\mathbb{E}[W(2^{-n})|\mathcal{F}_{(k-1)/2^n}], k = 1, 2, \cdots)$ is a martingale, by Doob's inequality,

$$\mathbb{E}\sup_t(\Delta A_t^n)^2\leqslant\mathbb{E}\Big[\sup_k(\mathbb{E}[W(2^{-n})|\mathcal{F}_{(k-1)/2^n}])^2\Big]\leqslant 4\mathbb{E}[W(2^{-n})^2].$$

Consequently, there is a subsequence (n_j) such that $\sup_t \Delta A_t^{n_j} \to 0$ a.s., as $j \to \infty$. Therefore, almost all paths of process (A_t) are continuous.

Finally, we prove that $(X_t - A_t)$ is a martingale. Since (X_t) and (A_t) are square-integrable continuous processes, we only need to prove that for $s, t \in \Delta_n, s < t, B \in \mathcal{F}_s$, we have

$$\mathbb{E}[(X_t - A_t)I_B] = \mathbb{E}[(X_s - A_s)I_B].$$

But since

$$\mathbb{E}[(X_t - A_t^m)I_B] = \mathbb{E}[(X_s - A_s^m)I_B], \quad \forall m\geqslant n,$$

letting $m \to \infty$ gives $\mathbb{E}[(X_t - A_t)I_B] = \mathbb{E}[(X_s - A_s)I_B]$. The theorem is proved. $\square$

4.1.8 Quadratic Variation Processes of Semimartingales

Let M be a continuous local martingale. Then M^2 is a continuous local submartingale. By Theorem 4.4 (Doob-Meyer decomposition theorem), there is a unique continuous increasing process with initial value zero, denoted as $\langle M \rangle$, such that $M^2 - \langle M \rangle$ is a continuous local martingale. We call $\langle M \rangle$ the *quadratic variation process* of M.

For two continuous local martingales M and N, we put

$$\langle M, N \rangle = \frac{1}{2}\Big(\langle M + N \rangle - \langle M \rangle - \langle N \rangle\Big).$$

Then $\langle M, N \rangle$ is a unique continuous finite variation process with initial value zero such that $MN - \langle M, N \rangle$ is a continuous local martingale. We call $\langle M, N \rangle$ the *quadratic covariation process* of M and N.

Theorem 4.5 *Let (B_t) be dimensional $(\mathcal{F}_t)$-standard Brownian motion. Then its quadratic variation process $\langle B \rangle_t = t$. Let (B_t) and (W_t) be two independent one-dimensional $(\mathcal{F}_t)$-standard Brownian motions. Then their quadratic covariation process $\langle B, W \rangle_t = 0$.*

Proof For $s < t$, $B_t - B_s$ is a normal random variable with mean 0 and variance $t - s$, and $\mathbb{E}[B_s B_t \mid \mathcal{F}_s] = B_s^2$. Thus, we have

$$\mathbb{E}[B_t^2 - B_s^2 \mid \mathcal{F}_s] = \mathbb{E}[(B_t - B_s)^2 \mid \mathcal{F}_s]$$
$$= \mathbb{E}[(B_t - B_s)^2] = (t - s).$$

This shows that $(B_t^2 - t)$ is a martingale, and hence $\langle B \rangle_t = t$. Similarly, we can prove that $(B_t W_t)$ is a martingale, and hence $\langle B, W \rangle_t = 0$. $\qquad\square$

Remark 1 Let $\Pi_n = \{t_0^n, \cdots, t_{k_n}^n\}$, $0 = t_0^n < t_1^n < \cdots < t_{k_n}^n = t$ be a sequence of partitions of $[0, t]$ such that the step sizes of the partitions $\delta(\Pi_n) = \max_i |t_i^n - t_{i-1}^n|$ tend to 0, when $n \to \infty$. Then when $n \to \infty$,

$$V_t(\Pi_n) = \sum_{i=1}^{k_n} (B_{t_i^n} - B_{t_{i-1}^n})^2$$

tends to t in L^2. In fact, since for $i \neq j$, $B_{t_i^n} - B_{t_{i-1}^n}$ and $B_{t_j^n} - B_{t_{j-1}^n}$ are mutually independent, we have

$$\mathbb{E}(V_t(\Pi_n) - t)^2 = \sum_{i=1}^{k_n} (t_i^n - t_{i-1}^n)^2 \mathbb{E}\left[\frac{(B_{t_i^n} - B_{t_{i-1}^n})^2}{t_i^n - t_{i-1}^n} - 1\right]^2 \leqslant t\delta(\Pi_n)\mathbb{E}[(Z^2 - 1)^2],$$

where Z is a standard normal random variable. This proves that $V_t(\Pi_n)$ converges to t in L^2.

Remark 2 If we further have $\sum_{n=1}^{\infty} \delta(\Pi_n) < \infty$, then the above convergence can be strengthened to a.s. convergence. In fact, for any $\varepsilon > 0$,

$$\sum_{n=1}^{\infty} \mathbb{P}[|V_t(\Pi_n) - t| > \varepsilon] \leqslant \frac{1}{\varepsilon^2} \sum_{n=1}^{\infty} \mathbb{E}(V_t(\Pi_n) - t)^2$$

$$\leqslant \frac{t\mathbb{E}[(Z^2 - 1)^2]}{\varepsilon^2} \sum_{n=1}^{\infty} \delta(\Pi_n) < \infty.$$

By Borel-Cantelli lemma, $\mathbb{P}[|V_t(\Pi_n) - t| > \varepsilon, \text{ i.o.}] = 0$. This proves that $[0, t]V_t(\Pi_n)$ a.s. converges to t. Here "i.o." is the shorthand for "infinitely often."

Assume $X_t = M_t + A_t$, where M is a continuous local martingale and A is a continuous finite variation process with initial value zero. Then we call X a continuous *semimartingale*. By Theorem 4.2, such a decomposition of a continuous semimartingale is unique, called the *canonical decomposition*. We call M the continuous local martingale part of X and $\langle M \rangle$ the *quadratic variation process* of X, which is denoted by $\langle X \rangle$.

For two continuous semimartingales X and Y with canonical decompositions $X_t = M_t + A_t$ and $Y_t = N_t + B_t$, we put $\langle X, Y \rangle = \langle M, N \rangle$, and call it the *quadratic covariation process* of X and Y. Clearly, for any stopping time T, we have $\langle X, Y \rangle^T = \langle X^T, Y \rangle = \langle X^T, Y^T \rangle$.

The next theorem gives a constructive representation of quadratic variation processes of continuous semimartingales. Its proof is from Kunita (1984) (see also Revuz and Yor (1999)).

Theorem 4.6 *Let $T > 0$ be a real number and X a continuous semimartingale on $[0, T]$. Let $\Pi_n = \{t_0^n, \cdots, t_{k_n}^n\}$, $0 = t_0^n < t_1^n < \cdots < t_{k_n}^n = T$ be a sequence of partitions of $[0, T]$ such that when $n \to \infty$ the step sizes of the partitions $\delta(\Pi_n) = \max_i |t_i^n - t_{i-1}^n|$ tend to 0. Then for any $0 < t \leqslant T$, when $n \to \infty$, the quadratic variation of X on $[0, t]$ w.r.t. Π_n*

$$V^{(n)}(X)_t = \sum_{i=1}^{k_n} (X_{t_i^n \wedge t} - X_{t_{i-1}^n \wedge t})^2$$

converges in probability to $\langle X \rangle_t$.

Proof First we assume that X is a bounded continuous martingale with initial value zero. For any $0 \leqslant s \leqslant u < v$, we have

$$\mathbb{E}[(X_v - X_u)^2 | \mathcal{F}_s] = \mathbb{E}[X_v^2 - X_u^2 | \mathcal{F}_s].$$

For $0 \leqslant s < t$, there is k such that $t_{k-1}^n < s \leqslant t_k^n$, and hence

$$\mathbb{E}[V^{(n)}(X)_t - V^{(n)}(X)_s | \mathcal{F}_s]$$

$$= \mathbb{E}\left[\sum_{i>k} (X_{t_i^n \wedge t} - X_{t_{i-1}^n \wedge t})^2 | \mathcal{F}_s \right] + \mathbb{E}[(X_{t_k^n \wedge t} - X_{t_{k-1}^n})^2 - (X_s - X_{t_{k-1}^n})^2 | \mathcal{F}_s]$$

$$= \mathbb{E}\left[\sum_{i>k} X_{t_i^n \wedge t}^2 - X_{t_{i-1}^n \wedge t}^2 | \mathcal{F}_s \right] + \mathbb{E}[(X_{t_k^n \wedge t}^2 - X_s^2) | \mathcal{F}_s]$$

$$= \mathbb{E}[X_t^2 - X_s^2 | \mathcal{F}_s].$$

This shows that $(X_t^2 - V^{(n)}(X)_t)$ is a continuous martingale with initial value zero.

For given n and m, $Y_t^{(n,m)} = V^{(n)}(X)_t - V^{(m)}(X)_t$ is a continuous martingale with initial value zero. Let $\Pi_{n,m} = \{s_0, \cdots, s_M : 0 = s_0 < s_1 < \cdots < s_M = T\}$ represent the partition of $[0, T]$ consisting of the partitioning points of Π_n and Π_m. The quadratic variations of X, $Y^{(n,m)}$, $V^{(n)}(X)$, $V^{(m)}(X)$ on $[0, t]$ w.r.t. $\Pi_{n,m}$ are denoted as $V^{(n,m)}(X)_t$, $V^{(n,m)}(Y)_t$, $V^{(n,m)}(V^{(n)})_t$, $V^{(n,m)}(V^{(m)})_t$, respectively. We first prove that sequence $V^{(n)}(X)_T$ converges in L^2. Assume that s_k^n is the rightmost point in Π_n satisfying $s_k^n \leqslant s_j < s_{j+1} \leqslant s_{k+1}^n$. Then

$$V^{(n)}(X)_{s_{j+1}} - V^{(n)}(X)_{s_j} = (X_{s_{j+1}} - X_{s_k^n})^2 - (X_{s_j} - X_{s_k^n})^2$$

$$= (X_{s_{j+1}} - X_{s_j})(X_{s_{j+1}} + X_{s_j} - 2X_{s_k^n}).$$

Thus, we have

$$V^{(n,m)}(V^{(n)})_T \leqslant \sup_j |X_{s_{j+1}} + X_{s_j} - 2X_{s_k^n}|^2 V^{(n,m)}(X)_T.$$

By Schwarz inequality, we get

$$\mathbb{E}[V^{(n,m)}(V^{(n)})_T] \leqslant \left(\mathbb{E}\left[\sup_j |X_{s_{j+1}} + X_{s_j} - 2X_{s_k^n}|^4 \right] \right)^{1/2} (\mathbb{E}[V^{(n,m)}(X)_T^2])^{1/2}.$$

Similarly, we have

$$\mathbb{E}[V^{(n,m)}(V^{(m)})_T] \leqslant \left(\mathbb{E}\left[\sup_j |X_{s_{j+1}} + X_{s_j} - 2X_{s_k^n}|^4 \right] \right)^{1/2} (\mathbb{E}[V^{(n,m)}(X)_T^2])^{1/2}.$$

Finally, we get

$$\mathbb{E}[(Y_T^{(n,m)})^2] = \mathbb{E}[V^{(n,m)}(Y)_T] \leqslant 2\mathbb{E}[V^{(n,m)}(V^{(n)})_T + V^{(n,m)}(V^{(m)})_T]$$

$$\leqslant 4\left(\mathbb{E}\left[\sup_j |X_{s_{j+1}} + X_{s_j} - 2X_{s_k^n}|^4 \right] \right)^{1/2} (\mathbb{E}[V^{(n,m)}(X)_T^2])^{1/2}.$$

By the continuity and boundedness of X, when $n, m \to \infty$, the first factor on the right side of the above formula tends to 0. Therefore, in order to prove that sequence $V^{(n)}(X)_T$ converges in L^2, it suffices to prove that $\mathbb{E}[V^{(n,m)}(X)_T^2]$ is uniformly bounded about (n, m). Let $|X_t| \leqslant C$, $t \in [0, T]$. We have $\mathbb{E}[V^{(n,m)}(X)_T] = \mathbb{E}[X_T^2] \leqslant C^2$. On the other hand, since

$$V^{(n,m)}(X)_T^2 = \left(\sum_j (X_{s_j} - X_{s_{j-1}})^2 \right)^2$$

$$= 2\sum_j (V^{(n,m)}(X)_T - V^{(n,m)}(X)_{s_j})(V^{(n,m)}(X)_{s_j} - V^{(n,m)}(X)_{s_{j-1}})$$

$$+ \sum_j (X_{s_j} - X_{s_{j-1}})^4,$$

by using equality $\mathbb{E}[V^{(n,m)}(X)_T - V^{(n,m)}(X)_{s_j}|\mathcal{F}_{s_j}] = \mathbb{E}[(X_T - X_{s_j})^2|\mathcal{F}_{s_j}]$, we obtain

$$\mathbb{E}[V^{(n,m)}(X)_T^2] = 2\sum_j \mathbb{E}[(X_T - X_{s_j})^2(V^{(n,m)}(X)_{s_j} - V^{(n,m)}(X)_{s_{j-1}})]$$

$$+ \sum_j \mathbb{E}[(X_{s_j} - X_{s_{j-1}})^4]$$

$$\leqslant 8C^2\mathbb{E}[V^{(n,m)}(X)_T] + 4C^2\mathbb{E}[V^{(n,m)}(X)_T] \leqslant 12C^4.$$

Thus, sequence $V^{(n)}(X)_T$ converges in L^2.

Now we prove that $V^{(n)}(X)_t$ is uniformly convergent in L^2 for $t \in [0, T]$. By Doob's inequality, we have

$$\mathbb{E}\Big[\sup_{s \leqslant T} |V^{(n)}(X)_t - V^{(m)}(X)_t|^2\Big] \leqslant 4\mathbb{E}[|V^{(n)}(X)_T - V^{(m)}(X)_T|^2].$$

Consequently, sequence $V^{(n)}(X)_t$ is uniformly convergent in L^2 for $t \in [0, T]$. Its limit is denoted by A_t. Obviously, for $s < t$, it holds $A_s \leqslant A_t$ a.s., and $(X_t^2 - A_t)$ is a martingale. Hence, process (A_t) has a cadlag version, still denoted by (A_t). In order to prove $A = \langle X \rangle$, we only need to prove that almost all paths of increasing process (A_t) are continuous. To this end, we chose a subsequence n_k, such that $\sum_{k=1}^{\infty} 2^k \mathbb{E}[|V^{(n_k)}(X)_T - A_T|] < \infty$. Since for each n, $(V^{(n)}(X)_t - A_t)$ is a martingale, by Doob's maximal inequality, we get

$$\sum_{n=1}^{\infty} \mathbb{P}\Big[\sup_t |V^{(n_k)}(X)_t - A_t| > 2^{-k}\Big] \leqslant \sum_{n=1}^{\infty} 2^k \mathbb{E}\Big[|V^{(n_k)}(X)_T - A_T|\Big] < \infty.$$

By Borel-Cantelli lemma, almost all paths of $V^{(n_k)}(X)$ uniformly converge to the paths of A on $[0, T]$, and hence almost all paths of A are continuous.

Let $X = M + V$ be a continuous semimartingale, where M is a continuous bounded martingale with initial value zero, V is a continuous finite variance process with initial value zero, and the total variation of V on $[0, T]$ (denoted by $\text{Var}_T(V)$) is uniformly bounded. Since

$$\Big|\sum_j (M_{s_j^{(n)} \wedge t} - M_{s_{j-1}^{(n)} \wedge t})(V_{s_j^{(n)} \wedge t} - V_{s_{j-1}^{(n)} \wedge t})\Big| \leqslant \sup_j |M_{s_j^{(n)} \wedge t} - M_{s_{j-1}^{(n)} \wedge t}| \text{Var}_t(V),$$

$$\Big|\sum_j (V_{s_j^{(n)} \wedge t} - V_{s_{j-1}^{(n)} \wedge t})^2\Big| \leqslant \sup_j |V_{s_j^{(n)} \wedge t} - V_{s_{j-1}^{(n)} \wedge t}| \text{Var}_t(V),$$

the difference of $V^{(n)}(X)_t$ and $V^{(n)}(M)_t$ converges to 0 in L^2, and thus $V^{(n)}(X)_t$ converges to $\langle X \rangle_t$ in L^2. Furthermore, we can take a sequence of stopping times $T_n \uparrow \infty$ such that M^{T_n} are continuous bounded martingales with initial value zero and V^{T_n} is a continuous finite variance process with initial value zero. Since $\lim_{n\to\infty} \mathbb{P}(T_n \wedge T = T) = 1$ and

$$V^{(n)}(X)_t^{T_n} - \langle X \rangle_t^{T_n} = V^{(n)}(X^{T_n})_t - \langle X^{T_n} \rangle_t,$$

it is easy to see that $V^{(n)}(X)_t$ converges in probability to $\langle X \rangle_t$. This reasoning is called "stopping time reasoning". The theorem is proved.   □

4.2 Stochastic Integrals w.t.t. Brownian Motion

Let $(\Omega, \mathcal{F}, (\mathcal{F}_t), \mathbb{P})$ be a stochastic basis; $(B_t)_{0 \leqslant t \leqslant T}$ be a $(\mathcal{F}_t)$-Brownian motion. In 1944, K. Itô first defined stochastic integrals $\int_0^t \theta(s)dB_s$ of a class of measurable adapted processes θ w.r.t. a Brownian motion. Here by measurability we mean the joint measurability w.r.t. product σ-algebra $\mathcal{B}([0,T]) \times \mathcal{F}_T$.

4.2.1 Wiener Integrals

We first consider the case where the integrand process is a deterministic function, starting with step functions on $[0, T]$. A step function f on $[0, T]$ has the following form: let $0 = t_0 < t_1 < \cdots < t_N = T$ be a finite partition of interval $[0, T]$,

$$f(t) = a_0, \ t \in [0, t_1]; \ f(t) = a_j, \ t \in (t_j, t_{j+1}], \ j = 1, \cdots, N-1,$$

where $\{a_j, 0 \leqslant j \leqslant N - 1\}$ are real numbers. Put

$$I(f) := \sum_{0 \leqslant j \leqslant N-1} a_j (B_{t_{j+1}} - B_{t_j}). \tag{4.2}$$

By the independent increment property of Brownian motion, we have

$$\sum_{0 \leqslant i, j \leqslant N-1} a_i a_j \mathbb{E}\big[(B_{t_{i+1}} - B_{t_i})(B_{t_{j+1}} - B_{t_j}) \big] = \sum_{0 \leqslant i \leqslant N-1} a_i^2 (t_{i+1} - t_i).$$

Thus, we have

$$\mathbb{E}\big[I(f)^2 \big] = \sum_{0 \leqslant j \leqslant N-1} a_j^2 (t_{j+1} - t_j) = \int_0^T f(s)^2 ds. \tag{4.3}$$

Since the set of all step functions on $[0, T]$ (denoted by H) constitutes a Hilbert space which is dense in $L^2[0, T]$, from (4.3) we know that the mapping $f \mapsto I(f)$ defined on H by (4.2) can be uniquely extended to a norm-preserving linear mapping from $L^2[0, T]$ to $L^2(\Omega, \mathcal{F}, \mathbb{P})$. We call $I(f)$ the *Wiener integral* of f w.r.t. Brownian motion (B_t) on $[0, T]$.

4.2.2 Itô Stochastic Integrals

4.2.2.1 Simple Measurable Adapted Process Case

Now we want to define integrals of the type $\int_0^t \theta(s) dB_s$ for a certain class of measurable adapted processes θ. We start with a simple integrand. By a *simple measurable adapted process* $(\theta(t))$ on $[0, T]$, we mean that there exists a partition of $[0, T]$: $0 = t_0 < t_1 <, \cdots, < t_N = T$, such that

$$\theta(t) = \xi_0, \quad t \in [0, t_1]; \quad \theta(t) = \xi_j, \quad t \in (t_j, t_{j+1}], \quad j = 1, \cdots, N-1,$$

where ξ_j is an $\mathcal{F}_{t_j}$-measurable square integrable random variable. Similarly to Wiener integrals, for such a process, it is natural to define its integral w.r.t. the Brownian motion on $[0, T]$ as follows:

$$I(\theta) := \sum_{0 \leqslant j \leqslant N-1} \xi_j (B_{t_{j+1}} - B_{t_j}). \tag{4.4}$$

Then we have

$$\mathbb{E}\Big[I(\theta)^2\Big] = \mathbb{E}\Big[\sum_{0 \leqslant j \leqslant N-1} \xi_j^2 (t_{j+1} - t_j)\Big] = \mathbb{E}\Big[\int_0^T \theta(s)^2 ds\Big]. \tag{4.5}$$

In fact,

$$\mathbb{E}\Big[\xi_i^2 (B_{t_{i+1}} - B_{t_i})^2\Big] = \mathbb{E}\Big[\xi_i^2 \mathbb{E}[(B_{t_{i+1}} - B_{t_i})^2 | \mathcal{F}_{t_i}]\Big] = \mathbb{E}\Big[\xi_i^2 (t_{i+1} - t_i)\Big].$$

For $i < j$, we have

$$\mathbb{E}\Big[\xi_i \xi_j (B_{t_{i+1}} - B_{t_i})(B_{t_{j+1}} - B_{t_j})\Big] = \mathbb{E}\Big[\xi_i \xi_j (B_{t_{i+1}} - B_{t_i})\mathbb{E}[(B_{t_{j+1}} - B_{t_j}) | \mathcal{F}_{t_j}]\Big] = 0.$$

Now we define indefinite integrals of simple measurable adapted processes, i.e., the upper bound t of integrals can vary in $[0, T]$. For any t with $t_k < t \leqslant t_{k+1}$, we put

$$I_t(\theta) := \int_0^t \theta(s) dB_s = \sum_{0 \leqslant j \leqslant k-1} \xi_j (B_{t_{j+1}} - B_{t_j}) + \xi_k (B_t - B_{t_k}).$$

Obviously, for any $t \in [0, T]$, the above expression can be rewritten as

$$I_t(\theta) = \sum_{0 \leqslant j \leqslant N} \xi_j (B_{t_{j+1} \wedge t} - B_{t_j \wedge t}). \tag{4.6}$$

It is a continuous adapted process. By (4.5) we have

$$\mathbb{E}\left[\left(\int_0^t \theta(s) dB_s\right)^2\right] = \mathbb{E}\left[\int_0^t \theta(s)^2 ds\right]. \tag{4.7}$$

In addition, since for any $0 \leqslant s < t$, $B_t - B_s$ is independent of $\mathcal{F}_s$, it is easy to prove that $(\int_0^t \theta(s) dB_s)$ is an $(\mathcal{F}_t)$-martingale.

We let $\mathcal{L}$ denote the set of all measurable adapted processes on $[0, T]$. Put

$$\mathcal{H}^2[0, T] = \left\{\theta : \theta \in \mathcal{L}, \ |\theta|_{\mathcal{H}}^2 := \mathbb{E}\left[\int_0^T \theta(s)^2 ds\right] < \infty\right\}.$$

Lemma 4.7 *Let $\theta \in \mathcal{H}^2[0, T]$. Then there is a sequence of simple measurable adapted processes (θ^n) such that*

$$\lim_{n \to \infty} \mathbb{E}\left[\int_0^T |\theta(s) - \theta^n(s)|^2 ds\right] = 0.$$

Proof We define the norm of θ by $|\theta|_{\mathcal{H}}$. Let $h_n(s) = (-n) \vee \theta(s) \wedge n$. Then it is easy to see that $\lim_{n \to \infty} |\theta - h_n|_{\mathcal{H}} = 0$. Thus, we may assume that θ is bounded. We take a sequence p_n of continuous probability density functions, supported in $[-1/n, 0]$, and put

$$g_n(t, \omega) = \int_0^t p_n(s - t)\theta(s, \omega) ds, \ t \in [0, T].$$

Then g_n is a bounded continuous adapted process and $\lim_{n \to \infty} |\theta - g_n|_{\mathcal{H}} = 0$. Thus, in order to prove the lemma, we can further assume that θ is a bounded continuous adapted process. In this case, we take arbitrarily a sequence of finite partitions of $[0, T]$: $\Pi_n = \{t_0^n, \cdots, t_{k_n}^n\}$, $0 = t_0^n < t_1^n < \cdots < t_{k_n}^n = T$, such that when $n \to \infty$ the step sizes of partitions $\delta(\Pi_n) = \max_i |t_i^n - t_{i-1}^n|$ tend to 0. Put $\theta_n(0, \omega) = \theta(0, \omega)$, and

$$\theta_n(t, \omega) = \sum_{j=0}^{k_n - 1} \theta(t_j^n, \omega) I_{(t_j^n, t_{j+1}^n]}(t), \ t > 0.$$

It is easily seen that we have $\lim_{n \to \infty} |\theta - \theta_n|_{\mathcal{H}} = 0$. The lemma is proved. $\square$

According to Lemma 4.7, we can use (4.7) to extend the definition of stochastic integrals to the integrands which are elements of $\mathcal{H}^2[0, T]$, and we have

$$\mathbb{E}\left[\left(\int_0^t \theta(s)dB_s\right)^2\right] = \mathbb{E}\left[\int_0^t \theta(s)^2 ds\right]. \tag{4.8}$$

The obtained indefinite integral $I_t(\theta) = \int_0^t \theta(s)dB_s$ is an $(\mathcal{F}_t)$-martingale. Since $(\mathcal{F}_t)$ is right continuous, then by Theorem 4.1, we can take a right continuous version of $(I_t(\theta))$.

Theorem 4.8 *If $\theta \in \mathcal{H}^2[0, T]$, then the right-continuous version of the indefinite integral $I_t(\theta)$ is a continuous martingale.*

Proof We take a sequence (θ^n) of simple measurable adapted processes such that

$$\mathbb{E}[(I_T(\theta^n) - I_T(\theta))^2] = \mathbb{E}\left[\int_0^T |\theta^n(s) - \theta(s)|^2 ds\right] \leqslant \frac{1}{n^2}, \quad n \geqslant 1. \tag{4.9}$$

Since $(|I_t(\theta^n) - I_t(\theta)|)$ is a nonnegative right-continuous submartingale, by Doob's inequality, we get

$$\mathbb{E}\left[\sum_{n=1}^{\infty} \sup_{0 \leqslant t \leqslant T} |I_t(\theta^n) - I_t(\theta)|^2\right] \leqslant 4 \sum_{n=1}^{\infty} \left(\mathbb{E}[|I_T(\theta^n) - I_T(\theta)|^2]\right) < \infty.$$

In particular, we have

$$\sum_{n=1}^{\infty} \sup_{0 \leqslant t \leqslant T} |I_t(\theta^n) - I_t(\theta)|^2 < \infty \quad \text{a.s.,}$$

which implies that almost all paths of the process $(I_t(\theta^n))$ uniformly converge to the process $(I_t(\theta))$ on $[0, T]$. Consequently, almost all paths of the process $(I_t(\theta))$ are continuous. $\qquad \square$

Remark We call continuous martingale $(I_t(\theta))$ the *stochastic integral* or *Itô integral* of θ w.r.t. Brownian motion (B_t), denoted also by $\int_0^t \theta(s)dB_s$ or $(\theta.B)_t$. Obviously, Itô integration is linear with respect to the integrand processes.

The following theorem shows that the Itô integral of a particular type of continuous process on a finite interval can be obtained as the limit of Riemann sum in L^2.

Theorem 4.9 *Let (B_t) be a one-dimensional $(\mathcal{F}_t)$-Brownian motion. Let*

$$\Pi_n = \{t_0^n, \cdots, t_{k_n}^n\}, 0 = t_0^n < t_1^n < \cdots < t_{k_n}^n = t,$$

be a sequence of partitions of $[0, t]$ such that when $n \to \infty$ the step sizes $\delta(\Pi_n)$ of partitions tend to 0. If θ is an adapted continuous process in $\mathcal{H}^2[0, t]$ satisfying $\mathbb{E}[\sup_{0 \leqslant s \leqslant t} |\theta(s)|^2] < \infty$, then when $n \to \infty$ the Riemann sums on the partitions (Π_n) $\sum_{i=1}^{k_n} \theta(t_{i-1}^n)(B_{t_i^n} - B_{t_{i-1}^n})$ tend to the Itô integral $\int_0^t \theta(s)dB_s$ in L^2.

Proof Put

$$\theta^n(s) = \sum_{i=1}^{k_n} \theta(t_{i-1}^n) I_{[t_{i-1} < s \leqslant t_i^n]}, \quad s \leqslant t.$$

It is easy to see

$$\lim_{n \to \infty} \mathbb{E}\left[\int_0^t |\theta(s) - \theta^n(s)|^2 ds \right] = 0,$$

and thus from (4.8) we deduce the conclusion of the theorem.   □

4.2.2.2 An Extension of Stochastic Integrals

Let us further generalize the definition of stochastic integrals. Let U and V be two stopping times with $U \leqslant V \leqslant T$. Put

$$]\!]U, V]\!] = \{(\omega, t) \in \Omega \times \mathbb{R}_+ : U(\omega) < t \leqslant V(\omega)\}.$$

We first prove that the Itô integral has the so-called locality: if $\theta \in \mathcal{H}^2[0, T]$ and S is a stopping time with $S \leqslant T$, then

$$(\theta.B)_{t \wedge S} = ((\theta I_{[0, S]\!]}).B)_t. \tag{4.10}$$

To this end, let U and V be two stopping times with $U \leqslant V \leqslant T$, ξ be an $\mathcal{F}_U$-measurable square integrable random variable, and $\theta = \xi I_{]\!]U, V]\!]}$. It is clear that $\theta \in \mathcal{H}^2[0, T]$. Now we first prove that for a θ of this form, we have

$$\int_0^t \theta(s)dB_s = \xi(B_{t \wedge V} - B_{t \wedge U}), \tag{4.10'}$$

which implies that (4.10') holds for such a θ. For each $n \geqslant 1$, there exists an integer k_n such that $k_n/2^n \leqslant T < (k_n + 1)/2^n$. Put

$$U_n = \sum_{k=1}^{k_n} \frac{k}{2^n} I_{[\frac{k-1}{2^n} \leqslant U < \frac{k}{2^n}]} + T I_{[k_n/2^n \leqslant U \leqslant T]},$$

$$V_n = \sum_{k=1}^{k_n} \frac{k}{2^n} I_{[\frac{k-1}{2^n} \leqslant V < \frac{k}{2^n}]} + T I_{[k_n/2^n \leqslant V \leqslant T]}.$$

Then U_n and V_n are stopping times, and $\theta^n := \xi I_{\rrbracket U_n, V_n \rrbracket}$ is a sequence of simple measurable adapted processes satisfying

$$\lim_{n \to \infty} \mathbb{E}\left[\int_0^T |\theta(s) - \theta^n(s)|^2 ds\right] = 0.$$

From (4.6) it is easily seen that

$$\int_0^t \theta^n(s) dB_s = \xi(B_{t \wedge V_n} - B_{t \wedge U_n}),$$

namely, (4.10') holds for θ^n. Thus, by (4.8) for $\theta = \xi I_{\rrbracket U, V \rrbracket}$ (4.10') also holds, so that (4.10) holds. Since for any simple measurable adapted process θ and any stopping time $S \leqslant T$, $\theta I_{\rrbracket 0, S \rrbracket}$ can be repressed as a linear combination of forms $\xi I_{\rrbracket U, V \rrbracket}$, (4.10) holds for simple measurable adapted process θ. Furthermore, by (4.8) we know that for any $\theta \in \mathcal{H}^2[0, T]$, (4.10) holds.

Now put

$$\mathcal{L}^2[0, T] = \left\{\theta : \theta \in \mathcal{L}, \int_0^T \theta(s)^2 ds < \infty\right\},$$

$$\mathcal{L}^1[0, T] = \left\{\theta : \theta \in \mathcal{L}, \int_0^T |\theta(s)| ds < \infty\right\}.$$

For $\theta \in \mathcal{L}^2[0, T]$, put

$$V_n = \inf\left\{t \in [0, T] : \int_0^t \theta(s)^2 ds \geqslant n\right\} \wedge T.$$

Then V_n are stopping times, $V_n \uparrow T$, and for $\omega \in \Omega$, there is $N(\omega)$ such that $\forall n \geqslant N(\omega)$, $V_n(\omega) = T$. Thus, by the "locality" of Itô integrals and by letting $(\theta.B)_{t \wedge V_n} = (\theta I_{\rrbracket 0, V_n \rrbracket}.B)_t$, we can uniquely define indefinite stochastic integral $((\theta.B)_t)$ of θ w.r.t. B on $[0, T]$. It is a continuous local martingale on $[0, T]$.

If $\theta \in \mathcal{L}^2[0, \infty)$, i.e., θ is a measurable adapted process satisfying $\int_0^t |\theta(s)|^2 ds < \infty$, $\forall t > 0$, put

$$V_n = \inf\left\{t \in [0, \infty] : \int_0^t \theta(s)^2 ds \geqslant n\right\} \wedge n,$$

then V_n is a stopping time and $V_n \uparrow +\infty$. By the "locality" of the Itô integrals, we can use a sequence of Itô integrals $(\theta I_{\rrbracket 0, V_n \rrbracket}).B$ to define uniquely stochastic integral $\theta.B$ of θ w.r.t. B on $[0, \infty)$. It is a local martingale on $[0, \infty)$.

Finally, let $B = (B^1, \cdots, B^d)^\tau$ be a d-dimensional $(\mathcal{F}_t)$-Brownian motion and $H = (H^1, \cdots, H^d)^\tau$ be a $\mathbb{R}^d$-valued process in $(\mathcal{L}^2[0, \infty))^d$. We can define the stochastic integral of H w.r.t. B (denoted by $H.B$) as follows:

$$\int_0^t H_s^\tau dB_s = \sum_{j=1}^d \int_0^t H_s^j dB_s^j.$$

The next theorem gives an expression of quadratic covariation processes of stochastic integrals.

Theorem 4.10 *Let* $\theta, \psi \in \mathcal{L}^2[0, \infty)$. *Then the quadratic covariation process of* $\theta.B$ *and* $\psi.B$ *is given by*

$$\langle \theta.B, \psi.B \rangle_t = \int_0^t \theta(s)\psi(s)ds. \tag{4.11}$$

Proof We may assume $\theta, \psi \in \mathcal{H}^2[0, T]$, $T > 0$. If θ and ψ are bounded simple measurable adapted processes, then we can prove that $(\theta.B)_t(\psi.B)_t - \int_0^t \theta(s)\psi(s)ds$ is a martingale. Thus, by the limit transition, it is easy to prove that for $\theta, \psi \in \mathcal{H}^2[0, T]$ $(\theta.B)_t(\psi.B)_t - \int_0^t \theta(s)\psi(s)ds$ is also a martingale, and thus (4.11) is proved.   □

4.3   Itô's Formula and Girsanov's Theorem

If a process has a form

$$X_t = X_0 + \int_0^t \theta^\tau(s)dB_s + \int_0^t \phi(s)ds,$$

where X_0 is $\mathcal{F}_0$-measurable, $\theta \in \mathcal{L}^2[0, \infty)^d$, $\phi \in \mathcal{L}^1[0, \infty)$, then we call it *Itô process*. By Theorem 4.2, a local continuous martingale with finite variation is identically equal to its initial value. Thus, the above decomposition (called *canonical decomposition*) of a Itô process is unique. In particular, if Itô process (X_t) is a continuous local martingale, then the finite variation term in the above decomposition vanishes. A simple example of Itô process is the so-called *generalized Brownian motion with drift*:

$$X_t = X_0 + \int_0^t b(s)ds + \int_0^t a(s)dB_s,$$

where X_0 is a constant, a and b are deterministic real functions, and (B_t) is a one-dimensional Brownian motion. It is easy to see that the transition density function of (X_t) is given by

$$p(s, x; t, y) = \left[2\pi \int_s^t a^2(u)du \right]^{-\frac{1}{2}} \exp\left\{ -\frac{\left(y - x - \int_s^t b(u)du \right)^2}{2\int_s^t a^2(u)du} \right\}.$$

4.3.1 Itô's Formula

Let (X_t) be an Itô process given by (4.11) and (H_t) be a measurable adapted process such that $H\theta \in (\mathcal{L}^2[0, T])^d$ and $H\phi \in \mathcal{L}^1[0, T]$. Then we can define the stochastic integral of H w.r.t. Itô process X as

$$\int_0^t H_s dX_s = \int_0^t H_s \theta(s) dB_s + \int_0^t H_s \phi(s) ds, \quad 0 \leqslant t \leqslant T.$$

The following theorem provides the change of variables formula for Itô processes, called *Itô's formula*, which is a powerful tool in Itô calculus.

Theorem 4.11 *Let* $B = (B^1, \cdots, B^d)^\tau$ *be a d-dimensional* $(\mathcal{F}_t)$-*Brownian motion and* $X = (X^1, \cdots, X^m)$ *be an* $\mathbb{R}^m$-*valued Itô processes with*

$$X_t^i = X_0^i + \sum_{j=1}^d \int_0^t \theta_j^i(s) dB_s^j + \int_0^t \phi^i(s) ds, \quad 1 \leqslant i \leqslant m. \tag{4.12}$$

If $F = F(x)$ *is a function on* $\mathbb{R}^m$ *and twice continuously differentiable w.r.t.* x, *then we have*

$$F(X_t) = F(X_0) + \sum_{i=1}^m \int_0^t \frac{\partial F}{\partial x_i}(X_s) dX_s^i$$

$$+ \frac{1}{2} \sum_{i,k=1}^m \int_0^t \frac{\partial^2 F}{\partial x_i x_k}(X_s) d\langle X^i, X^k \rangle_s, \tag{4.13}$$

where

$$\langle X^i, X^k \rangle_t = \sum_{j=1}^d \int_0^t \theta_j^i(s) \theta_j^k(s) ds$$

is the quadratic covariation processes of X^i *and* X^k *(see Theorem 4.10). Furthermore, if* $f = f(t, x)$ *is a function on* $\mathbb{R}_+ \times \mathbb{R}^m$ *such that it is twice differentiable w.r.t.* x *and once differentiable w.r.t.* t, *with continuous partial derivatives in* (t, x), *then*

$$f(t, X_t) = f(0, X_0) + \int_0^t \frac{\partial f}{\partial s}(s, X_s) ds + \sum_{i=1}^m \int_0^t \frac{\partial f}{\partial x_i}(s, X_s) dX_s^i$$

$$+ \frac{1}{2} \sum_{i,k=1}^m \int_0^t \frac{\partial^2 f}{\partial x_i x_k}(s, X_s) d\langle X^i, X^k \rangle_s. \tag{4.14}$$

We first prove a special case of Itô's formula, the so-called *integration by parts formula*.

Theorem 4.12 *Let* X *and* Y *be two Itô processes. Then*

$$X_t Y_t = X_0 Y_0 + \int_0^t X_s dY_s + \int_0^t Y_s dX_s + \langle X, Y \rangle_t. \tag{4.15}$$

Proof It suffices to prove (4.15) for the case $X = Y$, namely, prove

$$X_t^2 = X_0^2 + 2 \int_0^t X_s dX_s + \langle X \rangle_t. \tag{4.15'}$$

Let

$$\tau_n : 0 = t_0^n < t_1^n \cdots < t_{m(n)}^n = t$$

be a sequence of finite partitions of $[0, t]$ with $\delta(\tau_n) \to 0$, then

$$X_t^2 - X_0^2 = \sum_i (X_{t_{i+1}^n}^2 - X_{t_i^n}^2)$$

$$= 2 \sum_i X_{t_i^n} (X_{t_{i+1}^n} - X_{t_i^n}) + \sum_i (X_{t_{i+1}^n} - X_{t_i^n})^2$$

$$= 2((H^{(n)}.X)_t - X_0^2) + \sum_i (X_{t_{i+1}^n} - X_{t_i^n})^2 ,$$

where

$$H^{(n)} = X_0 I_{\llbracket 0 \rrbracket} + \sum_i X_{t_i^n} I_{\rrbracket t_i^n - t_{i+1}^n \rrbracket}.$$

By Theorem 4.6, when $n \to \infty$, $\sum_i (X_{t_{i+1}^n} - X_{t_i^n})^2$ converges in probability to $\langle X \rangle_t$. On the other hand, it is easy to prove that if process

$$X_t = X_0 + \int_0^t \theta(s) dB_s + \int_0^t \phi(s) ds$$

is uniformly bounded, and $\theta \in \mathcal{H}^2[0, t]$, $\phi \in \mathcal{L}^1[0, t]$, $\mathbb{E}\left[\int_0^t |\phi(s)|ds\right] < \infty$, then $(H^{(n)}.X)_t$ converges to $(X.X)_t$ in L^1. Thus, by using "stopping time reasoning" (see the proof of Theorem 4.6), we can further prove that in the general case, $H^{(n)}.X_t$ converges in probability to $(X.X)_t$, so we have (4.15'). The theorem is proved. □

Proof of Theorem 4.11 We may assume that $X^1, \cdots, X^d$ are all bounded Itô processes: $|X^j| \leqslant C(j = 1, \cdots, d)$, where C is a constant. Taking a sequence (F_n) of polynomials on $\mathbb{R}^d$ such that $F_n, D_j F_n, D_{ij} F_n$ uniformly converges to $F, D_j F, D_{ij} F, i, j = 1, \cdots, d$ on $[-C, C]^d$, respectively. If (4.13) holds for each F_n, then (4.13) holds also for F. Therefore, it suffices to prove (4.13) for a polynomial F on $\mathbb{R}^d$.

If $F(x^1, \cdots, x^d) = x^i x^j$, (4.13) becomes (4.15). By induction, we only need to prove that if (4.13) holds for a polynomial F, then (4.13) holds also for the following polynomial

$$G(x^1, \cdots, x^d) = x^i F(x^1, \cdots, x^d).$$

In fact, since

$$\langle X^i, F(X) \rangle_t = \sum_{k=1}^{m} \int_0^t \frac{\partial F}{\partial x_k}(X_s) d\langle X^i, X^k \rangle_s,$$

$$\frac{\partial G}{\partial x^k} = x^i \frac{\partial F}{\partial x^k}, \ \forall k \neq i; \quad \frac{\partial G}{\partial x^i} = F + x^i \frac{\partial F}{\partial x^i},$$

we have

$$G(X_t) - G(X_0) = X_t^i F(X_t) - X_0^i F(X_0)$$

$$= \int_0^t X_s^i dF(X_s) + \int_0^t F(X_s) dX_s^i + \langle X^i, F(X) \rangle_t$$

$$= \sum_{j=1}^d \int_0^t \frac{\partial G}{\partial x^j}(X_s) dX_s^j + \frac{1}{2} \sum_{j,k=1}^m \int_0^t \frac{\partial^2 G}{\partial x_j x_k}(X_s) d\langle X^j, X^k \rangle_s,$$

namely, (4.13) holds for G. The proof of (4.14) is left to the reader. ☐

Remark The simple proof given here is derived from Dellacherie and Meyer (1982); see also Revuz and Yor (1999).

4.3.2 Lévy's Martingale Characterization of Brownian Motion

Using Itô's formula we can prove the following *Lévy's martingale characterization* of Brownian motion.

Theorem 4.13 *Let (B_t) be an $\mathbb{R}^d$-valued $(\mathcal{F}_t)$-continuous adapted process with $B_0 = 0$. B is an $(\mathcal{F}_t)$-Brownian motion if and only if each (B_t^i) is a local martingale, and for all i, j, $(B_t^i B_t^j - \delta_{ij} t)$ is a local martingale.*

Proof The necessity is derived from Theorem 4.5. We are going to prove the sufficiency. Assume that for any $1 \leqslant i, j \leqslant d$, (B_t^i) and $(B_t^i B_t^j - \delta_{ij} t)$ are local martingales. Then $\langle B^i, B^j \rangle_t = \delta_{ij} t$. Let $u \in \mathbb{R}^d$. Put

$$Z_t = \exp \left\{ i \sum_{j=1}^d u_j B_t^j + \frac{|u|^2}{2} t \right\}.$$

By Itô's formula,

$$Z_t = 1 + \frac{|u|^2}{2} \int_0^t Z_s ds + i \sum_{j=1}^d u_j \int_0^t Z_s dB_s^j - \frac{1}{2} \sum_{j,k=1}^d u_j u_k \int_0^t Z_s d\langle B^j, B^k \rangle_s$$

$$= 1 + i \sum_{j=1}^d u_j \int_0^t Z_s dB_s^j.$$

Thus, (Z_t) is a complex valued local martingale. Since $|Z_t| \leqslant \exp\left\{\frac{|u|^2}{2}t\right\}$, (Z_t) is in fact a martingale. Therefore, $\forall s < t$,

$$\mathbb{E}\left[\exp\left\{i\sum_{j=1}^{d} u_j(B_t^j - B_s^j)\right\}\Big|\mathcal{F}_s\right] = \exp\left\{-\frac{|u|^2}{2}(t-s)\right\}.$$

This means that $B_t^j - B_s^j$ is independent of $\mathcal{F}_s$, and $B_t^j - B_s^j$ is normally distributed with mean 0 and variance $t - s$. On the other hand, for $i \neq j$, the fact that $B_t^i B_t^j$ is a martingale implies that $\mathbb{E}(B_t^i B_t^j) = 0$. Thus, B is an $(\mathcal{F}_t)$-Brownian motion. $\quad\square$

4.3.3 Reflection Principle of Brownian Motion

Let (B_t) be a one-dimensional Brownian motion and T be a stopping time. Put

$$\widetilde{B}_t = B_t I_{[t<T]} + (2B_T - B_t)I_{[t \geqslant T]} = 2B_{T \wedge t} - B_t.$$

By (4.10) we have

$$\widetilde{B}_t = \int_0^t H_s dB_s,$$

where $H = 2I_{[\![0,T]\!]} - 1$ is a predictable process and $|H| = 1$. By the integration by parts formula, we obtain

$$\widetilde{B}_t^2 - t = 2\int_0^t \widetilde{B}_s H_s dB_s.$$

Therefore, $(\widetilde{B}_t)$ and $(\widetilde{B}_t^2 - t)$ are locale martingales. Thus, by Lévy's martingale characterization of Brownian motion, $(\widetilde{B}_t)$ is a standard Brownian motion. This result is known as the *reflection principle* of Brownian motion.

In particular, let $x \in \mathbb{R}$, $T_x = \inf\{t \geqslant 0 : B_t = x\}$. Put

$$B_t^x = B_t I_{[t<T_x]} + (2x - B_t)I_{[t \geqslant T_x]},$$

then (B_t^x) is a standard Brownian motion.

4.3.4 Stochastic Exponentials and Novikov Theorem

The following theorem is a simple application of Itô's formula.

Theorem 4.14 *Let X be an Itô process with $X_0 = 0$. Put*

$$\mathcal{E}(X)_t = \exp\left\{X_t - \frac{1}{2}\langle X, X\rangle_t\right\}. \tag{4.16}$$

Then

$$\mathcal{E}(X)_t = 1 + \int_0^t \mathcal{E}(X)_s \, dX_s.$$

We call $\mathcal{E}(X)$ the stochastic exponential of X. Expression (4.16) of $\mathcal{E}(X)$ is called the Doléans (exponential) formula. In particular, if X is a local martingale, then $\mathcal{E}(X)$ is a local martingale and a strictly positive supermartingale.

Proof Applying Itô's formula to function e^x and Itô process $Y_t = X_t - X_0 - \frac{1}{2}\langle X, X\rangle_t$, we can verify that $\exp Y_t = 1 + \int_0^t \exp Y_s \, dX_s$.   □

The next theorem is due to Novikov (1972).

Theorem 4.15 (Novikov theorem) *Let $(B_t)_{0 \leqslant t \leqslant T}$ be a d-dimensional Brownian motion. If $\theta \in (\mathcal{L}^2[0, T])^d$ and satisfies the Novikov condition*

$$\mathbb{E}\left[\exp\left(\frac{1}{2}\int_0^T |\theta(s)|^2 ds\right)\right] < \infty, \tag{4.17}$$

then $\mathcal{E}(\theta.B)$ is a martingale (or equivalently, $\mathbb{E}[\mathcal{E}(\theta.B)_T] = 1$).

Proof The following simple proof was given in Yan (1980c). Let $0 < a < 1$. We have

$$\mathcal{E}(a\theta.B)_t = \exp\left\{a(\theta.B)_t - \frac{1}{2}a^2\int_0^t |\theta(s)|^2 ds\right\}$$

$$= \mathcal{E}(\theta.B)_t^a \exp\left\{\frac{a(1-a)}{2}\int_0^t |\theta(s)|^2 ds\right\}.$$

For any stopping time $\tau \leqslant T$ and any $A \in \mathcal{F}_T$, by Hölder inequality, we have

$$\mathbb{E}[I_A \mathcal{E}(a\theta.B)_\tau] \leqslant \left(\mathbb{E}[\mathcal{E}(\theta.B)_\tau]\right)^a \left(\mathbb{E}\left[I_A \exp\left\{\frac{a}{2}\int_0^\tau |\theta(s)|^2 ds\right\}\right]\right)^{1-a}$$

$$\leqslant \left(\mathbb{E}\left[I_A \exp\left\{\frac{1}{2}\int_0^T |\theta(s)|^2 ds\right\}\right]\right)^{1-a}.$$

This shows that $\mathcal{E}(a\theta.B)$ is a uniformly integrable martingale on $[0, T]$. In the first inequality above, we take $A = \Omega$ and $\tau = T$ and let a tend to 1; we get $\mathbb{E}[\mathcal{E}(\theta.B)_T] \geqslant 1$. Since $\mathcal{E}(\theta.B)$ is a nonnegative supermartingale, we must have $\mathbb{E}[\mathcal{E}(\theta.B)_T] = 1$, and $\mathcal{E}(\theta.B)$ is a martingale.   □

Remark

(1) Using the same proof as above, we get the following more general result: let M be a continuous local martingale with $M_0 = 0$. If $\mathbb{E}\left[\exp\left\{\frac{1}{2}\langle M\rangle_T\right\}\right] < \infty$, then $\mathcal{E}(M)_t, 0 \leqslant t \leqslant T$, is a martingale.

(2) From the proof of the theorem, we can see that the Novikov condition can be weakened to

$$\lim_{a \uparrow\uparrow 1} \left(\mathbb{E}\left[\exp\left\{\frac{a}{2}\int_0^T |\theta(s)|^2 ds\right\}\right]\right)^{1-a} = 1.$$

See (Yan (1980c).

(3) Let M be a continuous local martingale with $M_0 = 0$. Assume $\langle M\rangle_t \leqslant ct$. Then from the above result, we obtain the following so-called *exponential inequality* (see McKean (1969)): for any $a > 0$,

$$\mathbb{P}[\sup_{s \leqslant t} M_s \geqslant at] \leqslant \exp(-a^2 t/2c). \tag{4.18}$$

In fact, noting that for any $\alpha > 0$, it holds that $\mathbb{E}[\mathcal{E}(\alpha M)_t] = 1$, by Doob maximal inequality, we get

$$\mathbb{P}[\sup_{s \leqslant t} M_s \geqslant at] \leqslant \mathbb{P}\left[\sup_{s \leqslant t}\exp\left\{\alpha M_s - \frac{1}{2}\alpha^2\langle M\rangle_s\right\} \geqslant \exp\left(\alpha at - \frac{1}{2}\alpha^2\langle M\rangle_t\right)\right]$$

$$\leqslant \mathbb{P}\left[\sup_{s \leqslant t}\mathcal{E}(\alpha M)_s \geqslant \exp\left(\alpha at - \frac{1}{2}\alpha^2 ct\right)\right]$$

$$\leqslant \exp\left(-\alpha at + \frac{1}{2}\alpha^2 ct\right).$$

However, since $\inf_{\alpha>0}\left(-\alpha at + \frac{1}{2}\alpha^2 ct\right) = -\frac{a^2 t}{2c}$, we obtain (4.18).

4.3.5 Girsanov's Theorem

The following theorem describes the structure of a Brownian motion under an equivalent change of probability measure.

Theorem 4.16 (Girsanov's Theorem) *Let $(B_t)_{0 \leqslant t \leqslant T}$ be a d-dimensional $(\mathcal{F}_t)$-Brownian motion. If $\theta \in (\mathcal{L}^2[0, T])^d$ and $\mathbb{E}[\mathcal{E}(\theta.B)_T] = 1$, then $B_t^* = B_t -$*

$\int_0^t \theta(s)ds$ is a d-dimensional $(\mathcal{F}_t)$-Brownian motion under a new probability measure $\mathbb{P}^*$ with $\left.\frac{d\mathbb{P}^*}{d\mathbb{P}}\right|_{\mathcal{F}_T} = \mathcal{E}(\theta.B)_T$.

Proof We follow a proof given in Lamberton and Lapeyre (1996). We denote by L_t the martingale $\mathcal{E}(\theta.B)_t$. For each $1 \leqslant i \leqslant d$, we have

$$\langle B^{*i}, L \rangle_t = \langle B^i, L \rangle_t = \int_0^t L_s \theta^i(s)ds.$$

Using this fact and by the integration by parts formula,

$$B_t^{*i} L_t = \int_0^t L_s(dB_s^i - \theta^i(s)ds) + \int_0^t B_s^{*i}dL_s + \langle B^{*i}, L \rangle_t$$

$$= \int_0^t L_s dB_s^i + \int_0^t B_s^{*i}dL_s.$$

Hence, $(B_t^{*i} L_t)$ is a $\mathbb{P}$-local martingale. This means that B_t^{*i} is a local martingale under $\mathbb{P}^*$. Similarly, one can prove that $(B_t^{*i} B_t^{*j} - \delta_{ij}t)L_t$ is a local martingale under $\mathbb{P}$. This means that $B_t^{*i} B_t^{*j} - \delta_{ij}t$ is a local martingale under $\mathbb{P}^*$. Thus, by Theorem 4.13 (martingale characterization of Brownian motion), B^* is a Brownian motion under $\mathbb{P}^*$. $\square$

Remark As an application of Girsanov's theorem, we obtain the following formula:

$$\mathbb{E}[f(B^*.)g(B_T^*)] = \mathbb{E}[\mathcal{E}(-\theta.B)_T f(B.)g(B_T)],$$

where f is a Borel function on $C([0, T], \mathbb{R}^d)$ and g is a Borel function on $\mathbb{R}^d$. Here $C([0, T], \mathbb{R}^d)$ represents the set of all R^d-valued continuous functions on $[0, T]$. In fact, by Bayes' rule on conditional expectation (Theorem 4.11), we obtain

$$\mathbb{E}[f(B^*.)g(B_T^*)] = \mathbb{E}^*\left[\left(\left.\frac{d\mathbb{P}^*}{d\mathbb{P}}\right|_{\mathcal{F}_T}\right)^{-1} f(B^*.)g(B_T^*)\right]$$

$$= \mathbb{E}^*[\mathcal{E}(-\theta.B^*)_T f(B^*.)g(B_T^*)].$$

Since (B_t^*) is a Brownian motion under $\mathbb{P}^*$, we obtain the abovementioned formula.

4.4 Martingale Representation Theorem

The following is the *martingale representation theorem* for a Brownian motion. We present below a proof given by Revuz and Yor (1999).

Theorem 4.17 *Let (B_t) be a d-dimensional Brownian motion defined on $[0, T]$ and $(\mathcal{F}_t^B)$ be its natural filtration. Then (B_t) has the martingale representation property in the sense that for any local martingale w.r.t. $(\mathcal{F}_t^B)$, there exists a $\theta \in (\mathcal{L}^2[0, T])^d$ such that*

$$M_t = M_0 + \int_0^t \theta(s)dB_s, \quad 0 \leqslant t \leqslant T. \tag{4.19}$$

Moreover, such a θ is unique w.r.t. $\mathbb{P} \times \lambda$, where λ is the Lebesgue measure. In particular, any local $(\mathcal{F}_t^B)$-martingale is continuous.

Proof First of all we show that for any random variable $\xi \in L^2(\Omega, \mathcal{F}_T, \mathbb{P})$, there exists an $\mathbb{R}^d$-valued adapted process $f \in (\mathcal{H}^2[0, T])^d$ such that

$$\xi = \mathbb{E}[\xi] + \int_0^T f(s)dB_s, \quad \mathbb{E}\Big[(\xi - \mathbb{E}[\xi])^2\Big] = \sum_{i=1}^d \mathbb{E}\Big[\int_0^T f^i(s)^2 ds\Big].$$

It is easy to see that $\mathcal{G} = c + \{(f.B)_T : c \in \mathbb{R}, f \in (\mathcal{H}^2[0, T])^d\}$ is a closed subspace of Hilbert space $L^2(\Omega, \mathcal{F}_T, \mathbb{P})$. In order to show that $\mathcal{G} = L^2(\Omega, \mathcal{F}_T, \mathbb{P})$, it suffices to show that there exists a subspace $\mathcal{G}_0$ of $\mathcal{G}$ such that it is dense in $L^2(\Omega, \mathcal{F}_T, \mathbb{P})$. To this end, let $\mathcal{K}$ denote the set of all random variables of the form $\mathcal{E}(\theta.B)_T$ with θ being right-continuous $\mathbb{R}^d$-valued step functions, and let $\mathcal{G}_0$ denote the linear span of $\mathcal{K}$. Then by Theorem 4.14, $\mathcal{G}_0 \subset \mathcal{G}$. Now let $\mathcal{T}$ be a given partition of $[0, T]$: $0 = t_0 < t_1 < \cdots < t_n = T$, $\xi \in L^2(\Omega, \mathcal{F}_T^B, \mathbb{P})$, and is orthogonal to $\mathcal{K}$. We define

$$\varphi(z_1, \cdots, z_n) = \mathbb{E}\Big[\exp\Big\{\sum_{j=1}^n z_j^\tau(B_{t_j} - B_{t_{j-1}})\Big\}\xi\Big], \quad z_j \in \mathbb{C}^d, 1 \leqslant j \leqslant n.$$

Then φ must be identically equal to 0, because it is an analytic function and vanishes on $\mathbb{R}^d$. In particular,

$$\mathbb{E}\Big[\exp\Big\{i\sum_{j=1}^n \lambda_j^\tau(B_{t_j} - B_{t_{j-1}})\Big\}\xi\Big] = 0.$$

This shows that under mapping $\omega \to (B_{t_1}(\omega), \cdots, B_{t_j}(\omega) - B_{t_{j-1}}(\omega), \cdots)$, the image measure of $\xi \cdot \mathbb{P}$ is zero, because its Fourier transform is zero. Hence, $\xi = 0$ a.s. The above result means that $\mathcal{G}_0$ is dense in $L^2(\Omega, \mathcal{F}_T^B, \mathbb{P})$.

Let M be a square integrable $(\mathcal{F}_t^B)$-martingale on $[0, T]$. According to the above proved result, we have a representation (4.19) for M. In particular, any square integrable $(\mathcal{F}_t^B)$-martingale is continuous. Now let M be a uniformly integrable martingale. Since $L^2(\Omega, \mathcal{F}_T, \mathbb{P})$ is dense in $L^1(\Omega, \mathcal{F}_T, \mathbb{P})$, we can choose a sequence of square integrable $(\mathcal{F}_t^B)$-martingales $M^{(n)}$ such that

$$\sum_{n=1}^{\infty} 2^n \mathbb{E}\left[|M_T^{(n)} - M_T|\right] < \infty.$$

By Doob maximal inequality, we have

$$\sum_{n=1}^{\infty} \mathbb{P}\left[\sup_t |M_t^{(n)} - M_t| > 2^{-n}\right] \leqslant \sum_{n=1}^{\infty} 2^n \mathbb{E}\left[|M_T^{(n)} - M_T|\right] < \infty.$$

Thus, from Borel-Cantelli lemma, we know that almost all paths of $M^{(n)}$ uniformly converge to paths of M on $[0, T]$. Thus, almost all paths of M are continuous. Consequently, any local $(\mathcal{F}_t^B)$-martingale is continuous and is a local square integrable martingale. Therefore, any local $(\mathcal{F}_t^B)$-martingale has a representation (4.19). In fact, let M be a local $(\mathcal{F}_t^B)$-martingale on $[0, T]$. Then there is a sequence of stopping times $T_n \uparrow T$ with $\mathbb{P}(T_n = T) \to 1$ such that $M_t^{T_n} = M_0 + \int_0^t \theta^n(s)dB_s$. Let $\theta_t = \sum_n \theta_t^n I_{[T_{n-1} < t \leqslant T_n]}$, then we obtain (4.19). □

The following theorem is due to Fujisaki et al. (1972). We present below a constructive proof given by Lamberton and Lapeyre (1996).

Theorem 4.18 *Under the assumption of Theorem 4.16, if $(\mathcal{F}_t)$ is the natural filtration of (B_t), then (B_t^*) has the martingale representation property w.r.t. $(\mathcal{F}_t)$ under $\mathbb{P}^*$.*

Proof We denote by L_t the martingale $\mathcal{E}(\theta.B)_t$. Let M_t be a local martingale under $\mathbb{P}^*$ with $M_0 = 0$. Then by the Bayes' rule of conditional expectation (Theorem 1.11), we know that $M_t L_t$ is a local martingale under $\mathbb{P}$. Thus, by Theorem 4.17, there exists a measurable adapted process H_t such that

$$M_t L_t = \int_0^t H_s dB_s,$$

and thus,

$$M_t = L_t^{-1} \int_0^t H_s dB_s.$$

Since by Itô's formula,

$$L_t^{-1} = 1 + \int_0^t L_s^{-1}[-\theta(s)dB_s + \theta(s)^2 ds],$$

then using the integration by parts formula we obtain

$$M_t = \int_0^t (L_s^{-1} H_s - M_s \theta(s))dB_s^*.$$

The theorem is proved. □

4.5   Itô Stochastic Differential Equations

Let $(\Omega, \mathcal{F}, \mathbb{P}, (\mathcal{F}_t))$ be a stochastic basis. Let $(B_t)_{t \geqslant 0}$ be a d-dimensional $(\mathcal{F}_t)$-Brownian motion and $0 \leqslant t_0 \leqslant T$. Let $b : [t_0, T] \times \mathbb{R}^m \to \mathbb{R}^m$ and $\sigma : [t_0, T] \times \mathbb{R}^m \to M^{m,d}$ be Borel measurable maps, where $M^{m,d}$ is the set of all $m \times d$-matrices. An $\mathbb{R}^m$-valued continuous $(\mathcal{F}_t)$-adapted process X is called a solution to the following *Itô stochastic differential equation* (SDE)

$$dX_t = b(t, X_t)dt + \sigma(t, X_t)dB_t, \quad t \in [t_0, T], \quad X_{t_0} = \xi, \tag{4.20}$$

where $\xi = (\xi^1, \cdots, \xi^m)$ is $\mathcal{F}_{t_0}$-measurable, if X satisfies the stochastic integral equation

$$X_t^i = \xi^i + \int_{t_0}^t b^i(s, X_s)ds + \sum_{j=1}^d \int_{t_0}^t \sigma_j^i(s, X_s)dB_s^j, \quad 1 \leqslant i \leqslant m, \ t \in [t_0, T]. \tag{4.21}$$

Such a solution to (4.21) is called a *strong solution*, meaning that it is based on the paths of underlying Brownian motion (B_t). In particular, let $(\mathcal{F}_t^B)$ be the natural filtration of (B_t); if ξ is $\mathcal{F}_{t_0}^B$-measurable, a strong solution is adapted to the natural filtration of the Brownian motion (B_t).

There is another notion of solutions for SDE. We say that SDE (4.20) has a *weak solution* (X, B) with initial distribution μ, if there exists a Brownian motion (B_t) on a suitable stochastic basis and an adapted process (X_t) such that X_{t_0} has distribution μ and (4.20) holds. In financial mathematics we have no need for weak solutions.

If whenever (X, B) and (X', B') are two solutions of SDE (4.20) defined on the same stochastic basis with $B = B'$ and $X_{t_0} = X'_{t_0}$ a.s., X and X' are the same (i.e., indistinguishable), then we say that SDE (4.20) has *pathwise uniqueness*. If whenever (X, B) and (X', B') are two solutions of SDE (4.20) defined on the same stochastic basis or different ones with the same initial distribution, X and X' have same finite dimensional distributions, then we say SDE (4.20) has *uniqueness in law*. One can show that if SDE (4.20) has pathwise uniqueness, then it has also uniqueness in law and every solution to SDE (4.20) is a strong one (see Revuz and Yor (1999), Chap. IX, Theorem 1.7).

4.5.1   *Existence and Uniqueness of Solutions*

In the sequel, for $x \in \mathbb{R}^m$ and $\gamma \in M^{m,d}$, we define the following norms:

$$|x|^2 = \sum_{i=1}^m x_i^2, \quad |\gamma|^2 = \text{tr}(\gamma \gamma^T) = \sum_{j=1}^d \sum_{i=1}^m (\gamma^{ij})^2.$$

For notational simplicity, we let $t_0 = 0$ in (4.20).

The next theorem is a basic result about the existence and uniqueness of solutions to SDE (4.20). Its proof can be found in many books on stochastic analysis.

Theorem 4.19 *If b and σ satisfy the Lipschitz condition in x:*

$$|b(t, x) - b(t, y)| + |\sigma(t, x) - \sigma(t, y)| \leqslant K|x - y|, \tag{4.22}$$

and b satisfies the linear growth condition in x:

$$|b(t, x)| + |\sigma(t, x)| \leqslant K(1 + |x|), \tag{4.23}$$

where K is a constant, then (4.20) has a unique solution X. Moreover, if on $[0, T]$ b and σ satisfies the polynomial growth condition in x:

$$\sup_{0 \leqslant t \leqslant T} |b(t, x)| + |\sigma(t, x)| \leqslant C(1 + |x|^{2\mu}), \quad x \in \mathbb{R}^m, \tag{4.23'}$$

for some constant $C > 0$, $\mu \geqslant 1$ and $\mathbb{E}[|\xi|^{2\mu}] < \infty$, then we have

$$\mathbb{E}[\sup_{0 \leqslant t \leqslant T} |X_t|^{2\mu}] \leqslant K_1 + K_2 \mathbb{E}[|\xi|^{2\mu}].$$

Remark If b and σ satisfy only the *local Lipschitz condition* in the sense that for each positive constant L there is a constant K such that (4.22) is satisfied for x and y with $|x| \leqslant L$, $|y| \leqslant L$, then (4.20) still has a unique solution.

4.5.1.1 Linear SDE

If in (4.20) b and σ are linear functions in x:

$$b(t, x) = G(t)x + g(t); \quad \sigma(t, x) = \big(H_1(t)x + h_1(t), \cdots, H_d(t)x + h_d(t)\big),$$

where $G(t)$ and $H_i(t)$ are $m \times m$ matrix-valued measurable functions and $g(t)$ and $h_i(t)$ are $\mathbb{R}^m$-valued measurable functions, we call (4.20) a linear SDE.

The following theorem gives an expression for the solution of a linear SDE.

Theorem 4.20 *Assume that G, g, H_i, h_i are locally bounded measurable functions. Then linear SDE (4.20) has the unique solution:*

$$X_t = \Phi_t \left(\xi + \int_{t_0}^{t} \Phi_s^{-1} dY_s \right),$$

where

$$dY_t = \left(g(t) - \sum_{i=1}^{d} H_i(t)h_i(t) \right) dt + \sum_{i=1}^{d} h_i(t) dB_t^i,$$

Φ_t *is the solution to the homogeneous SDE*

$$d\Phi_t = G(t)\Phi(t)dt + \sum_{i=1}^{d} H_i(t)\Phi_t dB_t^i$$

with initial value $\Phi_0 = I$. *In particular, if c is a constant or a normal r.v., the solution to a linear SDE is a Gaussian process.*

Remark If $G(t) = G$ and $H_i(t) = H_i$, $1 \leqslant i \leqslant d$ do not depend on t and they are commutable in the sense:

$$GH_i = H_i G, \ H_i H_j = H_j H_i, \ \forall i, j,$$

then

$$\Phi_t = \exp\left\{\left(G - \sum_{i=0}^{d} H_i^2/2\right)t + \sum_{i=1}^{d} H_i B_t^i\right\}.$$

For a one-dimensional SDE (i.e., $m = d = 1$), the following result due to Yamada and Watanabe (1971) relaxes considerably the conditions in Theorem 4.19.

Theorem 4.21 *Assume $m = d = 1$. In order for (4.20) to have a unique solution, it suffices that b is continuous and Lipschitz in x and σ is continuous with the property*

$$|\sigma(t, x) - \sigma(t, y)| \leqslant \rho(|x - y|)$$

for all x and y and t, where $\rho : \mathbb{R}_+ \to \mathbb{R}_+$ is a strictly increasing function with $\rho(0) = 0$ and for any $\epsilon > 0$,

$$\int_{(0,\epsilon)} \rho^{-2}(x)dx = \infty.$$

4.5.2 Examples

In this subsection we present several useful processes which are solutions of SDEs.

4.5.2.1 Ornstein-Uhlenbeck Process

Consider the following SDE:

$$dX_t = -cX_t dt + \sigma dB_t, \ X_0 = \xi. \tag{4.24}$$

Its unique solution is given by

$$X_t = e^{-ct}\left(\xi + \sigma \int_0^t e^{cs} dB_s\right).$$

We call it the *Ornstein-Uhlenbeck process*. This SDE is called the *Langevin equation*, because it was originally introduced by Langevin (1908) to model the velocity of a physical Brownian particle. If ξ is a constant or a normal r.v., then (X_t) is a Gaussian process. If $\xi \sim N(0, \delta^2)$, and ξ is independent of (B_t), then for $s \leqslant t$ we have

$$E[X_t] = 0, \quad \text{cov}(X_s, X_t) = e^{-ct-cs}\left[\delta^2 + \frac{\sigma^2}{2c}(e^{2cs} - 1)\right],$$

$$\rho(X_t, X_s) = \sqrt{\frac{\delta^2 + \frac{\sigma^2}{2c}(e^{2cs} - 1)}{\delta^2 + \frac{\sigma^2}{2c}(e^{2ct} - 1)}}.$$

In particular, if $\sigma^2 = 2c\delta^2$, then $\rho(X_t, X_s) = e^{-c(t-s)}$, $X_t \sim N(0, \delta^2)$. In this case, X is a stationary Gaussian process and a homogenous Markov process. The transition density of X is given by

$$p(t; x, y) = \frac{1}{\sqrt{2\pi}\sigma\sqrt{\frac{1-e^{-2ct}}{2c}}} e^{-\frac{c(y-e^{-ct}x)^2}{\sigma^2(1-e^{-2c})}}.$$

4.5.2.2 Generalized Ornstein-Uhlenbeck Process

More generally, we consider the following SDE:

$$dX_t = -b(t)X_t dt + \sigma(t)dB_t, \quad X_0 = \xi, \tag{4.24'}$$

where $b(t)$ and $\sigma(t)$ are deterministic functions. The unique solution to the equation is given by

$$X_t = e^{-l(t)}\left(\xi + \int_0^t e^{l(s)}\sigma(s)dB_s\right),$$

where $l(t) = \int_0^t b(u)du$. It is called a *generalized Ornstein-Uhlenbeck process*. If ξ is a constant or a normal random variable, then (X_t) is a Gaussian process. If $\xi \sim N(0, \delta^2)$ and ξ is independent of (B_t), then for $s \leqslant t$ we have

$$E[X_t] = 0, \quad \text{cov}(X_s, X_t) = e^{-l(t)-l(s)}\left[\delta^2 + \int_0^s e^{2l(u)}\sigma^2(u)du\right],$$

$$\rho(X_t, X_s) = \sqrt{\frac{\delta^2 + \int_0^s e^{2l(u)}\sigma^2(u)du}{\delta^2 + \int_0^t e^{2l(u)}\sigma^2(u)du}}.$$

It is a Markov process, and its transition density is given by

$$p(s, x; t, y) = \frac{1}{\sqrt{2\pi}e^{-l(t)}\sqrt{\int_s^t e^{2l(u)}\sigma^2(u)du}} \exp\left\{-\frac{(y - e^{-(l(t)-l(s))}x)^2}{e^{-2l(t)}\int_s^t e^{2l(u)}\sigma^2(u)du}\right\}.$$

(4.25)

4.5.2.3 Bessel Process

Let $W(t) = (W_1(t), \cdots, W_d(t))$ be a d-dimensional Brownian motion. Let $X_t = \sum_{i=1}^d W_i(t)^2$. By Itô's formula,

$$dX_t = 2\sum_{i=1}^d W_i(t)dW_i(t) + d\,dt.$$

Put

$$B_t = \sum_{i=1}^d \int_0^t \sqrt{X_s^{-1}}W_i(s)dW_i(s).$$

Then B_t is a local martingale, and

$$\langle B, B \rangle_t = \sum_{i=1}^d \int_0^t X_s^{-1}W_i(s)^2 ds = t.$$

Thus by Lévy's martingale characterization of Brownian motion, B_t is a Brownian motion, and X_t satisfies the following SDE:

$$dX_t = 2(X_t)^{\frac{1}{2}}dB_t + d\,dt.$$

Now let $\delta > 0$ be a real number. Consider the following SDE:

$$dX_t = 2(X_t^+)^{\frac{1}{2}}dB_t + \delta dt,$$

(4.26)

with $X_0 = x > 0$, where $y^+ = \max\{y, 0\}$. It can be proved that (4.26) has a unique nonnegative strong solution (see Revuz and Yor 1994, 420–432). The solution (X_t) of Eq. (4.26) is called the *square of δ-dimensional Bessel process* started at x. If $\delta \geqslant 2$, the point is a polar, i.e., there is no $t > 0$ for which $X(t) = 0$, for almost all paths. If $0 < \delta < 1$, point 0 is an absorbing barrier of process R_t. If $1 \leqslant \delta < 2$, point 0 is an instantaneously reflected point.

Let $R_t = \sqrt{X_t}$. We call R_t a *Bessel process of dimension δ*. It is a diffusion process on $[0, \infty)$. Its transition density is given by

$$p^{(\nu)}(t; x, y) = \frac{y}{t}\left(\frac{y}{x}\right)^\nu \exp\left\{-\frac{x^2 + y^2}{2t}\right\} I_\nu\left(\frac{xy}{t}\right), \quad x, y > 0, \tag{4.27}$$

where $\nu = \frac{\delta}{2} - 1$, $I_\nu(x) = \left(\frac{x}{2}\right)^\nu \sum_{n=0}^\infty \frac{1}{n!\Gamma(\nu+n+1)}\left(\frac{x}{2}\right)^{2n}$ is the modified Bessel function of the first kind of index ν. If $\delta \geqslant 2$, then R_t never hits point 0. So by Itô's formula,

$$dR_t = \frac{1}{2}X_t^{-\frac{1}{2}}\left[2X_t^{\frac{1}{2}}dB_t + \delta dt\right] - \frac{1}{2}X_t^{-\frac{3}{2}}X_t dt$$

$$= dB_t + \frac{\delta - 1}{2}\frac{1}{R_t}dt.$$

But for $\delta < 2$, the situation is not simple. For example, if $\delta = 1$, it will involve the local time of the Brownian motion.

4.5.2.4 Scale and Time Changed Bessel Process

Consider the following SDE:

$$dX_t = \sigma dB_t + \frac{\varepsilon}{X_t}dt,$$

where σ and ε are constants. If we let $R_t = X_t/\sigma$, then

$$dR_t = dB_t + \frac{\varepsilon}{\sigma^2}\frac{1}{R_t}dt.$$

Thus, if $\varepsilon/\sigma^2 \geqslant 1/2$, then R_t is a δ-dimensional Bessel process, where $\delta = 2\varepsilon/\sigma^2 + 1$. In the sequel we assume that $\sigma \neq 0$, and $\varepsilon/\sigma^2 \geqslant 1/2$. Now we consider another SDE:

$$dY_t = \sigma dB_t + \frac{\varepsilon}{Y_t}dt - \eta Y_t dt, \tag{4.28}$$

where σ, ε, and η are constants. We shall show that this SDE has a weak solution Y_t, and it is a scale and time-changed Bessel process. In fact, let $Y_t = f(t)X_{\tau(t)}$, where f and τ are deterministic differential functions and τ is increasing with $\tau(0) = 0$. By Itô's formula,

$$dY_t = X_{\tau(t)} f'(t) dt + f(t) dX_{\tau(t)} = \frac{f'(t)}{f(t)} Y_t dt + f(t) [\sigma d B_{\tau(t)} + \frac{\varepsilon}{X_{\tau(t)}} \tau'(t) dt]$$

$$= \frac{f'(t)}{f(t)} Y_t dt + f(t)\sigma d B_{\tau(t)} + \frac{\varepsilon f(t)^2 \tau'(t)}{Y_t} dt.$$

In order that Y_t is a solution to (4.28), let

$$f(t) = e^{-\eta t}, \quad \tau(t) = \int_0^t f(s)^{-2} ds = \frac{e^{2\eta t} - 1}{2\eta}.$$

Put

$$W_t = \int_0^t \frac{1}{\sqrt{\tau'(s)}} d B_{\tau(s)}.$$

Then W_t is a local martingale w.r.t. $(\mathcal{F}_{\tau(t)})$, and $\langle W, W \rangle_t = t$. So W_t is a Brownian motion w.r.t.$(\mathcal{F}_{\tau(t)})$. Since $f(t)\sqrt{\tau'(t)} = 1$, we have

$$dY_t = \sigma d W_t + \frac{\varepsilon}{Y_t} dt - \eta Y_t dt.$$

Thus, Y_t is a weak solution to SDE (4.28).

4.5.2.5 Brownian Bridge (Process)

Let $Y_t = B_t - t B_1, t \in [0, 1]$. Then Y is a Gaussian process and $E[X_s X_t] = s/(1 - t)$, if $s \leqslant t$. It is called the *Brownian bridge (process)*, because it starts from 0 at time 0 and return to 0 at time 1, and between time 0 and 1 it behaves as a Brownian motion. Now we show how this process can be represented as a solution of a SDE. Consider the following SDE on $[0, 1)$:

$$dX_t = -\frac{X_t}{1 - t} dt + d B_t, \quad X_0 = 0. \tag{4.29}$$

Its unique solution is

$$X_t = (1 - t) \int_0^t \frac{1}{1 - s} d B_s, \quad t \in [0, 1). \tag{4.30}$$

Process X is a Gaussian process. It is easy to verify that for $0 \leqslant s \leqslant t < 1$ we have $E[X_s X_t] = s/(1 - t)$. Therefore, X has the same distributions as that of the Brownian bridge process on $[0, 1)$. In particular, we have $\lim_{t \to 1} X_t = 0$.

4.6 Itô Diffusion Processes

The strong solution to an Itô SDE (4.20) is called an *Itô diffusion process*, or *(Itô) diffusion*. We call $b(t, x)$ the *drift coefficient* and $\sigma(t, x)$ the *diffusion coefficient* of X. If b and σ only depend on x, we call X a *time-homogeneous diffusion*. In this case, if we denote by $X_t^{s,x}$ the solution X_t of (4.20) on $[s, \infty)$ with $X_s = x$, then two processes $(X_{s+h}^{s,x})$ and $(X_h^{0,x})$, $h \geq 0$, have the same finite dimensional distributions.

Note that by adding the time parameter to state variables, any time-inhomogeneous diffusion can be translated to a time-homogeneous one. In fact, for a time-inhomogeneous diffusion X and a given $s \geq 0$, if we let

$$Y_t = (s + t, X_t), \quad \widehat{b}(t, x) = (1, b(t, x)), \quad \widehat{\sigma}(t, x) = (0, \sigma(t, x)),$$

then we have

$$dY_t = \widehat{b}(Y_t)dt + \widehat{\sigma}(Y_t)dB_t. \tag{4.31}$$

So Y is a time-homogeneous diffusion.

Theorem 4.22 *Assume that X_t is the strong solution to Itô Eq. (4.20) and coefficients $b(t, x)$ and $\sigma(t, x)$ satisfy the conditions of Theorem 4.19. Then X_t is a diffusion process w.r.t. $\mathcal{F}_t^B$, where $\mathcal{F}_t^B$ is the natural filtration of Brownian motion B. Moreover, the transition probabilities of X are given by*

$$P(s, x, t, A) = P(X_t \in A | X_s = x) = P(X_t^{s,x} \in A), \tag{4.32}$$

and the generator of (X_t) is

$$\mathcal{A}_t = \frac{1}{2} \sum_{i,k=1}^{d} a_{ik}(t, x) \frac{\partial^2}{\partial x_i \partial x_k} + \sum_{i=1}^{d} b_i(t, x) \frac{\partial}{\partial x_i},$$

where $a(t, x) = \sigma(t, x)^\tau \sigma(t, x)$, i.e.,

$$a_{ik}(t, x) = \sum_{l=1}^{m} \sigma_i^l(t, x) \sigma_k^l(t, x).$$

Proof We denote by $\mathcal{G}_t$ the σ-algebra generated by $B_u - B_t$ for $u \geq t$. Let $A \in \mathcal{B}(\mathbb{R}^d)$. For a $s \geq t$, we put

$$g(x, \omega) = I_A(X_t^{s,x}(\omega)).$$

Then for any x, $g(x, \cdot)$ is $\mathcal{G}_t$-measurable. Consequently, $g(x, \cdot)$ is independent of $\mathcal{F}_s$. However, by the pathwise uniqueness,

$$X_t(\omega) = X_t^{t_0, \xi}(\omega) = X_t^{s, X_s(\omega)}(\omega),$$

so we have $g(X_s, \omega) = I_A(X_t(\omega))$. Finally, by Bayes' rule of conditional expectation (Theorem 1.11), we get

$$\mathbb{P}(X_t \in A | \mathcal{F}_s) = \mathbb{E}[g(X_s, \cdot) | \mathcal{F}_s] = \mathbb{E}[g(x, \cdot)]|_{x=X_s} = \mathbb{P}(X_t^{s,x} \in A)|_{x=X_s}.$$

This means $\mathbb{P}(X_t \in A | \mathcal{F}_s) = \mathbb{P}(X_t \in A | X_s)$. Thus, (4.32) holds. □

4.7 Feynman-Kac Formula

The following theorem provides a generalized version of the *Feynman-Kac formula*, which provides a probabilistic representation for the solution to the following *Cauchy problem*

$$-\frac{\partial u}{\partial t} + ku = \mathcal{A}_t u + g, \quad (t, x) \in [0, T) \times \mathbb{R}^d, \tag{4.33}$$

subject to terminal condition

$$u(T, x) = f(x), \quad x \in \mathbb{R}^d. \tag{4.34}$$

Here $\mathcal{A}_t$ is the generator of diffusion process (X_t) given in Theorem 4.22, $f : \mathbb{R}^d \to \mathbb{R}$, $k : [0, T] \times \mathbb{R}^d \to \mathbb{R}_+$, $g : [0, T] \times \mathbb{R}^d \to \mathbb{R}$ is a continuous function, and f and g satisfy the polynomial growth condition in x or are nonnegative.

Theorem 4.23 *Let u be a continuous real function on $[0, T] \times \mathbb{R}^d$ of class $C^{1,2}$, satisfying (4.33) and (4.34). Assume that u satisfies the polynomial growth condition in x:*

$$\sup_{0 \leqslant t \leqslant T} |u(t, x)| \leqslant C(1 + |x|^{2\mu}), \quad x \in \mathbb{R}^d, \tag{4.35}$$

for some constant $C > 0$, $\mu \geqslant 1$. Then u admits the representation

$$
\begin{aligned}
u(t, x) = \mathbb{E}^{t,x} \Big[& f(X_T) \exp\Big\{ -\int_t^T k(\theta, X_\theta) d\theta \Big\} \\
& + \int_t^T g(s, X_s) \exp\Big\{ -\int_t^s k(\theta, X_\theta) d\theta \Big\} ds \Big],
\end{aligned}
\tag{4.36}
$$

where $\{\mathbb{P}^{t,x}, t \geqslant 0, x \in \mathbb{R}^d\}$ is the family of probability measures associated with the Markov process (X_t). In particular, the solution to (4.33) and (4.34) satisfying the polynomial growth condition is unique.

Proof On $[t, T]$, put

$$M_s = u(s, X_s) \exp\left\{-\int_t^s k(\theta, X_\theta)d\theta + \int_t^s g(\tau, X_\tau)\exp\{-\int_t^\tau k(\theta, X_\theta)d\theta\}\, d\tau.\right.$$

By Itô's formula and using the fact that u is the solution to (4.33), we see that under each measure $\mathbb{P}^{t,x}$,

$$M_s = M_t + \int_t^s \exp\left\{-\int_t^\tau k(\theta, X_\theta)d\theta\right\}\sigma(\tau, X_\tau)\frac{\partial u}{\partial x}(\tau, X_\tau)dB_\tau.$$

Thus, M_s is a local martingale on $[t, T]$. Since u, f, and g satisfy the polynomial growth condition in x, and k is nonnegative, we have

$$\mathbb{E}[\sup_{t \leqslant s \leqslant T} |M_s|] \leqslant C(1+\mathbb{E}\sup_{t \leqslant s \leqslant T} |X_s|^{2\mu}) + C(T-t)(1+\mathbb{E}\sup_{t \leqslant s \leqslant T} |X_s|^{2\mu})$$

$$\leqslant C(1 + T - t)(1 + K_1 + K_2|x|^{2\mu}) < \infty.$$

Thus, (M_s) is a $\mathbb{P}^{t,x}$-martingale on $[t, T]$, and consequently, $u(t, x) = M_t = \mathbb{E}^{t,x}[M_T]$. $\qquad\square$

Remark Assume that v is a continuous real function on $[0, T] \times \mathbb{R}^d$, and $v \in C^{1,2}$ is the unique solution to the following Cauchy problem:

$$\frac{\partial v}{\partial t} + kv = \mathcal{A}_t v + g, \quad (t, x) \in (0, T) \times \mathbb{R}^d, \tag{4.37}$$

subject to initial condition

$$v(0, x) = f(x), \quad x \in \mathbb{R}^d. \tag{4.38}$$

In addition, we assume that v satisfies the polynomial growth condition in x. Similarly as above, we can show that v has the following representation:

$$v(t, x) = \mathbb{E}^{0,x}\left[f(X_t)\exp\left\{-\int_0^t k(\theta, X_\theta)d\theta\right\} \right.$$
$$\left. + \int_0^t g(t - s, X_s)\exp\left\{-\int_0^s k(t - \theta, X_\theta)d\theta\right\}ds\right]. \tag{4.39}$$

4.8 Snell Envelop (Continuous Time Case)

Let $(\Omega, \mathcal{F}, (\mathcal{F}_t), \mathbb{P})$ be a stochastic basis satisfying the usual condition.

Theorem 4.24 (Snell envelop) *Let (Z_t) be a progressively measurable process of class D. Put*

$$X_t = \operatorname{ess\,sup}_{\tau \in \mathcal{T}_t} \mathbb{E}[Z_\tau | \mathcal{F}_t], \tag{4.40}$$

where $\mathcal{T}_t$ is the set of all $(\mathcal{F}_t)$-stopping times taking values in $[t, \infty)$. Then (X_t) is the smallest supermartingale dominating (Z_t) and has a cadlag version. In addition, for $t > s$ we have

$$\mathbb{E}[X_t | \mathcal{F}_s] = \operatorname{ess\,sup}_{\tau \in \mathcal{T}_t} \mathbb{E}[Z_\tau | \mathcal{F}_s]. \tag{4.41}$$

Proof Taking $\tau_n \in \mathcal{T}_t$ such that

$$X_t = \sup_{n \geq 1} \mathbb{E}[Z_{\tau_n} | \mathcal{F}_t].$$

Put $A_n = \left[\mathbb{E}[Z_{\tau_{n-1}} | \mathcal{F}_t] \geqslant \mathbb{E}[Z_{\tau_n} | \mathcal{F}_t] \right]$,

$$\sigma_1 = \tau_1, \quad \sigma_n = \sigma_{n-1} I_{A_n} + \tau_n I_{A_n^c}, \quad n \geqslant 2.$$

Then $\sigma_n \in \mathcal{T}_t$, and by induction we get

$$\begin{aligned}
\mathbb{E}[Z_{\sigma_n} | \mathcal{F}_t] &= I_{A_n} \mathbb{E}[Z_{\sigma_{n-1}} | \mathcal{F}_t] + I_{A_n^c} \mathbb{E}[Z_{\tau_n} | \mathcal{F}_t] \\
&= \max\{\mathbb{E}[Z_{\sigma_{n-1}} | \mathcal{F}_t], \mathbb{E}[Z_{\tau_n} | \mathcal{F}_t]\} \\
&= \max_{i \leqslant n} \mathbb{E}[Z_{\tau_i} | \mathcal{F}_t].
\end{aligned}$$

Since

$$X_t = \lim_{n \to \infty} \max_{1 \leqslant i \leqslant n} \mathbb{E}[Z_{\tau_i} | \mathcal{F}_t],$$

by the monotone convergence theorem of conditional expectation, we obtain

$$\begin{aligned}
\mathbb{E}[X_t | \mathcal{F}_s] &= \lim_{n \to \infty} \mathbb{E}[Z_{\sigma_n} | \mathcal{F}_s] \leqslant \operatorname{ess\,sup}_{\tau \in \mathcal{T}_t} \mathbb{E}[Z_\tau | \mathcal{F}_s] \\
&\leqslant \operatorname{ess\,sup}_{\tau \in \mathcal{T}_s} \mathbb{E}[Z_\tau | \mathcal{F}_s] = X_s,
\end{aligned}$$

which implies that

$$\mathbb{E}[X_t | \mathcal{F}_s] \leqslant \operatorname{ess\,sup}_{\tau \in \mathcal{T}_t} \mathbb{E}[Z_\tau | \mathcal{F}_s].$$

But the opposite inequality holds obviously, so (4.41) is proved. Besides, the above inequality shows that (X_t) is the smallest supermartingale dominating (Z_t). Using the fact that (Z_t) is of class D, it is easy to prove that $(\mathbb{E}[X_t])$ is right continuous, and thus by Theorem 4.1, (X_t) has a cadlag version. $\square$

Chapter 5
The Black-Scholes Model and Its Modifications

In Chap. 3, we have studied discrete-time financial market models, which are suitable for qualitative research and statistical analysis. However, for theoretical research, continuous-time models are proved to be a convenient framework, because one can use stochastic analysis tools in studying of pricing and hedging of contingent claims and portfolio selection. Using stochastic analysis tools can often lead to explicit solutions or analytical expressions.

In the early 1970s, Black and Scholes (1973) made a breakthrough in option pricing. They gave an explicit pricing formula for European options. However, empirical data show deviations from the Black-Scholes option pricing formula, which are mainly embodied in two aspects: (1) the implied volatility of a call option at different strike prices increases as the deviation between the strike price and the current price of the stock increases, presenting a U shape, and the U shape of the implied volatility is not symmetric; (2) the empirical distribution of the logarithmic returns has a large deviation from the normal distribution, with a fat tail phenomenon. In the past 40 years, option pricing in theoretical and empirical researches progressed rapidly. The same is true in practical applications.

This chapter provides the main ideas and results on option pricing in the Black-Scholes model and its modifications. In Sect. 5.1, we introduce the martingale method for pricing and hedging of contingent claims and derive the Black-Scholes formulas for European options. A brief discussion on the problem of pricing American options is also included in this section. Several illustrative examples of option pricing are presented in Sect. 5.2. In Sect. 5.3, we present practical uses of the Black-Scholes formulas. Finally, in Sect. 5.4, we introduce some modifications of the Black-Scholes model to correct biases in the Black-Scholes formulas for option pricing.

© Springer Nature Singapore Pte Ltd. and Science Press 2018
J.-A. Yan, *Introduction to Stochastic Finance*, Universitext,
https://doi.org/10.1007/978-981-13-1657-9_5

5.1 Martingale Method for Option Pricing and Hedging

In this section we introduce the Black-Scholes model and present a martingale method for the pricing and hedging of contingent claims.

5.1.1 The Black-Scholes Model

Consider a security market which consists of two instruments: a risky asset, called simply the asset, and a risk-free asset (i.e., bank account). Assume that the asset pays no dividends. According to the discussion in the approximately continuous trading case of binomial tree model, it is reasonable to assume that the continuous-time price process of the asset satisfies the following Itô SDE

$$dS_t = S_t(\mu dt + \sigma dB_t), \tag{5.1}$$

where $S_0 > 0$, μ and σ are constants and (B_t) is a Brownian motion defined on a filtered probability space $(\Omega, \mathcal{F}, (\mathcal{F}_t), \mathbb{P})$. We assume that $(\mathcal{F}_t)$ is the natural filtration of Brownian motion (B_t). This model originated in the seminal paper of Black and Scholes (1973) and is called the Black-Scholes *model*. Such a process (S_t) is called a *geometric Brownian motion*. It is also called a *log-normal process*, because by Theorem 4.14,

$$S_t = S_0 \exp\left\{(\mu - \frac{\sigma^2}{2})t + \sigma B_t\right\}, \tag{5.2}$$

so that $\log(S_t)$ is normally distributed. We call μ the *(instantaneous) expected rate of return* and σ the *volatility* of S. Here one should beware that the continuously compounded rate of return $\log(S_t/S_0)$ is different from the expected rate of return. The value process (β_t) of the bank account is assumed to satisfy

$$d\beta_t = r\beta_t dt, \tag{5.3}$$

where r is the constant interest rate. In the sequel, we always assume $\beta_0 = 1$, so that $\beta_t = e^{rt}$.

We assume that the market is *frictionless* in the sense that there are no transaction costs, no bid/ask spread, no restrictions on trade such as margin requirements or short sale restrictions, there are no taxes, borrowing and lending are at the same interest rate, and asset shares are divisible. In addition, we assume that trading in assets takes place continuously in time. A *trading strategy* is a pair of $\mathcal{F}_t$-adapted processes $\{a, b\}$ such that $a \in \mathcal{L}^2$ and $b \in \mathcal{L}^1$, where $a(t)$ denotes the number of units of the asset held at time t, $b(t)\beta_t$ is the total money invested in the bank at time t, and

$$\mathcal{L}^2 = \left\{\theta : \forall t > 0, \int_0^t \theta(s)^2 ds < \infty\right\},$$

$$\mathcal{L}^1 = \left\{ \theta : \forall t > 0, , \int_0^t |\theta(s)| ds < \infty \right\}.$$

So the *wealth* V_t at time t of a portfolio $\{a(t), b(t)\}$ is given by

$$V_t = a(t)S_t + b(t)\beta_t.$$

A trading strategy $\{a, b\}$ is said to be *self-financing* if the change of its wealth is only due to the changes in the assets prices weighted by the portfolio, meaning that for all t we have

$$dV_t = a(t)dS_t + b(t)d\beta_t,$$

or equivalently,

$$dV_t = rV_t dt + a(t)[dS_t - rS_t dt]. \tag{5.4}$$

Conversely, for an adapted process $a \in \mathcal{L}^2$ and $x > 0$, the solution (V_t) to (5.4) with $V_0 = x$ is the wealth process of self-financing strategy $\{a, b\}$, where $b(t) = e^{-rt}[V_t - a(t)S_t]$.

Let $\{a, b\}$ be a self-financing strategy and (V_t) be its wealth process. This strategy is called an *arbitrage strategy*, if $V_0 = 0$ and there is $t > 0$ such that $V_t \geqslant 0$ and $\mathbb{P}(V_t > 0) > 0$. It is called *admissible*, if its wealth process is nonnegative. It is called *allowable*, if there exists a positive constant c such that $V_t \geqslant -c(e^{rt} + S_t)$, for all $t \in [0, T]$.

5.1.2   Equivalent Martingale Measures

Let (X_t) be a process. We put $\widetilde{X}_t = X_t e^{-rt}$ and call $(\widetilde{X}_t)$ the *discounted process*. The starting point of the martingale method for option pricing is the following observation:

Lemma 5.1 *A trading strategy* $\{a, b\}$ *is self-financing if and only if its discounted wealth process* $(\widetilde{V}_t)$ *satisfies*

$$d\widetilde{V}_t = a(t)d\widetilde{S}_t. \tag{5.5}$$

Proof Let $\{a, b\}$ be a self-financing strategy. Then by (5.4), we have

$$d\widetilde{V}_t = -V_t r e^{-rt} dt + e^{-rt} dV_t$$
$$= a(t)[S_t d(e^{-rt}) + e^{-rt} dS_t]$$
$$= a(t)d\widetilde{S}_t,$$

which gives (5.5). Similarly, we can prove that (5.5) implies (5.4). $\square$

We will show that there exists a unique probability measure $\mathbb{P}^*$, equivalent to $\mathbb{P}$, such that the process $\widetilde{S}_t$ is a $\mathbb{P}^*$-martingale. To this end, we rewrite (5.1) as

$$d\widetilde{S}_t = \widetilde{S}_t[(\mu - r)dt + \sigma dB_t].$$

If we put $\frac{d\mathbb{P}^*}{d\mathbb{P}}|_{\mathcal{F}_T} = \mathcal{E}(-\theta.B)_T$ with $\theta(t) = \theta = (\mu - r)/\sigma$, then by Girsanov's theorem (Theorem 4.16), $B_t^* = B_t + \theta t$ is a $\mathbb{P}^*$-Brownian motion and

$$d\widetilde{S}_t = \widetilde{S}_t \sigma dB_t^*. \tag{5.6}$$

Thus $(\widetilde{S}_t)$ is a $\mathbb{P}^*$-martingale. It is easy to see that such a probability measure is unique.

Let $T > 0$. An $\mathcal{F}_T$-measurable nonnegative random variable is called a *European contingent claim* at time T. A result of Dudley (1977) shows that in the Black-Scholes model, for any European contingent claim ξ at time T, there is an adapted process $a \in \mathcal{L}^2$ such that $\widetilde{\xi} = \int_0^T a(s)d\widetilde{S}_s$. This means that if self-financing strategies are not subject to certain restrictions, then there exist arbitrage opportunities in the market.

According to Lemma 5.1, the discounted wealth process of a self-financing strategy is a local martingale under $\mathbb{P}^*$. Thus, for an allowable self-financing strategy, its discounted wealth process $(\widetilde{V}_t)$ is a $\mathbb{P}^*$-supermartingale, because it can be expressed as the difference of a nonnegative local $\mathbb{P}^*$-martingale and a $\mathbb{P}^*$-martingale. In fact, if $\widetilde{V}_t \geqslant -c(1 + \widetilde{S}_t)$ with $c > 0$, then

$$\widetilde{V}_t = \widetilde{V}_t + c(1 + \widetilde{S}_t) - c(1 + \widetilde{S}_t).$$

From this fact, it is easy to see that the market has no arbitrage for allowable strategies. It should be emphasized that when we say that the market has "no arbitrage," we need to point out that the market has no arbitrage for what kind of self-financing strategy.

We call probability measure $\mathbb{P}^*$ the *equivalent martingale measure* for the market. By contrast, $\mathbb{P}$ is called the *objective* or *physical* probability measure. On the other hand, (4.6) can be rewritten as

$$dS_t = S_t[rdt + \sigma dB_t^*]. \tag{5.7}$$

This means that under probability measure $\mathbb{P}^*$, the expected rate of return of the risky asset is equal to the risk-free interest rate r. For this reason, the equivalent martingale measure $\mathbb{P}^*$ is also called the *risk-neutral probability measure*.

Note that (5.7) can be expressed as

$$dS_t = S_t[(r + \sigma\theta)dt + \sigma dB_t],$$

where $\theta = (\mu - r)/\sigma$. We call θ the *market price of risk* or *risk premium* of the risky asset. It represents the excess rate of return above the risk-free rate of return per unit of extra risk. To further explain the economic meaning of the market price of risk, we envisage another risky asset in the market, which pays no dividends and its price process (W_t) is the following Itô process

$$dW_t = W_t[\mu_t dt + \sigma_t dB_t].$$

Thus, the discounted price process $(\widetilde{W}_t)$ satisfies

$$dW_t = \widetilde{W}_t[(\mu_t - r)dt + \sigma_t dB_t]$$
$$= \widetilde{W}_t[(\mu_t - r - \theta\sigma_t)dt + \sigma_t dB_t^*].$$

So in order for the market to have no arbitrage, $(\widetilde{W}_t)$ must be a $\mathbb{P}^*$-martingale. Consequently, we must have $\sigma_t^{-1}(\mu_t - r) = \theta$. This means that in an arbitrage-free market, the assets having the same sources of uncertainty must have the same market price of risk.

5.1.3 Pricing and Hedging of European Contingent Claims

Let ξ be a European contingent claim at time T. Let $\{a, b\}$ be an admissible self-financing strategy and (V_t) be its wealth process. If $V_T = \xi$, we say that this strategy hedges (or replicates) ξ. The following theorem shows that in the Black-Scholes model, any European contingent claim that is integrable under $\mathbb{P}^*$ can be replicated by an admissible self-financing strategy.

Theorem 5.2 *Let ξ be a European contingent claim which is integrable under $\mathbb{P}^*$. Then there exists a unique admissible self-financing strategy $\{a, b\}$ replicating ξ, and its wealth process (V_t) is given by*

$$V_t = \mathbb{E}^*\big[e^{-r(T-t)}\xi|\mathcal{F}_t\big], \tag{5.8}$$

In addition, there is $H \in \mathcal{L}^2[0, T]$ such that $d\widetilde{V}_t = H_t dB_t^*$ and

$$a(t) = H_t/(\sigma\widetilde{S}_t). \tag{5.9}$$

In particular, if $V_t = F(t, S_t)$, $F \in C^{1,2}([0, T) \times \mathbb{R}_+)$, then $a(t) = F_x(t, S_t)$.

Proof Let V_t be given by (5.8). Since $(\widetilde{V}_t)$ is a $\mathbb{P}^*$- martingale, and $(\mathcal{F}_t)$ is also the natural filtration of (B_t^*), by the martingale representation theorem of Brownian motion, there exists an $H \in \mathcal{L}^2$ such that

$$\widetilde{V}_t = V_0 + \int_0^t H_s dB_s^*, \quad t \in [0, T].$$

Put

$$a(t) = H_t/(\sigma \widetilde{S}_t), \quad b(t) = \widetilde{V}_t - a(t)\widetilde{S}_t. \tag{5.10}$$

Then $\{a, b\}$ is a hedging strategy for ξ, and (V_t) is its wealth process. By (5.6) and (5.10), we have

$$a(t)d\widetilde{S}_t = a(t)\widetilde{S}_t \sigma dB_t^* = H_t dB_t^* = d\widetilde{V}_t.$$

Thus, by Lemma 5.1, the strategy $\{a, b\}$ is self-financing and admissible.

Let $\{a, b\}$ be an admissible self-financing strategy replicating ξ. Then by (5.5) and (5.6), we have

$$d\widetilde{V}_t = a(t)\sigma \widetilde{S}_t dB_t^*,$$

from which we get immediately (5.9). This shows that $\{a, b\}$ is the unique admissible self-financing strategy replicating ξ.

Now assume that $V_t = F(t, S_t)$, where $F \in C^{1,2}([0, T) \times \mathbb{R}^+)$. We claim that $a(t) = F_x(t, S_t)$. In fact, by Itô's formula,

$$d\widetilde{V}_t = d\left(e^{-rt}F(t, S_t)\right) = e^{-rt}F_x(t, S_t)dS_t + \text{``}dt\text{''} \text{ term}$$

$$= F_x(t, S_t)d\widetilde{S}_t + \text{``}dt\text{''} \text{ term}$$

$$= F_x(t, S_t)\widetilde{S}_t \sigma dB_t^*.$$

By Theorem 4.2, the "dt" term vanishes, because $(\widetilde{V}_t)$ and $\widetilde{S}_t$ are both $\mathbb{P}^*$-martingales. Thus from (5.9), we get $a(t) = F_x(t, S_t)$. □

Let ξ be a European contingent claim which is integrable under $\mathbb{P}^*$. By Theorem 5.2, if we define V_t as the price at time t of contingent claim ξ, then there exists no arbitrage opportunity for both the seller and buyer of the contingent claim. This method of pricing contingent claims is called *pricing by arbitrage* or *arbitrage pricing*. Equation (5.8) is called the *risk-neutral valuation formula*.

Theorem 5.3 *If a European contingent claim is of the form $\xi = f(S_T)$, then its price process can be expressed as $V_t = F(t, S_t)$, where*

$$F(t, x) = e^{-r(T-t)} \int_{-\infty}^{\infty} f\left(xe^{(r-\sigma^2/2)(T-t)+\sigma y\sqrt{T-t}}\right) \frac{e^{-y^2/2}}{\sqrt{2\pi}} dy. \tag{5.11}$$

Proof We express S_T as

$$S_T = S_t \exp\{(r - \sigma^2/2)(T - t) + \sigma(B_T^* - B_t^*)\}.$$

Since S_t is measurable w.r.t. $\mathcal{F}_t$ and $B_T^* - B_t^*$ is independent of $\mathcal{F}_t$, by (5.8) and Theorem 1.10, we have

$$V_t = \mathbb{E}^*\left[e^{-r(T-t)}f\left(x\exp\{(r-\sigma^2/2)(T-t)+\sigma(B_T^*-B_t^*)\}\right)\right]\Big|_{x=S_t},$$

from which we get $V_t = F(t, S_t)$. □

Theorem 5.4 *If $f(x) = (x - K)^+$, then $\xi = (S_T - K)^+$ is a European call option. As a particular case of formula (5.11), the price process of ξ is given by $V_t = C(t, S_t)$, where*

$$C(t, x) = xN(d_1) - Ke^{-r(T-t)}N(d_2), \tag{5.12}$$

$N(z)$ is the standard normal distribution function:

$$N(z) = \int_{-\infty}^{z} \frac{1}{\sqrt{2\pi}}\exp\left\{-\frac{y^2}{2}\right\}dy,$$

$$d_1 = \frac{\ln(x/K)+\left(r+\frac{1}{2}\sigma^2\right)(T-t)}{\sigma\sqrt{T-t}}, \quad d_2 = \frac{\ln(x/K)+\left(r-\frac{1}{2}\sigma^2\right)(T-t)}{\sigma\sqrt{T-t}}. \tag{5.13}$$

Formula (5.12) is called the Black-Scholes *formula* for European call option $\xi = (S_T - K)^+$. By the put-call parity formula, we can get the corresponding Black-Scholes formula for a European put option. It means that the price process of European put option $\xi = (S_T - K)^-$ is given by $V_t = P(t, S_t)$, where

$$P(t, x) = Ke^{-r(T-t)}N(-d_2) - xN(-d_1). \tag{5.14}$$

Form Theorem 5.3 we see that an important feature of the Black-Scholes model is that the expected rate of return on the asset does not enter Black-Scholes formulas. Note that if we replace constant μ in (5.1) by an adapted process $(\mu(t))$ such that $\mathbb{E}[\exp\{\frac{1}{2}\int_0^T(\theta(s))^2ds\}] < \infty$, where $\theta(t) = (\mu(t) - r)/\sigma$, then by letting

$$\frac{d\mathbb{P}^*}{d\mathbb{P}}\Big|_{\mathcal{F}_T} = \mathcal{E}(-\theta.B)_T, \quad B_t^* = B_t + \int_0^t \theta(s)ds,$$

(5.6) is still valid. Consequently, the Black-Scholes formulas are applicable to this case. This remarkable fact was first observed by Merton (1973b). We will present the Merton's generalization of the Black-Scholes model in Chap. 7, Sect. 7.3.1.

5.1.4 Pricing of American Contingent Claims

Now we turn to the problem of pricing American contingent claims. We assume that the asset pays no dividends. In the continuous-time case, an *American contingent claim* is naturally defined as an adapted nonnegative process $(h_t)_{0 \leqslant t \leqslant T}$. For simplicity, we only consider an American contingent claim of the form $h_t = g(S_t)$, where g is called a *reward function*. For a call, $g(x) = (x - K)^+$, and for a put, $g(x) = (K - x)^+$.

Similar to the discrete-time case, for pricing American contingent claims, we should introduce the *trading-consumption strategy*. By such a strategy, we mean a trading strategy $\phi = \{a, b\}$ as well as an adapted continuous nondecreasing process C_t with $C_0 = 0$ (called *cumulative consumption process*), such that the wealth process (V_t) of the trading strategy ϕ (i.e., $V_t(\phi) = a(t)S_t + b(t)\beta_t$) satisfies the following equation:

$$a(t)S_t + b(t)\beta_t = a(0)S_0 + b(0) + \int_0^t a(s)dS_s + \int_0^t b(s)d\beta_s - C_t, \quad \forall t \in [0, T].$$

It is easy to see that for any trading-consumption strategy (ϕ, C), the following process

$$e^{-rt}V_t(\phi) + \int_0^t e^{-rs}dC_s$$

is a martingale under $\mathbb{P}^*$. A trading-consumption strategy (ϕ, C) is called a *superhedging strategy* for American contingent claim (h_t), if for all $t \in [0, T]$, $V_t(\phi) \geqslant h_t$ a.s.

Let $\mathcal{T}_{t,T}$ be the set of all stopping times taken values in $[t, T]$. Put

$$\Phi(t, x) = \sup_{\tau \in \mathcal{T}_{t,T}} \mathbb{E}^* \left[e^{-r(\tau - t)} g\left(x \exp\{(r - (\sigma^2/2))(\tau - t) + \sigma(B_\tau^* - B_t^*)\}\right) \right],$$

where we assume $g(x) \leqslant A + Bx$ so that $\Phi(t, x)$ is well defined. Since

$$S_t \exp\{(r - (\sigma^2/2))(\tau - t) + \sigma(B_\tau^* - B_t^*)\} = S_\tau,$$

by Theorem 1.11 we have

$$e^{-rt}\Phi(t, S_t) = \operatorname*{ess\,sup}_{\tau \in \mathcal{T}_{t,T}} \mathbb{E}^* \left[e^{-r\tau} g(S_\tau) | \mathcal{F}_t \right].$$

By Theorem 4.24, process $e^{-rt}\Phi(t, S_t)$ is the smallest supermartingale such that $\Phi(t, S_t)$ dominates process $g(S_t)$ for all $t \in [0, T]$.

The following theorem is the main result concerning the pricing American contingent claims. For a proof we refer the reader to Karatzas and Shreve (1998), p. 376–378.

Theorem 5.5 *There is a trading-consumption strategy ϕ such that ϕ super-hedges $(g(S_t))$ and $V_t(\phi) = \Phi(t, S_t)$, $\forall t \in [0, T]$. Moreover, for any trading-consumption strategy ψ super-hedging $(g(S_t))$, we have $V_t(\psi) \geqslant \Phi(t, S_t)$ for all $t \in [0, T]$.*

We call $\Phi(t, S_t)$ the *seller's price* at time t of the American contingent claim.

The following theorem shows that the American call option and the European call option have the same price at any time t, if the underlying asset pays no dividend.

Theorem 5.6 *In the call option case, i.e., $g(x) = (x - K)^+$, we have $\Phi(t, x) = C(t, x)$, where C is defined by (5.12).*

Proof We only consider the case $t = 0$, other cases being similar. Since $\widetilde{S}_t$ is martingale under $\mathbb{P}^*$, for any stopping time τ taking values in $[0, T]$, we have

$$\widetilde{S}_\tau - e^{-rT} K = \mathbb{E}^*[e^{-rT}(S_T - K)|\mathcal{F}_\tau] \leqslant \mathbb{E}^*[e^{-rT}(S_T - K)^+|\mathcal{F}_\tau],$$

from which we get

$$(\widetilde{S}_\tau - e^{-r\tau} K)^+ \leqslant (\widetilde{S}_\tau - e^{-rT} K)^+ \leqslant \mathbb{E}^*[e^{-rT}(S_T - K)^+|\mathcal{F}_\tau].$$

Taking expectations in two sides, we obtain

$$\mathbb{E}^*[(\widetilde{S}_\tau - e^{-r\tau} K)^+] \leqslant \mathbb{E}^*[e^{-rT}(S_T - K)^+].$$

This implies the wanted result. □

Now we turn to the optimal exercise problem on American contingent claims. Let $\tau \in \mathcal{T}_{0,T}$. If one exercises the American contingent claim at the stopping time τ, the expected discounted payoff under $\mathbb{P}^*$ is given by

$$V_0^\tau = \mathbb{E}^*\left[e^{-r\tau} g(S_\tau)\right].$$

Put

$$\tau^* = \inf\left\{t \in [0, T] : \Phi(t, S_t) = g(S_t)\right\}. \tag{5.15}$$

Then τ^* is a stopping time which maximizes V_0^τ within $\mathcal{T}_{0,T}$. So it is reasonable to consider τ^* as the optimal exercise time for the American contingent claim. For an American put option, we set

$$c(t) = \sup\left\{x \in \mathbb{R}_+ : \Phi(t, x) = (K - x)^+\right\}, \quad t \leqslant T. \tag{5.16}$$

Then

$$\tau^* = \inf\left\{t \in [0, T] : S_t \leqslant c(t)\right\}. \tag{5.17}$$

The function $c(t)$ is a nondecreasing function of t. It is called the *critical price* (or *optimal stopping boundary*). Put

$$\mathcal{C} = \{(t, x) \in [0, T] \times \mathbb{R}_+ : x > c(t)\} = \{(t, x) \in [0, T] \times \mathbb{R}_+ : \Phi(t, x) > (K-x)^+\}.$$

We call $\mathcal{C}$ the *continuation region*. What we know about the critical price are the following: c is a C^∞-function on $(0, T)$, $\lim_{t \to T} c(t) = K$ and

$$K - c(t) \sim K\sqrt{(T - t)\ln(T - t)},$$

as t approaches the expiry date T (i.e., $T - t \to 0$). If we put $b(t) = c(T - t)$, $0 \leqslant t \leqslant T$, then

$$\lim_{t \to \infty} b(t) = \frac{2rK}{2r + \sigma^2}.$$

From an analytical point of view, Φ satisfies the following *free boundary condition*:

$$\Phi(t, c(t)) = (K - c(t))^+, \quad \frac{\partial \Phi}{\partial x}(t, c(t)) = -1. \tag{5.18}$$

Thus, the problem of pricing and optimal exercising American put options is reduced to solving a free boundary problem for a partial differential equations (PDE). We refer the reader to Wilmott et al. (1993) for a detailed treatment of this problem.

5.2   Some Examples of Option Pricing

We illustrate below the martingale method for option pricing through several examples.

5.2.1   *Options on a Stock with Proportional Dividends*

If a stock pays no dividends, then under the risk-neutral probability measure, its price process (S_t) is modeled by (5.7). Now we assume that the stock pays proportional dividends with dividend rate q. Then the ex-dividend price at time t of this stock is $X_t = e^{-qt}S_t$. So under the risk-neutral probability $\mathbb{P}^*$, we have

$$dX_t = X_t\big((r - q)dt + \sigma dB_t^*\big). \tag{5.19}$$

Thus, by (5.8) the time t price of a call option on this stock with dividend rate q and strike price K at maturing time T is given by

$$C_t = e^{-r(T-t)}\mathbb{E}^*[(X_T - K)^+|\mathcal{F}_t] = e^{-q(T-t)}e^{-(r-q)(T-t)}\mathbb{E}^*[(X_T - K)^+|\mathcal{F}_t].$$

However, from (5.19) we see that $\mathbb{P}^*$ would be the risk-neutral probability for (X_t) if the interest rate were $r - q$. So according to (5.8), we have $C_t = e^{-q(T-t)}C_t'$, where C_t' is the time t price of a call option with strike price K at maturing time T in a market with interest rate $r - q$. Thus, by Black-Scholes formula (5.12), we get

$$C_t = e^{-q(T-t)}[X_t N(d_1') - Ke^{-(r-q)(T-t)}N(d_2')]$$
$$= e^{-q(T-t)}X_t N(d_1') - Ke^{-r(T-t)}N(d_2'),$$

where

$$d_1' = \frac{\ln(X_t/K) + \left(r - q + \frac{1}{2}\sigma^2\right)(T-t)}{\sigma\sqrt{T-t}}, \quad d_2' = d_1' - \sigma\sqrt{T-t}.$$

5.2.2 Foreign Currency Option

Consider a contract which gives its owner the right to buy M units of a foreign currency at a pre-specified exchange rate K and date T. This contract is called foreign currency option (see Garman and Kohlhagen (1983)). We assume that both the domestic and foreign risk-free interest rates are nonnegative constants, say r^d and r^f, respectively, and the exchange rate Q satisfies the following equation:

$$dQ_t = Q_t(\mu dt + \sigma dB_t), \tag{5.20}$$

where μ and σ are constants. From the domestic point of view, there are two basic assets: one is the domestic bank account (regarded as a risk-free asset), whose price process is

$$d\beta_t^d = r^d \beta_t^d dt;$$

the other is the foreign bank account in domestic term (regarded as a risky asset), whose price process is $S_t := \beta_t^f Q_t$, where

$$d\beta_t^f = r^f \beta_t^f dt.$$

By Itô's formula, we get

$$dS_t = S_t[(r^f + \mu)dt + \sigma dB_t].$$

Now the contract can be regarded as a European call option in this domestic market, whose payoff at expiry date T is $\xi = M(Q_T - K)^+$.

We take β_t^d as a numeraire and let $\widetilde{S}_t = (\beta_t^d)^{-1} S_t$. Then

$$d\widetilde{S}_t = \widetilde{S}_t[(r^f - r^d + \mu)dt + \sigma dB_t].$$

Consequently, if we put $\frac{d\mathbb{P}^*}{d\mathbb{P}}|_{\mathcal{F}_T} = \mathcal{E}(-\theta.B)_T$ with $\theta(t) = \theta = (r^f - r^d + \mu)/\sigma$, then by Girsanov's theorem $B_t^* = B_t + \theta t$ is a $\mathbb{P}^*$-Brownian motion, and $(\widetilde{S}_t)$ is a $\mathbb{P}^*$-martingale. We call $\mathbb{P}^*$ the *domestic martingale measure*. By (5.8), the price at time t of option ξ is given by

$$V_t = \mathbb{E}^*[e^{-r^d(T-t)}M(Q_T - K)^+|\mathcal{F}_t]. \tag{5.21}$$

We can rewrite (5.20) and (5.21) as

$$dQ_t = Q_t[(r^d - r^f)dt + \sigma dB_t^*],$$

$$V_t = Me^{-r^f(T-t)}\mathbb{E}^*[e^{-(r^d-r^f)(T-t)}(Q_T - K)^+|\mathcal{F}_t],$$

respectively. Thus, by the Black-Scholes formula, we obtain immediately the pricing formula for option ξ:

$$V_t = Me^{-r^f(T-t)}C(t, Q_t), \tag{5.22}$$

where $C(t, x)$ is given by (5.12) with r replaced by $r^d - r^f$. We remark that just like the Black-Scholes formula, parameter μ does not enter the valuation formula for currency option, even when μ is time dependent and random.

5.2.3   Compound Option

A *compound option* is an option whose underlying asset is an option contract. There are four main types of compound options: a call or put based on a call option, and a call or put based on a put option. In the following, we take a call based on a call as an example. This compound call gives the holder, subject to a certain time $T_1 < T$ prior to the expiry T, at an agreed contract price, the right to buy the underlying call option. Assume that the strike prices of the call and the compound call are K and K_1, respectively, and the price process of the underlying call is $C(t, S_t)$. Then the return of the compound call at the executing time T_1 is

$$V(T_1) = (C(T_1, S_{T_1}) - K_1)^+,$$

where $C(t, x)$ is given by (5.12). The pricing formula of the compound call is first given by Geske (1979) using the Fourier integral method. Here we give a brief introduction to how to deduce the pricing formula using a probability method. From (5.8) and Theorem 1.11, the price $V(t)$ of the compound call at time t with $t \leqslant T_1$ is given by

$$V(t) = e^{-r(T_1-t)} \mathbb{E}^*[(C(T_1, S_{T_1}) - K_1)^+ | \mathcal{F}_t] = C_1(t, S_t)$$

where

$$C_1(t, x) = e^{-r(T_1-t)} \mathbb{E}^*[(C(T_1, x S_t^{-1} S_{T_1}) - K_1)^+].$$

By some calculations (see, e.g., Kwok 1998), one can prove:

$$C_1(t, x) = x N_2 \left(D_1(t, x) + \sigma\sqrt{T_1 - t}, D(t, x) + \sigma\sqrt{T - t}; \sqrt{\frac{T_1 - t}{T - t}} \right)$$

$$- K e^{-r(T-t)} N_2 \left(D_1(t, x), D(t, x); \sqrt{\frac{T_1 - t}{T - t}} \right)$$

$$- K_1 K e^{-r(T_1-t)} N(D_1(t, x)),$$

where $N(z)$ is the standard normal distribution function, and

$$N_2(y, z; \rho) = \frac{1}{2\pi\sqrt{1 - \rho^2}} \int_{-\infty}^{y} \int_{-\infty}^{z} \exp\left\{ -\frac{1}{2}\left(\frac{u^2 - 2\rho u v + v^2}{1 - \rho^2}\,du\,dv\right)\right\}.$$

$$D_1(t, x) = \frac{\ln(x/S^*) + \left(r - \frac{1}{2}\sigma^2\right)(T_1 - t)}{\sigma\sqrt{T_1 - t}},$$

$$D(t, x) = \frac{\ln(x/K) + \left(r - \frac{1}{2}\sigma^2\right)(T - t)}{\sigma\sqrt{T - t}},$$

S^* being the unique solution to

$$C_1(T_1, S^*) - K_1 = 0.$$

5.2.4 Chooser Option

A *chooser option* gives the option holder the following right: at a predetermined future time $T_c < T$, the holder may choose one between a call option and a put options having the same strike price K. Therefore, the return at the executing time of a chooser option options is

$$V(T_c) = \max\{C(T_c, S_{T_c})), P(T_c, S_{T_c})\},$$

where $C(t, x)$ and $P(t, x)$ are given by (5.12) and (5.14), respectively. The price $V(t)$ of the chooser option at time t is as follows: for $t \geqslant T_c$, $V(t) = V(T_c)$; for $t \leqslant T_c$,

$$V(t) = e^{-r(T-t)}\mathbb{E}^*[V(T_c)|\mathcal{F}_t]$$
$$= e^{-r(T-t)}g(t, S_t),$$

where

$$g(t, x) = \mathbb{E}^*[\max\{C(T_c, xS_t^{-1}S_{T_c}), P(T_c, xS_t^{-1}S_{T_c})\}].$$

5.3 Practical Uses of the Black-Scholes Formulas

5.3.1 Historical and Implied Volatilities

Note that the only unknown parameter in the Black-Scholes formulas is the volatility. One might use historic data of the asset prices to calculate the standard deviation of asset returns as an estimate of the volatility, known as a *historical volatility*. On the other hand, assuming that the asset price process obeys a geometric Brownian motion and the market is arbitrage-free, we can take the quoted prices of options on the same asset with different maturities and/or strike prices to deduce, from the Black-Scholes formula, some estimates of the volatility. By taking some weighted average over these implied volatilities, we obtain an estimate of the volatility, which can be considered as a forecast of the future volatility. Such an estimate is called an *implied volatility*. Empirical studies have shown that an implied volatility is better suited than a historical volatility for predicting an asset's future volatility. The pricing of other options can be based on an implied volatility. So extracting an implied volatility is one of the primary uses of the Black-Scholes formula by practitioners, including arbitragers and brokerage houses.

5.3.2 Delta Hedging and Analyses of Option Price Sensitivities

Black-Scholes formula provides useful measures of the sensitivity of an option price to various parameter changes. These sensitivity measures prove to be effective tools for monitoring an option position risk exposure. By definition, the *delta* of an option measures the change in the option price for a unit change in the underlying asset price. The sensitivity of delta to changes in the value of the underlying asset is called *gamma*. The sensitivity of the option price to changes in the time to maturity (resp., in the volatility, in the interest rate) is called *theta* (resp., *vega, rho*).

From the Black-Scholes formulas, we see that the price of a call option depends on S_t, K, σ, r, and $T - t$, the time to maturity of the option. Since we have $N'(x) = 1/\sqrt{2\pi}e^{-\frac{1}{2}x^2}$, it is easy to prove that

$$xN'(d_1) = Ke^{-r(T-t)}N'(d_2),$$

where d_1 and d_2 are given by (5.13). From the above equation, it is easy to prove

$$\Delta = \frac{\partial C}{\partial x} = N(d_1) > 0,$$

$$\Gamma = \frac{\partial^2 C}{\partial x^2} = \frac{1}{x\sigma\sqrt{T-t}}N'(d_1) > 0,$$

$$V = \frac{\partial C}{\partial \sigma} = x\sqrt{T-t}N'(d_1) > 0,$$

$$\rho = \frac{\partial C}{\partial r} = (T-t)e^{-r(T-t)}KN(d_2) > 0,$$

$$\theta = \frac{\partial C}{\partial t} = -\frac{x\sigma}{2\sqrt{T-t}}N'(d_1) - Kre^{-r(T-t)}N(d_2) < 0.$$

By Theorem 5.2, we obtain immediately a hedging strategy $\{a(t), b(t)\}$ for the call option, where $a(t)$ is given by the delta $\frac{\partial C}{\partial x}(t, S_t)$ at time t of the option. So this hedging is also called the *delta hedging*. In practice, due to transaction costs, one should rebalance the portfolio only when the position's delta has moved noticeably from its target level. To this point, the option's gamma helps us to know how frequently one should rebalance the portfolio.

The theta of a call option is always negative. This suggests that a long position in a call will lose its time value with the passage of time. This loss can only be avoided by setting a theta-neutral position consisting of short and long holdings in options that have the same theta. If one believes that the volatility is not constant, one should also take vega into consideration.

Another sensitivity measure of an option price is the so-called *elasticity* or *lambda*, denoted by λ. It also refers to the *leverage* of the option's position. It measures the percentage change in the option price for 1% change in the underling asset's price. In the Black-Scholes setting, we have

$$\lambda = \frac{\partial C}{\partial x}\frac{x}{C} = \frac{xN(d_1)}{C}.$$

From (5.12), we see that $\lambda > 1$ always holds. This phenomenon is called the *leverage effect*. For a put option, $\lambda = -\frac{\partial P}{\partial x}\frac{x}{P}$, where P stands for the price of the put. It is always negative, but not necessary less than -1. It means that a put option does not necessarily have a leverage effect.

5.4   Capturing Biases in Black-Scholes Formulas

For a European option, if the discounted exercise price is less than, or greater than, or equal to the current price of the stock, the option is called *in-the-money*, or *out-of-the-money*, or *at-the-money*, respectively. If the Black-Scholes model fits well the real market, then the implied volatility would be the same for all options with different strike prices or maturities. However, empirical evidence shows that the implied volatilities for call options with different strike prices tend to rise for deeply in- or out-of-money options, referred to as the *volatility smile*, and the graph of implied volatilities is lopsided, referred to as the *skewness*. Statistical analysis of high-frequency data further shows that the empirical distribution of the standardized innovations of the log-price is not normal: it has fat-tails, which are measured by the so-called *excess kurtosis*. Here the kurtosis stands for the fourth moment around the mean, which is equal to 3 for the normal distribution. So the excess kurtosis of a fat-tailed distribution is its kurtosis mines 3. In addition, some commodity markets also exhibit volatility clustering and/or mean reversion property. All these biases suggest that the Black-Scholes model has to be modified.

Below we shall present several modifications: level-dependent volatility model (including constant elasticity of variance (CEV) model as a particular case), stochastic volatility model, variance gamma (VG) model, stochastic alpha-beta-rho (SABR) model, and GARCH models. Further discussion of some models will be given in later chapters.

5.4.1   CEV Model and Level-Dependent Volatility Model

Observed stock price data manifests the so-called *leverage effect*, meaning that there is a negative correlation between volatility and price level. Taking the leverage effect into consideration, Cox (1975) and Cox and Ross (1976) proposed a *constant elasticity of variance model* (abbreviated as CEV *model*):

$$dS_t = \mu S_t dt + \sigma S_t^{\frac{\alpha}{2}} dB_t,$$

where $0 < \alpha \leqslant 2$ is a constant, known as the *elasticity factor*. For $\alpha = 2$, it reduces to a geometric Brownian motion. If $0 < \alpha < 2$, the CEV process has an absorbing barrier 0. In this model, the volatility is endogenously determined by the stock price level. The option pricing under CEV models will be discussed in Chap. 7.

The CEV model can only capture the negative skew in the graph of implied volatilities. In order to capture volatility smiles, we model stock price process (S_t) by a diffusion process:

$$dS_t = S_t[\mu(t, S_t)dt + \sigma(t, S_t)dB(t)], \qquad (5.23)$$

where μ and σ are assumed to be sufficiently smooth such that there exists a unique solution to (5.23). This model is called the *level-dependent volatility model*. For notational simplicity, we assume that the interest rate for the savings account is 0. Put

$$\eta_T = \exp\left\{-\int_0^T \frac{\mu}{\sigma}(t, S_t)dB_t - \frac{1}{2}\int_0^T \left(\frac{\mu}{\sigma}(t, S_t)\right)^2 dt\right\}.$$

Assume $\mathbb{E}[\eta_T] = 1$. We define a new probability measure $\mathbb{P}^*$ on $(\Omega, \mathcal{F}_T)$ by $d\mathbb{P}^*/d\mathbb{P} = \eta_T$. Then by Girsanov's theorem, $B_t^* = B_t + \int_0^t \frac{\mu}{\sigma}(u, S_u)du$ is a Brownian motion under $\mathbb{P}^*$, and we have

$$dS_t = S_t\sigma(t, S_t)dB^*(t), \qquad (5.24)$$

Under certain condition on $\sigma(t, x)$, S_t will be a $\mathbb{P}^*$-martingale, so that $\mathbb{P}^*$ is the risk-neutral measure.

Breeden and Litzenberger (1978) remarked that if we denote by $f(T, x)$ the density function of the distribution of S_T under the risk measure $\mathbb{P}^*$, and by $G(T, K)$ the price at time 0 of the call option with maturity T and strike price K, then

$$G(T, K) = \mathbb{E}^*[(S_T - K)^+] = \int_0^\infty (x - K)^+ f(T, x)dx.$$

Thus, by differentiating twice w.r.t. K, we obtain the following equality:

$$\frac{\partial^2}{\partial K^2}G(T, K) = f(T, K). \qquad (5.25)$$

Based on (5.25) Dupire (1997) further shows that "volatility function" $\sigma(t, x)$ can be inferred from family $G(T, K)_{T>t, K>0}$ of call option prices. In fact, integrating twice Kolmogorov's forward equation on K

$$\frac{\partial}{\partial T}f(T, K) = \frac{1}{2}\frac{\partial^2}{\partial K^2}\left(\sigma^2(T, K)K^2 f(T, K)\right),$$

and using (5.25), we obtain

$$\frac{\partial}{\partial T}G(T, K) = \frac{1}{2}\sigma^2(T, K)K^2\frac{\partial^2}{\partial K^2}G(T, K) + a_1 K + a_2.$$

Assume that $\lim_{T \to \infty} \frac{\partial}{\partial T} G(T, K) = 0$ and $\frac{\partial^2}{\partial K^2} G(T, K) > 0$, then we must have $a_1 = 0$, $a_2 = 0$, and we obtain the following formula for volatility function $\sigma(t, x)$:

$$\sigma^2(T, K) = \frac{2\frac{\partial}{\partial T} G(T, K)}{K^2 \frac{\partial^2}{\partial K^2} G(T, K)}. \tag{5.26}$$

The biggest advantage of this model is that when the market is complete, volatility function $\sigma(t, x)$ can be estimated by using a parametric form for the volatility function or by an interpolation algorithm based on the observed call option prices with different strikes and maturities. The estimated volatility function is called an *implied volatility function*. However, empirical tests showed that implied volatility functions are often unstable over time. So the endogenously determined volatility model is also not quite suited to modeling the stock price process.

5.4.2 Stochastic Volatility Model

To overcome the shortcoming of level-dependent model, we model the volatility as a stochastic process dependent of an exogenous random factor. This means that we model stock price process (S_t) by

$$dS_t = S_t[\mu_t dt + \sigma_t dB_t], \tag{5.27}$$

but volatility σ_t itself is modeled by a diffusion:

$$d\sigma_t = b(\sigma_t)dt + a(\sigma_t)dW_t, \tag{5.28}$$

where B_t and W_t are two correlated Brownian motions with the correlation coefficient $\rho \in (-1, 1)$ (i.e., $\mathbb{E}[B_t W_t] = \rho t$, or equivalently, $[B, W]_t = \rho t$). Usually, ρ is negative. This reflects the negative correlation between volatility and price. Such a model is called a *stochastic volatility model*. If we let Z_t be a Brownian motion independent of B_t and put $W_t = \rho B_t + \sqrt{1 - \rho^2} Z_t$, then we can rewrite (5.28) as

$$d\sigma_t = b(\sigma_t)dt + \rho a(\sigma_t)dB_t + \sqrt{1 - \rho^2} a(\sigma_t)dZ_t. \tag{5.29}$$

This model was first proposed by Hull and White (1987), in which μ_t is a constant, and the square of the volatility, denoted by v_t, is modeled by a geometric Brownian motion:

$$dv_t = v_t[bdt + \delta dW_t].$$

Heston (1993) proposed another stochastic volatility model, in which v_t follows a *mean-reverting process*

$$dv_t = (\theta - \kappa v_t)dt + \delta\sqrt{v_t}dW_t,$$

where θ and κ are two positive constants, κ is the reversion rate, and θ/κ is the reversion level.

It turned out that stochastic volatility models are well suited to capture smiles and skewness in implied volatility. For more general stochastic volatility models, we refer the reader to Hofmann et al. (1992) and Frey (1997).

Now we come to option pricing. Since the volatility is not a tradable asset, markets with stochastic volatility model are incomplete. Consequently, many contingent claims cannot be priced by arbitrage (see below Chap. 7). In this case, option pricing will depend on the risk preferences of investors. They may choose utility functions as preferences or specify the market price of volatility risk.

Consider a stochastic volatility model (5.27), (5.28), and (5.29), where $\mu_t = \mu(\sigma_t)$. Assume that $Z_t^* = Z_t + \int_0^t \lambda(\sigma_s)ds$ is a Brownian motion under a risk-neutral probability measure $\mathbb{P}^*$. Then we have

$$dS_t = S_t[rdt + \sigma_t dB_t^*], \tag{5.30}$$

where $B_t^* = B_t + \int_0^t \frac{\mu(\sigma_s)-r}{\sigma_s}ds$ is a Brownian motion under $\mathbb{P}^*$. Since $[Z^*, B^*]_t = [Z, B]_t = 0$, Z_t^* and B_t^* are two independent Brownian motions. From (5.29) and (5.30), we get

$$d\sigma_t = \tilde{b}(\sigma_t)dt + \rho a(\sigma_t)dB_t^* + \sqrt{1-\rho^2}a(\sigma_t)dZ_t^*, \tag{5.31}$$

where

$$\tilde{b}(\sigma_t) = b(\sigma_t) - \frac{\rho a(\sigma_t)(\mu(\sigma_t)-r)}{\sigma_t} - \sqrt{1-\rho^2}a(\sigma_t)\lambda(\sigma_t). \tag{5.32}$$

If μ_t does not depend on σ_t and two Brownian motions B_t and W_t are independent (i.e., $\rho = 0$), then it is easy to show that the price of a European option on the stock is equal to the Black-Scholes price integrated over the probability distribution of the average squared volatility. For the derivation of the result, we refer to Hull and White (1987).

For the general case, we anticipate that the price process V_t of a European option $f(S_T)$ written on the stock can be expressed as $F(t, S_t, \sigma_t)$ with $F(t, x, y)$ being a $C^{1,2,2}$-function on $[0, T) \times (0, \infty) \times \mathbb{R}$. Then we have

$$F(t, x, y) = e^{-r(T-t)}\mathbb{E}^*[f(S_T) \mid S_t = x, \sigma_t = y]. \tag{5.33}$$

By Feynman-Kac formula, $F(t, x, y)$ satisfies the following PDE:

$$-rF + F_t + \frac{1}{2}a^2 F_{yy} + rxF_x + \rho xya F_{xy} + \frac{1}{2}x^2 y^2 F_{xx} + \tilde{b}F_y = 0, \qquad (5.34)$$

subject to the terminal condition $F(T, x, y) = f(x)$. In some particular cases, a closed-form expression for the option price is available.

5.4.3 SABR Model

Hagan et al. (2002) proposed a *stochastic alpha-beta-rho model* (abbreviated as *SABR model*), in which the stock price (S_t) is described by (5.27), where

$$\mu_t \equiv 0, \quad \sigma_t = \alpha_t S_t^{\beta-1},$$

(α_t) obeys the following geometric Brownian motion:

$$d\alpha_t = v\alpha_t dW_t.$$

Here $\beta \in [0, 1]$ and $v > 0$ are constants, B_t and W_t are two correlated Brownian motions with the correlation coefficient ρ. In fact, v is the volatility of process (α_t). If $v = 0$, the SABR model is reduced to a CEV model with elasticity factor 2β.

The SABR model is very popular in the financial industry, especially in the foreign exchange market and the interest rate market. This is mainly due to the fact that it uses only four parameters $(\alpha_0, v, \rho, and \beta)$ to better fit the various types of implied volatility curve observed in the market. In addition, Hagan et al. (2002) derives explicit asymptotic expansions of European option prices and implied volatilities using singular perturbation techniques, which facilitate the related calculation under the SABR model. Option pricing and related calculations under the SABR model can be further referred to Glasserman and Wu (2011).

5.4.4 Variance-Gamma (VG) Model

One possible explanation for random volatility is that due to nonhomogeneous information flow and varying volume of trading, the asset return process is a Brownian motion with drift, only when the time parameter t is interpreted as an *intrinsic clock* rather than calendar time. In other words, the asset return process is given by a random time change of a Brownian motion with drift:

$$X(t) = \sigma B(A(t)) + \theta A(t),$$

where A_t is an increasing process, independent of S. This process A is called a *subordinator*.

Madan and Seneta (1990) proposed to choose a gamma process as the subordinator. The *gamma process* $\Gamma(t, \nu)$ is an increasing Lévy process (i.e., a process of independent and stationary increments) whose increments over time intervals $(t, t + h)$ have the gamma density function $f_h(g)$ with mean h and variance νh:

$$f_h(g) = \frac{g^{h/\nu - 1} \exp(-g/\nu)}{(\nu)^{h/\nu} \Gamma(\frac{h}{\nu})}, \quad g > 0,$$

where $\Gamma(x)$ is the gamma function. The *variance-Gamma (VG) process* is given by

$$X(t; \sigma, \nu, \theta) = \theta G(t, \nu) + \sigma B(G(t, \nu)).$$

It turns out that a gamma process is a pure jump process. The first four central moments of random variable $X(t)$ are as follows:

$$\mathbb{E}[X(t)] = \theta t,$$
$$\mathbb{E}[(X(t) - \theta t)^2] = (\theta^2 \nu + \sigma^2) t,$$
$$\mathbb{E}[(X(t) - \theta t)^3] = (2\theta^3 \nu^2 + 3\sigma^2 \theta \nu) t,$$
$$\mathbb{E}[(X(t) - \theta t)^4] = (3\theta^4 \nu + 12\sigma^2 \theta^2 \nu^2 + 6\sigma^4 \nu^3) t$$
$$+ (3\sigma^4 + 6\sigma^2 \theta^2 \nu + 3\theta^4 \nu^2) t^2.$$

We observe that the skewness is zero if $\theta = 0$, and the fourth central moment divided by the squared second central moment is $3(1 + \nu)$, and so ν is the ratio of the excess kurtosis of $X(t)$.

Now we assume that under the historical probability measure, the log stock price is driven by a VG process $X(t; \sigma, \nu, \theta)$:

$$\ln S_t = \ln S_0 + \alpha t + \theta G(t, \nu) + \sigma B(G(t, \nu)).$$

Then under the risk-neutral probability measure, we may write the stock price process as:

$$S_t = S_0 \exp\left[rt + \frac{t}{\nu} \ln(1 - \theta \nu - \sigma^2 \nu/2) + \theta G(t, \nu) + \sigma B(G(t, \nu)) \right],$$

where r is the constant interest rate of the money market account. The skewness and kurtosis of the stock return distribution are controlled, respectively, by parameters θ and ν. When ν tends to zero, we obtain the Black-Scholes model. We refer the reader to Madan et al. (1998) for a closed form solution to option pricing in this model (see also Madan (2001)).

5.4.5 GARCH Model

Engle (1982) introduced for the first time the *autoregressive conditional heteroscedasticity model* (abbreviated as *ARCH model*) in an attempt to explain a stylized fact (called *volatility clustering*) observed in financial markets that there is a succession of periods with high and low volatilities. If we denote by P_n the sequence of asset prices, observed at discrete times, then an ARCH(p) model assumes that the (adjusted) innovation of log-price $h_n = \ln P_n - \ln P_{n-1} - v$ obeys

$$h_n = \sigma_n \varepsilon_n, \tag{5.35}$$

where (ε_n) is a sequence of i.i.d. random variables with the standard normal distribution and the *conditional variance* σ_n^2 possesses an autoregressive structure with random coefficients:

$$\sigma_n^2 = \alpha_0 + \sum_{i=1}^{p} \alpha_i h_{n-i}^2. \tag{5.36}$$

According to this model, the distribution of the "innovation" of the standardized log-price is normal. This does not meet the empirical data. In order to overcome this shortcoming, Bollerslev (1986) proposed a class of generalized ARCH or GARCH models. In a GARCH(p, q) model, we have

$$\sigma_n^2 = \alpha_0 + \sum_{i=1}^{p} \alpha_i h_{n-i}^2 + \sum_{j=1}^{q} \beta_j \sigma_{n-j}^2. \tag{5.37}$$

It turned out that GARCH models are discrete-time approximations of stochastic volatility models. For further discussion about this result, one can see Nelson (1990) and Bollerslev et al. (1994). For the option pricing under GARCH models see Duan (2005).

Chapter 6
Pricing and Hedging of Exotic Options

Usually we call an option whose payoff at the exercise or expiry time depends only on the current price of the underlying asset (such as European option, American option, compound option, etc., studied in Chap. 5) a *vanilla option* (here "vanilla" stands for "ordinary"). Any option that is not vanilla is called an *exotic option*. Exotic options are widely used in investment and risk management by banks, corporations, and institutional investors. Typical exotic options are the *barrier options*, *Asian options*, *lookback options*, and *reset options*. They are all *path dependent* in the sense that their payoffs at exercise or expiry times depend on the history of the underlying asset. For an analysis of many exotic options we refer Zhang (2006).

In this chapter we will study pricing and hedging of the abovementioned exotic options in the Black-Scholes setting. We consider only European-style options with notations of Chap. 5.

6.1 Running Extremum of Brownian Motion with Drift

We will need the following result about the joint distribution of the Brownian motion with drift at time t and its running maximum or minimum on interval $[0, T]$.

Theorem 6.1 *Let B be -dimensional Brownian motion, $\alpha, \lambda \in \mathbb{R}, \sigma > 0$. Then for $x \geqslant \alpha, y \leqslant x$, we have*

$$
\begin{aligned}
\mathbb{P}&\left(\alpha + \sigma B_t + \lambda t \geqslant y, \max_{s \leqslant t}(\alpha + \sigma B_s + \lambda s) \leqslant x\right) \\
&= N\left(\tfrac{x-\alpha-\lambda t}{\sigma\sqrt{t}}\right) - N\left(\tfrac{y-\alpha-\lambda t}{\sigma\sqrt{t}}\right) \\
&\quad - e^{2\lambda(x-\alpha)/\sigma^2}\left[N\left(\tfrac{-x+\alpha-\lambda t}{\sigma\sqrt{t}}\right) - N\left(\tfrac{-(2x-\alpha-y)-\lambda t}{\sigma\sqrt{t}}\right)\right],
\end{aligned}
\tag{6.1}
$$

© Springer Nature Singapore Pte Ltd. and Science Press 2018
J.-A. Yan, *Introduction to Stochastic Finance*, Universitext,
https://doi.org/10.1007/978-981-13-1657-9_6

$$\mathbb{P}\Big(-\alpha + \sigma B_t + \lambda t \leqslant -y, \min_{s\leqslant t}(-\alpha + \sigma B_s + \lambda s) \leqslant -x\Big)$$
$$= N\Big(\tfrac{-x+\alpha-\lambda t}{\sigma\sqrt{t}}\Big) + e^{-2\lambda(x-\alpha)/\sigma^2}\Big[N\Big(\tfrac{-x+\alpha+\lambda t}{\sigma\sqrt{t}}\Big) - N\Big(\tfrac{-(2x-\alpha-y)+\lambda t}{\sigma\sqrt{t}}\Big)\Big]. \tag{6.2}$$

Proof We only need to prove the result for the maximum case (6.1). For notational simplicity we assume that $\alpha = 0$ and $\sigma = 1$. Put $T_x = \inf\{t \geqslant 0 : B_t = x\}$ and

$$\widetilde{B}_t = B_t I_{[t<T_x]} + (2x - B_t)I_{[t\geqslant T_x]}.$$

By the reflection principle for Brownian motion (see Sect. 4.3.3 of Chap. 4), $(\widetilde{B}_t)$ is a Brownian motion. Since for all $s \leqslant t$ $\widetilde{B}_s = B_s$ on $[t \leqslant T_x]$, we have

$$[\max_{s\leqslant t}\widetilde{B}_s \geqslant x, y \leqslant \widetilde{B}_t \leqslant x] = [2x - y \geqslant B_t \geqslant x] \subset [t \geqslant T_x].$$

Consequently, by Girsanov's theorem, we have

$$\mathbb{P}\Big(B_t + \lambda t \geqslant y, \max_{s\leqslant t}(B_s + \lambda s) \leqslant x\Big)$$
$$= \mathbb{P}\Big(\widetilde{B}_t + \lambda t \geqslant y, \max_{s\leqslant t}(\widetilde{B}_s + \lambda s) \leqslant x\Big)$$
$$= \mathbb{P}(y \leqslant \widetilde{B}_t + \lambda t \leqslant x) - \mathbb{P}\Big(\max_{s\leqslant t}(\widetilde{B}_s + \lambda s) \geqslant x, y \leqslant \widetilde{B}_t + \lambda t \leqslant x\Big)$$
$$= \mathbb{P}(y \leqslant \widetilde{B}_t + \lambda t \leqslant x) - \mathbb{E}\Big[\exp\Big\{\lambda\widetilde{B}_t - \frac{\lambda^2}{2}t\Big\}I_{[\max_{s\leqslant t}\widetilde{B}_s \geqslant x, y \leqslant \widetilde{B}_t \leqslant x]}\Big]$$
$$= \mathbb{P}(y \leqslant B_t + \lambda t \leqslant x) - \mathbb{E}\Big[\exp\Big\{\lambda(2x - B_t) - \frac{\lambda^2}{2}t\Big\}I_{[2x-y\geqslant B_t \geqslant x]}\Big]$$
$$= \mathbb{P}(y \leqslant B_t + \lambda t \leqslant x) - e^{2\lambda x}\mathbb{E}\Big[\exp\Big\{-\lambda B_t - \frac{\lambda^2}{2}t\Big\}I_{[2x-y\geqslant B_t \geqslant x]}\Big]$$
$$= \mathbb{P}(y \leqslant B_t + \lambda t \leqslant x) - e^{2\lambda x}\mathbb{P}(-(2x - y) \leqslant -B_t + \lambda t \leqslant -x)$$
$$= N\Big(\frac{x - \lambda t}{\sqrt{t}}\Big) - N\Big(\frac{y - \lambda t}{\sqrt{t}}\Big) - e^{2\lambda x}\Big[N\Big(\frac{-x - \lambda t}{\sqrt{t}}\Big) - N\Big(\frac{-(2x - y) - \lambda t}{\sqrt{t}}\Big)\Big].$$

The theorem is proved.   □

Let B be a one-dimensional Brownian motion. Let $\alpha, \lambda \in \mathbb{R}, \sigma > 0$. Put

$$X_t = \alpha + \sigma B_t + \lambda t, \quad m_t = \min\{X_s : 0 \leqslant s \leqslant t\}, \quad M_t = \max\{X_s : 0 \leqslant s \leqslant t\}.$$

Below we follow the method of Freedman (1983) to study the joint distribution of (B_t, m_t, M_t). To this end we fist prove a lemma.

Lemma 6.2 *Let* $y \in \mathbb{R}$, *H be a Borel set. Put* $\tau_y = \inf\{t \geqslant 0 : B_t = y\}$, $r_y H = \{2y - x : x \in H\}$. *Let* $a, b \in \mathbb{R}$, $a < 0 < b$.

(1) *If* $H \subset (-\infty, a]$, *then*

$$\mathbb{P}(B_t \in H, \tau_b < \tau_a) = \mathbb{P}(B_t \in r_b H, \tau_b < \tau_a).$$

(2) *If* $H \subset [b, \infty)$, *then*

$$\mathbb{P}(B_t \in H, \tau_a < \tau_b) = \mathbb{P}(B_t \in r_a H, \tau_a < \tau_b).$$

Proof We only prove (1); the proof of (2) is similar. Obviously,

$$[B_t \in H, \tau_b < \tau_a] \subset [\tau_b \leqslant t], \quad [B_t \in r_b H] \subset [\tau_b \leqslant t].$$

For stopping time τ_b, applying the reflection principle for Brownian motion gives

$$\mathbb{P}(B_t \in H, \tau_b < \tau_a) = \mathbb{P}(2b - B_t \in H, \tau_b < \tau_a) = \mathbb{P}(B_t \in r_b H, \tau_b < \tau_a).$$

Thus (1) is proved.   □

Theorem 6.3 *Assume* $a < \alpha < b$, $a < x < b$. *For any Borel set* $J \subset [a, b]$, *let* $J(\alpha, \sigma) = \left\{ \frac{x - \alpha}{\sigma} : x \in J \right\}$. *Then*

$$\mathbb{P}(X_t \in J, a < m_t \leqslant M_t < b) = \int_{J(\alpha, \sigma)} \exp\left\{ \frac{\lambda}{\sigma} y - \frac{\lambda^2}{2\sigma^2} \right\} k(y) dy, \qquad (6.3)$$

where

$$k(y) = \sum_{n=-\infty}^{+\infty} \left[\varphi\left(y; -\frac{2n(b-a)}{\sigma}, \sqrt{t}\right) - \varphi\left(y; \frac{2a - 2n(b-a)}{\sigma}, \sqrt{t}\right) \right], \qquad (6.4)$$

$$\varphi(y; \mu, \delta) = \frac{1}{\delta\sqrt{2\pi}} \exp\left\{ -\frac{(y-\mu)^2}{2\delta^2} \right\}.$$

Proof We may assume $\alpha = 0$. First consider the case of standard Brownian motion, namely, $\sigma = 1$ and $\lambda = 0$. For any Borel set $J \subset [a, b]$,

$$\mathbb{P}(B_t \in J, a < m_t \leqslant M_t < b) = \mathbb{P}(B_t \in J) - \mathbb{P}(B_t \in J, \tau_b < \tau_a, \tau_b \leqslant t)$$

$$-\mathbb{P}(B_t \in J, \tau_a < \tau_b, \tau_a \leqslant t) = T_1 - T_2 - T_3.$$

Since $[B_t \in r_a J] \subset [\tau_a \leqslant t]$, for stopping time τ_a, applying the reflection principle for Brownian motion gives

$$T_3 = \mathbb{P}(B_t \in J, \tau_a < \tau_b, \tau_a \leqslant t) = \mathbb{P}(B_t \in r_a J, \tau_a < \tau_b)$$

$$= \mathbb{P}(B_t \in r_a J) - \mathbb{P}(B_t \in r_a J, \tau_b < \tau_a).$$

Since $r_a J \subset (-\infty, a]$, then $r_b r_a J \subset [b, \infty), \cdots$. Repeatedly applying Lemma 6.2 gives

$$T_3 = \mathbb{P}(B_t \in r_a J) - \mathbb{P}(B_t \in r_b r_a J) + \mathbb{P}(B_t \in r_a r_b r_a J) - \cdots.$$

Since

$$(r_b r_a)^n x = x + 2n(b - a), \quad (r_a r_b)^n x = x - 2n(b - a),$$

we have

$$\mathbb{P}(B_t \in (r_b r_a)^n J) = \mathbb{P}(B_t - 2n(b - a) \in J) = \int_J \varphi(y; -2n(b - a), \sqrt{t}) dy,$$

$$\mathbb{P}(B_t \in r_a (r_b r_a)^n J) = \mathbb{P}(-B_t + 2a - 2n(b-a) \in J) = \int_J \varphi(y; 2a - 2n(b-a), \sqrt{t}) dy,$$

from which it follows $T_3 = \int_J k_3(y) dy$, where

$$k_3(y) = \sum_{n=0}^{\infty} \varphi(y; 2a - 2n(b - a), \sqrt{t}) - \sum_{n=1}^{\infty} \varphi(y; -2n(b - a), \sqrt{t}).$$

Similarly, we can prove $T_2 = \int_J k_2(y) dy$, where

$$k_2(y) = \sum_{n=-\infty}^{-1} \varphi(y; 2a - 2n(b - a), \sqrt{t}) - \sum_{n=-\infty}^{-1} \varphi(y; -2n(b - a), \sqrt{t}).$$

So in the case of Brownian motion, we have

$$\mathbb{P}(X_t \in H, a < m_t \leqslant M_t < b) = \int_J k_1(y) dy, \tag{6.5}$$

where

$$k_1(y) = \sum_{n=-\infty}^{+\infty} [\varphi(y; -2n(b - a), \sqrt{t}) - \varphi(y; 2a - 2n(b - a), \sqrt{t})]. \tag{6.6}$$

Finally, by Girsanov's theorem, it is easy to derive (6.3) from (6.5) for the Brownian motion case. The detailed proof is left to the reader as an exercise. $\square$

6.2 Barrier Options

Barrier options are options that are either worthless ("out") or established ("in"), when the price of the underlying asset crosses a particular level ("barrier"). Common examples of single-barrier options are "down-and-out", "down-and-in", "up-and-out," and "up-and-in" options for calls or puts. A *double-knock binary option* or *up-and-down out binary option* is the simplest *double-barrier option*. It is characterized by two barriers, L (lower barrier) and U (upper barrier): the option knocks out if either barrier is touched. Otherwise, the option gives at maturity the usual binary payoff.

Barrier options have become increasingly popular over the last few years. Since their price is less expensive than the standard options, barrier options may be an appropriate hedge in a number of situations. For instance, a down-and-in put with a low barrier offers an inexpensive protection against a downward move of the underlying asset. Pricing barrier options in the Black-Scholes setting is not difficult, as will be seen below. Closed-form solutions for all types of single-barrier options have been offered by Goldman et al. (1979) (see also Rubinstein (1992)). Two different analytic expressions for double-barrier options have been worked out by Geman and Yor (1996) and Hui (1996), using Laplace transform and partial differential equation methods, respectively. We will use a probabilistic approach of Bjork (1998) to convert the pricing of barrier options into the pricing of ordinary options.

6.2.1 Single-Barrier Options

We only consider a down-and-out call option with strike price K, maturity T, and a *lower barrier* $L < S_0$; the other cases can be treated in a similar way. The payoff at expiry of this option is the same as that for call option (i.e., $(S_T - K)^+$), provided that S_t never falls below L before T. Therefore,

$$\xi = \begin{cases} (S_T - K)^+, & \text{if for all } t \in [0, T], S_t > L, \\ 0, & \text{if for certain } t \in [0, T], S_t \leqslant L, \end{cases}$$

Namely, $\xi = (S_T - K)^+ I_{[\min_{0 \leqslant t \leqslant T} S_t > L]}$. Let $\widetilde{C}(t, S_t)$ to express the price of this barrier option at time t. Since under the risk-neutral probability measure $\mathbb{P}^*$,

$$\ln S_t = \ln S_0 + \left(r - \frac{\sigma^2}{2}\right) t + \sigma B_t^*,$$

from formula (6.1), we derive easily:

$$\widetilde{C}(t, x) = C(t, x) - \left(\frac{x}{L}\right)^{-(k-1)} C(t, X^2/x), \tag{6.7}$$

where $k = \frac{2r}{\sigma^2}$. More generally, we can consider down-and-out contingent claim $\xi = g(S_T)I_{[\inf_{0\leqslant t\leqslant T} S_t > L]}$, where g is a nonnegative Borel function. Using formula (6.1), we can give its pricing formula.

6.2.2   Double-Barrier Options

Now we consider a double-barrier knockout call option with strike price K, maturity T, lower barrier $L < S_0$, and upper barrier $U > S_0$. The option knocks out if before the maturity T either barrier is touched. Otherwise, the option gives at maturity the usual call option payoff (i.e., $(S_T - K)^+$). Thus,

$$\xi = \begin{cases} (S_T - K)^+, & \text{if for all } t \in [0, T], U > S_t > L, \\ 0, & \text{if for certain } t \in [0, T], S_t \geqslant U, \text{ or } S_t \leqslant L, \end{cases}$$

i.e., $\xi = (S_T - K)^+ I_{[U > \max_{0\leqslant t\leqslant T} S_t \geqslant \min_{0\leqslant t\leqslant T} S_t > L]}$. Using (6.3) we can give the pricing formula for the double-barrier knockout call option, and the concrete expression is omitted.

6.3   Asian Options

An *Asian option* is an option whose payoff depends on a suitably defined average of prices of the underlying asset. There are two kinds of average: geometric and arithmetic. For each kind of average, there are two types of options: the *average strike* and *average rate* options. They are similar to a European vanilla option. The only difference is that in the payoff: for the former, the strike price is replaced by the average, and for the latter, the asset price is replaced by the average. The sampling for the average can be discrete or continuous. In this section we only consider the average rate options with continuous sampling.

There are two types of average rate call options whose payoffs are equal to

$$\xi_1 = \left(\exp\left\{\frac{1}{T}\int_0^T \ln(S_u)du\right\} - K\right)^+$$

and

$$\xi_2 = \left(\frac{1}{T}\int_0^T S_u du - K\right)^+,$$

respectively. The first one uses the geometric average, while the second one uses the arithmetic average. We denote by $C_t^{(i)}$ the price at time t of ξ_i, $i = 1, 2$.

6.3.1 Geometric Average Asian Options

Consider the case of geometric average. Let $\mathbb{P}^*$ be the equivalent martingale measure for $\widetilde{S}$. Then

$$C_t^{(1)} = e^{-r(T-t)}\mathbb{E}^*[\xi_1|\mathcal{F}_t].$$

Put

$$I_t = \int_0^t \ln(S_u)du.$$

We have

$$\xi_1 = \left(\exp\left\{\frac{1}{T}I_t + \frac{1}{T}\int_t^T \ln(S_uS_t^{-1})du + \frac{T-t}{T}\ln S_t\right\} - K\right)^+,$$

$$= (X_tY_t - K)^+,$$

where

$$X_t = e^{I_t/T}S_t^{(T-t)/T}, \quad Y_t = \exp\left\{\frac{1}{T}\int_t^T \ln S_u S_t^{-1}du\right\}. \tag{6.8}$$

Noting that

$$S_t = S_0 \exp\left\{\left(r - \frac{\sigma^2}{2}\right)t + \sigma B_t^*\right\},$$

we get

$$Y_t = \exp\left\{\frac{1}{T}\int_t^T\left[\left(r - \frac{\sigma^2}{2}\right)(u-t) + \sigma(B_u^* - B_t^*)\right]du\right\} = e^{r^*(T-t)+Z_t},$$

where

$$r^* = \left(r - \frac{\sigma^2}{2}\right)\frac{T-t}{2T}, \quad Z_t = \frac{1}{T}\int_t^T \sigma(B_u^* - B_t^*)du. \tag{6.9}$$

Since Z_t is independent of $\mathcal{F}_t$, and X_t is $\mathcal{F}_t$-measurable, by Theorem 1.11, we obtain

$$C_t^{(1)} = e^{-r(T-t)}F(t, X_t),$$

where

$$F(t, x) = \mathbb{E}^*[(xe^{r^*(T-t)+Z_t} - K)^+].$$

Noting that Z_t is a Gaussian random variable with mean zero and variance $\sigma^{*2}(T - t)$, where

$$\sigma^{*2} = \frac{\sigma^2(T - t)^2}{3T^2}, \tag{6.10}$$

we have

$$F(t, x) = e^{r^*(T-t)} \int_{-\infty}^{\infty} \left(xe^{\sigma^*\sqrt{T-t}y} - Ke^{-r^*(T-t)}\right)^+ \frac{1}{\sqrt{2\pi}} e^{-\frac{y^2}{2}} dy \tag{6.11}$$

$$= xe^{(r^*+\frac{\sigma^{*2}}{2})(T-t)} N(d_1^*) - KN(d_2^*),$$

where

$$d_1^* = \frac{\ln(x/K) + (r^* + \sigma^{*2})(T - t)}{\sigma^*\sqrt{T - t}}, \quad d_2^* = \frac{\ln(x/K) + r^*(T - t)}{\sigma^*\sqrt{T - t}}.$$

Finally, we obtain

$$\tilde{C}_t^{(1)} = e^{-rT} F(t, X_t), \quad X_t = e^{I_t/T} S_t^{(T-t)/T}.$$

In order to get the hedging strategy for the geometric average rate call option, we first compute $F_x(t, x)$. By the first equality in (6.11), it is easy to see that

$$F_x(t, x) = e^{(r^*+\frac{\sigma^{*2}}{2})(T-t)} N(d_1^*). \tag{6.12}$$

In addition, we have

$$dX_t = e^{I_t/T} \frac{T-t}{T} S_t^{-t/T} S_t \sigma dB_t^* + \text{``}dt\text{'' term}$$

$$= \frac{T-t}{T} X_t \sigma dB_t^* + \text{``}dt\text{'' term}.$$

Since $\tilde{C}_t^{(1)}$ is a $\mathbb{P}^*$-martingale, from Theorem 4.2, we must have

$$d\tilde{C}_t^{(1)} = e^{-rT} F_x(t, X_t) \frac{T-t}{T} X_t \sigma dB_t^*.$$

Let $a(t)$ denote the number of units of the risky asset held at time t of the hedging strategy. By (5.9) and (6.12), we get

$$a(t) = e^{-r(T-t)} F_x(t, X_t) \frac{T-t}{T} \frac{X_t}{S_t}$$

$$= e^{\left(-r+r^*+\frac{\sigma^{*2}}{2}\right)(T-t)} \frac{X_t}{S_t} N\left(\frac{\ln(X_t/K) + (r^* + \sigma^{*2})(T - t)}{\sigma^*\sqrt{T - t}}\right) \frac{T-t}{T}.$$

6.3.2 Arithmetic Average Asian Options

Now we turn to the pricing of an arithmetic average rate call option. Geman and Yor (1993) worked out the Laplace transform of the price. Rogers and Shi (1995) have derived a PDE for the price. But both approaches are difficult to perform numerically. Vecer (2001, 2002) obtained a simple PDE for the price which is easy to solve numerically. A similar formulation of the pricing PDE is obtained independent by Hoogland and Neumann (2001).

6.3.2.1 PDE Method

First of all, we follow Rogers and Shi (1995) to derive the PDE satisfied by the pricing function of an arithmetic average rate call option. We have

$$C_t^{(2)} = \mathbb{E}^* \left[e^{-r(T-t)} \left(\frac{1}{T} \int_0^T S_u du - K \right)^+ \Big| \mathcal{F}_t \right].$$ (6.13)

Put

$$M_t = \mathbb{E}^* \left[\left(\int_0^T S_u du - TK \right)^+ \Big| \mathcal{F}_t \right].$$ (6.14)

Since $\int_t^T S_t^{-1} S_u du$ is independent of $\mathcal{F}_t$, we have

$$M_t = S_t \mathbb{E}^* \left[\left(\int_t^T S_t^{-1} S_u du - S_t^{-1} \left(TK - \int_0^t S_u du \right) \right)^+ \Big| \mathcal{F}_t \right] = S_t f(t, Y_t),$$

where

$$f(t, x) = \mathbb{E}^* \left[\left(\int_t^T S_t^{-1} S_u du - x \right)^+ \right],$$ (6.15)

$$Y_t = S_t^{-1} \left(TK - \int_0^t S_u du \right).$$

For the case of $x \leqslant 0$, we have

$$f(t, x) = \mathbb{E}^* \left[\left(\int_t^T S_t^{-1} S_u du - x \right) \right] = \int_t^T e^{r(u-t)} du - x = r^{-1}(e^{r(T-t)} - 1) - x.$$

Now consider the case of $x \geqslant 0$. Since

$$dS_t^{-1} = -S_t^{-2} dS_t + S_t^{-3} d\langle S, S \rangle_t = S_t^{-1}[(\sigma^2 - r)dt - \sigma dB_t^*],$$ (6.16)

we get

$$dY_t = \left(TK - \int_0^t S_u du\right)dS_t^{-1} - dt = Y_t[(\sigma^2 - r)dt - \sigma dB_t^*] - dt.$$

Thus, we have

$$d\langle Y, Y\rangle_t = Y_t^2 \sigma^2 dt,$$

$$d\langle S, f(\cdot, Y.)\rangle_t = f_x(t, Y_t)d\langle S, Y\rangle_t = -f_x(t, Y_t)S_t Y_t \sigma^2 dt.$$

In the sequel, by notation $A \sim B$, we mean that $A - B$ is a $\mathbb{P}^*$-local martingale. By Itô's formula,

$$dM_t = S_t df(t, Y_t) + f(t, Y_t)dS_t + d\langle S, f(\cdot, Y.)\rangle_t$$

$$= S_t\left[f_t(t, Y_t)dt + f_x(t, Y_t)dY_t + \frac{1}{2}f_{xx}(t, Y_t)d\langle Y, Y\rangle_t\right]$$

$$+ f(t, Y_t)dS_t + d\langle S, f(\cdot, Y.)\rangle_t$$

$$\sim S_t\left(f_t - (1 + rY_t)f_x + \frac{1}{2}\sigma^2 Y_t^2 f_{xx} + rf\right)(t, Y_t)dt.$$

Since (M_t) is a $\mathbb{P}^*$-martingale, from Theorem 4.2, we know that the right side of the above equation must be equal to zero. This induces the following PDE:

$$f_t - (1 + rx)f_x + \frac{\sigma^2 x^2}{2}f_{xx} + rf = 0, \quad x \geqslant 0. \tag{6.17}$$

According to (6.15), the first boundary condition is

$$f(T, x) = 0,$$

and the second boundary condition is

$$f(t, 0) = r^{-1}(e^{r(T-t)} - 1),$$

which is due to the fact that

$$\mathbb{E}^*[S_t^{-1}S_u] = \mathbb{E}^*\left[\exp\left\{\sigma(B_u^* - B_t^*) - \left(\frac{\sigma^2}{2} - r\right)(u - t)\right\}\right] = e^{r(u-t)}.$$

In addition, since

$$f_x(t, x) = -\mathbb{P}^*\left(\int_t^T S_t^{-1}S_u du \geqslant x\right),$$

we have the third boundary condition:

$$\lim_{x \to \infty} f_x(t, x) = 0.$$

Unfortunately, we have not yet found an analytical solution to Eq. (6.17), and it is difficult to implement the numerical calculation of the equation. Especially, when σ is very small, the error of the numerical calculation will be very big. However, Rogers and Shi (1995) give a relatively lower bound for the solution.

In the following we introduce a partial differential equation given by Vecer (2001), which is convenient for the numerical solution to the option pricing function. Put $\xi = \frac{1}{T} \int_0^T S_u du - K$. Suppose that the self-financing strategy replicating ξ holds the stock share at time t is $a(t)$ and its wealth process is (V_t). Then from Theorem 5.2, we know

$$V_t = \mathbb{E}^* \left[e^{-r(T-t)} \left(\frac{1}{T} \int_0^T S_u du - K \right) | \mathcal{F}_t \right], \tag{6.18}$$

$$d\widetilde{V}_t = a(t) \sigma \widetilde{S}_t d B_t^*. \tag{6.19}$$

By (6.18) we get

$$\widetilde{V}_t = \frac{e^{-rT}}{T} \left\{ \int_0^t S_u du + \mathbb{E}^* \left[\int_t^T S_u du | \mathcal{F}_t \right] \right\} - e^{-rT} K$$

$$= \frac{e^{-rT}}{T} \left\{ \int_0^t S_u du + \widetilde{S}_t \int_t^T e^{ru} du \right\} - e^{-rT} K$$

$$= \frac{e^{-rT}}{T} \int_0^t S_u du + \frac{1}{rT}(1 - e^{-r(T-t)}) \widetilde{S}_t - e^{-rT} K.$$

Since $(\widetilde{V}_t)$ is a local martingale, $d\widetilde{S}_t = \sigma \widetilde{S}_t d B_t^*$, by Theorem 4.2, we must have

$$d\widetilde{V}_t = \frac{1}{rT}(1 - e^{-r(T-t)}) \sigma \widetilde{S}_t d B_t^*.$$

Thus, we get

$$a(t) = \frac{1}{rT}(1 - e^{-r(T-t)}), \tag{6.20}$$

and

$$V_0 = a(0) S_0 - e^{-rT} K.$$

By (6.19),

$$dV_t = a(t) \sigma S_t d B_t^* + r V_t dt.$$

Put $Z_t = V_t S_t^{-1}$. From the above equation and (6.16), we obtain

$$
\begin{aligned}
dZ_t &= S_t^{-1} dV_t + V_t dS_t^{-1} + d[V, S^{-1}]_t \\
&= a(t)\sigma dB_t^* + r Z_t dt + Z_t[(\sigma^2 - r)dt - \sigma dB_t^*] - a(t)\sigma^2 dt \\
&= \sigma(a(t) - Z_t)d(B_t^* - \sigma t).
\end{aligned}
$$

Set

$$
D_t = \frac{d\widetilde{\mathbb{P}}}{d\mathbb{P}}\bigg|_{\mathcal{F}_t} = \exp\left\{\sigma B_t^* - \frac{1}{2}\sigma^2 t\right\} = e^{-rt}\frac{S_t}{S_0}.
$$

Then $\widetilde{B}_t = B_t^* - \sigma t$ is a $\widetilde{\mathbb{P}}$-Brownian motion, and

$$
dZ_t = \sigma(a(t) - Z_t)d\widetilde{B}_t.
$$

Put

$$
g(t, x) = \widetilde{\mathbb{E}}[(Z_T)^+ | Z_t = x],
$$

i.e., $\widetilde{\mathbb{E}}[(Z_T)^+ | \mathcal{F}_t] = g(t, Z_t)$. By Itô's formula,

$$
\begin{aligned}
dg(t, Z_t) &= \left(g_t(t, Z_t) + \frac{1}{2}g_{zz}(t, Z_t)\sigma^2(a(t) - Z_t)^2\right)dt \\
&\quad + g_z(t, Z_t)\sigma(a(t) - Z_t)d\widetilde{B}_t.
\end{aligned}
$$

Since $g(t, Z_t)$ is a $\widetilde{\mathbb{P}}$-martingale, by Theorem 4.2, the "dt" term in the above equation must be zero. Therefore, $g(t, x)$ satisfies the following PDE:

$$
\begin{cases} g_t(t, x) + \frac{1}{2}\sigma^2(a(t) - x)^2 g_{xx}(t, x) = 0, \\ g(T, x) = x^+. \end{cases}
\tag{6.21}
$$

In addition, g satisfies another two boundary conditions:

$$
\lim_{x \to -\infty} g(t, x) = 0, \quad \lim_{x \to \infty} g(t, x) = \infty.
$$

On the other hand, using $C_t^{(2)}$ to repress the price of arithmetic average rate call option at time t, by Bayes' rule of conditional expectation (Theorem 1.11), we have

$$
\begin{aligned}
C_t^{(2)} &= e^{-r(T-t)}\mathbb{E}^*[(V_T)^+ | \mathcal{F}_t] = e^{-r(T-t)} D_t \widetilde{\mathbb{E}}[D_T^{-1}(V_T)^+ | \mathcal{F}_t] \\
&= S_t \widetilde{\mathbb{E}}[(Z_T)^+ | \mathcal{F}_t] \\
&= S_t g(t, Z_t) = S_t g(t, V_t S_t^{-1}),
\end{aligned}
$$

where

$$V_t = \frac{e^{-r(T-t)}}{T} \int_0^t S_u du + \frac{1}{rT}(1 - e^{-r(T-t)})S_t - e^{-r(T-t)}K.$$

6.3.2.2 Probabilistic Method

Now we will follow closely Yang et al. (2003) to give an analytic expression for the option pricing using a probabilistic method.

The price at time t of an arithmetic average rate call option is given by

$$C_t^{(2)} = \mathbb{E}^*\left[e^{-r(T-t)}\left(\frac{1}{T}\int_0^T S_u du - K\right)^+ \Big| \mathcal{F}_t\right]. \tag{6.22}$$

The following lemma and theorem are due to Yang et al. (2003).

Lemma 6.4 *Let $t > 0$ and a, b be constants. The probability density function of random variable $\int_0^t \exp(au + bB_u)\,du$ is given by*

$$p(t, a, b; x) =$$
$$M(x) \int_0^{+\infty} \int_0^{+\infty} L(v) y^{\frac{2a}{b^2}} \exp\left\{-\frac{2}{b^2 x}(y^2 + 2y\cosh(v) + 1)\right\} dy\,dv, \tag{6.23}$$

where

$$M(x) = 8(\pi b^3 x^2 \sqrt{2\pi t})^{-1} \exp\left\{\frac{4\pi^2 - (at)^2}{2b^2 t}\right\},$$

and

$$L(v) = \sin\left(\frac{4\pi v}{b^2 t}\right)\sinh(v)\exp\left\{-\frac{2v^2}{b^2 t}\right\}.$$

Proof Define

$$U_t(a, b; x) = \mathbb{P}\left(\int_0^t \exp(au + bB_u)du \geqslant x\right).$$

Let $v = b^2 u/4$, and noting that $\left(\frac{b}{2}B_{4v/b^2}\right)_{v \geqslant 0}$ is a Brownian motion, we obtain

$$U_t(a, b; x) = \mathbb{P}\left(\int_0^{\frac{b^2 t}{4}} \exp\left\{2\left(B_v + \frac{2av}{b^2}\right)\right\}dv \geqslant \frac{b^2 x}{4}\right).$$

We define a new probability measure $\widetilde{\mathbb{P}}$ by

$$\left.\frac{d\widetilde{\mathbb{P}}}{d\mathbb{P}}\right|_{\mathcal{F}_t} = \exp\left\{-\frac{2a}{b^2}B_t - \frac{2a^2}{b^4}t\right\}.$$

Put

$$\widetilde{B}_u = B_u + \frac{2au}{b^2}, \quad A = \left\{\omega : \int_0^{\frac{b^2t}{4}} \exp\{2\widetilde{B}_u(\omega)\}du \geqslant \frac{b^2x}{4}\right\}.$$

Then $(\widetilde{B}_u)$ is a Brownian motion under $\mathbb{P}^*$, and by Girsanov's theorem, we get

$$U_t(a, b; x) = \int_\Omega I_A \exp\left\{\frac{2a}{b^2}\widetilde{B}_{\frac{b^2t}{4}} - \frac{2a^2}{b^4} \times \frac{b^2t}{4}\right\}d\widetilde{\mathbb{P}}. \tag{6.24}$$

If we denote by $f_t(x, y)$ the joint probability density function of random vector $\left(\int_0^t \exp\{2B_u\}du, B_t\right)$ under $\mathbb{P}$, then by (6.24) we have

$$U_t(a, b; x) = \exp\left\{-\frac{a^2t}{2b^2}\right\}\int_{\frac{b^2x}{4}}^\infty \int_{-\infty}^\infty \exp\left\{\frac{2ay}{b^2}\right\}f_{\frac{b^2t}{4}}(u, y)dudy. \tag{6.25}$$

However, from (6.c) in Yor (1992) we know that $f_t(x, y)$ is given by

$$f_t(x, y) = \exp\left\{\frac{2xyt+\pi^2x-t-t\exp\{2y\}}{2xt}\right\}\bigg/x^2\sqrt{2\pi^3t}$$

$$\times \int_0^\infty \exp\left\{-\frac{z^2}{2t} - \frac{\exp\{y\}}{x}\cosh(z)\right\}\sinh(z)\sin\left(\frac{\pi z}{t}\right)dz,$$

from which and (6.25) it follows (6.23).   □

Theorem 6.5 *For any $t \in [0, T]$, we have*

$$C_t^{(2)} = \frac{1}{T}e^{-r(T-t)}S_th(t, A_t), \tag{6.26}$$

where

$$h(t, y) = \mathbb{E}^*\left[\left(\int_t^T S_t^{-1}S_udu - y\right)^+\right], \quad A_t = S_t^{-1}\left(KT - \int_0^t S_udu\right). \tag{6.27}$$

Furthermore,

$$h(t, y) = \int_y^{+\infty}(x - y)p\left(T - t, r - \frac{\sigma^2}{2}, \sigma; x\right)dx, \tag{6.28}$$

where $p(t, a, b; x)$ is given by (6.23).

Proof From (6.27) we get

$$\xi_2 = \frac{1}{T} S_t (X_t - A_t)^+,$$

where

$$X_t = \int_t^T S_t^{-1} S_u du = \int_t^T \exp\left\{\sigma(B_u^* - B_t^*) + \left(r - \frac{1}{2}\sigma^2\right)(u - t)\right\} du.$$

Since S_t and A_t are $\mathcal{F}_t$-measurable and X_t is independent of $\mathcal{F}_t$ under $\mathbb{P}^*$, from Theorem 1.11, we deduce (6.26). By Lemma 6.4, under $\mathbb{P}^*$ the density function of X_t is $p(T - t, r - \sigma^2/2, \sigma; x)$, from which we obtain (6.28). $\qquad\square$

Theorem 6.6 *The hedging strategy $a(t)$ for ξ_2 is as follows:*

$$a(t) = \frac{e^{-r(T-t)}}{T}\left[h(t, A_t) - h_x(t, A_t)A_t\right], \tag{6.29}$$

where A_t is defined by (6.27) and $h(t, y)$ is given by (6.28).

Proof From (6.16) we get

$$dA_t = \left(TK - \int_0^t S_u du\right) dS_t^{-1} - dt = A_t\left[(\sigma^2 - r)dt - \sigma dB_t^*\right] - dt. \tag{6.30}$$

Put $M_t = S_t g(t, A_t)$. Since $\widetilde{C}_t^{(2)} = \frac{e^{-rT}}{T} M_t$ is a $\mathbb{P}^*$-martingale, by (6.30) and Theorem 4.2, we must have

$$d\widetilde{C}_t^{(2)} = \frac{1}{T}e^{-rT}dM_t = \frac{1}{T}e^{-rT}\left[h(t, A_t) - h_x(t, A_t)A_t\right]S_t \sigma dB_t^*.$$

Thus, from (5.9) we deduce (6.29). $\qquad\square$

6.4 Lookback Options

A *lookback option* is an option whose payoff depends on the maximum or minimum of the realized asset prices over the option's life. Like Asian options, there are two types of options: the *lookback strike* and *lookback rate*, in both call and put varieties. They are similar to a European vanilla option, the only difference being that in the payoff. For the former, the strike price is replaced by the sampled maximum or minimum and, for the latter, the asset price is replaced by the sampled maximum or minimum. The sampling of the underlying asset price can be discrete or continuous. In this section we only consider the continuous sampling.

6.4.1   Lookback Strike Options

The payoff of a lookback strike call (resp. put) is defined by

$$\xi = S_T - \min_{0 \leqslant s \leqslant T} S_s, \quad \eta = \max_{0 \leqslant s \leqslant T} S_s - S_T.$$

We denote the prices of a call and a put at time t by C_t and P_t, respectively. By Theorem 5.2, we have

$$C_t = e^{-r(T-t)} \mathbb{E}^*[\xi \mid \mathcal{F}_t], \quad P_t = e^{-r(T-t)} \mathbb{E}^*[\eta \mid \mathcal{F}_t]. \tag{6.31}$$

We are going to deduce an explicit expression for P_t. We let $\lambda = r - \frac{1}{2}\sigma^2$ and set

$$W_t = \max_{0 \leqslant s \leqslant t} S_s, \quad L_t = \max_{t \leqslant s \leqslant T} S_s.$$

Then W_t is $\mathcal{F}_t$-measurable. Since

$$S_t^{-1} L_t = \exp\left\{ \max_{t \leqslant s \leqslant T} \left(\sigma(B_s^* - B_t^*) + \lambda(s - t) \right) \right\},$$

$S_t^{-1} L_t$ is independent of $\mathcal{F}_t$. By using these notations, we have

$$P_t = e^{-r(T-t)} \mathbb{E}^*[W_T - S_T | \mathcal{F}_t] = e^{-r(T-t)} S_t \mathbb{E}^*[\max(S_t^{-1} W_t, S_t^{-1} L_t) | \mathcal{F}_t] - S_t$$

$$= e^{-r(T-t)} S_t \mathbb{E}^*[\max(x, S_t^{-1} L_t)]|_{x = S_t^{-1} W_t} - S_t .$$

Put

$$G(t, x) = \mathbb{E}^*[\max(x, S_t^{-1} L_t)], \quad R(t, x) = \mathbb{P}\left(\max_{s \leqslant t} (\sigma B_s + \lambda s) \leqslant x \right).$$

Then by (6.1), we get

$$G(t, x) = \mathbb{E}^*\left[\exp\{\max\left(\ln x, \max_{t \leqslant s \leqslant T}[\sigma(B_s^* - B_t^*) + \lambda(s - t)] \right)\} \right]$$

$$= \mathbb{E}\left[\exp\{\max\left(\ln x, \max_{0 \leqslant s \leqslant T - t} (\sigma B_s + \lambda s) \right)\} \right] \tag{6.32}$$

$$= x R(T - t, \ln x) + \int_{\ln x}^{\infty} e^y R_x(T - t, y) dy,$$

where

$$R(t, x) = N\left(\frac{x - \lambda t}{\sigma \sqrt{t}} \right) - e^{2\lambda x / \sigma^2} N\left(\frac{-x - \lambda t}{\sigma \sqrt{t}} \right).$$

After computations we get

$$P_t = S_t(-1 + N(d_3)(1 + \sigma^2/2r))$$

$$+ W_t e^{-r(T-t)} \left(N(d_1) - \sigma^2/2r (S_t^{-1} W_t)^{(2r/\sigma^2)-1} N(d_2) \right),$$

(6.33)

where

$$d_1 = \frac{\ln(W_t/S_t) - \left(r - \frac{1}{2}\sigma^2\right)(T-t)}{\sigma\sqrt{T-t}},$$

$$d_2 = \frac{-\ln(W_t/S_t) - \left(r - \frac{1}{2}\sigma^2\right)(T-t)}{\sigma\sqrt{T-t}},$$

$$d_3 = \frac{-\ln(W_t/S_t) + \left(r + \frac{1}{2}\sigma^2\right)(T-t)}{\sigma\sqrt{T-t}}.$$

Similarly, from (6.2) we obtain an explicit expression for C_t.

The explicit expressions for P_t and C_t were first derived by Goldman et al. (1979). For further results, see Conze and Viswanathan (1991).

Now we derive the hedging strategy for the lookback strike put option. Since $\widetilde{P}_t$ is a $\mathbb{P}^*$-martingale and

$$\widetilde{P}_t = e^{-rT} S_t G(t, S_t^{-1} W_t) - \widetilde{S}_t,$$

by Theorem 4.2, we must have

$$d\widetilde{P}_t = -e^{-rT} S_t G_x(t, S_t^{-1} W_t) W_t S_t^{-1} \sigma dB_t^*$$

$$+ e^{-rT} G(t, S_t^{-1} W_t) S_t \sigma dB_t^* - \widetilde{S}_t \sigma dB_t^*$$

$$= e^{-rT} [G(t, S_t^{-1} W_t) S_t - G_x(t, S_t^{-1} W_t) W_t] \sigma dB_t^* - \widetilde{S}_t \sigma dB_t^*.$$

By (6.32) we have

$$G_x(t, x) = R(T - t, \ln x).$$

Therefore, from (5.9) we get

$$a(t) = e^{-r(T-t)} \left[G(t, S_t^{-1} W_t) - \frac{W_t}{S_t} R(T - t, \ln(S_t^{-1} W_t)) \right] - 1.$$

6.4.2   Lookback Rate Options

We only consider the sampled maximum and call option case. In this case, the payoff
of a lookback rate call option is defined by

$$\xi = (\max_{0 \leqslant s \leqslant T} S_s - K)^+.$$

We denote by C_t its price at time t. By Theorem 5.2, we have

$$C_t = e^{-r(T-t)} \mathbb{E}^*[\xi \mid \mathcal{F}_t]. \tag{6.34}$$

Using the notations of the previous subsection and letting $K_t = \max(W_t, K)$, we
have

$$\begin{aligned}
\mathbb{E}^*[\xi \mid \mathcal{F}_t] &= \mathbb{E}^*[\max(W_T, K) - K \mid \mathcal{F}_t] \\
&= \mathbb{E}^*[\max(K_t, L_t) - K_t \mid \mathcal{F}_t] + K_t - K \\
&= \mathbb{E}^*[(L_t - K_t)^+ \mid \mathcal{F}_t] + K_t - K \\
&= S_t \mathbb{E}^*[(S_t^{-1} L_t - S_t^{-1} K_t)^+ \mid \mathcal{F}_t] + K_t - K \\
&= S_t \mathbb{E}^*[(S_t^{-1} L_t - x)^+] \mid_{x = S_t^{-1} K_t} + K_t - K \\
&= S_t H(t, S_t^{-1} K_t) + K_t - K,
\end{aligned}$$

where

$$\begin{aligned}
H(t, x) &= \mathbb{E}^*[(S_t^{-1} L_t - x)^+] = \mathbb{E}^*[\max(x, S_t^{-1} L_t) - x] \\
&= G(t, x) - x = -x(1 - R(T - t, \ln x)) + \int_{\ln x}^{\infty} e^y R_x(T - t, y) dy.
\end{aligned} \tag{6.35}$$

From the same computations as in the previous subsection, we get

$$\begin{aligned}
C_t = {}& S_t N(\tilde{d}_3)(1 + \sigma^2/2r) \\
& + K_t e^{-r(T-t)} \left(N(\tilde{d}_1) - \sigma^2/2r (S_t^{-1} K_t)^{(2r/\sigma^2)-1} N(\tilde{d}_2) \right) - e^{-r(T-t)} K,
\end{aligned} \tag{6.36}$$

where

$$\tilde{d}_1 = \frac{\ln(K_t/S_t) - \left(r - \frac{1}{2}\sigma^2\right)(T - t)}{\sigma\sqrt{T - t}},$$

$$\tilde{d}_2 = \frac{-\ln(K_t/S_t) - \left(r - \frac{1}{2}\sigma^2\right)(T - t)}{\sigma\sqrt{T - t}},$$

$$\tilde{d}_3 = \frac{-\ln(K_t/S_t) + \left(r + \frac{1}{2}\sigma^2\right)(T - t)}{\sigma\sqrt{T - t}}.$$

Now we derive the hedging strategy for the lookback rate call option. Since $\widetilde{C}_t$ is a $\mathbb{P}^*$-martingale, and

$$\widetilde{C}_t = e^{-rT} S_t H(t, S_t^{-1} K_t) + e^{-rt}(K_t - K),$$

by Theorem 4.2, we must have

$$
\begin{aligned}
d\widetilde{C}_t &= e^{-rT} H(t, S_t^{-1} K_t) S_t \sigma d B_t^* \\
&\quad -e^{-rT} S_t H_x(t, S_t^{-1} K_t) K_t S_t^{-1} \sigma d B_t^* \\
&= e^{-rT} [H(t, S_t^{-1} K_t) S_t - H_x(t, S_t^{-1} K_t) K_t] \sigma d B_t^*.
\end{aligned}
$$

From (6.35) we get

$$H_x(t, x) = R(T - t, \ln x) - 1.$$

Therefore, from (5.9) we obtain

$$a(t) = e^{-r(T-t)} \left[H(t, S_t^{-1} K_t) - \frac{K_t}{S_t}(R(T - t, \ln(S_t^{-1} K_t)) - 1) \right].$$

6.5 Reset Options

A *reset option* is a lookback strike call-type option. Unlike a lookback strike call, if, on predetermined observation moments, the minimum of observed underlying asset prices falls in predetermined ranges, the strike price will be adjusted to the predetermined strike prices. Specifically, if n observation moments $0 < t_1 < \cdots < t_n < T$ and n times strike prices $(K_1, \cdots, K_m)$ are predetermined, the payoff of the reset option at time maturity T is $C(T) = (S_T - K^*)^+$, where

$$
K^* = \begin{cases}
K_0, & \text{if } \min[S_{t_1}, \cdots, S_{t_n}] > D_1, \\[2mm]
K_i, & \text{if } D_i \geqslant \min[S_{t_1}, \cdots, S_{t_n}] > D_{i+1}, \quad i = 1, \cdots, m-1, \\[2mm]
K_m, & \text{if } D_m \geqslant \min[S_{t_1}, \cdots, S_{t_n}].
\end{cases} \tag{6.37}
$$

Here $D_1 > D_2 > \cdots > D_m$ are also previously given. Gray and Whaley (1999) first gave an explicit pricing formula for the reset option in the case of $m = 1$. For the derivation of the explicit option pricing formula in the general case, see Liao and Wang (2003).

Chapter 7
Itô Process and Diffusion Models

In this chapter we will introduce a general framework of financial market. In the first section we present some basic concepts and fundamental results on martingale methods in the pricing and hedging of European contingent claims under the Itô process setting. In Sect. 7.2 we show that within the diffusion process framework, the pricing and hedging of European contingent claims can be done through a PDE approach. In Sect. 7.3 we present two probabilistic methods for closed-form pricing of European options and illustrate these methods through examples. In the fourth section we briefly address the problem of pricing American contingent claims in diffusion models.

7.1 Itô Process Models

7.1.1 Self-Financing Trading Strategies

We fix a finite time horizon T. Let $B = (B^1, \cdots, B^d)$ be a Brownian motion on a complete probability space $(\Omega, \mathcal{F}, \mathbb{P})$. We denote by $(\mathcal{F}_t)$ the natural filtration of (B_t) and by $\mathcal{L}$ the set of all measurable $(\mathcal{F}_t)$-adapted processes. We adopt the same notations $\mathcal{L}^1$ and $\mathcal{L}^2$ as defined in Chap. 4. For an Itô process X with the canonical decomposition

$$X_t = X_0 + \int_0^t a(t)^\tau dB_t + \int_0^t b(t)dt,$$

we put

$$\mathcal{L}^2(X) = \left\{ \theta \in \mathcal{L} : \theta a \in (\mathcal{L}^2)^d, \ \theta b \in \mathcal{L}^1 \right\}. \tag{7.1}$$

© Springer Nature Singapore Pte Ltd. and Science Press 2018
J.-A. Yan, *Introduction to Stochastic Finance*, Universitext,
https://doi.org/10.1007/978-981-13-1657-9_7

We consider a financial market which consists of $m + 1$ assets. The price process (S_t^i) of each asset i is assumed to be a strictly positive Itô process. Since its logarithm is also an Itô process, we can represent (S_t^i) as

$$dS_t^i = S_t^i \left[\sigma^i(t)dB_t + \mu^i(t)dt \right], \quad S_0^i = p_i, \quad 0 \leqslant i \leqslant m. \tag{7.2}$$

We call $\mu = (\mu^0, \cdots, \mu^m)$ the *appreciation rate vector* and σ the *volatility matrix*.

We specify arbitrarily one of the assets, say, asset 0, as the numeraire asset. We set $\gamma_t \hat{=} (S_t^0)^{-1}$ and call γ_t the *deflator* at time t. By Itô's formula we have

$$d\gamma_t = -\gamma_t \left[\sigma^0(t)dB_t + (\mu^0(t) - |\sigma^0(t)|^2)dt \right].$$

Put $S_t = (S_t^1, \cdots, S_t^m)$, $\tilde{S}_t = (\tilde{S}_t^1, \cdots, \tilde{S}_t^m)$, where $\tilde{S}_t^i = \gamma_t S_t^i$. Then we have

$$d\tilde{S}_t^i = \tilde{S}_t^i \left[a^i(t)dB_t + b^i(t)dt \right], \quad 1 \leqslant i \leqslant m, \tag{7.3}$$

where

$$a^i(t) = \sigma^i(t) - \sigma^0(t); \quad b^i(t) = \mu^i(t) - \mu^0(t) + |\sigma^0(t)|^2 - \sigma^i(t)\sigma^0(t).$$

In particular, if asset 0 is a bank account with interest rate process $(r(t))$, then

$$a^i(t) = \sigma^i(t), \quad b^i(t) = \mu^i(t) - r(t).$$

A *trading strategy* is a pair $\phi = \{\theta^0, \theta\}$ of $\mathcal{F}_t$-adapted processes, where $\theta^0(t) \in \mathcal{L}^2(S^0)$,

$$\theta(t) = (\theta^1(t), \cdots, \theta^m(t))^\tau, \quad \theta^i \in \mathcal{L}^2(S^i), \quad \forall 1 \leqslant i \leqslant m,$$

where $\theta^i(t)$ represents the numbers of units of asset i held at time t. The wealth V_t of a trading strategy $\phi = \{\theta^0, \theta\}$ at time t is

$$V_t = \theta^0(t)S^0(t) + \theta(t)S(t). \tag{7.4}$$

Its deflated wealth at time t is $\tilde{V}_t = V_t \gamma_t$. A trading strategy $\phi = \{\theta^0, \theta\}$ is called *self-financing* if

$$V_t = V_0 + \int_0^t \theta^0(u)dS_u^0 + \int_0^t \theta(u)^\tau dS_u. \tag{7.5}$$

A self-financing strategy is called *admissible* if its wealth process is nonnegative. Put $S_t^{m+1} = \sum_{i=0}^m S_t^i$. As in the discrete-time case, a strategy $\phi = \{\theta^0, \theta\}$ is called *allowable*, if there exists a positive constant c such that the wealth process (V_t) is bounded from below by $-cS_t^{m+1}$.

Similar to the Black-Scholes model (see Lemma 5.1), we have the following characterization of the self-financing strategy.

Lemma 7.1 *A trading strategy* $\phi = \{\theta^0, \theta\}$ *is self-financing if and only if*

$$d\widetilde{V}_t = \theta(t)d\widetilde{S}_t. \tag{7.6}$$

Proof Assume that $\phi = \{\theta^0, \theta\}$ is a self-financing strategy. We rewrite dS_t and $d\gamma_t$ as

$$dS_t = \sigma_S(t)dB_t + \mu_S(t)dt,$$

$$d\gamma_t = \sigma_\gamma(t)dB_t + \mu_\gamma(t)dt.$$

Applying Itô's formula to the product $V_t\gamma_t$, we obtain (noting that $d(S_t^0\gamma_t) = 0$)

$$\begin{aligned} d\widetilde{V}_t &= V_t d\gamma_t + \gamma_t dV_t + d\langle V, \gamma \rangle_t \\ &= (\theta(t) \cdot S_t)d\gamma_t + \gamma_t\theta(t)^\tau dS_t + [\theta(t)^\tau\sigma_S(t)]\sigma_\gamma(t)dt + \theta^0(t)d(S_t^0\gamma_t) \\ &= \theta(t)[S_t d\gamma_t + \gamma_t dS_t + d\langle S, \gamma \rangle_t] = \theta(t)d\widetilde{S}_t. \end{aligned}$$

Similarly, we can prove the "sufficiency" part (i.e., (7.6) implies (7.5)). □

7.1.2 Equivalent Martingale Measures and No Arbitrage

Assume that asset j is taken as the numeraire asset. Let $\mathbb{Q}$ be a probability measure equivalent to the "objective" probability measure $\mathbb{P}$. If the deflated price process is a (vector-valued) $\mathbb{Q}$-martingale, we call $\mathbb{Q}$ an *equivalent martingale measure* for the market. We denote by $\mathcal{M}^j$ the set of all equivalent martingale measures. Here the superscript j indicates that asset j is taken as the numeraire.

The next theorem shows that for any pair (i, j) of indices from $\{0, \cdots, m\}$, there exists a bijection from $\mathcal{M}^i$ onto $\mathcal{M}^j$.

Theorem 7.2 *Let* $j \in \{0, 1, \cdots, m\}$. *For* $\mathbb{P}^* \in \mathcal{M}^0$, *we define a probability measure* $\mathbb{Q}$ *by*

$$\frac{d\mathbb{Q}}{d\mathbb{P}^*} = \frac{S_0^0}{S_0^j}(S_T^0)^{-1}S_T^j. \tag{7.7}$$

We denoted $\mathbb{Q}$ *by* $h_j(\mathbb{P}^*)$. *Then* h_j *is a bijection from* $\mathcal{M}^0$ *onto* $\mathcal{M}^j$.

Proof We let $\gamma'_t = (S^j_t)^{-1}$ and put

$$\widehat{S}^i_t = \gamma'_t S^i_t, \quad 0 \leqslant i \leqslant m.$$

Let $\mathbb{P}^* \in \mathcal{M}^0$. We define a probability measure $\mathbb{Q}$ by (7.7). Since $(S^0_t)^{-1} S^j_t$ is a $\mathbb{P}^*$-martingale, we must have

$$M_t := \mathbb{E}^* \left[\frac{d\mathbb{Q}}{d\mathbb{P}^*} \bigg| \mathcal{F}_t \right] = \frac{S^0_0}{S^j_0} (S^0_t)^{-1} S^j_t.$$

According to the following fact

$$M_t \widehat{S}^i_t = M_t \gamma'_t S^i_t = \frac{S^0_0}{S^j_0} \widetilde{S}^i_t,$$

we know that $\mathbb{Q}$ is an equivalent martingale measure for the market with asset j as the numeraire asset, i.e., $\mathbb{Q} \in \mathcal{M}^j$.   □

A market is said to have *arbitrage* opportunity if there exists an allowable self-financing strategy such that its initial wealth V_0 is zero but its terminal wealth V_T is nonnegative and $\mathbb{P}(V_T > 0) > 0$.

In the following we specify asset 0 as the numeraire. By Lemma 7.1, the deflated wealth process of any self-financing strategy is a local $\mathbb{Q}$-martingale for each $\mathbb{Q} \in \mathcal{M}^0$. As a consequence, for any $\mathbb{Q} \in \mathcal{M}^0$, the wealth process of any allowable self-financing strategy is a $\mathbb{Q}$-supermartingale.

Theorem 7.3 *If in the market there exist equivalent martingale measures, namely, $\mathcal{M}^0 \neq \emptyset$, then the market has no arbitrage. In this case, let $\mathbb{Q} \in \mathcal{M}^0$, then $\frac{d\mathbb{Q}}{d\mathbb{P}} \big|_{\mathcal{F}_T} = \mathcal{E}(-\psi.B)_T$, $\psi \in (\mathcal{L}^2[0, T])^d$, and ψ is a solution to the following linear equation:*

$$a(t)\psi(t) = b(t), \quad dt \times d\mathbb{P}\text{-a.e. a.s. on } [0, T] \times \Omega. \tag{7.8}$$

Proof Let $\mathbb{P}^* \in \mathcal{M}^0$, and $\phi = \{\theta^0, \theta\}$ be an allowable self-financing strategy with initial wealth zero. As mentioned above, its deflated wealth process $(\widetilde{V}_t)$ is a $\mathbb{P}^*$-supermartingale. Therefore, we must have $\mathbb{E}^*[\widetilde{V}_T] \leqslant 0$. So the market has no arbitrage.

Let $\mathbb{Q} \in \mathcal{M}^0$, and put

$$M_t = \mathbb{E} \left[\frac{d\mathbb{Q}}{d\mathbb{P}} \bigg| \mathcal{F}_t \right].$$

Then (M_t) is a $\mathbb{P}$-martingale. By the martingale representation theorem for Brownian motion, there exists $\phi \in (\mathcal{L}^2)^d$ such that $dM_t = \phi(t) dB_t$. Put $\psi(t) = -\phi(t)/M_t$. Then $M = \mathcal{E}(-\psi.B)$, and by Girsanov's theorem, $B^*_t = B_t + \int_0^t \psi(s) ds$

is a Brownian motion under $\mathbb{Q}$. Moreover, by Theorem 4.18, (B_t^*) has also the martingale representation property w.r.t. $(\mathcal{F}_t)$ under $\mathbb{Q}$. Thus, there exists some $\sigma^* \in (\mathcal{L}^2)^{m \times d}$ such that

$$d\widetilde{S}_t = \sigma^*(t)dB_t^* = \sigma^*(t)(dB_t + \psi(t)dt).$$

According to the uniqueness of the representation of Itô process $(\widetilde{S}_t)$ and the invariance of the stochastic integral under a change of probability, from (7.3) we know that $\sigma^*(t) = \widetilde{S}_t a(t)$, $dt \times d\mathbb{P}$-a.e. a.s., and consequently, $a(t)\psi(t) = b(t)$, $dt \times d\mathbb{P}$-a.e. a.s. So $(\psi(t))$ is a solution to Eq. (7.8). $\qquad\square$

If a market has an equivalent martingale measure, it will be called a *fair market*. It is natural to raise a question: under which conditions on coefficients a and b of diffusion process $(\widetilde{S}_t)$ the market is fair? The following theorem gives a partial answer to the question.

Theorem 7.4 *If (7.8) has a solution* $\psi \in (\mathcal{L}^2)^d$ *satisfying*

$$\mathbb{E}\left[\exp\left\{\frac{1}{2}\int_0^T |\psi(t)|^2 dt\right\}\right] < \infty \qquad (7.9)$$

and

$$\mathbb{E}\left[\exp\left\{\frac{1}{2}\int_0^T |a^i(t) - \psi(t)|^2 dt\right\}\right] < \infty, \quad 1 \leqslant i \leqslant m, \qquad (7.10)$$

then probability measure $\mathbb{Q}$ *with Radon-Nikodym derivative*

$$\frac{d\mathbb{Q}}{d\mathbb{P}} = \mathcal{E}(-\psi.B)_T$$

belongs to $\mathcal{M}^0$*. In particular,* $\mathcal{M}^0 \neq \emptyset$.

Proof Assume that ψ is a solution to (7.8) and verifies (7.9). By Novikov theorem (Theorem 4.11) $\mathcal{E}(-\psi.B)$ is a $\mathbb{P}$-martingale. So we can define a probability measure $\mathbb{Q}$ such that $\frac{d\mathbb{Q}}{d\mathbb{P}} = \mathcal{E}(-\psi.B)_T$. In order to prove that $\mathbb{Q} \in \mathcal{M}^0$, i.e., $(\widetilde{S}_t)$ is a $\mathbb{Q}$-martingale, it suffices to prove that the product $\mathcal{E}(-\psi.B)\widetilde{S}$ is a $\mathbb{P}$-martingale. By (7.3) and (4.16) we have

$$\mathcal{E}(-\psi.B)_t\widetilde{S}_t^i = \widetilde{S}_0^i \exp\left\{\int_0^t (a^i(s) - \psi(s))dB_s - \frac{1}{2}\int_0^t |a^i(s) - \psi(s)|^2 ds\right\}.$$

Once again by (7.10) and Novikov theorem, $\mathcal{E}(-\psi.B)\widetilde{S}^i$ is a $\mathbb{P}$-martingale. The theorem is proved. $\qquad\square$

According to the theorem we propose the following definition.

Definition 7.5 If a satisfies (7.9) and Eq. (7.8) has a solution ψ satisfying (7.10), then the market is called a *standard market*.

According to Theorems 7.4 and 7.3, any standard market is fair. The following theorem provides a sufficient condition for the existence of a unique equivalent martingale measure.

Theorem 7.6 *Assume that $m \geqslant d$, $a(t)^{\tau} a(t)$ are non-degenerated on $[0, T] \times \Omega$ w.r.t. $dt \times d\mathbb{P}$, a.e. a.s., where $a(t)^{\tau}$ stands for the transpose of $a(t)$. Put $\psi(t) = (a(t)^{\tau} a(t))^{-1} a(t)^{\tau} b(t)$. If ψ satisfies (7.8), a and ψ satisfy (7.9) and (7.10), then there exists a unique equivalent martingale measure $\mathbb{P}^*$ for the market. Moreover, we have*

$$\mathbb{E}\Big[\frac{d\mathbb{P}^*}{d\mathbb{P}} \Big| \mathcal{F}_t\Big] = \exp\Big\{ - \int_0^t \psi(s) dB_s - \frac{1}{2} \int_0^t |\psi(s)|^2 ds \Big\}, \quad 0 \leqslant t \leqslant T.$$

Proof Under the assumptions of the theorem, the market is standard, so by Theorem 7.4 there exists an equivalent martingale measure. To prove the uniqueness of equivalent martingale measure, let $\mathbb{Q}$ be an equivalent martingale measure. There exists a $\theta \in (\mathcal{L}^2)^d$ such that $\frac{d\mathbb{Q}}{d\mathbb{P}} = \mathcal{E}(-\theta.B)_T$. By Theorem 7.3, we have $a(t)\theta(t) = b(t)$. Consequently, multiplying $(a^{\tau}(t)a(t))^{-1}a^{\tau}(t)$ to both sides of the equation, we get $\theta(t) = \psi(t)$. The uniqueness is thus proved. $\square$

Remark If $m = d$, put $\psi(t) = a(t)^{-1}b(t)$, then $\psi(t) = a(t)^{-1}b(t)$ automatically satisfies (7.8).

Definition 7.7 Let $\mathbb{Q} \in \mathcal{M}^0$. A European contingent claim ξ (i.e., a nonnegative $\mathcal{F}_T$-measurable random variable) is called $\mathbb{Q}$-*replicatable* (or $\mathbb{Q}$-*attainable*), if there exists an admissible self-financing strategy such that its terminal wealth is equal to ξ and its deflated wealth process is a $\mathbb{Q}$-martingale. Such a trading strategy is called a $\mathbb{Q}$-*hedging strategy* for ξ.

Now assume that there exists a unique equivalent martingale measure $\mathbb{P}^*$ for the market. According to a theorem due to Jacod and Yor (1977), the uniqueness of equivalent martingale measure implies the martingale representation property under the equivalent measure. From this result we can prove that the market is *complete* in the sense that any European contingent claim ξ with the deflated value $\gamma_T \xi$ being $\mathbb{P}^*$-integrable is $\mathbb{P}^*$-replicatable. In our particular case we can directly prove this result, as will be shown below.

Theorem 7.8 *If the conditions in Theorem 6.5 are satisfied, then the market is complete.*

Proof Let $\mathbb{P}^*$ be the unique equivalent martingale measure. By Theorem 7.6 we have $\frac{d\mathbb{P}^*}{d\mathbb{P}} = \mathcal{E}(-\psi.B)_T$, where

$$\psi(t) = (a^{\tau}(t)a(t))^{-1}a^{\tau}(t)b(t).$$

Put $B_t^* = B_t + \int_0^t \psi(s)ds$, then (B_t^*) is a $\mathbb{P}^*$-Brownian motion. Let ξ be a nonnegative contingent claim such that $\mathbb{E}^*[\gamma_T \xi] < \infty$. We put

$$V_t = \gamma_t^{-1} \mathbb{E}^*[\gamma_T \xi \mid \mathcal{F}_t].$$

Then the deflated value process $(\widetilde{V}_t)$ is a $\mathbb{P}^*$-martingale. Since by Theorem 4.14 (B_t^*) also has the martingale representation property w.r.t. $(\mathcal{F}_t)$ under $\mathbb{P}^*$, there exist two processes, H and K, in $(\mathcal{L}^2)^{m \times d}$ such that

$$d\widetilde{V}_t = H_t dB_t^* = H_t(dB_t + \psi(t)dt),$$

$$d\widetilde{S}_t = K_t dB_t^* = K_t(dB_t + \psi(t)dt).$$

Let C_t be a matrix with $c_{i,j}(t) = \delta_{i,j}\widetilde{S}_t^i, 1 \leqslant i, j \leqslant m$. From (7.3) we know $K_t = C_t a(t), a(t)\psi(t) = b(t)$. Thus we have

$$(a^\tau(t)a(t))^{-1}a^\tau(t)C_t^{-1}d\widetilde{S}_t = dB_t + \psi(t)dt.$$

Put

$$\theta(t) = H_t(a^\tau(t)a(t))^{-1}a^\tau(t)C_t^{-1},$$

$$\theta^0(t) = S_t^{0^{-1}}[V_t - \theta(t) \cdot S_t].$$

Then $\phi = \{\theta^0, \theta\}$ is a hedging strategy for ξ, whose wealth process is (V_t). By definition, the market is complete. □

Remark If $m = d$ and the market is standard, then the market is complete if and only if $a(t, \omega)$ is non-singular, for $(t, \omega) \in [0, T] \times \Omega$, a.e. a.s. See Karatzas (1997).

7.1.3 Pricing and Hedging of European Contingent Claims

In this subsection we study the problem of pricing and hedging of European contingent claims. We assume that the market is fair.

Let ξ be a contingent claim. One raises naturally a question: what is a "fair" price process of ξ? Assume that $\gamma_T \xi$ is $\mathbb{P}^*$-integrable for some $\mathbb{P}^* \in \mathcal{M}^0$. We put

$$V_t = \gamma_t^{-1} \mathbb{E}^*[\gamma_T \xi \mid \mathcal{F}_t]. \tag{7.11}$$

If we consider (V_t) as the price process of an asset, then $\mathbb{P}^*$ is still an equivalent martingale measure in the market augmented with this asset, because the deflated price process of this asset is a $\mathbb{P}^*$-martingale. So it seems that (V_t) can be considered as a candidate for a"fair" price process of ξ. However this definition of the "fair" price depends on the choice of equivalent martingale measure. Determining an equivalent martingale measure is equivalent to specifying the corresponding market price of risk. However, it is the market itself who determines the market price of risk. Consequently, in practice, based on statistical methods one uses market data on prices of existing contingent claims to estimate the market price of risk.

The next theorem shows that for replicatable contingent claims, their fair prices can be uniquely determined.

Theorem 7.9 *Let* $\mathbb{P}^*, \mathbb{Q} \in \mathcal{M}^0$ *and* ξ *be a* $\mathbb{P}^*$- *and* $\mathbb{Q}$-*replicatable contingent claim. Let* (V_t) *(resp.* (U_t)*) be the wealth process of a* $\mathbb{P}^*$- *(resp.* $\mathbb{Q}$-*)hedging strategy for* ξ. *Then* (V_t) *and* (U_t) *are the same. Moreover,* V_t *is given by (7.11).*

Proof Put $\widetilde{V}_t = \gamma_t V_t$, $\widetilde{U}_t = \gamma_t U_t$. Then $(\widetilde{V}_t)$ is a $\mathbb{P}^*$-martingale and $\mathbb{Q}$-supermartingale, and $(\widetilde{U}_t)$ is a $\mathbb{Q}$-martingale and a $\mathbb{P}^*$-supermartingale. Note that $U_T = V_T = \xi$ we have

$$\mathbb{E}^*[\widetilde{V}_T | \mathcal{F}_t] = \widetilde{V}_t \geqslant \mathbb{E}_{\mathbb{Q}}[\widetilde{V}_T | \mathcal{F}_t] = \widetilde{U}_t.$$

Thus we have $V_t \geqslant U_t$ a.s. Similarly, we have $U_t \geqslant V_t$ a.s. Hence $V = U$. The last assertion of the theorem is obvious. $\square$

Remark According to Theorem 7.9, for a $\mathbb{P}^*$-replicatable contingent claim ξ, it is natural to define its fair price at time t by (7.11). We call this method of pricing the *risk-neutral pricing* (or *arbitrage pricing*).

Now assume that the conditions in Theorem 7.6 are satisfied. So there exists a unique equivalent martingale measure $\mathbb{P}^*$ for the market, and by Theorem 7.8 the market is complete. Let ξ be a European contingent claim such that $\gamma_T \xi$ is $\mathbb{P}^*$-integrable. From the proof of Theorem 7.8, there exists actually a $\mathbb{P}^*$-hedging strategy for ξ. So in this case, the fair price process of ξ is given by (7.11).

In general, if a contingent claim ξ is not replicatable, we can not define its fair price process. In this case we need new kinds of trading strategies. Similar to the discrete-time case, a *strategy with consumption* is a trading strategy $\phi = \{\theta^0, \theta, C\}$ with the property that for all $t \in [0, T]$,

$$\theta^0(t)S_t^0 + \theta(t) \cdot S_t = \theta^0(0)S_0^0 + \theta(0) \cdot S_0 + \int_0^t \theta^0(u)dS_u^0 + \int_0^t \theta(u)dS_u - C_t,$$

where (C_t) is an adapted continuous nondecreasing process null at $t = 0$. C_t represents the cumulative consumption up to time t. In contrast, a *strategy with reinvestment* is a trading strategy $\phi = \{\theta^0, \theta, R\}$ with the property that for all $t \in [0, T]$,

$$\theta^0(t)S_t^0 + \theta(t) \cdot S_t = \theta^0(0)S_0^0 + \theta(0) \cdot S_0 + \int_0^t \theta^0(u)dS_u^0 + \int_0^t \theta(u)dS_u + R_t,$$

where (R_t) is an adapted continuous nondecreasing process null at $t = 0$, which represents the cumulative reinvestment up to time t. For a strategy ϕ of these two kinds, we denote by $V_t(\phi)$ the wealth at time t of ϕ, namely, $V_t(\phi) = \theta^0(t)S_t^0 + \theta(t) \cdot S_t$. Similar to the discrete-time case, the deflated wealth process of a strategy with consumption is a local $\mathbb{Q}$-supermartingale for any equivalent martingale measure $\mathbb{Q}$. In contrast, the deflated wealth process of a strategy with reinvestment is a local $\mathbb{Q}$-submartingale under any equivalent martingale measure $\mathbb{Q}$. We denote by $\mathcal{G}_c$ the set of all admissible strategies with consumption and by $\mathcal{G}_r$ the set of all admissible strategies with reinvestment.

The following definition seems to be reasonable.

Definition 7.10 Let ξ be a European contingent claim such that ξ is $\mathbb{Q}$-integrable for some $\mathbb{Q} \in \mathcal{M}^0$. We put

$$V_t^s = \operatorname{essinf}\{V_t(\phi) : \phi \in \mathcal{G}_c, V_T(\phi) \geqslant \xi\},$$

$$V_t^b = \operatorname{esssup}\{V_t(\phi) : \phi \in \mathcal{G}_r, V_T(\phi) \leqslant \xi\}.$$

We call V_t^s and V_t^b the *seller's price* and *buyer's price* at time t of ξ, respectively.

Note that there exists a version of V^s (resp. V^b) such that V^s (resp. V^b) is a $\mathbb{Q}$-supermartingale (resp. local $\mathbb{Q}$-submartingale).

We refer the reader to Karatzas (1997) or Musiela and Rutkowski (1997) for an account of this subject. Note that our definition of buyer's price is a little different from that given in the above two books.

7.1.4 Change of Numeraire

The following theorem shows that the risk-neutral pricing is invariant under the change of numeraire.

Theorem 7.11 *Let $j \in \{0, 1, \cdots, m\}$, h_j be a bijection from $\mathcal{M}^0$ onto $\mathcal{M}^j$ defined by (7.7). If $\mathbb{P}^* \in \mathcal{M}^0$, and ξ is a $\mathbb{P}^*$-replicatable contingent claim, then ξ is a $h_j(\mathbb{P}^*)$-replicatable contingent claim, and its "fair" price process is invariant under the change of numeraire.*

Proof Let ξ be a $\mathbb{P}^*$-replicatable contingent claim. Put $\gamma_t' = (S_t^j)^{-1}$, we have

$$\mathbb{E}_{\mathbb{Q}}[\gamma_T'\xi] = \mathbb{E}^*[M_T\gamma_T'\xi] = \frac{S_0^0}{S_0^j}\mathbb{E}^*[\gamma_T\xi] = (S_0^j)^{-1}V_0.$$

This implies that a $\mathbb{P}^*$-hedging strategy for ξ is also a $\mathbb{Q}$-hedging strategy for ξ. So ξ is a $\mathbb{Q}$-replicatable contingent claim. Moreover, by Bayes' rule of conditional expectation (Theorem 1.11), we have

$$(\gamma_t')^{-1} \mathbb{E}_{\mathbb{Q}}[\gamma_T' \xi \,|\, \mathcal{F}_t] = (\gamma_t')^{-1} M_t^{-1} \mathbb{E}^*[M_T \gamma_T' \xi \,|\, \mathcal{F}_t]$$

$$= \mathbb{E}^*[\gamma_T \xi \,|\, \mathcal{F}_t].$$

This proves that the "fair" price process of ξ is invariant under the change of numeraire. $\square$

Now we illustrate by an example how to use the martingale method and the change of numeraire to do option pricing.

Consider a market consisting of $d + 1$ assets with $d \geqslant 2$: a bank account with interest rate r and d stocks whose price processes (S_t^i), $i = 1, 2, \cdots, d$, satisfy the following Itô's equation:

$$dS_t^i = S_t^i[\mu^i dt + \sum_{j=1}^{d} \sigma_{ij} dB_t^j], \ i = 1, 2, \cdots, d,$$

where μ^i, σ_{ij} are constants and $(B_t^1, B_t^2, \cdots, B_t^d)$ is a d-dimensional Brownian motion. We consider the so-called *exchange option*, which gives its owner the right but no obligation to exchange stock 2 for stock 1 at time T. The payoff of this option is $\xi = (S_T^1 - S_T^2)^+$. The exchange option was first studied by Margrabe (1978) in the Black-Scholes economy. The change of numeraire approach to this problem is due to Davis (1994). We follow closely Karatzas (1997).

Assume that matrix (σ_{ij}) is positive definite. Put

$$Z_0(t) = \exp\left\{-\sum_{i=1}^{d} \theta^i B_t^i - \frac{1}{2} \sum_{i=1}^{d} (\theta^i)^2 t\right\}, \quad 0 \leqslant t \leqslant T,$$

and

$$\frac{d\mathbb{P}^*}{d\mathbb{P}} = Z_0(T),$$

where $(\theta^1, \cdots, \theta^d)^\tau = A^{-1}(\mu^1 - r, \cdots, \mu^d - r)^\tau$. Then (Z_t) is a $\mathbb{P}$-martingale and $\mathbb{P}^*$ is the unique martingale measure for the market. Put

$$B_t^{i*} = B_t^i + \theta^i t, \ \ S_t^{i*} = e^{-rt} S_t^i, \ i = 1, 2, \cdots, d.$$

Then $(B_t^{1*}, B_t^{2*}, \cdots, B_t^{d*})^\tau$ is a d-dimensional Brownian motion under $\mathbb{P}^*$, and

$$dS_t^{i*} = S_t^{i*}\left[\sum_{j=1}^{d} \sigma_{ij} dB_t^{j*}\right].$$

By (7.11) the price process of the exchange option is given by

$$V_t = \gamma_t^{-1} \mathbb{E}^*[\gamma_T (S_T^1 - S_T^2)^+ | \mathcal{F}_t],$$

where $\gamma_t = e^{-rt}$. Now we take asset 2 as a numeraire, then the associated martingale measure $\mathbb{Q}$ is given by

$$\frac{d\mathbb{Q}}{d\mathbb{P}^*} = \frac{\gamma_T}{S_0^2} S_T^2 = \frac{1}{S_0^2} S_T^{2*}.$$

Put

$$\widetilde{B}_t^i = B_t^{i*} - \sigma_{2i} t, \quad X_t = \frac{S_t^1}{S_t^2} = \frac{S_t^{1*}}{S_t^{2*}}.$$

Then $(\widetilde{B}_t^1, \widetilde{B}_t^2, \cdots, \widetilde{B}_t^d)^\tau$ is a d-dimensional Brownian motion under $\mathbb{Q}$, and

$$dX_t = X_t d\left[\sum_{j=1}^d (\sigma_{1j} - \sigma_{2j}) \widetilde{B}_t^j \right] = X_t \sigma dW_t,$$

where $\sigma = \sqrt{\sum_{j=1}^d (\sigma_{1j} - \sigma_{2j})^2}$, (W_t) is a one-dimensional Brownian motion under $\mathbb{Q}$. Thus, we have

$$V_t = S_t^2 \mathbb{E}_{\mathbb{Q}} \left[\left(\frac{S_T^1}{S_T^2} - 1 \right)^+ | \mathcal{F}_t \right] = S_t^2 \mathbb{E}_{\mathbb{Q}}[(X_T - 1)^+ | \mathcal{F}_t]$$

$$= S_t^2 [X_t N(d_1) - N(d_2)] = S_t^1 N(d_1) - S_t^2 N(d_2),$$

where

$$d_1 = \frac{\ln(X_t) + \frac{1}{2}\sigma^2(T-t)}{\sigma\sqrt{T-t}}, \quad d_2 = \frac{\ln(X_t) - \frac{1}{2}\sigma^2(T-t)}{\sigma\sqrt{T-t}}.$$

Remark that in the above pricing formula, the parameters μ^i and r do not appear.

7.1.5 Arbitrage Pricing Systems

In a fair market, if the deflated wealth process of an admissible self-financing strategy is strictly positive martingale under certain $\mathbb{P}^* \in \mathcal{M}_0$, we call such a strategy *regular strategy*. If we regard the wealth process of a regular strategy as a price process of an asset, then we can also take it as a numeraire and deduce all related martingale measures.

A pair of a numeraire and a related martingale measure is called an *arbitrage pricing system*. In an arbitrage pricing system, the fair price of a contingent claim is defined as the mathematical expectation of the deflated contingent claim with respect to the martingale measure. Two arbitrage pricing systems, $(S, \mathbb{Q})$ and $(S', \mathbb{Q}')$, are said to be equivalent, if for any bound contingent claim, they determine the same fair price, i.e., it holds that $d\mathbb{Q}'/d\mathbb{Q}|_{\mathcal{F}_T} = S_0/S_0' \times S_T'/S_T$.

Now we assume that asset 0 is a bank account with interest rate $r(t)$, and Eq. (7.8) has a solution $\eta \in (\mathcal{L}^2)^d$. We call η a *market price of risk (process)*. Note that in this case, Eq. (7.8) is reduced to

$$\sigma(t)\eta(t) = \mu(t) - r(t)1_m, \quad dt \times d\mathbb{P}\text{--a.e. a.s. on } [0, T] \times \Omega, \tag{7.8'}$$

where $1_m = (1, \cdots, 1)^\tau$. So the economic meaning of a market price of risk is that η provides a proportional relationship between the mean rates of change of prices $\mu(t) - r(t)1_m$ and the amounts σ of "risk" in asset price changes. If $\mathbb{Q}$ is an equivalent martingale measure and $\frac{d\mathbb{Q}}{d\mathbb{P}} = \mathcal{E}(-\eta.B)_T$, then η is just a market price of risk.

Let $\{\pi(t) = (\pi^{(1)}(t), \pi^{(2)}(t), \cdots, \pi^{(m)}(t))\}_{t \in [0, T]}$ be a portfolio process of a self-financing strategy, where $\pi^{(i)}(t)$ is the proportion of wealth invested in the i-th stock at time t, $i = 1, 2, \cdots, m$. The remaining proportion $1 - \sum_{i=1}^{m} \pi^{(i)}(t)$ is invested in the bond at time t. The corresponding wealth process X^π is determined by

$$dX_t^\pi = X_t^\pi \left(1 - \sum_{i=1}^{m} \pi^{(i)}(t)\right) r(t)dt + X_t^\pi \sum_{i=1}^{m} \pi^{(i)}(t) \left(\mu^{(i)}(t)dt + \sum_{j=1}^{d} \sigma^{(ij)}(t)dB_t^{(j)}\right)$$

$$= X_t^\pi \left((r(t) + \pi(t)\widetilde{\mu}(t)^\tau)dt + \pi(t)\sigma(t)dB_t^\tau\right),$$

where $\widetilde{\mu}(t) = \mu(t) - r(t)1_m$. Without loss of generality, we assume that $X_0^\pi = 1$. The explicit solution is

$$X_t^\pi = \exp\left\{\int_0^t \left(r(s) + \pi(s)\widetilde{\mu}(s)^\tau - \frac{1}{2}(\pi(s)\sigma(s))(\pi(s)\sigma(s)^\tau)\right)ds + \int_0^t \pi_s\sigma_s dB_s^\tau\right\}.$$

We will show that if we take the wealth process of a regular strategy as numeraire, then there is a natural family of equivalent martingale measures associated with the market prices of risk. Moreover, all arbitrage pricing systems with different numeraires and martingale measures associated with the same market price of risk are equivalent.

Let $\mathcal{Y}$ denote the collection of all adapted row vector processes $\eta(t)$ satisfying Eq. (7.8'), and such that the process $M_t = \exp\left\{-\frac{1}{2}\int_0^t \eta(s)\eta(s)\tau ds - \int_0^t \eta(s)dB_s^\tau\right\}$, $0 \leqslant t \leqslant T$, is a martingale. If $\mathcal{Y}$ is non-empty, then the market is fair. For example, if there is a solution $\mathbf{y}$ of Eq. (7.8') such that $\mathbb{E}\left[\exp\{\frac{1}{2}\int_0^T \eta(s)\eta(s)\tau ds\right] <$

∞, then by Novikov's criterion for the uniform integrability of exponential martingale, $\mathcal{Y}$ is non-empty. In the following we assume that $\mathcal{Y}$ is non-empty. The following theorem is due to Luo et al. (2002).

Theorem 7.12 *Assume that*

$$\mathbb{E}\left[\exp\left\{ \frac{1}{2} \int_0^T \pi(t)\sigma(t)(\pi(t)\sigma(t))^\tau dt \right\} \right] < \infty.$$

Let $\eta \in \mathcal{Y}$. Put

$$\widehat{\pi}_t := \mathbf{y}_t - \pi_t \sigma_t, \qquad t \in [0, T].$$

Define a new probability measure $\mathbb{P}_{\pi,\eta}$ on $(\Omega, \mathcal{F}_T)$ by

$$\frac{d\mathbb{P}_{\pi,\eta}}{d\mathbb{P}} = \exp\left\{ -\frac{1}{2} \int_0^T \widehat{\pi}(t)\widehat{\pi}(t)^\tau dt - \int_0^T \widehat{\pi}(t)dB_t^\tau \right\}.$$

Then $(X^\pi, \mathbb{P}_{\pi,\eta})$ is an arbitrage pricing system. Moreover, if π' is another portfolio satisfying the same condition as that for π, then two arbitrage pricing systems $(X^\pi, \mathbb{P}_{\pi,\eta})$ and $(X^{\pi'}, \mathbb{P}_{\pi',\eta})$ are equivalent.

Proof We define a new probability measure on $(\Omega, \mathcal{F}_T)$ by

$$\frac{d\mathbb{Q}_\eta}{d\mathbb{P}} = \exp\left\{ -\frac{1}{2} \int_0^T \eta(t)\eta(t)^\tau dt - \int_0^T \eta(t)dB_t^\tau \right\}. \tag{7.12}$$

Then $(\beta, \mathbb{Q}_\eta)$ is an arbitrage pricing system, and $\beta_t^{-1} X_t^\pi$ is a $\mathbb{Q}_\eta$-martingale, where $\beta_t = \exp\{\int_0^t r(s)ds\}$ is the price process of the bond. In particular, π is a portfolio of regular strategy. Since it is easy to verify that

$$\frac{X_t^\pi}{\beta_t} = \frac{d\mathbb{P}_{\pi,y}}{d\mathbb{P}}\bigg|_{\mathcal{F}_t} \bigg/ \frac{d\mathbb{Q}_\eta}{d\mathbb{P}}\bigg|_{\mathcal{F}_t} = \frac{d\mathbb{P}_{\pi,\eta}}{d\mathbb{Q}_\eta}\bigg|_{\mathcal{F}_t},$$

by Bayes' rule on conditional expectation, we know that $(X^\pi, \mathbb{P}_{\pi,\eta})$ is an arbitrage pricing system, which is equivalent to the arbitrage pricing system $(\beta, \mathbb{Q}_\eta)$. Consequently, another conclusion is trivially true. $\qquad\square$

Remark Process $\eta = (\eta(t))$ is interpreted as the market price of risk. For a given portfolio π, we have associated a natural family of arbitrage pricing systems $(X^\pi, \mathbb{P}_{\pi,\eta})$ to market prices of risk in $\mathcal{Y}$. If $m = d$ and the matrix σ_t is invertible, for any $t \in [0, T]$, then Eq. (7.8') has a unique solution. In this case the martingale measure is unique and the market is complete.

7.2   PDE Approach to Option Pricing

In this and the next sections, we assume that the market consists of $m + 1$ assets, one of which is a bank account. We denote by S_t^0 the value process of the bank account and $S_t = (S_t^1, \cdots, S_t^m)$ the price processes of the other assets. We take (S_t^0) as the numeraire. Assume that the interest rate process is of the form $r(t, S_t)$ where $r : \mathbb{R}_+ \times \mathbb{R}^m \to \mathbb{R}_+$ is Borel measurable and (S_t) is a diffusion process. Moreover, we assume that there exists a unique equivalent martingale measure $\mathbb{P}^*$ for $(\tilde{S}_t)$. Then under $\mathbb{P}^*$, (S_t) can be expressed as:

$$dS_t^i = S_t^i \left[\sigma^i(t, S_t) dB_t^* + r(t, S_t) dt \right], \quad S_0^i = p_i, \quad 1 \leqslant i \leqslant m, \tag{7.13}$$

where $\sigma : \mathbb{R}_+ \times \mathbb{R}^m \to M^{m,d}$ is Borel measurable, and (B_t^*) is a d-dimensional Brownian motion under $\mathbb{P}^*$. If $r(t, x)x$ and matrix $(x^i \sigma_j^i(t, x))$ are Lipschitz in x and satisfy the linear growth condition in x, then according to Theorem 4.19, (7.13) has a unique solution.

When dimension m is lower, say, 1 or 2, a Monte-Carlo approach to computing option prices is useful. However, if dimension m is higher than 3, a PDE method is more suitable. We show how option pricing is related to a parabolic PDE.

Let a European option ξ be of the form $g(S_T)$ with a nonnegative Borel function g on $\mathbb{R}_+^d$. Assume that $\mathbb{E}^*[\gamma_T g(S_T)] < \infty$. By (7.11), the price at time t of contingent claim ξ is given by

$$V_t = \mathbb{E}^* \left[e^{-\int_t^T r(s, S_s) ds} g(S_T) \,\Big|\, \mathcal{F}_t \right]. \tag{7.14}$$

If V_t can be expressed as $V_t = F(t, S_t)$, then by the Markovian property of the diffusion (S_t), we should have

$$F(t, x) = \mathbb{E}^{*t, x} [e^{-\int_t^T r(s, S_s) ds} g(S_T)]. \tag{7.15}$$

Consequently, under some technical conditions on coefficients r and σ^i, $F(t, x)$ solves the following parabolic PDE:

$$-\frac{\partial u}{\partial t} + ru = \mathcal{A}_t u, \quad (t, x) \in [0, T) \times \mathbb{R}_+^d, \tag{7.16}$$

subject to terminal condition $u(T, x) = g(x)$, where

$$\mathcal{A}_t = \frac{1}{2} \sum_{i,k=1}^d a_{ik}(t, x) \frac{\partial^2}{\partial x_i \partial x_k} + \sum_{i=1}^d r(t, x) x_i \frac{\partial}{\partial x_i},$$

$a(t, x) = x\sigma(t, x)(x\sigma(t, x))^{\tau}$. The classical Black-Scholes differential equation is a special case of (7.16). Usually, one uses the finite difference method to solve this PDE numerically. See Wilmott et al. (1993) and Jiang (2003).

7.3 Probabilistic Methods for Option Pricing

In this section we present a probabilistic method, due to Goldenberg (1991), for closed-form pricing of European options, where the price processes of the underlying assets follow a diffusion process. The method uses linear and nonlinear time and scale changes to reduce complex diffusion processes to known processes, thereby generating option pricing formulas and unifying existing results.

7.3.1 Time and Scale Changes

Consider a market consisting of two assets: a bank account with a deterministic interest rate function (r_t) and a stock. Assume that under the unique equivalent martingale measure $\mathbb{P}^*$, the stock price process is modeled by a diffusion:

$$dS_t = S_t\left[\sigma(t, S_t)dB_t^* + r_t dt\right], \quad S_0 = p, \qquad (7.17)$$

where $\sigma(t, x)$ is a Borel function on $\mathbb{R}^+ \times \mathbb{R}$ and (B_t^*) is a standard Brownian motion under $\mathbb{P}^*$. We put

$$P(t, s) = e^{-\int_t^s r_u du}, \quad t \leqslant s,$$

which is the time t price of a unit discount bond maturing at time s. The price at time t of a call option $(X_s - K)^+$ with maturity s and strike price K is given by $V_t = C(t, S_t; s, K)$, where

$$C(t, x; s, K) = P(t, s)\mathbb{E}^*[(S_s - K)^+ | S_t = x].$$

Assume that $C(t, x; s, K)$ has an explicit expression. Our objective below is to show how to use time and scale changes to deduce the pricing formula for a call option based on another risky asset, starting from $C(t, x; s, K)$.

Theorem 7.13 *Let τ be a strictly increasing differentiable function with $\tau(0) = 0$. Put $\mathcal{G}_t = \mathcal{F}_{\tau(t)}$. Assume that under stochastic basis $(\Omega, (\mathcal{G}_t), \mathcal{F}, \mathbb{P})$, the price process of a risky asset can be expressed as $Y_t = f(t)S_{\tau(t)}$, where f is a strictly positive function with $f(0) = 1$, and (S_t) is a diffusion process satisfying (7.17). Let*

$$P_Y(t, T) = \exp\left\{-\int_t^T r_Y(u)du\right\}$$

be the value at time t of the unit discount bond in the market, where $(r_Y(t))$ is a deterministic interest rate function. Then the price at time t of call option $(Y_T - K)^+$ is given by

$$\widehat{V}_t = \frac{P_Y(t, T)f(T)}{P(\tau(t), \tau(T))}C\left(\tau(t), \frac{Y_t}{f(t)}; \tau(T), \frac{K}{f(T)}\right). \tag{7.18}$$

Proof We have

$$\widehat{V}_t = P_Y(t, T)\mathbb{E}^*[(Y_T - K)^+|\mathcal{F}_{\tau(t)}]$$

$$= \frac{P_Y(t, T)f(T)}{P(\tau(t), \tau(T))}P(\tau(t), \tau(T))\mathbb{E}^*\left[(S_{\tau(T)} - \frac{K}{f(T)})^+\Big|\mathcal{F}_{\tau(t)}\right]$$

$$= \frac{P_Y(t, T)f(T)}{P(\tau(t), \tau(T))}C\left(\tau(t), \frac{Y_t}{f(t)}; \tau(T), \frac{K}{f(T)}\right).$$

Thus, (7.18) is proved. $\qquad\qquad\qquad\qquad\qquad\qquad\qquad\qquad\qquad\qquad\qquad\square$

7.3.2   Option Pricing in Merton's Model

Now we illustrate the formula (6.17) through an example. In Merton's generalization of the Black-Scholes model, the bank account earns a time-dependent interest rate $r(t)$ and the underlying asset pays dividends and has time-dependent expected rate of return, volatility, and dividend yield. We denote them by $\mu(t), \sigma(t)$ and $q(t)$, respectively, and assume that $r(t), \mu(t), \sigma(t)$, and $q(t)$ are deterministic functions of t. Under the risk-neutral measure $\mathbb{P}^*$, the ex-dividend stock price process satisfies

$$dY_t = Y_t\Big[(r(t) - q(t))dt + \sigma(t)dW_t\Big]. \tag{7.19}$$

Let (S_t) be as in Black-Scholes model, and let $Y_t = f(t)S_{\tau(t)}$. By Itô's formula,

$$dY_t = f'(t)S_{\tau(t)}dt + f(t)dS_{\tau(t)}$$

$$= f'(t)S_{\tau(t)}dt + f(t)S_{\tau(t)}\Big[\sigma dB^*_{\tau(t)} + r\tau'(t)dt\Big]$$

$$= Y_t\Big[\left(\frac{f'(t)}{f(t)} + r\tau'(t)\right)dt + \sigma\sqrt{\tau'(t)}dW_t\Big],$$

where, by Lévy's martingale characterization theorem for Brownian motion,

$$W_t = \int_0^t \frac{1}{\sqrt{\tau'(s)}}dB^*_{\tau(s)}$$

is a $\mathbb{P}^*$-Brownian motion w.r.t. $(\mathcal{F}_{\tau(t)})$. In order that Y_t satisfies (7.19), clearly we only need to take $\tau(t) = \sigma^{-2} \int_0^t \sigma(u)^2 du$, and

$$f(t) = \exp\left\{-r\tau(t) + \int_0^t (r(u) - q(u))du\right\}.$$

In the present case, we have $P_Y(t, T) = \exp\left\{-\int_t^T r(s)ds\right\}$.

Now we consider a European call option $\xi = (Y_T - K)^+$. Then by (7.18), it is easy to see that its price at time t is equal to $C(t, Y_t)$, where $C(t, x)$ is given by a generalized Black-Scholes formula:

$$C(t, x) = \tilde{x}N(\tilde{d}_1) - Ke^{-(T-t)\tilde{r}}N(\tilde{d}_2). \tag{7.18'}$$

In this formula,

$$\tilde{x} = xe^{-\int_t^T q(s)ds}, \quad \tilde{r} = \frac{1}{T-t}\int_t^T r(s)ds,$$

and $\tilde{d}_1$ and $\tilde{d}_2$ have the same expressions as (5.13), except that x, r and σ^2 therein are replaced by $\tilde{x}$, $\tilde{r}$ and $\frac{1}{T-t}\int_t^T \sigma^2(s)ds$, respectively. We refer the reader to Wilmott, Dewynne, Howison (1993) for the derivation of this formula using a PDE approach.

7.3.3 General Nonlinear Reduction Method

Let (Y_t) be a price process of an asset under the risk-neutral probability measure $\mathbb{P}^*$ satisfying the following SDE:

$$dY_t = rY_t dt + \sigma(Y_t)dB_t^*,$$

where (B_t^*) is a standard Brownian motion under $\mathbb{P}^*$. If we know the transition density function of diffusion process (Y_t), then we can give a pricing formula for options with this as the underlying asset. The following gives the so-called nonlinear reduction method.

Let (X_t) be a diffusion process with a known transition density function satisfying

$$dX_t = \mu(X_t)dt + \overline{\sigma}(X_t)dB_t^*.$$

Process (Y_t) is called reducible to (X_t) if there is a twice continuously differentiable and monotonic function g such that $g(Y_t) = X_t$. By Itô's formula,

$$dg(Y_t) = (rg'(Y_t)Y_t + \frac{1}{2}g''(Y_t)\sigma^2(Y_t))dt + g'(Y_t)\sigma(Y_t)dB_t^*.$$

Thus, if

$$ryg'(y) + \frac{1}{2}g''(y)\sigma^2(y) = \mu(g(y)),$$

$$g'(y)\sigma(y) = \overline{\sigma}(g(y)),$$

then $g(Y_t) = X_t$.

Theorem 7.14 *Let $p_X(t, x; T, z)$ be the transition density function of (X_t). Then the transition density function $\overline{p}_Y(t, y; T, z)$ of the diffusion process (Y_t) is given by*

$$\overline{p}_Y(t, y; T, z) = p_X(t, g(y); T, g(z))g'(z). \tag{7.20}$$

Proof We have

$$\mathbb{E}^*[h(Y_T)|Y_t = y] = \mathbb{E}^*[h(g^{-1}(X_T))|X_t = g(y)]$$

$$= \int h(g^{-1}(w))p_X(t, g(y); T, w)dw$$

$$= \int h(z)p_X(t, g(y); T, g(z))g'(z)dz.$$

Thus, we get (7.20).   □

7.3.4   Option Pricing Under the CEV Model

In this subsection, we give a concrete example on how to use the nonlinear reduction method to make option pricing.

We assume that the stock price process is modeled by a CEV process (see Chap. 5, Sect. 5.4):

$$dS_t = \mu S_t dt + \sigma S_t^{\frac{\alpha}{2}} dB_t, \tag{7.21}$$

where $0 < \alpha < 2$ is a constant, known as the elasticity factor. If $\alpha = 2$ in (7.21), this model reduces to the Black-Scholes model.

Under risk-neutral probability $\mathbb{P}^*$ (7.21) becomes

$$dS_t = r S_t dt + \sigma S_t^{\frac{\alpha}{2}} dB_t^*. \tag{7.22}$$

Now we show how to derive the option pricing formula for the $\alpha = 1$ case by using Theorem 7.14. First we consider the $r = 0$ case. The corresponding process is denoted by (Y_t), i.e.,

$$dY_t = \sigma Y_t^{\frac{1}{2}} dB_t^*.$$

Let $X_t = 2\frac{\sqrt{Y_t}}{\sigma}$. Then

$$dX_t = -\frac{1}{2X_t}dt + dB_t^*.$$

Note that (X_t) is a Bessel process with transition density function:

$$p_X(t, x; T, y) = \frac{x}{T-t} \exp\left\{-\frac{x^2 + y^2}{2(T-t)}\right\} I_{-1}\left(\frac{xy}{T-t}\right),$$

where I_{-1} is the modified Besssel function of index -1. The call option price at time t for the process (Y_t) is

$$C_Y(t, Y_t; T, K) = \int_{g^{-1}(y)>K} [g^{-1}(y) - K] p_X(t, g(Y_t); T, y)dy,$$

where $g(x) = 2\frac{\sqrt{x}}{\sigma}$. Now put $S_t = f(t)Y_{\tau(t)}$, where

$$f(t) = e^{rt}, \quad \tau(t) = r^{-1}(1 - e^{-rt}).$$

Then (S_t) satisfies

$$dS_t = rS_t dt + \sigma S_t^{\frac{1}{2}} dW_t,$$

where (W_t) is a Brownian motion. Thus, by Theorem 7.13, we get

$$C_S(t, S_t; T, K) = e^{rT}e^{-r(T-t)}C_Y\left(\tau(t), \frac{S_t}{f(t)}; \tau(T), \frac{K}{f(T)}\right)$$

$$= S_t \sum_{n=0}^{\infty} \left[\frac{e^{-S_t'}(S_t')^n(n+1)G(n+2, K')}{(n+1)!}\right]$$

$$-Ke^{r(T-t)}\sum_{n=0}^{\infty}\left[\frac{e^{-S_t'}(S_t')^{n+1}G(n+1, K')}{(n+1)!}\right],$$

where

$$G(m, x) = \frac{1}{(m-1)!}\int_x^{\infty} e^{-z}z^{m-1}dz,$$

$$S_t' = \frac{2re^{r(T-t)}S_t}{\sigma^2(e^{r(T-t)} - 1)}, \quad K' = \frac{2rK}{\sigma^2(e^{r(T-t)} - 1)}.$$

For the general case of $0 < \alpha < 2$, put $X_t = \sigma^{-1} \left(1 - \frac{\alpha}{2}\right)^{-1} S_t^{1-\frac{\alpha}{2}}$. Then by Itô's formula,

$$dX_t = r\left(1 - \frac{\alpha}{2}\right)X_t dt - \frac{\alpha}{4\left(1 - \frac{\alpha}{2}\right)X_t}dt + dB_t^*.$$

The process (X_t) is a scale and time changed Bessel process, whose transition density function is known. So we can still give an analytic expression for option pricing.

7.4 Pricing American Contingent Claims

Now we address the problem of pricing American contingent claims in the diffusion model setting. We will use the same notations adopted in the previous section. Recall that an American contingent claim is defined as an adapted nonnegative process $(h_t)_{0 \leqslant t \leqslant T}$. For simplicity, we only consider an American contingent claim of the form $h_t = g(t, S_t)$. If $m = 1$, for an American call, we have $g(t, x) = (x - K)^+$, and for an American put, $g(t, x) = (K - x)^+$.

Let $\mathcal{T}_{t,T}$ be the set of all stopping times taking values in $[t, T]$. Put

$$\Phi(t, x) = \sup_{\tau \in \mathcal{T}_{t,T}} \mathbb{E}^{*t,x}\left[e^{-\int_t^\tau r(s, S_s)ds} g(\tau, S_\tau)\right], \tag{7.23}$$

where function g is assumed to be good enough such that $\Phi(t, x)$ is well defined. It is not difficult to prove that process $\gamma_t \Phi(t, S_t)$ is the supermartingale that dominates the process $\gamma_t g(t, S_t)$ for all $t \in [0, T]$.

The following two theorems are the main results concerning pricing American contingent claims. For proofs we refer the reader to Karatzas (1988).

Theorem 7.15 *There exists a trading strategy with consumption ϕ such that ϕ super-hedges $(g(t, S_t))$ and its wealth process $V_t(\phi)$ is given by $V_t(\phi) = \Phi(t, S_t)$, $\forall t \in [0, T]$. Moreover, for any trading strategy with consumption ψ which super-hedges $(g(t, S_t))$, we have $V_t(\psi) \geqslant \Phi(t, S_t)$ for all $t \in [0, T]$.*

We call $\Phi(0, S_0)$ the *seller's price* at time 0 of the American contingent claim.

Theorem 7.16 *Under some technical conditions, $\Phi(t, x)$ solves the following system of partial differential inequalities:*

$$\frac{\partial u}{\partial t} + \mathcal{A}_t u - ru \leqslant 0, \ \ u \geqslant g, \tag{7.24}$$

$$\left(\frac{\partial u}{\partial t} + \mathcal{A}_t u - ru\right)(g - u) = 0, \tag{7.25}$$

$$u(T, x) = g(T, x), \ \ x \in \mathbb{R}^m. \tag{7.26}$$

Now we turn to the optimal exercise problem on American contingent claims. Let $\tau \in \mathcal{T}_{0,T}$. If one exercises the American contingent claim at stopping time τ, the initial value of the payoff $g(\tau, S_\tau)$ is given by

$$V_0^\tau = \mathbb{E}^* \left[e^{-\int_0^\tau r(s, S_s)ds} g(\tau, S_\tau) \right].$$

Put

$$\tau^* = \inf \left\{ t \in [0, T] : \Phi(t, S_t) = g(t, S_t) \right\}. \tag{7.27}$$

Then τ^* is a stopping time which maximizes V_0^τ within $\mathcal{T}_{0,T}$. So it is reasonable to consider τ^* as the optimal exercise time for the American contingent claim.

Chapter 8
Term Structure Models for Interest Rates

In the Black-Scholes model, it was assumed that the interest rate is a constant or a deterministic function. For short-dated options on stock-like assets, it is an acceptable approximation. However, for pricing interest rate derivatives or interest rate risk management, it is an unreasonable assumption. Therefore, one of the major topics in finance theory is the modeling of random interest rates and the pricing of interest rate derivatives.

The so-called interest rate model is a mathematical description about the dependence relationship of the yield on bonds from bond maturity. This relationship is called the *interest rate term structure*. In accordance with the interest rate term structure modeling process to distinguish, they can be broadly divided into two types: the first type is the *equilibrium model* which is derived on the basis of market equilibrium conditions. The Vasicek model (Vasicek 1997) and CIR model (Cox et al. 1985) belong to this type. The second type is the *no-arbitrage model*, which is created by the no-arbitrage condition that must be satisfied between the relevant bonds. The Hull-White model (Hull and White 1990), HJM model (Heath et al. 1987, 1992), and BGM model (Brace et al. 1997) belong to this type.

On the other hand, based on the descriptions of the term structure of interest rates, they can be divided into short-term interest rate model (in abbreviation, short rate model), forward rate model, and state price density model. These three approaches are pioneered by Vasicek (1997), Heath et al. (1987, 1992), and Flesaker and Hughston (1996), respectively.

In this chapter, according to the above three methods, we introduce some models for term structure of interest rates, including a variety of single-factor short rate models, HJM model, and a variant of HJM model (*BGM model*), proposed by Brace et al. (1997). The pricing of interest rate derivatives is also briefly discussed.

© Springer Nature Singapore Pte Ltd. and Science Press 2018
J.-A. Yan, *Introduction to Stochastic Finance*, Universitext,
https://doi.org/10.1007/978-981-13-1657-9_8

8.1   The Bond Market

8.1.1   Basic Concepts

Throughout the chapter, we fix a time horizon $[0, T]$. We consider a financial market, called the *bond market*, which consists of a bank account and discount bonds with all possible maturities $s : 0 < s \leqslant T$. By a *discount bond* (or *zero-coupon bond*), we mean a financial security which pays no dividends and is sold at a price lower than the face value paid at maturity. In the following we call a discount bond maturing at time s an s- *bond*, denote its price at time t by $P(t, s)$, and assume that $P(s, s)$ is equal to 1 (i.e., one unit of cash).

The *yield-to-maturity* (or simply, *yield*) at time $t < s$ of an s-bond is defined as

$$Y(t, s) = -\frac{\ln P(t, s)}{s - t}.$$

It is an average of returns of the s-bond over the time interval $[t, s]$. The difference in yields at different maturities reflects market beliefs about future interest rates. A *yield curve* at time t is the graph of $Y(t, s)$ against maturities s. The dependence of the yield curve on the time to maturity, $s - t$, is called the *term structure of interest rates*. The *short rate* r_t at time t is defined as $\lim_{s>t, s \to t} Y(t, s)$, when the limit exists a.s.

In the sequel, we assume that short rate r_t exists for all $t \in [0, T]$, and process (r_t), which is $(\mathcal{F}_t)$-adapted, admits a measurable version. Moreover we assume $\int_0^T r_t dt < \infty$. Put

$$R_t = \exp\left\{ \int_0^t r_u du \right\}.$$

We call process (R_t) the price process of the *savings account* (or *money market account*).

Assume $P(t, s)$ is differentiable w.r.t. s, put

$$f(t, s) = -\frac{\partial \ln P(t, s)}{\partial s} = -\frac{\partial P(t, s)/\partial s}{P(t, s)}.$$

Then

$$P(t, s) = \exp\left\{ -\int_t^s f(t, u)du \right\}. \tag{8.1}$$

We call $f(t, s)$ the *(instantaneous) forward rates* with maturity s, contracted at t. Intuitively, $f(t, s)$ is the interest rate over the infinitesimal time interval $[s, s + ds]$, as seen from time t. A *forward rate curve* at time t is the graph of $f(t, s)$ against maturities s. It is another measure of future values of interest rates.

8.1.2 Bond Price Process

Let $B = (B^1, \cdots, B^d)$ be a d-dimensional Brownian motion on a complete probability space $(\Omega, \mathcal{F}, \mathbb{P})$. We denote by $(\mathcal{F}_t)$ the natural filtration of (B_t). If $P(t, s)_{t \leqslant s}$, for $s \leqslant T$, are known deterministic smooth functions, then under no-arbitrage condition, $P(t, s)$ must have the form

$$P(t, s) = \exp\left\{-\int_t^s r_u du\right\},$$

where r_t is the short rate at time t. In this case the bond prices are completely determined by the short rates.

However, in the uncertain world, this is no longer true. Assume that we are given a short rate process (r_t), which is a measurable $(\mathcal{F}_t)$-adapted nonnegative process. If $\mathbb{P}^*$ is any probability measure equivalent to $\mathbb{P}$, then we can model an arbitrage-free bond market by letting

$$P(t, s) = \mathbb{E}^*\left[e^{-\int_t^s r_u du} \,\Big|\, \mathcal{F}_t\right], \quad t \leqslant s \leqslant T \tag{8.2}$$

be bond prices. By so doing, $\mathbb{P}^*$ is automatically a martingale measure for the market. So different equivalent probability measures lead to different models for bond market.

Assume that we are given a filtered probability space $(\Omega, \mathcal{F}, (\mathcal{F}_t), \mathbb{P})$, where the filtration $(\mathcal{F}_t)$ is the natural filtration associated to a d-dimensional (standard) Brownian motion (B_t). Let (r_t) be a short rate process which is $(\mathcal{F}_t)$-adapted measurable process. How to select a risk-neutral measure for the bond market? We will show that it consists in specifying a market price of risk process. In fact, according to the martingale representation theorem, the Radon-Nikodym derivative of a martingale measure $\mathbb{P}^*$ w.r.t. $\mathbb{P}$ has the form

$$\frac{d\mathbb{P}^*}{d\mathbb{P}} = \exp\left\{-\int_0^T \lambda_u^\tau dB_u - \frac{1}{2}\int_0^T |\lambda_u|^2 du\right\}.$$

On the other hand, by (8.2) and the martingale representation theorem, the s-bond price process $P(t, s)$ should have the form

$$d_t P(t, s) = P(t, s)\left[r_t dt + v(t, s)^\tau dB_t^*\right], \quad t \leqslant s \leqslant T,$$

where

$$B_t^* = B_t + \int_0^t \lambda_u du$$

is a d-dimensional Brownian motion under $\mathbb{P}^*$. Thus, under the objective probability measure $\mathbb{P}$, we have

$$d_t P(t, s) = P(t, s)\Big[m(t, s)dt + v(t, s)^\tau dB_t\Big], \quad t \leqslant s \leqslant T,$$

where

$$m(t, s) = r_t + \lambda_t^\tau v(t, s).$$

This shows that process (λ_t) is the market price of risk process for all bonds with different maturities. For the economic meaning of the market price of risk, see Chap. 7, Sect. 7.1.2. For simplicity, we assume that short rate process (r_t) is modeled, under objective probability measure $\mathbb{P}$, by a diffusion process

$$dr_t = \mu_0(t, r_t)dt + \sigma(t, r_t)^\tau dB_t, \quad t \leqslant T, \tag{8.3}$$

where $\mu_0(t, x)$ and $\sigma(t, x)$ are real and $\mathbb{R}^d$-valued Borel functions, respectively.

Now we only consider those equivalent probability measures $\mathbb{P}^*$ whose Radon-Nikodym derivatives w.r.t. $\mathbb{P}$ have the form

$$\frac{d\mathbb{P}^*}{d\mathbb{P}} = \exp\Big\{ -\int_0^T \lambda(u, r_u)^\tau dB_u - \frac{1}{2}\int_0^T |\lambda(u, r_u)|^2 du \Big\}, \tag{8.4}$$

where $\lambda(t, x)$ is a $\mathbb{R}^d$-valued Borel function on $[0, T] \times \mathbb{R}$. Consequently, selecting such a probability measure consists of specifying a function $\lambda(t, x)$. The latter can be determined by solving a utility-based general equilibrium economy and by using the market data, because $\lambda(t, r_t)_{0 \leqslant t \leqslant s}$ is the market price of risk process. Once $\lambda(t, x)$ is specified, the short rate process (r_t), modeled by (8.3), can be remodeled, in the "risk-neutral" world, as:

$$dr_t = \mu(t, r_t)dt + \sigma(t, r_t)^\tau dB_t^*, \quad t \leqslant T, \tag{8.5}$$

where $\mu(t, x) = \mu_0(t, x) - \sigma(t, x)\lambda(t, x)$. In this case, by the Feynman-Kac formula, we know that, under some regularity conditions, the s-bond price process can be expressed as $P(t, s) = F(t, r_t; s)$, where $F(t, x; s)$ is a $C^{1,2}$-function on $[0, T] \times \mathbb{R}$, for any fixed $s \in (0, T]$, and is the unique solution to the following PDE

$$F_t(t, x; s) + \mu(t, x)F_x(t, x; s) + \frac{1}{2}\sigma(t, x)^2 F_{xx}(t, x; s) - xF(t, x; s) = 0, \tag{8.6}$$

subject to the terminal condition $F(s, x; s) = 1$.

8.2 Short Rate Models

The simplest term structure models are short rate models, which are described by an Itô process or a diffusion process. By specifying a market price of risk (or, equivalently, an equivalent martingale measure), we can reformulate short rate models in a risk-neutral world. Such a model is consistent with the initial value of the short rate but in general cannot fit the initial term structure (e.g., the initial bond prices with different maturities or initial forward rate). For certain short rate models (e.g., Hull-White's extensions of the Vasicek model and the CIR model), one can adjust the deterministic drift function in such a way that the model can fit perfectly the initial term structure.

In this section, for simplicity of notations, we assume that objective probability $\mathbb{P}$ is itself a risk-neutral measure.

8.2.1 One-Factor Models and Affine Term Structures

In this subsection, we assume that under risk-neutral measure $\mathbb{P}^*$, the short rate process (r_t) is modeled by (8.5), where B_t^* is a one-dimensional Brownian motion. Since the only state variable in (8.5) is the short rate, we call such a model *one-factor model*. In general, the number of factors in a model is the same as the number of the fundamental sources of uncertainty in the model, i.e., the dimensionality of Brownian motion employed in the model.

In the past 40 years, researchers have proposed a variety of concrete models of (8.5) and studied the pricing of interest rate derivatives under these models. The standard one-factor models include Merton (1973a), Vasicek (1977), Marsh and Rosenfeld (1983), Constantinides and Ingersoll (1984), Cox et al. (1985), Ho and Lee (1986), Black et al. (1990), Black and Karasinski (1991), Pearson and Sun (1994), and other models. Ait-Sahalia (1996) tested many of these models empirically.

Below is the table that lists these widely recognized models.

Interest rate models	References
$dr_t = \alpha dt + \sigma dB_t$	Merton (1973a)
$dr_t = \alpha(t)dt + \sigma dB_t$	Ho and Lee (1986)
$dr_t = ar_t dt + \sigma r_t dB_t$	Dothan (1978)
$dr_t = \beta(\alpha - r_t)dt + \sigma dB_t$	Vasicek 1977
$dr_t = (a(t) - b(t)r_t)dt + \sigma(t)dB_t$	Hull and White (1990)
$d\ln r_t = (a(t) - b(t)\ln r_t)dt + \sigma(t)dB_t$	Black and Karasinski (1991)
$\left(b(t) = -\frac{\sigma'(t)}{\sigma(t)}\right)$	Black et al. (1990)
$dr_t = (\alpha r_t^{\delta-1} + \beta r_t)dt + \sigma r_t^{\delta/2}dB_t$	Marsh and Rosenfeld (1983)
$(\delta = 3, \beta = 0)$	Constantinides and Ingersoll (1984)
$dr_t = \beta(\alpha - r_t)dt + \sigma\sqrt{r_t}dB_t$	Cox-Ingersoll-Ross (CIR) (1985)
$dr_t = (\alpha(t) - \beta(t)r_t)dt + \sigma(t)r_t^{1/2}dB_t$	Hull-White (extended CIR) (1990)
$dr_t = \beta(\alpha - r_t)dt + \sigma(\gamma + r_t)^{1/2}dB_t$	Pearson and Sun (1994)

Under certain conditions on μ and σ^2 in (8.5), the following theorem gives an analytic expression for the bond price process in terms of short rate process.

Theorem 8.1 *Let the short rate process be modeled by (8.5) under the risk-neutral measure. If μ and σ^2 are affine functions in x, i.e.,*

$$\mu(t, x) = \alpha(t)x + \beta(t), \ \sigma(t, x) = \sqrt{\gamma(t)x + \delta(t)},$$

then

$$P(t, s) = e^{A(t,s) - B(t,s)r_t},$$

where $B(t, s)$ and $A(t, s)$ are deterministic functions of s and t and satisfy the following equations:

$$B_t(t, s) = -\alpha(t)B(t, s) + \frac{1}{2}\gamma(t)B(t, s)^2 - 1,$$

$$A_t(t, s) = \beta(t)B(t, s) - \frac{1}{2}\delta(t)B(t, s)^2, \tag{8.7}$$

subject to the boundary condition: $A(s, s) = B(s, s) = 0$. Here we denote $\partial_t B(t, s)$ by $B_t(t, s)$.

Proof (see (Björk 1997)) Assume that the solution to (8.6) has the form $F(t, x; s) = e^{A(t,s) - B(t,s)x}$, then under the assumption on μ and σ, for all x, we have

$$A_t(t, s) - \beta(t)B(t, s) + \frac{1}{2}\delta(t)B(t, s)^2$$

$$- \left\{1 + B_t(t, s) + \alpha(t)B(t, s) - \frac{1}{2}\gamma(t)B(t, s)^2\right\}x = 0.$$

This shows that $B(t, s)$ and $A(t, s)$ satisfy the stated equations (8.7), and the boundary condition $F(s, x; s) = 1$ implies $A(s, s) = B(s, s) = 0$. By reversing the above reasoning, we get the desired result. □

Remark 1 If a short rate model is such that the bond price is of the form (8.6), then the yield curve $Y(t, s)$ at time t is a linear function of the short rate r_t:

$$Y(t, s) = \frac{1}{s - t}[B(t, s)r_t - A(t, s)].$$

Thus, such a short rate model is said to possess an *affine term structure*. In this case, possible shapes of the yield curve are upward-slopping, downward-slopping, and slightly humped. We refer the reader to Duffie (1996) for a detailed discussion on the subject.

Remark 2 It has been shown in Brown and Schaefer (1994) and Duffie and Kan (1996) that a short rate model described by a diffusion with an affine term structure is necessarily of the type stated in Theorem 8.1. Filipović (2001) further gives a complete characterization of the affine term structure models based on a general nonnegative Markov short rate process.

We give bellow some examples of term structures for short rate models.

The Vasiceck model   The Vasiceck model is

$$dr_t = \beta(\alpha - r_t)dt + \sigma B_t.$$

Then Eq. (8.7) becomes

$$B_t(t, s) - \beta B(t, s) = -1, \quad A_t(t, s) = \alpha \beta B(t, s) - \frac{1}{2}\sigma^2 B^2(t, s),$$

subject to boundary condition $A(s, s) = B(s, s) = 0$. Its solution is given by

$$B(t, s) = \frac{1}{\beta}\left\{1 - e^{-\beta(s-t)}\right\},$$

$$A(t, s) = \frac{\{B(t, s) - s + t\}\left(\alpha\beta^2 - \frac{1}{2}\sigma^2\right)}{\beta^2} - \frac{\sigma^2 B^2(t, s)}{4\beta}.$$

The Ho-Lee model   The Ho-Lee model is

$$dr_t = \alpha(t)dt + \sigma dB_t.$$

Equation (8.7) becomes

$$B_t(t, s) = -1, \quad A_t(t, s) = \alpha(t)B(t, s) - \frac{1}{2}\sigma^2 B^2(t, s),$$

subject to boundary condition $A(s, s) = B(s, s) = 0$. Its solution is given by

$$B(t, s) = s - t,$$

$$A(t, s) = \int_t^s \alpha(u)(u - s)du + \frac{\sigma^2}{2} \cdot \frac{(s - t)^3}{3}.$$

Now in order that $\alpha(t)$ fits the observed initial term structure $\{P^*(0, s), s \geqslant 0\}$, i.e., such that $P(0, s) = P^*(0, s)$, we must have

$$\alpha(t) = \frac{\partial f^*(0, t)}{\partial t} + \sigma^2 t,$$

where $f^*(0, t) = -\frac{\partial \ln P^*(0,t)}{\partial t}$ are the observed forward rates. Finally, we obtain

$$P(t, s) = \frac{P^*(0, s)}{P^*(0, t)} \exp\left\{(s-t)f^*(0, t) - \frac{\sigma^2}{2}t(s-t)^2 - (s-t)r_t\right\}.$$

The Hull-White model The Hull-White model (Hull and White 1993b) is

$$dr_t = (a(t) - b(t)r_t)dt + \sigma(t)dB_t,$$

where $a(t)$, $b(t)$, and $\sigma(t)$ are nonrandom positive functions of t. Then Eq. (8.7) becomes

$$B_t(t, s) = b(t)B(t, s) - 1, \quad A_t(t, s) = a(t)B(t, s) - \frac{1}{2}\sigma(t)^2 B^2(t, s),$$

subject to the boundary condition: $A(s, s) = B(s, s) = 0$. If $b(t) = b, \sigma(t) = \sigma$, the solution to the above equation is

$$B(t, s) = \frac{1}{b}\left\{1 - e^{-b(s-t)}\right\},$$

$$A(t, s) = \int_t^s \left\{\frac{1}{2}\sigma^2 B^2(u, s) - a(u)B(u, s)\right\}du.$$

In the present case, in order that $a(t)$ fits initial term structure $\{P^*(0, s), s \geqslant 0\}$ (i.e., $P(0, s) = P^*(0, s)$), $a(t)$ must satisfy the equation

$$f^*(0, s) = e^{-bs}r(0) + \int_o^s e^{-b(s-u)}a(u)du - \frac{\sigma^2}{2b^2}(1 - e^{-bs})^2,$$

where $f^*(0, t) = -\frac{\partial \ln P^*(0,t)}{\partial t}$ are the observed forward rates. Put $g(t) = \frac{\sigma^2}{2b^2}(1 - e^{-bs})^2$. The solution is given by

$$a(s) = f_s^*(0, s) - g'(s) + b(f^*(0, s) - g(s)).$$

Finally, we obtain

$$P(t, s) = \frac{P^*(0, s)}{P^*(0, t)} \exp\left\{B(t, s)f^*(0, t) - \frac{\sigma^2}{4b}B^2(t, s)(1 - e^{-2bt}) - B(t, s)r_t\right\}.$$

The CIR model The disadvantage of the Vasicek model is the possibility of negative interest rates. In order to avoid this defect, Cox et al. (1985) proposed the following CIR model:

$$dr_t = \beta(\alpha - r_t)dt + \sigma\sqrt{r_t}dB_t.$$

For CIR model, Eq. (8.7) becomes

$$B_t(t, s) = \beta B(t, s) + \frac{1}{2}\sigma^2 B(t, s)^2 - 1, \quad A_t(t, s) = \beta\alpha B(t, s),$$

subject to boundary condition $A(s, s) = B(s, s) = 0$. Its solution is given by

$$B(t, s) = \frac{\sinh\gamma(s - t)}{\gamma\cosh\gamma(s - t) + \frac{1}{2}\beta\sinh\gamma(s - t)},$$

$$A(t, s) = \frac{2\beta\alpha}{\sigma^2}\ln\left\{\frac{\gamma e^{\beta(s-t)/2}}{\gamma\cosh\gamma(s - t) + \frac{1}{2}\beta\sinh\gamma(s - t)}\right\},$$

where $\gamma = \frac{1}{2}(\beta^2 + 2\sigma^2)^{1/2}$.

8.2.2 Functional Approach to One-Factor Models

Each model listed in Sect. 8.2.1. can be studied separately and has its own desirable properties. The calibration of model parameters are typically done using numerical and statistical methods or Monte Carlo simulation.

Luo et al. (2012) propose a functional approach to model short rates: first it postulates a Markov state variable which is a generalized Ornstein-Uhlenbeck process or a scale and time-changed Bessel process and then models the short rate as a nonlinear function of the underlying state variable. This approach turns out to be simple and effective: it not only incorporates all of the above tabulated examples but also has advantages in numerical computations.

For notational simplicity, we assume that the objective probability measure $\mathbb{P}$ is itself a martingale measure. Let $\{B_t\}_{t\geqslant 0}$ be a one-dimensional Brownian motion under $\mathbb{P}$. In the following, $\eta(t)$ and $\sigma(t)$ are nonnegative deterministic functions of t, η, $\sigma > 0$, and $\epsilon \geqslant \frac{1}{2}\sigma^2$ are constants. We consider a generalized Ornstein-Uhlenbeck process X_t and a scale and time-changed Bessel process Y_t, which satisfy the following SDE, respectively (see Chapt. 4, Sect. 4.5.2):

$$dX_t = -\eta(t)X_t dt + \sigma(t)dB_t, \tag{8.8}$$

$$dY_t = \left(-\eta Y_t + \frac{\epsilon}{Y_t}\right)dt + \sigma dB_t. \tag{8.9}$$

We model the interest rate r_t as a function of time t and the state variable X_t or Y_t, namely,

$$r_t = f(t, X_t), \quad \text{or } r_t = f(t, Y_t) \quad t \geqslant 0,$$

where f is an appropriate function. As we have shown in Chap. 4 that the state process is Markovian and their transition density functions are explicitly known. The interest rate processes are also Markovian and their transition density functions are also known.

When $f(t, x)$ is twice continuously differentiable in x and continuously differentiable in t, applying Itô's formula, we obtain the SDE for r_t:

$$
\begin{aligned}
dr_t &= df(t, X_t) \\
&= f_t(t, X_t)dt + f_x(t, X_t)dX_t + \tfrac{1}{2}f_{xx}(t, X_t)d[X]_t \\
&= \left\{ f_t(t, X_t) + \tfrac{\sigma(t)^2}{2}f_{xx}(t, X_t) - \eta(t)X_t f_x(t, X_t) \right\}dt \\
&\quad + \sigma(t)f_x(t, X_t)dB_t,
\end{aligned} \tag{8.10}
$$

or

$$
\begin{aligned}
dr_t &= df(t, Y_t) \\
&= \left\{ f_t(t, Y_t) + \tfrac{\sigma^2}{2}f_{xx}(t, Y_t) + \left(\tfrac{\varepsilon}{Y_t} - \eta Y_t \right)f_x(t, Y_t) \right\}dt \\
&\quad + \sigma f_x(t, Y_t)dB_t.
\end{aligned} \tag{8.11}
$$

If we model the interest rate r_t as a function of the single state variable X_t or Y_t, then we write simply $r_t = f(X_t)$ or $r_t = f(Y_t)$. In this setting, a good model depends on the judicious choice of f, $\eta(t)$, $\sigma(t)$ or η, ϵ and σ. If the range of f is confined to positive real values, the usual drawback of obtaining a negative interest rate such as that occurring in the Vasicek model is avoided.

To demonstrate the usefulness and simplicity of this approach, we show that by specifying f as some functions, we recover all the interest rate models tabulated above. In the cases of the Merton, the Dothan, and the Ho-Lee models, interest rates are trivially functions of a single Brownian motion. So we omit the discussion on them.

First, we construct short-term interest rate models by using process (X_t).

Example 1 The Black-Karasinski model (Black and Karasinski 1991) is

$$
d\ln r_t = (a(t) - \eta(t)\ln r_t)dt + \sigma(t)dB_t.
$$

If we let $f(t, x) = \exp\{g(t, x)\}$, i.e., $r_t = \exp\{g(t, X_t)\}$, then applying (8.10) to function g instead of f gives

$$
d\ln r_t = \left\{ g_t(t, X_t) + \frac{\sigma(t)^2}{2}g_{xx}(t, X_t) - \eta(t)X_t g_x(t, X_t) \right\}dt + \sigma(t)g_x(t, X_t)dB_t. \tag{8.12}
$$

So if we take $\eta(t) = b(t)$ in (8.8) and let $g(t, x) = x + \int_0^t a(s)ds$, i.e., $\ln r_t = X_t + \int_0^t a(s)ds$ with $X_0 = \ln r(0)$, then by (8.12) we recover the Black-Karasinski model. In particular, if we let $b(t) = -\frac{\sigma'(t)}{\sigma(t)}$, then we obtain the Black-Derman-Toy model (see Black et al. 1990).

Example 2 The Vasicek model is

$$dr_t = \beta(\alpha - r_t)dt + \sigma dB_t.$$

By comparing the above equation with (8.8), it is easy to see that if we let $\eta(t) = \beta$, $\sigma(t) = \sigma$, and $r_t = X_t + \alpha$ with $X_0 = r(0) - \alpha$, then we recover the Vasicek model.

Example 3 The Hull-White model is

$$dr_t = (a(t) - b(t)r_t)dt + \sigma(t)dB_t.$$

Let $l(t) = \int_0^t b(u)du$. Then we have

$$d(e^{l(t)}r_t) = e^{l(t)}(a(t)dt + \sigma(t)dB_t).$$

Thus, we get

$$r_t = e^{-l(t)}\left(r_0 + \int_0^t e^{l(u)}a(u)du + \int_0^t e^{l(u)}\sigma(u)dB_u\right).$$

If we let $\eta(t) = b(t)$, $g(t) = e^{-l(t)}\int_0^t e^{l(u)}a(u)du$, and $r_t = X_t + g(t)$ with $X_0 = r_0$, then we recover the Hull-White model.

We show below that the remaining three models listed in the table can be recovered by choosing f as a function of the single state variable Y_t, namely, $r_t = f(Y_t)$. The only exception is the Marsh-Rosenfeld model with $\delta = 2$, which can be easily modelled as $r_t = f(t, Y_t)$ (see Example 5).

Example 4 The CIR model (Cox et al. 1985) is

$$dr_t = \beta(\alpha - r_t)dt + \sigma r_t^{1/2}dB_t.$$

Let $f(x) = \frac{1}{4}x^2$ and put $r_t = \frac{1}{4}Y_t^2$ with $Y_0 = 2\sqrt{r(0)}$. By (8.11) we have

$$dr_t = \left(\frac{1}{4}\sigma^2 + \frac{\epsilon}{2} - 2\eta r_t\right)dt + \sigma\sqrt{r_t}dB_t. \tag{8.13}$$

Identifying $\beta = 2\eta$ and $\alpha = (\sigma^2 + 2\epsilon)/8\eta$, we recover the CIR model.

Remark In a special set of parameters, namely, when $\beta = \sigma^2/4\alpha$, the CIR model can be mapped from the Ornstein-Uhlenbeck process. This observation is also noted by, e.g., Rogers (1995) and Maghsoodi (1996). This special case corresponds to $\epsilon = 0$ in our model. In general cases, the CIR model has to be mapped from the scale and time-changed Bessel process given by (8.9).

Example 5 The Marsh-Rosenfeld model (Marsh and Rosenfeld 1983) is

$$dr_t = (\alpha r_t^{\delta-1} + \beta r_t)dt + \sigma r_t^{\delta/2}dB_t.$$

We consider separately the cases $\delta < 2$, $\delta > 2$ and $\delta = 2$. If $\delta \neq 2$, we let $f(x) = ax^{\frac{2}{2-\delta}}$ and put $r_t = aY_t^{\frac{2}{2-\delta}}$. Then (8.11) becomes

$$dr_t = \left\{ (\delta\sigma^2 + 2(2-\delta)\varepsilon)\frac{a^{2-\delta}}{(2-\delta)^2}r_t^{\delta-1} - \frac{2\eta}{2-\delta}r_t \right\}dt + \frac{2\sigma a^{1-\delta/2}}{2-\delta}r_t^{\delta/2}dB_t.$$

(8.14)

First, assume $\delta < 2$. In this case, we should take a such that $\frac{2a^{1-\delta/2}}{2-\delta} = 1$, i.e., $a = \left(\frac{2-\delta}{2}\right)^{\frac{2}{2-\delta}}$. By identifying η and ε through equation $(\delta\sigma^2 + 2(2-\delta)\varepsilon)\frac{a^{2-\delta}}{(2-\delta)^2} = \alpha$ and $-\frac{2\eta}{2-\delta} = \beta$, that is, putting $\eta = -\frac{\beta(2-\delta)}{2}$ and $\varepsilon = \frac{4\alpha-\delta\sigma^2}{2(2-\delta)}$, we recover the Marsh-Rosenfeld model. In order that condition $\varepsilon \geqslant \frac{1}{2}\sigma^2$ to be satisfied, one should assume that $\alpha \geqslant \frac{1}{2}\sigma^2$.

Next, assume $\delta > 2$. In this case, if we solve the equation $\frac{2a^{1-\delta/2}}{2-\delta} = 1$, we will obtain a negative a. This is not reasonable for an interest rate model. In order to overcome this difficulty, we should consider the following state process:

$$dY_t = \left(-\eta Y_t + \frac{\varepsilon}{Y_t}\right)dt - \sigma dB_t,$$

and solve the equation $\frac{2a^{1-\delta/2}}{2-\delta} = -1$ to obtain $a = \left(\frac{\delta-2}{2}\right)^{\frac{2}{2-\delta}}$. The parameters η and ε are determined as above. In order for the condition $\varepsilon \geqslant \frac{1}{2}\sigma^2$ to be satisfied, one should assume that $\alpha \leqslant \frac{1}{2}\sigma^2$. In particular, if $\delta = 3$ and $\beta = 0$, we recover the Constantinides-Ingersoll model (Constantinides and Ingersoll 1984).

Finally, we consider the case $\delta = 2$. In this case, the Marsh-Rosenfeld model becomes

$$dr_t = (\alpha + \beta)r_t dt + \sigma r_t dB_t.$$

If we choose $f(t, x) = \exp\left\{x + \left(\alpha + \beta - \frac{\sigma^2}{2}\right)t\right\}$, i.e., $r_t = \exp\{Y_t + (\alpha + \beta - \frac{\sigma^2}{2})t\}$ with $X_0 = \ln r(0)$, we recover the model.

Example 6 The Pearson-Sun model (Pearson and Sun 1994) is

$$dr_t = \beta(\alpha - r_t)dt + \sigma(\gamma + r_t)^{1/2}dB_t.$$

Let $f(x) = \frac{1}{4}x^2 - \gamma$ and put $r_t = \frac{1}{4}Y_t^2 - \gamma$ with $X_0 = 2\sqrt{\gamma + r(0)}$. Then (8.11) becomes

$$dr_t = \left(\frac{\sigma^2}{4} + \frac{\varepsilon}{2} - 2\eta\gamma - 2\eta r_t\right)dt + \sigma(r_t + \gamma)^{1/2}dB_t.$$

By identifying $\eta = \frac{1}{2}\beta$, $\varepsilon = 2\beta(\alpha + \gamma) - \frac{\sigma^2}{2}$, we recover the Pearson-Sun model.

In practice, we need to estimate the parameters of the model under an unknown martingale measure. Instead, we use the historical data of the short rate to estimate the parameters of the model under the objective probability, and then we calculate the values for a set of traded bonds and options based on estimated parameters and compare them with market values. We adjust values of parameters and repeated the same procedure until the model fits well the historical data. The freedom of choosing f allows fitting of any current distribution curve.

When we calculate the price of a contingent claim based on the proposed interest rate model, there is a unified numerical approach based on simple lattice approximation. For example, consider an interest rate-dependent contingent claim $C_T = g(r_T)$ at time T. Its price at time t is given by

$$C_t = E\left[e^{-\int_t^T r_u du} C_T \middle| \mathcal{F}_t\right],$$

where $\{\mathcal{F}_t\}$ is the filtration of the interest rate. We have assumed that the original probability is already a martingale measure. Since the distribution of the state variable is known, so is the distribution of interest rate r_t. Thus, the above price can be easily calculated via direct lattice approximation. That is, we construct a sequence of discrete-time, discrete-state processes $(r_t^{(n)})$, which converges weakly to (r_t), such that

$$E\left[e^{-\int_t^T r_u^{(n)} du} C_T \middle| \mathcal{F}_t\right]$$

is calculated by the backward induction (see Hull and White (1990, 1993a, 1996)), and Schmidt (1997)). Then taking limit as $n \to \infty$, we obtain the value of C_t.

8.2.3 Multifactor Short Rate Models

The one-factor short rate models presented above provide explicit expressions for the bond prices. However these models do not fit the real interest rate movement well. A more realistic short rate model should include some other economic variables, such as the long-term interest rate (or long rate), the yields on a fixed number of bonds, and the short rate volatility, among others. The sources of uncertainty are represented by a multidimensional Brownian motion. Such a model is called a *multifactor model*.

Two-factor models are described by two-dimensional diffusions. Fong and Vasicek (1991) modify the Vasicek model by allowing the variance of the short rate to follow a stochastic process and present the following model:

$$dr_t = \alpha(\bar{r} - r_t)dt + \sqrt{v(t)} dB_t,$$

$$dv(t) = \gamma(\bar{v} - v(t))dt + \xi\sqrt{v(t)} dW_t,$$

where (B_t) and (W_t) are two correlated Brownian motions in a risk-neutral world, γ is the intensity of mean reversion, $\bar{v}$ is the long-run average level of v, and ξ is the volatility parameter for v. It turns out that this model possesses an affine term structure in the sense that the yield curve at time t is a linear function of the rate r_t and its variance $v(t)$. The drawback of the Vasicek model that the short rate can be negative remains for the Fong-Vasicek model and can be even worse. The model of Longstaff and Schwartz (1992a) is similar to the Fong-Vasicek model.

Another two-factor model was proposed by Hull and White (1994). They added a random parameter u in their extension of Vasicek's model as follows:

$$dr_t = \big(a(t) + u(t) - b(t)r_t\big)dt + \sigma(t)dB_t,$$

where $a(t), b(t)$, and $\sigma(t)$ are nonrandom positive functions of t and $u(t)$ following the stochastic process:

$$du(t) = -c(t)dt + \delta(t)dW_t,$$

where $c(t)$ and $\delta(t)$ are nonrandom positive functions of t.

A three-factor model was proposed by Chen (1996). In this model, in addition to the short rate, two other factors are the short-term mean rate and the short rate volatility.

In the 1990s, there have been many papers dealing with the so-called *higher-dimensional squared-Gauss-Markov process model*, which is described as

$$dX_t = (a_t + C_t X_t)dt + \sigma_t dB_t^*,$$

$$r_t = \frac{1}{2}|X_t|^2,$$

where (B_t^*) is a d-dimensional Brownian motion under the risk-neutral probability measure $\mathbb{P}^*$, and σ, C are $\mathbb{R}^d \times \mathbb{R}^d$-valued functions on $\mathbb{R}^+$, and a is an $\mathbb{R}^d$-valued function. This model has the advantage that it leads to an explicit formula for bond prices. In fact, it is easy to prove that the s-bound price at time t is given by

$$P(t, s) = \exp\left\{ -\frac{1}{2}X_t^\tau Q_t X_t + b_t^\tau X_t - \gamma_t \right\},$$

where Q_t solves the matrix *Riccati equation*

$$I + Q_t C_t + (Q_t C_t)^\tau + Q_t - Q_t \sigma \sigma^\tau Q_t^\tau = 0, \quad Q_t(s, s) = 0,$$

and b_t, γ_t solve

$$\frac{db_t}{dt} - Q_t a_t - (Q_t \sigma_t \sigma_t^\tau - C_t^\tau)b_t = 0, \quad b_t(s, s) = 0,$$

$$\frac{d\gamma_t}{dt} = b_t^\tau a_t - \frac{1}{2}\mathrm{tr}(\sigma_t^\tau Q_t \sigma_t) + \frac{1}{2}b_t^\tau \sigma_t \sigma_t^\tau b_t, \quad \gamma_t(s, s) = 0.$$

We refer the reader to Rogers (1995) and Duffie and Kan (1996) for details on this class of models.

8.2.4   Forward Rate Models: The HJM Model

Heath, Jarrow, and Morton proposed in 1987 another way to model the term structure (see Heath et al. 1992). They choose the forward rate curve as their (infinite dimensional) state variable. In this way, the HJM model fits automatically the current yield curve. A discrete-time analog of the HJM model in the form of a binomial tree was proposed by Ho and Lee (1986).

For each fixed maturity s, the HJM model for forward interest rates is described, in the risk-neutral world, by an Itô process:

$$f(t, s) = f(0, s) + \int_0^t \alpha(u, s)du + \int_0^t \sigma(u, s)dB_u^*, \quad t \leqslant s, \tag{8.15}$$

where (B_t^*) is a d-dimensional Brownian motion under an equivalent martingale measure $\mathbb{P}^*$, $\{\alpha(t, s) : 0 \leqslant t \leqslant s\}$ and $\{\sigma(t, s) : 0 \leqslant t \leqslant s\}$ are measurable adapted processes valued in $\mathbb{R}$ and $\mathbb{R}^d$, respectively, such that (8.15) is well defined as an Itô process and the initial forward curve, $f(0, s)$, is deterministic and satisfies the condition that $\int_0^T f(0, u)du < \infty$.

We now show that in (8.15), $\alpha(u, s)$ is in fact uniquely determined by $\sigma(u, s)$ (see below (8.18)). Assume that (r_t) is the interest rate process determined by forward interest rate process $f(t, s)$. Put

$$W_t = \mathbb{E}^*\left[e^{-\int_0^s r_u du}\middle| \mathcal{F}_t\right] = e^{-\int_0^t r_u du} P(t, s).$$

Since $(W_t)_{0 \leqslant t \leqslant s}$ is a strictly positive martingale, by the martingale representation theorem for Brownian motion, there exists an $\mathbb{R}^d$-valued adapted process $(H(u, s)_{u \leqslant s})$ such that

$$W_t = W_0 \exp\left\{\int_0^t H(u, s)dB_u^* - \frac{1}{2}\int_0^t |H(u, s)|^2 du\right\}, \tag{8.16}$$

i.e.,

$$\ln W_t = \ln W_0 + \int_0^t H(u, s)dB_u^* - \frac{1}{2}\int_0^t |H(u, s)|^2 du.$$

However, by (8.1) and (8.15), we get

$$\ln W_t = -\int_0^t r_u du - \int_t^s f(t, u)du$$

$$= -\int_0^t r_u du - \int_t^s \left[f(0, u) + \int_0^t \alpha(\tau, u)d\tau + \int_0^t \sigma(\tau, u)dB_\tau^*\right]du.$$

Thus, under some technical conditions ensuring the applicability of a stochastic Fubini's theorem (cf. Protter 2004), by comparing the martingale parts of the two expressions of $\ln W_t$, we obtain

$$H(t, s) = -\int_t^s \sigma(t, u)du. \tag{8.17}$$

Since $P(t, s) = W_t e^{\int_0^t r_u du}$, by (8.16) we get

$$d_t P(t, s) = P(t, s)\left[r_t dt + H(t, s)dB_t^*\right].$$

On the other hand, by (8.15) and (8.1), applying Itô's formula, we obtain

$$d_t P(t, s) = P(t, s)\left[\left(r_t - \int_t^s \alpha(t, u)du + \frac{1}{2}|H(t, u)|^2\right)dt + H(t, s)dB_t^*\right].$$

By comparing the above two equalities and using (8.17), we get

$$\int_t^s \alpha(t, u)du = \frac{1}{2}\left|\int_t^s \sigma(t, u)du\right|^2.$$

Thus, we have

$$\alpha(t, s) = \sigma(t, s) \cdot \int_t^s \sigma(t, u)du, \tag{8.18}$$

where "$\cdot$" stands for the inner product in $\mathbb{R}^d$. From (8.15) and (8.18), we obtain

$$f(t, s) = f(0, s) + \int_0^t \sigma(v, s) \cdot \int_v^s \sigma(v, u)du\, dv + \int_0^t \sigma(v, s)dB_v^*, \tag{8.19}$$

$$r_t = f(0, t) + \int_0^t \sigma(v, t) \cdot \int_v^t \sigma(v, u)du\, dv + \int_0^t \sigma(v, t)dB_v^*. \tag{8.20}$$

Consider an example where $d = 1$ and $\sigma(t, s)$ is a constant σ. By (8.20), we obtain

$$dr_t = \Phi(t)dt + \sigma dB_t^*,$$

where

$$\Phi(t) = \frac{\partial f(0, t)}{\partial t} + \sigma^2 t.$$

This is the Ho-Lee model. In this case, for $u > t$,

$$r_u = r_t + \int_t^u \Phi(v)dv + \sigma(B_u^* - B_t^*).$$

Since $B_u^* - B_t^*$ is independent of $\mathcal{F}_t$, by (8.2), we have

$$P(t, s) = \exp\left\{ -r_t(s - t) - \int_t^s \Phi(v)(s - v)dv + \frac{1}{6}\sigma^2(s - t)^3 \right\}.$$

Finally, we get

$$f(t, s) = r_t + \int_t^s \Phi(v)dv - \frac{1}{2}\sigma^2(s - t)^2.$$

8.2.4.1 Short Rate Models in HJM Terms

Now following Baxter (1997), we show that any short rate model described by a diffusion process can be expressed in HJM terms. In fact, assume that the short rate r_t, in the risk-neutral world, satisfies

$$dr_t = \mu(t, r_t)dt + v(t, r_t)dB_t^*, \quad t \leqslant T.$$

If we put

$$g(t, x, s) = -\ln \mathbb{E}^*\left[\exp\left\{ -\int_t^s r_u du \right\} \middle| r_t = x \right], \tag{8.21}$$

then $\int_t^s f(t, u)du = -\ln P(t, s) = g(t, r_t, s)$, $f(t, s) = \frac{\partial g}{\partial s}(t, r_t, s)$. Thus, by Itô's formula, we get

$$d_t f(t, s) = \frac{\partial^2 g}{\partial x \partial s}(\mu(t, r_t)dt + v(t, r_t)dB_t^*) + \frac{\partial^2 g}{\partial t \partial s}dt + \frac{1}{2}\frac{\partial^3 g}{\partial^2 x \partial s}v^2(t, r_t)dt.$$

This is an HJM model with parameter

$$\sigma(u, s) = v(u, r_u)\frac{\partial^2 g}{\partial x \partial s}(u, r_u, s). \tag{8.22}$$

The initial forward curve $f(0, T)$ is given by

$$f(0, T) = \frac{\partial g}{\partial T}(r(0), 0, T). \tag{8.23}$$

How to compute g? Under suitable conditions, the Feynman-Kac formula assures that for any fixed $s > 0$, $F(t, x, s) = e^{-g(t,x,s)}$ solves the PDE

$$F_t(t, x, s) + \mathcal{A}_t F(t, x, s) = 0, \quad (t, x) \in [0, s) \times \mathbb{R}^d \tag{8.24}$$

subject to the boundary condition

$$F(s, x, s) = 1, \quad x \in \mathbb{R}^d, \tag{8.25}$$

where

$$\mathcal{A}_t f(x) = \sum_{i=1}^{d} \mu_i(t, x) f_{x_i}(x) + \frac{1}{2} \sum_{i,j=1}^{d} \sigma_{i,j}(t, x)^2 f_{x_i x_j}(x), \quad f \in C^2(\mathbb{R}^d).$$

$$(8.26)$$

8.3   Forward Price and Futures Price

A *forward contract* is an agreement between two parties whereby the seller (known as in a short position) agrees to deliver to the buyer (known as in a long position) on a specified date, and at a specified price, known as *delivery price* or *forward price*, a specified quantity of an underlying asset or a European option. A *futures contract* is similar to a forward contract. There are, however, three distinctions between futures and forward contracts. The first is that forward contracts are "tailored" to the specific needs of particular buyers and sellers, while futures contracts are *standardized* with regard to the quality of the asset, time to maturity, price quotation, and delivery procedure. Another distinction is that futures contracts are traded on exchanges, while forward contracts are traded in "over-the-counter" markets. The most important distinction lies in their respective price settlement procedures. There is no cash transfer between the two parties of a forward contract until its delivery date. So a forward contract has a market value at any time t before its delivery date, the initial value being equal to zero. However, futures contracts have a *daily settlement* (or so-called *marking-to-market*) procedure that requires the buyer and seller to adjust their position daily according to the gains or losses due to the futures price changes. Here the *futures price* is the anticipated future unit price of the underlying asset. Futures prices change continuously in such a way that they make the market value of a futures contract always equal to zero. Due to the daily settlement, the futures price and the forward price are generally not the same.

8.3.1   Forward Price

Consider a forward contract with maturity date T written on one unit of a risky asset (or a European option) whose price process is (S_t). Assume that the short rate process (r_t) is bounded and the discounted price process $(\widetilde{S}_t)$ is a martingale under the equivalent martingale measure $\mathbb{P}^*$. Let $F(t, T)$ be the forward price at time t of the underlying asset. By definition, it is the delivery price of the forward contract such that the contract has zero value if the contract is initiated at time t. Since the payoff at maturity T of this contract is equal to $S_T - F(t, T)$, we must have

$$0 = \mathbb{E}^* \left[\exp \left\{ - \int_t^T r_s ds \right\} (S_T - F(t, T) \Big| \mathcal{F}_t \right].$$

Consequently,

$$F(t, T) = \frac{\mathbb{E}^* \left[\exp \left\{ - \int_t^T r_s ds \right\} S_T \Big| \mathcal{F}_t \right]}{\mathbb{E}^* \left[\exp \left\{ - \int_t^T r_s ds \right\} \Big| \mathcal{F}_t \right]},$$

which gives

$$F(t, T) = \frac{S_t}{P(t, T)}. \tag{8.27}$$

One should beware that the value at time s with $t < s \leqslant T$ of a forward contract initiated at time t is no longer zero.

8.3.2   Futures Price

Consider a futures contract with maturity T written on one unit of a risky asset or a European option whose price process is (S_t). Let $\Phi(t, T)$ be the futures price at time t of the underlying asset. Assume that the settlements during the time period $(t, T]$ take place at the time $t_1 < t_2 \cdots < t_N = T$. Since the value at time t of the futures contract is zero, we must have

$$0 = \mathbb{E}^* \left[\sum_{i=1}^N \exp\{- \int_0^{t_i} r_s ds\}(\Phi(t_i, T) - \Phi(t_{i-1}, T)) \mid \mathcal{F}_t \right],$$

where $t_0 = t$, $\Phi(T, T) = S_T$. In order to get an approximation of the value $\Phi(t, T)$, we consider a continuous settlement which is purely fictitious. In this case, we should have

$$0 = \mathbb{E}^* \left[\int_t^T Y_s d\Phi(s, T) \Big| \mathcal{F}_t \right],$$

where

$$Y_s = \exp \left\{ - \int_0^s r_\tau d\tau \right\}.$$

It means that the stochastic integral $\int_0^t Y_s d\Phi(s, T)$ is a $\mathbb{P}^*$-martingale. Since there are constants $k_1, k_2 > 0$ such that $k_2 \leqslant Y \leqslant k_2$, $\Phi(t, T)$ is also a martingale. Therefore we have

$$\Phi(t, T) = \mathbb{E}^*[\Phi(T, T) \mid \mathcal{F}_t] = \mathbb{E}^*[S_T \mid \mathcal{F}_t]. \tag{8.28}$$

From (8.21) and (8.28), we see that if the short rate r_t is a deterministic function, then the forward price and the futures price are the same.

8.4 Pricing Interest Rate Derivatives

An *interest rate derivative* is a financial contract whose payoffs are contingent on future interest rates or bond prices. In order to price an interest rate derivative, we need to know not only the dynamics of the interest rates but also the associated market price of risk (the excess return required for an investor to be able to bear an extra unit of risk) or an equivalent martingale measure (risk-neutral measure). Since the existence of a risk-neutral measure is, under purely technical conditions, equivalent to the absence of arbitrage, we can assume without loss of generality that the martingale measure $\mathbb{P}^*$ is given.

This section introduces three methods of pricing interest rate derivatives.

8.4.1 PDE Method

First of all we assume that the interest rate process obeys a one-factor model specified by (8.5). We are going to show that in this case the value of an interest derivative can be expressed in terms of the solution to a PDE. Consider an interest rate derivative with maturity $\tau \leqslant T$, which has dividend rate $h(t, r_t)$ at any time $t \leqslant \tau$ and terminal payoff $g(\tau, r_\tau)$. By the definition of the equivalent martingale measure $\mathbb{P}^*$, the value at time t of the derivative is given by

$$F(t, r_t) = \mathbb{E}^* \left[\int_t^\tau \phi_{t,s} h(s, r_s) ds + \phi_{t,\tau} g(\tau, r_\tau) \Big| \mathcal{F}_t \right], \qquad (8.29)$$

where $\phi_{t,s} = \exp\{-\int_t^s r_u du\}$. Under certain conditions, the Feynman-Kac formula ensures that F satisfies the PDE

$$F_t(t, x) + \mathcal{A}_t F(t, x) - h(t, x) + x F(t, x) = 0, \quad (t, x) \in [0, \tau) \times \mathbb{R}^d, \qquad (8.30)$$

subject to the boundary condition

$$F(\tau, x) = g(\tau, x), \quad x \in \mathbb{R}^d, \qquad (8.31)$$

where $\mathcal{A}_t$ is given by (8.26). In particular, the price of the T-bond at t is $P(t, T) = f(t, r_t)$, where f is the solution to equation (8.30), $h = 0$ and $f(\tau, x) = 1$.

Now we show, based on the functional approach to interest rate modeling, how one can price various interest rate derivatives. This approach has an advantage in numerical computations. For clarity, we only consider the case where the interest rate process r_t has the form $f(t, Y_t)$, where Y_t satisfies SDE (8.9).

8.4.1.1 Bonds

The price at time t of the T-bond is given by $G(t, T) = v(t, r_t)$, where

$$v(t, x) = \mathbb{E}\left[e^{-\int_t^T f(s, Y_s)ds} \big| Y_t = x\right], \quad t < T,$$

Y_t satisfies SDE (8.9). By the Feynman-Kac formula, v satisfies the following PDE:

$$\left\{\frac{\partial}{\partial t} + \frac{1}{2}\sigma^2 \frac{\partial^2}{\partial x^2} + \left(-\eta x + \frac{\varepsilon}{x}\right)\frac{\partial}{\partial x} - f\right\}v = 0,$$

$$v(T, x) = 1.$$

8.4.1.2 Swaps

An *interest-rate swap* is a contract between two *counter parties* (referred to as A and B), to exchange a series of cash payments. A agrees to pay B interest at a fixed rate (called *swap rate*) and receive interest at a *floating rate*. The most common floating rate index is LIBOR, the London Interbank Offered Rate. The same notional principal is used in determining the size of the payments, and there is no exchange of principal. From A's point of view, a swap is nothing but a derivative which pays dividends at a rate $h(t, r_t) = r_t - r^*$, where r^* is the fixed interest rate agreed upon at time zero. Let Z be the notional principal and $Zv(t, x)$ be the value of the T-maturity swap for the amount of Z starting at (t, x). Then v is determined by the equation

$$\left\{\frac{\partial}{\partial t} + \frac{1}{2}\sigma^2 \frac{\partial^0}{\partial x^2} + \left(-\eta x + \frac{\varepsilon}{x}\right)\frac{\partial}{\partial x} - f\right\}v + f - r^* = 0,$$

$$v(T, x) = 0.$$

8.4.1.3 Caps and Floors

Very often, a loan may have a proviso such that the interest rate charged cannot go above (below) a specified value r^{cap} (r^{flr}), known as *cap (floor)*. Let Z be the amount of the loan to be paid back at time T. Let $Zv(t, x)$ be the value of the capped loan starting at (t, x). Then v is determined by the following equation:

$$\left\{\frac{\partial}{\partial t} + \frac{1}{2}\sigma^2 \frac{\partial^2}{\partial x^2} + \left(-\eta x + \frac{\varepsilon}{x}\right)\frac{\partial}{\partial x} - f\right\}v + f + \min(r, r^{\text{cap}}) = 0,$$

$$v(T, x) = 1.$$

The equation for determining a floor is the same as above except that "min" is replaced by "max" and r^{cap} is replaced by r^{flr}.

8.4.1.4 Swaptions, Captions, Floortions, and Other Derivatives

More generally, consider an interest rate derivative which pays $h(r_t)$ at maturity time T. Options on swaps (*swaptions*), options on caps (*captions*), and options on floors (*floortions*) are examples of these types of interest rate derivative securities. From our interest rate model, the payoff can be expressed as $h(f(T, X_T))$. Thus the price at time t of the derivative is given by $v(t, r_t)$, where

$$v(t, x) = E\left(e^{-\int_t^T f(s, X_s)ds} h(f(T, X_T)) \middle| X_t = x\right), t < T.$$

From the Feynman-Kac formula, v satisfies the following PDE:

$$\left(\frac{\partial}{\partial t} + \frac{1}{2}\sigma^2 \frac{\partial^2}{\partial x^2} + \left(-\eta x + \frac{\varepsilon}{x}\right)\frac{\partial}{\partial x} - f\right)v = 0$$

$$v(T, x) = h(f(T, x)).$$

In most cases, one does not have analytical expressions for pricing of interest rate derivatives. Numerical computation is the main tool. The functional approach to interest rate modelling offers a great advantage in numerical computations. In fact, all partial differential equations listed in the previous section have the form:

$$(D - f)v + F(f) = 0,$$

where

$$D = \frac{\partial}{\partial t} + \frac{1}{2}\sigma^2 \frac{\partial^2}{\partial x^2} + \left(-\eta x + \frac{\varepsilon}{x}\right)\frac{\partial}{\partial x}$$

is a rather simple differential operator. The term $F(f)$ does not involve a differential operator. It is well known, from the mathematical theory of numerical analysis, that to check the stability condition of a numerical scheme for the above equation, one only needs to check the stability of the numerical scheme for the case $F(f) = 0$. Furthermore, for many numerical schemes, such as the forward-time central-space scheme, the backward-time central-space scheme, and the Crank-Nicolson scheme, the lower order term $-fv$ does not affect the stability condition. Therefore, we only need to check whether such scheme is stable for the equation

$$Dv = 0$$

Note that the above equation does not depend on f. Therefore, such a class of numerical schemes will be stable for all functional forms of f. Consequently, interest rate derivatives based on the various interest rate models given in Sect. 7.2.2 can be computed using the same numerical method. For different models, i.e., different functions f, one only needs to make a one-line change in the computer program. It significantly simplifies our task in numerical computations and adds considerable power in choosing a desirable functional form for an interest rate model.

8.4.2 Forward Measure Method

Now we assume that the derivative is written on a bond or the term structure of interest rates is represented by the HJM model. In this case we can take the T-bond as a numeraire. More precisely, let $P(t, T)$ be the bond price at time t. We define $\alpha_t = P(t, T)/P(0, T)$ as the normalized T-bond with $\alpha_0 = 1$. We take (α_t) as the numeraire. We denote by β_t the value process of the bank account. Now we seek a probability measure $\mathbb{Q}$, called T-*forward measure*, which is equivalent to $\mathbb{P}^*$ such that the value process of the bank account deflated by the numeraire (α_t) is a $\mathbb{Q}$-martingale on $[0, T]$. To this end we define a probability measure $\mathbb{Q}$ on $(\Omega, \mathcal{F}_T)$ by

$$\frac{d\mathbb{Q}}{d\mathbb{P}^*} = \frac{\alpha_T}{\beta_T} = \frac{1}{P(0, T)\beta_T}.$$

Since the associated $\mathbb{P}^*$-martingale is obviously

$$L_t = \mathbb{E}^* \left[\frac{d\mathbb{Q}}{d\mathbb{P}^*} \, \middle| \, \mathcal{F}_t \right] = \frac{P(t, T)}{P(0, T)\beta_t} = \frac{\alpha_t}{\beta_t},$$

(β_t/α_t) is a $\mathbb{Q}$-martingale.

If an interest derivative with maturity T has only terminal payoff ξ, and ξ is $\mathbb{Q}$-integrable, then by Bayes' rule, its value at time t is given by

$$V_t = \beta_t \mathbb{E}^*[\beta_T^{-1}\xi \mid \mathcal{F}_t] = \beta_t L_t \mathbb{E}_{\mathbb{Q}}[L_T^{-1}\beta_T^{-1}\xi \mid \mathcal{F}_t] = P(t, T)\mathbb{E}_{\mathbb{Q}}[\xi \mid \mathcal{F}_t].$$

In other words, by (8.27), under the T-forward measure $\mathbb{Q}$, the T-forward price $F(t, T)$ at time t of ξ is given by

$$F(t, T) = \mathbb{E}_{\mathbb{Q}}[\xi \mid \mathcal{F}_t].$$

In particular, the T-forward price process $F(t, T)$ is a martingale under the T-forward measure $\mathbb{Q}$.

8.4.3 Changing Numeraire Method

Assume we are given a financial market with a short rate of interest r and asset price process $S(t)$. We want to compute the price at time 0 of call option $(S(T) - K)^+$ under an equivalent martingale measure P^*.

The following theorem is due to Björk (1998).

Theorem 8.2 *If under T-forward measure $\mathbb{Q}$ process $Z_T(t) = S(t)/P(t, T)$ is given by*

$$dZ_T(t) = Z_T(t)\sigma_T(t)dB_T(t),$$

where $B_T(t)$ is a d-dimensional Q-Brownian motion and $\sigma_T(t)$ is deterministic row-vector function, then the P^*-price at time 0 of call option $(S(T) - K)^+$ is given by

$$V_0 = S(0)N(d_1) - KP(0, T)N(d_2),$$

where

$$d_2 = \frac{\ln\left(\frac{S(0)}{KP(0,T)}\right) - \frac{1}{2}\sigma^2}{\sigma}, \quad d_1 = d_2 + \sigma, \quad \sigma^2 = \int_0^T \|\sigma_T(t)\|^2 dt.$$

Proof We have

$$V_0 = \mathbb{E}^*[\beta_T^{-1}(S(T) - K)I_{S(T) \geqslant K}]$$

$$= \mathbb{E}^*\left[\beta_T^{-1}S(T)I_{S(T) \geqslant K}\right] - K\mathbb{E}^*\left[\beta_T^{-1}I_{S(T) \geqslant K}\right].$$

In the first term, we use S as the numeraire and denote the corresponding martingale by Q^S with $dP^*/dQ^S = \beta_T S(0)/S(T)$, and for the second term, we use T-forward measure Q, and then we get

$$V_0 = S(0)Q^S(S(T) \geqslant K) - KP(0, T)Q(S(T) \geqslant K).$$

Since

$$Z_T(T) = \frac{S(0)}{P(0, T)} \exp\left\{\frac{1}{2}\int_0^T \|\sigma_T(t)\|^2 dt + \int_0^T \sigma_T(t)dB_T(t)\right\},$$

we have

$$\mathbb{Q}(S(T) \geqslant K) = \mathbb{Q}(Z_T(T) \geqslant K) = \mathbb{Q}(\ln Z_T(T) \geqslant \ln K) = N(d_2).$$

On the other hand, if we let

$$Y_T(t) = \frac{P(t, T)}{S(t)} = \frac{1}{Z_T(t)},$$

then $Y_T(T) = \frac{1}{S(T)}$, $Y_T(t)$ is a Q^S-martingale. Thus, we have

$$dY_T(t) = Y_T(t)\delta_T(t)dB_S(t),$$

where $B_S(t)$ is a d-dimensional Brownian motion under Q^S. Since $Y_T(t) = \frac{1}{Z_T(t)}$, by Itô's formula, we must have $\delta_T(t) = -\sigma_T(t)$. Consequently,

$$Y_T(T) = \frac{S(0)}{P(0, T)} \exp\left\{\frac{1}{2}\int_0^T \|\sigma_T(t)\|^2 dt - \int_0^T \sigma_T(t)dB_S(t)\right\}.$$

Thus, we have

$$Q^S(S(T) \geqslant K) = Q^S\left(\frac{1}{S(T)} \leqslant \frac{1}{K}\right) = Q^S\left(Y_T(T) \leqslant \frac{1}{K}\right) = N(d_1),$$

where $d_1 = d_2 + \sigma$. $\square$

As an example of application, we get the following formula for Hull-White bond call option (see Björk 1998, Proposition 19.16).

Theorem 8.3 *In the following Hull-White model*

$$dr_t = (a(t) - b)dt + \sigma dB_t,$$

the price at time 0 of European call option $(P(T_1, T_2) - K)^+$ *with maturity* T_1 *on the* T_2-*bond is given by*

$$V_0 = P(0, T_2)N(d_1) - KP(0, T_1)N(d_2),$$

where

$$d_2 = \frac{\ln\left(\frac{P(0,T_2)}{KP(0,T_1)}\right) - \frac{1}{2}\sigma^2}{\sigma}, \quad d_1 = d_2 + \Sigma,$$

and

$$\Sigma^2 = \frac{\sigma^2}{2b^3}\{1 - e^{-2bT_1}\}\left\{1 - e^{-b(T_2-T_1)}\right\}^2.$$

Proof In Sect. 8.2.2, we obtain an affine term structure

$$P(t, s) = e^{A(t,s) - B(t,s)r_t},$$

where A and B are deterministic functions and $B(t, s)\frac{1}{b}\{1 - e^{b(s-t)}\}$. Put $Z(t) = \frac{P(t,T_2)}{P(t,T_1)}$, then

$$Z(t) = \exp\{A(t, T_2) - A(t, T_1) - [B(t, T_2) - B(t, T_1)]r_t\}.$$

By Itô's formula,

$$dZ(t) = Z(t)\{\cdots\}dt + Z(t)\sigma_Z(t)dB_t,$$

where

$$\sigma_Z(t) = -\sigma[B(t, T_2) - B(t, T_1)] = \frac{\sigma}{b}e^{bt}[e^{-bT_2} - e^{-bT_1}].$$

Thus, σ_Z is a deterministic function, and we may apply Theorem 8.2 to get the result. $\square$

8.5   The Flesaker-Hughston Model

Flesaker and Hughston (1996) proposed a new approach to the term structure modeling of interest rates. The key point of this approach stems from the following observation on (8.2). Let s-bond price process $P(t, s)$ be defined by (8.2). Set

$$\eta_t = \frac{d\mathbb{P}^*}{d\mathbb{P}}\bigg|_{\mathcal{F}_t}, \quad 0 \leqslant t \leqslant T.$$

Then by the Bayes' rule of conditional expectation,

$$P(t, s) = A_t^{-1}\mathbb{E}[A_s \mid \mathcal{F}_t], \quad s \geqslant t, \tag{8.32}$$

where

$$A_s = \eta_s \exp\left\{ -\int_0^s r_\tau d\tau \right\}. \tag{8.33}$$

Since (η_t) is a $\mathbb{P}$-martingale, (A_t) is a $\mathbb{P}$-supermartingale. Note that the expression (8.33) is nothing but the product decomposition of supermartingale (A_t). Now assume that (A_t) is a strictly positive and bond price $P(t, s)$ is modeled by (8.32). If in (8.33) (η_t) is a $\mathbb{P}$-martingale and (r_t) is a nonnegative process, then the corresponding short rate process must be (r_t), and probability measure $\mathbb{P}^*$ with density process η^{-1} is an equivalent martingale measure for price processes of bonds with different maturities. As an example (due to Flesaker and Hughston 1996), let

$$A_t = f(t) + g(t)M_t, \quad t \in [0, T], \tag{8.34}$$

where $f, g : [0, T] \to \mathbb{R}_+$ are strictly positive decreasing functions with $f(0) + g(0) = 1$ and (M_t) is a strictly positive martingale defined on a filtered probability space $(\Omega, \mathcal{F}, (\mathcal{F}_t), \mathbb{P})$, with $M_0 = 1$. Then it follows immediately from (8.32) that

$$P(t, s) = \frac{f(s) + g(s)M_t}{f(t) + g(t)M_t}, \quad t \in [0, s]. \tag{8.35}$$

This model fits easily the initial curve: it suffices to choose f and g such that

$$P(0, s) = f(s) + g(s), \quad s \in [0, T]. \tag{8.36}$$

In order to get an explicit expression for the short rate, we assume that $(\mathcal{F}_t)$ is the natural filtration of a Brownian motion (B_t). Since (M_t) is a strictly positive martingale, it must be of the form $M_t = \mathcal{E}(\sigma.B)_t$, where $\sigma = (\sigma_t)$ is an adapted measurable process and $\mathcal{E}(\sigma.B)_t$ is the Doléans exponential of σ (Theorem 4.14). In fact, assume $M_t = 1 + (H.B)_t$, then $M_t = \mathcal{E}(N)_t$, where $N = M^{-1}.M$; thus we have $\sigma = M^{-1}H$.

Let $A_t = \eta_t C_t$ be the product decomposition of supermartingale A, where η is a strictly positive local martingale and C is a strictly positive decreasing process with $\eta_0 = C_0 = 1$ (cf. Yan 2002b). Then η must be of the form $\eta_t = \mathcal{E}(\gamma.B)_t$. Thus by Itô's formula,

$$\eta_t dC_t + C_t \eta_t \gamma_t dB_t = dA_t = f'(t)dt + M_t g'(t)dt + g(t)M_t \sigma_t dB_t. \qquad (8.37)$$

By comparing the "dB_t" terms and the remaining terms on both sides of (8.37), we find that

$$\gamma_t = \frac{\sigma_t g(t)M_t}{f(t) + g(t)M_t},$$

$$dC_t = C_t \frac{f'(t) + g'(t)M_t}{f(t) + g(t)M_t} dt. \qquad (8.38)$$

Consequently, if η is a martingale and we define a probability measure $\mathbb{P}^*$ by $\frac{d\mathbb{P}^*}{d\mathbb{P}} = \eta_T^{-1}$, then $\mathbb{P}^*$ is the unique probability measure such that

$$P(t,s) = C_t^{-1} \mathbb{E}^*[C_s \mid \mathcal{F}_t]. \qquad (8.39)$$

C_t can be solved from (8.38), and the result is

$$C_t = \exp\left\{ \int_0^t \frac{f'(\tau) + g'(\tau)M_\tau}{f(\tau) + g(\tau)M_\tau} d\tau \right\}.$$

Then from (8.39), we can derive an explicit expression for the short rate process (r_t):

$$r_t = -\frac{f'(t) + g'(t)M_t}{f(t) + g(t)M_t}. \qquad (8.40)$$

In particular, it is readily verifiable that $r_t = f(t,t)$, where $f(t,s)$ is the forward interest rates.

The main advantage of the Flesaker-Hughston model is that we can use directly upermartingale A to express the price at time t of a interest rate derivative ξ with maturity $s < T$ as

$$V_t = A_t^{-1} \mathbb{E}[\xi A_s \mid \mathcal{F}_t], \quad \forall t \in [0, s]. \qquad (8.41)$$

This equation enables us to obtain closed-form expressions for the prices of some interest rate derivatives, such as caps and swaptions. We refer the reader to Rutkowski (1997) for this subject.

8.6   BGM Models

Compared with the HJM model, practical models which have been more widely applied in the financial markets are Brace et al. (1997), Miltersen et al. (1997), and the BGM *model* proposed by Jamshidian (1997). In this case, we are modeling forward-LIBOR which has actual market transactions and is observable. Therefore, the BGM model is also known as the LIBOR *market model*. In this sense, we can also regard the BGM model as a discrete version of the HJM model.

We consider a set composed of a list of discrete times, $0 = T_0 < T_1 < \cdots < T_{N+1}$, and denote $\delta_i = (T_{i+1} - T_i)/$per year, $\forall i \in [0, N]$. In practice, (T_i) usually corresponds to the interest rate reset date of the interest rate derivatives, with $T_{i+1} - T_i$ usually being 3, 6, or 12 months. We define T_i-forward-LIBOR at time t as follows:

$$L(t, T_i, T_{i+1}) = \frac{P(t, T_i) - P(t, T_{i+1})}{\delta_i P(t, T_{i+1})}, \tag{8.42}$$

namely, $L(t, T_i, T_{i+1})$ represents the single phase of interest rate acting on the time period $[T_i, T_{i+1}]$. For simplicity, we denote $L(t, T_i, T_{i+1})$ by $L_i(t)$. It is different from the HJM model for instantaneous forward rates modelled under a usual equivalent martingale measure. We consider now the forward risk-neutral measure, i.e., the risk-neutral measure corresponding to the numeraire of zero coupon bond. We first consider the most remote forward risk-neutral measure $\mathbb{Q}^{N+1}$. Then by the definition of the above forward-LIBOR, $L_N(t)$ is a martingale under $\mathbb{Q}^{N+1}$. We assume that it obeys a diffusion process without drift term as follows:

$$dL_N(t) = L_N(t)\sigma_N(t)dZ^{N+1}(t), \quad t \leqslant T_N, \tag{8.43}$$

where $Z^{N+1}(t)$ is an m-dimensional standard Brownian motion under $\mathbb{Q}^{N+1}$ and $\sigma_N(t)$ is an m-dimensional function. The BGM model can represent different volatility term structures by using different forms of $\sigma_N(t)$. It can also reflect the actual market volatility smile or skew phenomenon. At this point, the BGM model has a lot of flexibility. Next, we consider the forward risk-neutral measure $\mathbb{Q}^N$. Similarly, $L_{N-1}(t)$ is a martingale under $\mathbb{Q}^N$. We model it as follows:

$$dL_{N-1}(t) = L_{N-1}(t)\sigma_{N-1}(t)dZ^N(t), \quad t \leqslant T_{N-1}, \tag{8.44}$$

where $Z^N(t)$ is an m-dimensional standard Brownian motion under $\mathbb{Q}^N$ and $\sigma_{N-1}(t)$ is an m-dimensional function. By the measure transformation technique, we can easily get the following relationship between two neighboring risk-neutral measures $\mathbb{Q}^{N+1}$ and $\mathbb{Q}^N$:

$$\left.\frac{d\mathbb{Q}^N}{d\mathbb{Q}^{N+1}}\right|_{\mathcal{F}_t} = \frac{1 + \delta_N L_N(t)}{1 + \delta_N L_N(0)}. \tag{8.45}$$

Furthermore by the Girsanov measure transformation formula (at this time we need to impose certain technical conditions on the volatility coefficient to ensure the Girsanov measure transformation formula holds), the following $Z^N(t)$ is an m-dimensional standard Brownian motion under $\mathbb{Q}^N$:

$$Z^N(t) = Z^{N+1}(t) - \int_0^t \frac{\delta_N L_N(s)\sigma_N(s)}{1 + \delta_N L_N(s)}ds. \qquad (8.46)$$

By analogy, we can model $L_i(t)$ under forward risk-neutral measure $\mathbb{Q}^{i+1}$ and obtain recursively a relationship between different forward risk-neutral measures and one between corresponding standard Brownian motions. In this way we can consider problems by switching to the same risk-neutral measure, when necessary.

Let us take an interest rate cap as an example. Consider the pricing problem under the BGM model. In order to obtain a closed-form expression, we assume that the above volatility coefficients are deterministic functions. Let $T_i (i = 1, \cdots, N)$ be the reset date of the interest rate cap, M be its nominal principal, and R^* be the upper limit of interest rate. We know that the interest rate cap can be decomposed into the sum of N caplets. By the risk-neutral pricing formula, the value of the ith caplet at time $t \leqslant T_i$ is given by

$$\text{Caplet}_i(t) = M\delta_i P(t, T_{i+1}) E_{\mathbb{Q}^{i+1}}[(L_i(T_i) - R^*)_+ |\mathcal{F}_t]. \qquad (8.47)$$

According to the previous model assumptions, we know that $L_i(T_i)$ obeys log-normal distribution under $\mathbb{Q}^{i+1}$ and satisfies

$$\text{Var}_{\mathbb{Q}^{i+1}}[\ln L_i(T_i)|\mathcal{F}_t] = \int_t^T \sigma_i^2(u)du, \quad E_{\mathbb{Q}^{i+1}}[\ln L_i(T_i)|\mathcal{F}_t]$$
$$= \ln L_i(t) - \tfrac{1}{2}\int_t^T \sigma_i^2(u)du. \qquad (8.48)$$

Put

$$v_i^2 = \frac{1}{T_i - t}\int_t^{T_i} \sigma_i^2(u)du, \qquad (8.49)$$

by (8.47), (8.48), and (8.49), we obtain a closed-form pricing formula for ith caplet:

$$\text{Caplet}_i(t) = M\delta_i P(t, T_{i+1})\{L_i(t)\Phi(d_{1i}) - R^*\Phi(d_{2i})\}, \qquad (8.50)$$

where

$$d_{1i} = \frac{\ln(L_i(t)/R^*) + v_i^2(T_i - t)/2}{v_i\sqrt{T_i - t}}, \quad d_{2i} = d_{1i} - v_i\sqrt{T_i - t}. \qquad (8.51)$$

In the actual market, the upper and lower limits of the interest rate and the exchange option prices are quoted through the implied volatility corresponding to the Black-Scholes formula. Then (8.49) actually gives a relationship between the implied volatility and the volatility coefficient used in the BGM model. Because the BGM model with the abovementioned pricing formula fits the empirical data well, calibration with the actual financial market quotation becomes easy to operate, making the model more popular than others over the past decade.

Chapter 9
Optimal Investment-Consumption Strategies in Diffusion Models

In complete markets with diffusion models, the expected utility maximization problem has been studied by many authors (see a review in Karatzas 1989). Karatzas et al. (1991) studied this problem in incomplete markets. They considered a market composed of a bond and d stocks, the latter being driven by an m-dimensional Brownian motion. They used some virtual stocks to expand the original market into a complete market. Under certain additional conditions, they proved that one can wisely choose virtual stocks, such that in the resulting optimal portfolio for the solution to utility maximization problem in the completed market, virtual stocks are superfluous. Thus, this solution is also the optimal one in the original incomplete market.

In this chapter, we first introduce investment-consumption strategies in diffusion models and then study the expected utility maximization problem in Sect. 9.2. The employed approach is the so-called martingale method (see Dana and Jeanblanc 2003). In Sect. 9.3, we study risk-mean portfolio selection problems, including the weighted mean-variance model and the Markowitz' mean-variance portfolio problem, in the range of L^2-allowable trading strategies.

9.1 Market Models and Investment-Consumption Strategies

We first introduce the market model. Let $W(t) = (W^1(t), \cdots, W^m(t))^\tau$ be an m-dimensional standard Brownian motion defined on the stochastic basis $(\Omega, \mathcal{F}, (\mathcal{F}_t), \mathbb{P})$, where $(\mathcal{F}_t)$ is the natural filtration of Brownian motion $W(\cdot)$. Suppose that there are $d + 1$ assets available for trading in the market, the first is a risk-free asset (commonly referred to as bond), whose price process, denoted by $S_0(t)$, obeys the following equation:

© Springer Nature Singapore Pte Ltd. and Science Press 2018
J.-A. Yan, *Introduction to Stochastic Finance*, Universitext,
https://doi.org/10.1007/978-981-13-1657-9_9

$$\begin{cases} dS_0(t) = S_0(t)r(t)dt, \quad t \in [0, T], \\ \\ S_0(0) = 1, \end{cases}$$

where $T > 0$ is a fixed terminal time and $r(\cdot)$ is the short-term interest rate, which is an $(\mathcal{F}_t)$ progressively measurable process satisfying

$$\int_0^T |r(s)|ds < \infty \quad \text{a.s.}$$

The remaining assets are d risky assets (commonly referred to as stocks). Each price process $S_i(t), i = 1, \cdots, d$, satisfies the following equation:

$$\begin{cases} dS_i(t) = S_i(t)\left[\mu_i(t)dt + \sum_{j=1}^m \sigma_{ij}(t)dW^j(t)\right], \quad t \in [0, T], \\ \\ S_i(0) = s_i > 0, \end{cases}$$

where $\mu_i(\cdot)$ and $\sigma_{i\cdot}(\cdot) := (\sigma_{i1}, \cdots, \sigma_{im})^\tau$ are the return and the volatility vector for i-th stock, respectively, which are $(\mathcal{F}_t)$ progressively measurable processes satisfying

$$\int_0^T \left[|\mu_i(t)| + \sum_{j=1}^m (\sigma_{ij}(t))^2\right]dt < \infty \quad \text{a.s.}$$

Define the *excess return vector*

$$b(t) := (\mu_1(t) - r(t), \ldots, \mu_d(t) - r(t))^\tau = \mu(t) - r(t)\mathbf{1},$$

where $\mathbf{1}$ represents the d-dimensional column vector with all components 1, and the *volatility matrix*

$$\sigma(t) := (\sigma_{ij}(t))_{d \times m}.$$

We consider an investor with initial wealth x. In Chap. 7, we introduced the concept of self-financing trading strategy. A trading strategy is an $(\mathcal{F}_t)$-adapted vector process $\varphi(t) = (\varphi_0(t), \cdots, \varphi_d(t))^\tau$, where each component $\varphi_i(t)$ expresses the amount of assets held i by the investor at time t. If its wealth process $X(t)$ is

$$X(t) = x + \sum_{i=0}^d \int_0^t \varphi_i(s)dS_i(s), \quad \forall t \in [0, T],$$

then such a trading strategy is called self-financing. In this chapter, we have another way to express trading strategies. We consider an investor who invests wealth $\pi_i(t)$ in asset i at time t. Apparently, $\pi_i(t) = \varphi_i(t)S_i(t)$. Set

$$\bar{\pi}(t) = (\pi_0(t), \pi_1(t), \cdots, \pi_d(t))^\tau.$$

We often call $\bar{\pi}(\cdot)$ a trading strategy. Obviously, $\bar{\pi}(\cdot)$ is self-financing if and only if

$$dX(t) = \pi_0(t)r(t)dt + \sum_{i=1}^{d} \pi_i(t)\left[\mu_i(t)dt + \sum_{j=1}^{m} \sigma_{ij}(t)dW^j(t)\right], \quad \forall t \in [0, T].$$

It is also obvious that the above condition can be rewritten as

$$dX(t) = X(t)r(t)dt + \sum_{i=1}^{d} \pi_i(t)\left[(\mu_i(t)-r(t))dt + \sum_{j=1}^{m} \sigma_{ij}(t)dW^j(t)\right], \forall t \in [0, T].$$

Usually, we use vector and matrix notations; the result is

$$dX(t) = X(t)r(t)dt + \pi(t)^\tau(\mu(t) - r(t)\mathbf{1})dt + \pi(t)^\tau \sigma(t)dW(t), \quad \forall t \in [0, T]. \tag{9.1}$$

Here, vector

$$\pi(t) = (\pi_1(t), \cdots, \pi_d(t))^\tau$$

indicates the amount of money invested in various stocks. Equation (9.1) shows that in a self-financing trading strategy, how the instantaneous change $dX(t)$ of the wealth depends on the amounts $\pi(t)$ invested in the stocks. Equation (9.1) is often referred to as the *dynamic budget constraint* or *dynamic budget equation*. Also it is worth noting that for a self-financing trading strategy, its wealth process $X(\cdot)$ entirely depends on initial wealth x and amount $\pi(\cdot)$ invested in the stocks, because in addition to investing in the stocks, the rest of the money is invested in the bond.

In reality of investments, in addition to trading in the securities market, investors also conduct consumer activities. In order to characterize investors' consumption behavior, we will introduce a nonnegative progressively measurable process: consumption (rate) process $c(\cdot)$, satisfying $\int_0^T c(s)ds < \infty$. Then $\int_0^t c(s)ds$ expresses the total wealth of cumulative consumption by the investor from time 0 to time t. So we have an *investment-consumption strategy* (φ, c). It is called self-financing if it satisfies

$$X(t) = x + \sum_{i=0}^{d} \int_0^t \varphi_i(s)dS_i(s) - \int_0^t c(s)ds, \quad \forall t \in [0, T].$$

Of course, we can also use (π, c) to represent an investment-consumption strategy. Clearly, $(\bar{\pi}, c)$ is self-financing if and only if $\forall t \in [0, T]$,

$$dX(t) = X(t)r(t)dt + \pi(t)^{\tau}(\mu(t) - r(t)\mathbf{1})dt + \pi(t)^{\tau}\sigma(t)dW(t) - c(t)dt. \qquad (9.2)$$

Equation (9.2) is also commonly known as the dynamic budget constraint equation. Similarly, for a self-financing investment-consumption strategy, its wealth process $X(\cdot)$ entirely depends on initial wealth x, amount $\pi(\cdot)$ invested in the stocks, and consumption rate $c(\cdot)$. In view of this, in the future as long as self-financing investment-consumption strategy is considered, we only need to specify the initial wealth x, the amount $\pi(\cdot)$ invested in the stocks, and the consumption rate process $c(\cdot)$.

9.2 Expected Utility Maximization

We consider two kinds of utility functions. One is about the consumer's time-dependent utility function:

$$U : [0, T] \times (0, \infty) \to (-\infty, \infty)$$

$$(t, c) \mapsto U(t, c).$$

We assume that for any $t \in [0, T]$, $U(t, \cdot)$ is strictly concave, strictly increasing, and continuously differentiable on $(0, \infty)$ and satisfies the following Inada *conditions*:

$$\frac{\partial U}{\partial c}(t, 0) := \lim_{c \downarrow 0} \frac{\partial U}{\partial c}(t, c) = \infty, \qquad \frac{\partial U}{\partial c}(t, \infty) := \lim_{c \uparrow \infty} \frac{\partial U}{\partial c}(t, c) = 0.$$

Another utility function is about the terminal wealth:

$$V : (0, \infty) \to (-\infty, \infty)$$

$$x \mapsto V(x).$$

We assume that on $(0, \infty)$, V is strictly concave, strictly increasing, and continuously differentiable and satisfies the following Inada conditions:

$$V'(0) := \lim_{x \downarrow 0} V'(x) = \infty, \qquad V'(\infty) := \lim_{x \uparrow \infty} V'(x) = 0.$$

It is easy to verify that $\ln x$ and x^{α}/α ($\alpha < 1, \alpha \neq 0$) both are utility functions satisfying the Inada conditions.

For a self-financing investment-consumption strategy (π, c), the wealth process satisfies dynamic budget Eq. (9.2). The expected utility that this strategy brings to investors is

$$\mathbb{E}\left[\int_0^T U(t, c(t))dt + V(X(T))\right].$$

For a self-financing trading strategy π, the wealth process meets dynamic budget Eq. (9.1). The expected utility that this strategy brings to investors is

$$\mathbb{E}[V(X(T))].$$

The investor's goal is, within a certain range, to select a self-financing strategy so that the expected utility is maximized.

Below we will specify the scope of selecting self-financing strategies.

Definition 9.1 A self-financing trading strategy π is called admissible, if at any time $t \in [0, T]$, we have wealth $X(t) \geqslant 0$ a.s. A self-financing investment-consumption strategy (π, c) is called admissible if at any time $t \in [0, T]$, we have $c(t) \geqslant 0$ a.s., and $X(t) \geqslant 0$ a.s.

Given initial wealth x, the set of all self-financing trading strategies π is denoted by $\mathcal{A}(x)$; the set of all self-financing investment-consumption strategies (π, c) is denoted by $\mathcal{A}_c(x)$, where the subscript c represents "with the consumption."

The investor's goal is

$$\max_{(\pi,c)\in\mathcal{A}_c(x)} \mathbb{E}\left[\int_0^T U(t, c(t))dt + V(X(T))\right] \tag{9.3}$$

subject to (9.2),

or

$$\max_{\pi\in\mathcal{A}(x)} \mathbb{E}[V(X(T))] \tag{9.4}$$

subject to (9.1).

Since solutions to those problems are similar, we only present how to solve problem (9.3).

In problem (9.3), the budget constraint (9.2) is *dynamic* in the sense that it describes how the wealth evolves in every moment. Noting that in the target of problem (9.3), only consumption $c(\cdot)$ and terminal wealth $X(T)$ are emerged. Such problem (9.3) can be broken down into two steps:

The first step is to characterize what consumption process $c(\cdot)$ and terminal wealth $X(T)$ are optimal, that is, to answer "what should the optimal outcome be?" The second step is to select a trading strategy $\pi(\cdot)$ to achieve the optimal consumption process $c(\cdot)$ and the terminal wealth $X(T)$, that is, to answer "specifically how to achieve the optimal outcome?" These are the two steps of the "martingale method" in the utility maximization problem. Namely, first consider the "ending problem," and then reversely consider the "realization problem."

We first consider the first step: the ending problem. To this end, we need to portray what outcomes are achievable and then find the optimal outcome within those achievable ones. Therefore, we must first find those consumption process $c(\cdot)$ and terminal wealth $X(T)$ which can be achieved by admissible investment-consumption strategies.

When the market is incomplete, portraying such achievable outcomes is difficult. Below, we only consider the complete market case, i.e., assuming that $d = m$ and market parameters $r(t)$, $\mu(t)$, and $\sigma(t)$ satisfy the following conditions:

(1) There exists $\delta > 0$, such that

$$\sigma(t)\sigma(t)^\tau \geqslant \delta I_m, \quad \forall t \in [0, T],$$

where I_m is an $m \times m$ unit matrix;

(2) $\mathbb{E}\left[e^{\frac{1}{2}\int_0^T |\theta(s)|^2 ds}\right] < \infty$, where $\theta(t) = \sigma(t)^{-1}b(t), t \in [0, T]$.

For such a complete market, there is a unique risk-neutral measure $\mathbb{Q}$, $\frac{d\mathbb{Q}}{d\mathbb{P}}\Big|_{\mathcal{F}_t} = Z(t)$, where

$$Z(t) = \exp\left\{-\int_0^t \theta(s)^\tau dW(s) - \frac{1}{2}\int_0^t |\theta(s)|^2 ds\right\}. \tag{9.5}$$

Under $\mathbb{Q}$, the stock price process satisfies the following SDE:

$$dS_i(t) = S_i(t)\left[r(t)dt + \sum_{j=1}^m \sigma_{ij}(t)d\widetilde{W}^j(t)\right],$$

where $\widetilde{W}(t) := W(t) + \int_0^t \theta(s)ds$, $0 \leqslant t \leqslant T$ is a Brownian motion under probability measure $\mathbb{Q}$. The wealth process X of self-financing investment-consumption strategy (π, c) satisfies

$$d\left(\frac{X(t)}{S_0(t)}\right) = \frac{1}{S_0(t)}\pi^\tau(t)\sigma(t)d\widetilde{W}(t) - \frac{c(t)}{S_0(t)}dt, \quad t \in [0, T].$$

Let

$$\begin{aligned}
\rho(t) &= \exp\left\{-\int_0^t r(s)ds\right\} Z(t) \\
&= \exp\left\{-\int_0^t \theta(s)^\tau dB(s) - \int_0^t \left(r(t) + \frac{1}{2}|\theta(s)|^2\right)ds\right\}.
\end{aligned} \tag{9.6}$$

We call ρ a *pricing kernel* or a *state price density*. It is obvious that $0 < \rho(T) < \infty$ a.s. and $0 < \mathbb{E}[\rho(T)] < \infty$. Then for all $i = 1, \cdots, d$ it holds that

$$d(\rho(t)S_i(t)) = \rho(t)S_i(t) \sum_{j=1}^{m} [\sigma_{ij}(t) - \theta_j(t)]dW^j(t), \quad \forall t \in [0, T],$$

and consequently,

$$\rho(t)S_i(t) = \mathbb{E}[\rho(T)S_i(T)|\mathcal{F}_t], \quad \forall t \in [0, T].$$

In addition, the wealth process X of the self-financing investment-consumption strategy (π, c) satisfies

$$d(\rho(t)X(t)) = \rho(t)\big[\pi^\tau(t)\sigma(t) - X(t)\theta(t)^\tau\big]dW(t) - \rho(t)c(t)dt, \quad \forall t \in [0, T]. \tag{9.7}$$

Lemma 9.2 *For any $(\pi, c) \in \mathcal{A}_c(x)$, we have*

$$\mathbb{E}\left[\int_0^T \rho(t)c(t)dt + \rho(T)X(T)\right] \leqslant x.$$

Proof From (9.7), we know that $\int_0^t \rho(s)c(s)ds + \rho(t)X(t)$ is a nonnegative local martingale, hence a supermartingale. So the conclusion of the lemma can be obtained. □

Lemma 9.3 *For any given nonnegative consumption process $c(\cdot)$ and nonnegative $\mathcal{F}_T$-measurable random variables ξ, if they satisfy*

$$\mathbb{E}\left[\int_0^T \rho(t)c(t)dt + \rho(T)\xi\right] = x,$$

then there is a trading strategy $\pi(\cdot)$ such that $(\pi, c) \in \mathcal{A}_c(x)$ and $X(T) = \xi$.

Proof Let

$$M(t) = \mathbb{E}\left[\int_0^T \rho(t)c(t)dt + \rho(T)\xi \,\middle|\, \mathcal{F}_t\right], \quad t \in [0, T].$$

Then $M(\cdot)$ is a martingale. By the martingale representation theorem, there is an m-dimensional progressively measurable process $\phi(\cdot)$ such that

$$M(t) = M(0) + \int_0^t \phi(s)^\tau dW(s), \quad t \in [0, T].$$

Put

$$\pi(t) = (\sigma(t)^{-1})^{\tau}\left(\frac{\phi(t)}{\rho(t)} + \theta(t)X(t)\right),$$

where

$$X(t) = \frac{M(t) - \int_0^t \rho(s)c(s)ds}{\rho(t)}.$$

Then it is easy to verify that $(\pi, c) \in \mathcal{A}_c(x)$ and its wealth process is $X(\cdot)$ satisfying $X(T) = \xi$.   □

Lemma 9.4 *For any* $(\pi, c) \in \mathcal{A}_c(x)$, *if*

$$\mathbb{E}\left[\int_0^T \rho(t)c(t)dt + \rho(T)X(T)\right] < x,$$

then there is a trading strategy $\widetilde{\pi}(\cdot)$ *such that* $(\widetilde{\pi}, c) \in \mathcal{A}_c(x)$ *and its terminal wealth* $\widetilde{X}(T) > X(T)$.

Proof Assume $(\pi, c) \in \mathcal{A}_c(x)$ satisfies $\mathbb{E}\left[\int_0^T \rho(t)c(t)dt + \rho(T)X(T)\right] < x$. Then there is $\varepsilon > 0$ such that

$$\mathbb{E}\left[\int_0^T \rho(t)c(t)dt + \rho(T)(X(T) + \varepsilon)\right] = x.$$

By Lemma 9.3, there is a trading strategy $\widetilde{\pi}(\cdot)$ such that $(\widetilde{\pi}, c) \in \mathcal{A}_c(x)$ and the corresponding terminal wealth $\widetilde{X}(T) = X(T) + \varepsilon > X(T)$.   □

Lemma 9.4 shows that the optimal solution to the problem (9.3) must satisfy

$$\mathbb{E}\left[\int_0^T \rho(t)c(t)dt + \rho(T)X(T)\right] = x.$$

From Lemma 9.3 again, problem (9.3) is equivalent to the following one:

$$\max_{c(\cdot), \xi}\ \mathbb{E}\left[\int_0^T U(t, c(t))dt + V(\xi)\right]$$

subject to $c(t) \geqslant 0,\ \xi \geqslant 0,\ c(\cdot)$ is progressively measurable, $\xi \in \mathcal{F}_T,$ (9.8)

$$\mathbb{E}\left[\int_0^T \rho(t)c(t)dt + \rho(T)\xi\right] = x.$$

In the above problem, constraint $\mathbb{E}\left[\int_0^T \rho(t)c(t)dt + \rho(T)\xi\right] = x$ replaces the dynamic budget constraint in problem (9.3). The former is usually called the *static budget constraint*. It portrays the consumption rate process and terminal wealth (c, ξ), which can be reached from the initial capital x.

Let us solve problem (9.8) by using Lagrange's multiplier method. Consider the following problem:

$$\max_{c(\cdot), \xi} \quad \mathbb{E}\left[\int_0^T U(t, c(t))dt + V(\xi)\right] - \lambda\left(\mathbb{E}\left[\int_0^T \rho(t)c(t)dt + \rho(T)\xi\right] - x\right)$$
$$\text{subject to} \quad c(t) \geqslant 0, \; \xi \geqslant 0, \; c(\cdot) \text{ is progressively measurable}, \; \xi \in \mathcal{F}_T.$$
$$(9.9)$$

It is easy to prove that if there is λ such that the optimal solution (c, ξ) of problem (9.9) is just to meet the static budget constraint

$$\mathbb{E}\left[\int_0^T \rho(t)c(t)dt + \rho(T)\xi\right] = x,$$

then (c, ξ) must be the optimal solution to problem (9.8). In addition, we will find such a $\lambda > 0$.

Problem (9.9) is apparently equivalent to the following problem:

$$\max_{c(\cdot), \xi} \quad \mathbb{E}\left[\int_0^T (U(t, c(t)) - \lambda\rho(t)c(t))dt + V(\xi) - \lambda\rho(T)\xi\right] \qquad (9.10)$$
$$\text{s.t.} \quad c(t) \geqslant 0, \; \xi \geqslant 0, \; c(\cdot) \text{ is progressively measurable}, \; \xi \in \mathcal{F}_T.$$

But problem (9.10) can be reduced to the following one: for every $\omega \in \Omega$,

$$\max_{c(\cdot), \xi} \quad \int_0^T (U(t, c(t)) - \lambda\rho(t)c(t))dt + V(\xi) - \lambda\rho(T)\xi \qquad (9.11)$$
$$\text{s.t.} \quad c(t) \geqslant 0, \xi \geqslant 0.$$

To make notational simplicity, we have omitted ω in problem (9.11). For any $\lambda > 0$, the optimal solution (c^*, ξ^*) of problem (9.11) satisfies obviously the following first-order condition:

$$\begin{cases} \frac{\partial U}{\partial c}(t, c^*(t)) = \lambda\rho(t), & \forall t \in [0, T], \\ \\ V'(\xi^*) = \lambda\rho(T). \end{cases} \qquad (9.12)$$

For each t, let $I(t, \cdot)$ be the inverse function of $\frac{\partial U}{\partial c}(t, \cdot)$, and then it is easy to show that $I(t, \cdot)$ is a continuous, strictly decreasing function on $(0, \infty)$ and satisfies

$$I(t, 0) := \lim_{y \downarrow 0} I(t, y) = \infty, \quad I(t, \infty) := \lim_{y \uparrow \infty} I(t, y) = 0.$$

In addition, let J be the inverse function of V', and then it is easy to know that J is a continuous, strictly decreasing function on $(0, \infty)$ and satisfies

$$J(0) := \lim_{y \downarrow 0} J(y) = \infty, \quad J(\infty) := \lim_{y \uparrow \infty} J(y) = 0.$$

Thus, the first-order condition (9.12) can be written as

$$\begin{cases} c^*(t) = I(t, \lambda \rho(t)), & \forall t \in [0, T], \\ \\ \xi^* = J(\lambda \rho(T)). \end{cases} \qquad (9.13)$$

For any $\lambda > 0$, as defined above, (ξ^*, c^*) are automatically strictly positive. That is, the constraint in problem (9.11) is actually ineffective.

Thus, from the above discussion, we know that for any $\lambda > 0$, the optimal solution to problem (9.9) is given by (9.13). In order to finally solve the problem (9.8), we still need to set a good λ. Actually, λ is given by the following equation:

$$\mathbb{E}\left[\int_0^T \rho(t) I(t, \lambda \rho(t)) dt + \rho(T) J(\lambda \rho(T)) \right] = x. \qquad (9.14)$$

We summarize the conclusions of the first step: we first solve Eq. (9.14) and set a good λ, and then the optimal solution to problem (9.9) is given by (9.13).

Finally, we come to complete the second step: the "realization problem." Actually, this problem has been solved in the proof of Lemma 9.3, so we do not repeat it here.

Example 9.5   Assume $U(t, c) = e^{-\beta t} \ln c, \forall t \in [0, T]$, where $\beta \geqslant 0$. In addition, assume $V(x) = \ln x$. Find the solution to problem (9.3).

Solution   We have

$$I(t, y) = e^{-\beta t} y^{-1}, \quad J(y) = y^{-1}.$$

Then Eq. (9.14) is just

$$\int_0^T e^{-\beta t} \frac{1}{\lambda} dt + \frac{1}{\lambda} = x.$$

Its solution is

$$\lambda = \frac{1 + \beta - e^{-\beta T}}{\beta x}.$$

By (9.13), the optimal solution to problem (9.9) is

$$c^*(t) = \frac{\beta e^{-\beta t} x}{\rho(t)(1 + \beta - e^{-\beta T})}, \quad \xi^* = \frac{\beta x}{\rho(T)(1 + \beta - e^{-\beta T})}.$$

This completes the first step. Below, in accordance with the procedure of the proof for Lemma 9.3, we complete the proof of the second step. It is easy to get

$$\int_0^T \rho(t)c^*(t)dt + \rho(T)\xi^* = x.$$

Thus,

$$M(t) = \mathbb{E}\left[\int_0^T \rho(t)c^*(t)dt + \rho(T)\xi^* \,\Big|\, \mathcal{F}_t\right] = x.$$

Consequently, $\phi(t) = 0$, and

$$\pi^*(t) = (\sigma(t)^{-1})^\tau \theta(t) X(t),$$

where

$$X(t) = \frac{\beta + e^{-\beta t} - e^{-\beta T}}{1 + \beta - e^{-\beta T}} \cdot \frac{x}{\rho(t)}.$$

It can be seen, at every moment t,

$$\frac{\pi^*(t)}{X(t)} = (\sigma(t)^{-1})^\tau \theta(t).$$

This vector represents the optimal ratio of investment in each stock, i.e., the ratio of the funds $\pi_i^*(t)$ invested in various stocks i in the total wealth $X(t)$. We found that the optimal trading strategy is very simple. That is, at each time t, we arrange the total wealth to each stock in proportion to the vector $(\sigma(t)^{-1})^\tau \theta(t)$. In addition, we can get

$$\frac{c^*(t)}{X(t)} = \frac{\beta e^{-\beta t}}{\beta + e^{-\beta t} - e^{-\beta T}}.$$

This is the optimal consumption proportion of the total wealth.

9.3   Mean-Risk Portfolio Selection

In this section, we study a *risk portfolio selection problem*, which is a natural generalization of Markowitz' mean-variance portfolio theory. This section is based on Jin, Yan, and Zhou (2005) and Zhou (2008).

9.3.1   General Framework for Mean-Risk Models

Now we assume that $d = m$, and market parameters $r(t)$, $\mu(t)$, and $\sigma(t)$ meet the conditions in Sect. 9.2. Then the market is complete, and there is a unique risk-neutral measure $\mathbb{Q}$ with $d\mathbb{Q}/d\mathbb{P}|_{\mathcal{F}_t} = Z_t$, where $Z(t)$ is given by (9.5).

Definition 9.6 A self-financing trading strategy is called L^2-*allowable* if it satisfies $\mathbb{E}\left[\int_0^T |\pi(t)|^2 dt\right] < \infty$.

We use $\mathcal{A}_2(x)$ to express the set of all L^2-allowable trading strategies with initial wealth x.

We have

$$S_i(t) = \rho(t)^{-1} E(\rho(T)S_i(T)|\mathcal{F}_t), \quad \forall t \in [0, T],$$

where ρ is the pricing kernel. Besides, for L^2-allowable trading strategy π, its wealth process $X(t)$ satisfies

$$X(t) = \rho(t)^{-1} E(\rho(T)X(T)|\mathcal{F}_t), \quad \forall t \in [0, T].$$

Consider a more general mean-risk portfolio selection problem:

$$
\begin{cases}
\pi^* = \arg\min_{\pi} \mathbb{E}[f(X(T) - \mathbb{E}[X(T)])], \\[2mm]
\pi \in \mathcal{A}_2(x_0), \ \mathbb{E}[X(T)] = z,
\end{cases}
\tag{9.15}
$$

where $x_0, z \in \mathbb{R}$, $f : \mathbb{R} \to \mathbb{R}$ is a given measurable function. Markowitz' mean-variance portfolio corresponds to the case of $f(x) = x^2$ in (9.15).

As the market is complete, the mean-risk portfolio selection problem (9.15) can be reduced to the following static optimization problem:

$$
\begin{cases}
\xi^* = \arg\min_{\xi} \mathbb{E}[f(\xi - z)], \\[2mm]
\xi \in L^2(\mathcal{F}_T, \mathbb{R}), \quad \mathbb{E}[\xi] = z, \quad \mathbb{E}[\rho(T)\xi] = x_0,
\end{cases}
\tag{9.16}
$$

and a replicating contingent claims issue. Assume that ξ^* is the solution to problem (9.16). By (9.1), we know that the replicating problem boils down to the following backward stochastic differential equation (BSDE):

$$\begin{cases} dX(t) = [r(t)X(t)dt + Z(t)^\tau \sigma(t)^{-1}b(t)]dt + Z(t)^\tau dW(t), \\ X(T) = \xi^*. \end{cases} \tag{9.17}$$

Once we get its solution (X, Z), we just put

$$\pi(t) = (\sigma(t)^{-1})^\tau Z(t). \tag{9.18}$$

But in general mean-risk models, BSDE (9.17) has usually no analytical solution.

9.3.2 Weighted Mean-Variance Model

This section considers a most simple extension of Markowitz' mean-variance model, the weighted mean-variance model. Let $Y = \xi - z$. Consider the following static optimization problem:

$$\begin{cases} Y^* = \arg\min_Y \alpha \mathbb{E}[Y_+^2] + \mathbb{E}[Y_-^2], \\ Y \in L^2(\mathcal{F}_T, \mathbb{R}), \quad \mathbb{E}[Y] = 0, \quad \mathbb{E}[\rho Y] = y_0, \end{cases} \tag{9.19}$$

where $\alpha > 0$ is a previously given positive constant, $\rho = \rho(T)$, $y_0 = x_0 - z\mathbb{E}[\rho]$. To solve this optimization problem, we use the Lagrange multiplier method. First, for any pair of multiplier (λ, μ), we solve the following problem:

$$\min_{Y \in L^2(\mathcal{F}_T, \mathbb{R})} \mathbb{E}[\alpha Y_+^2 + Y_-^2 - 2(\lambda - \mu\rho)Y]. \tag{9.20}$$

Lemma 9.7 *The unique solution to problem* (9.20) *is* $Y^* = \frac{(\lambda - \mu\rho)_+}{\alpha} - (\lambda - \mu\rho)_-$.

Proof For any $Y \in L^2(\mathcal{F}_T, \mathbb{R})$, we have

$$\alpha Y_+^2 + Y_-^2 - 2(\lambda - \mu\rho)Y$$

$$= \alpha(Y_+^2 - 2\frac{\lambda - \mu\rho}{\alpha}Y_+) + Y_-^2 + 2(\lambda - \mu\rho)Y_-$$

$$= \alpha(Y_+ - \frac{\lambda - \mu\rho}{\alpha})^2 - \frac{(\lambda - \mu\rho)^2}{\alpha} + (Y_- + \lambda - \mu\rho)^2 - (\lambda - \mu\rho)^2$$

$$\geq -\frac{(\lambda - \mu\rho)_+^2}{\alpha} - (\lambda - \mu\rho)_-^2$$

$$= \alpha(Y_+^*)^2 + (Y_-^*)^2 - 2(\lambda - \mu\rho)Y^*.$$

This shows that Y^* is the unique solution to problem (9.20). $\square$

The proof of the next theorem is left as an exercise for the reader.

Theorem 9.8 *The unique solution to problem (9.19) is given as*

$$Y^* = \frac{(\lambda - \mu\rho)_+}{\alpha} - (\lambda - \mu\rho)_-,$$

where (λ, μ) is the unique solution to the following set of equations:

$$\begin{cases} \frac{\mathbb{E}[(\lambda - \mu\rho)_+]}{\alpha} - \mathbb{E}[(\lambda - \mu\rho)_-] = 0, \\ \\ \frac{\mathbb{E}[\rho(\lambda - \mu\rho)_+]}{\alpha} - \mathbb{E}[\rho(\lambda - \mu\rho)_-] = y_0. \end{cases}$$

For the weighted mean-variance model, BSDE (9.17) has no analytical solution. But when $\alpha = 1$, the weighted mean-variance model degenerates into Markowitz' mean-variance model. Then the only solution to the problem (9.19) is $Y^* = \lambda - \mu\rho$, where (λ, μ) is the unique solution to the following set of linear equations:

$$\mathbb{E}[\rho(\lambda - \mu\rho)] = y_0, \quad \mathbb{E}[\lambda - \mu\rho] = 0,$$

i.e.,

$$\lambda = -\frac{y_0\mathbb{E}[\rho]}{\mathrm{Var}(\rho)}, \quad \mu = -\frac{y_0}{\mathrm{Var}(\rho)}. \tag{9.21}$$

In the following we assume that $r(t)$, $\mu_i(t)$, and $\sigma_{i,j}(t)$ are bounded measurable functions. First, we can expect to find $X(t)$ by calculating the conditional mathematical expectation and then find the optimal investment strategy π^*.

Theorem 9.9 *Assume that $r(t)$, $\mu_i(t)$ and $\sigma_{i,j}(t)$ are bounded measurable functions. Then the optimal trading strategy of Markowitz' mean-variance portfolio problem is given by*

$$\pi^*(t) = -(\sigma(t)^\tau)^{-1}\theta(t)\left(X^*(t) - (\lambda - z)e^{-\int_t^T r(s)ds}\right), \tag{9.22}$$

$$X^*(t) = (\lambda - z)e^{-\int_t^T r(s)ds} - \mu e^{-\int_t^T (2r(s) - |\theta(s)|^2)ds}\rho(t), \tag{9.23}$$

where (λ, μ) is given by (9.21) and

$$\mathbb{E}[\rho] = \exp\left\{-\int_0^T r(s)ds\right\}, \quad \mathbb{E}[\rho^2] = \exp\left\{\int_0^T (|\theta(s)|^2 - 2r(s))ds\right\}. \quad (9.24)$$

Proof Since $\rho(t) = \exp\left\{-\int_0^t r(s)ds\right\} Z(t)$, $(Z(t))$ is a martingale defined by (9.5), we obtain immediately (9.24) and have

$$\mathbb{E}[\rho|\mathcal{F}_t] = e^{-\int_t^T r(s)ds}\rho(t), \quad \mathbb{E}[\rho^2|\mathcal{F}_t] = e^{-\int_t^T (2r(s)-|\theta(s)|^2)ds}\rho(t)^2.$$

Because of $X^*(T) = Y^* - z = \lambda - \mu\rho - z$, we have

$$X^*(t) = \rho(t)^{-1}\mathbb{E}\left[(\lambda - \mu\rho - z)\rho|\mathcal{F}_t\right].$$

Thereby immediately, we get (9.23). Consequently,

$$dX^*(t) = \mu e^{-\int_t^T (2r(s)-|\theta(s)|^2)ds}\rho(t)\theta(t)^\tau dW(t) + \text{``}dt\text{''} \text{ term}.$$

In comparison of this equation with (9.2), we obtain (9.22).   □

Chapter 10
Static Risk Measures

The financial market faces risks arising from many types of uncertain losses, including market risk, credit risk, liquidity risk, operational risk, etc. In 1988, the Basel Committee on Banking Supervision proposed measures to control credit risk in banking. A risk measure called the *value-at-risk*, acronym VaR, became, in the 1990s, an important tool of risk assessment and management for banks, securities companies, investment funds, and other financial institutions in asset allocation and performance evaluation. The VaR associated with a given confidence level for a venture capital is the upper limit of possible losses in the next certain period of time. In 1996 the Basel Committee on Banking Supervision endorsed the VaR as one of the acceptable methods for the bank's internal risk measure. However, due to the defects of VaR, a variety of new risk measures came into being. This chapter focuses on the representation theorems for static risk measures. For an overview of the subject we refer to Song and Yan (2009b).

10.1 Coherent Risk Measures

Assume that all possible states and events that may occur at the terminal time are known, namely, a measurable space $(\Omega, \mathcal{F})$ is given. The financial position (here refers to the wealth deducted investment cost) is usually described by a measurable function ξ on $(\Omega, \mathcal{F})$. If we assume that a probability measure $\mathbb{P}$ is given on measurable space $(\Omega, \mathcal{F})$, the financial positions is usually described by a random variables ξ on $(\Omega, \mathcal{F}, \mathbb{P})$. In order to facilitate the notation and description, we use $X = -\xi$ to denote the *potential loss* at the terminal time of trading. Here the potential loss is relative to a reference point in terms. If X is a negative value, it indicates a surplus.

© Springer Nature Singapore Pte Ltd. and Science Press 2018
J.-A. Yan, *Introduction to Stochastic Finance*, Universitext,
https://doi.org/10.1007/978-981-13-1657-9_10

10.1.1 Monetary Risk Measures and Coherent Risk Measures

A risk measure is a numerical value $\rho(X)$ to quantify the risk of a potential loss X. If we denote the set of potential losses to be considered by $\mathcal{G}$, a risk measure ρ is a map from $\mathcal{G}$ to $\mathbb{R}$. In this chapter we take always $L^\infty(\Omega, \mathcal{F})$ or $L^\infty(\Omega, \mathcal{F}, \mathbb{P})$ as the set of all potential losses $\mathcal{G}$ or $\mathcal{G}(\mathbb{P})$, where the former is the set of all bounded $\mathcal{F}$ measurable functions on $(\Omega, \mathcal{F})$, endowed with the uniform norm $||\cdot||_\infty$, and the latter is the set of equivalence classes of the former under probability $\mathbb{P}$. In the former case, the states and the probabilities of the possible events are unknown or are not consensus in the market, and then the risk measure is called *model-free*. In the latter case, the risk measure is called *model-dependent*. In the model-dependent case, naturally, we always assume that the risk measure ρ satisfies the following property: If $X = Y$, $\mathbb{P}$-a.s., then $\rho(X) = \rho(Y)$.

We only consider a single period of uncertainty, namely, we only investigate *static risk measures*.

Definition 10.1 A map ρ from $\mathcal{G}$ to $\mathbb{R}$ is called a *monetary risk measure*, abbreviated as *risk measure*, if it satisfies two conditions:

(1) *monotonicity*: For all $X, Y \in \mathcal{G}$ satisfying $X \leqslant Y$, it holds that $\rho(Y) \leqslant \rho(X)$.
(2) *translation invariance*: For all $X \in \mathcal{G}$ and any real number α, it holds that

$$\rho(X + \alpha) = \rho(X) + \alpha.$$

Remark Any monetary risk measure ρ is Lipschitz continuous w.r.t. the uniform norm $||\cdot||_\infty$ on $\mathcal{G}$:

$$|\rho(X) - \rho(Y)| \leqslant ||X - Y||_\infty. \tag{10.1}$$

In fact, clearly, we have $X \leqslant Y + ||X - Y||_\infty$, so by the monotonicity and the transition invariance, we get

$$\rho(X) \leqslant \rho(Y) + ||X - Y||_\infty.$$

By the position symmetry of X and Y, (10.1) follows.

Let $X \in \mathcal{G}$ be a given potential loss. For $\alpha \in (0, 1)$, the *VaR with confidence level* $1 - \alpha$ is defined by

$$\mathrm{VaR}_\alpha(X) = \inf\{x \in \mathbb{R} : \mathbb{P}[X > x] \leqslant \alpha\}.$$

For $\alpha = 0$, we have

$$\mathrm{VaR}_0(X) = \lim_{\alpha \to 0} \mathrm{VaR}_\alpha(X) = \mathrm{ess\,sup}(X).$$

Obviously, VaR_α is a monetary risk measure. However, one can construct an example to show that VaR_α does not satisfy the following property.

(3) *subadditivity*: For all $X_1, X_2 \in \mathcal{G}$, it holds that $\rho(X_1 + X_2) \leqslant \rho(X_1) + \rho(X_2)$. For example, Let X and Y be two independent and identically distributed random variables, satisfying

$$\mathbb{P}(X = 0) = 0.95, \quad \mathbb{P}(X = 1) = 0.05,$$

then

$$\text{VaR}_{0.05}(X) = \text{VaR}_{0.05}(Y) = 0.$$

However, since

$$\mathbb{P}(X + Y = 0) = 0.9025, \quad \mathbb{P}(X + Y = 1) = 0.095, \quad \mathbb{P}(X + Y = 2) = 0.0025,$$

we have $\text{VaR}_{0.05}(X + Y) = 1$.

Many financial economists argue that a reasonable measure of risk should satisfy subadditivity. Embrechts et al. (2000) pointed out that only for elliptical distributions of X, the VaR has subadditivity. In order to overcome this shortcoming of the VaR, Artzner et al. (1997, 1999) and Delbaen (2002) introduced the notion of *coherent risk measure* $\rho : \mathcal{G} \to \mathbb{R}$.

A monetary risk measure is called a coherent risk measure, if it has subadditivity and the following property.

(4) *Positive homogeneity*: For all $\lambda \geqslant 0$ and any $X \in \mathcal{G}$, it holds that $\rho(\lambda X) = \lambda \rho(X)$.

From the positive homogeneity of a coherent risk measure ρ, we know $\rho(0) = 0$. Thus, for all $X \in \mathcal{G}$, we have $0 = \rho(X - X) \leqslant \rho(X) + \rho(-X)$, and, consequently,

$$\rho(X) \geqslant -\rho(-X). \tag{10.2}$$

Example 1 Put

$$\rho_{\max}(X) = \sup_{\omega \in \Omega} X(\omega), \quad \forall X \in \mathcal{G}.$$

It is easy to verify that $\rho_{\max}$ is a coherent risk measure, called the *worst-case risk measure*.

Example 2 For a $\lambda \in (0, 1]$, set

$$\mathcal{Q}_\lambda = \{Q \in \mathcal{M}_1(\mathbb{P}) : d\mathbb{Q}/d\mathbb{P} \leqslant 1/\lambda\}.$$

One can verify that as defined below

$$\pi_\lambda(X) := \sup_{Q \in \mathcal{Q}_\lambda} E_{\mathbb{Q}}(X), \quad \forall X \in \mathcal{G},$$

is a coherent risk measure.

We will see later (Theorem 10.20) that π_λ has the following representation:

$$\pi_\lambda(X) = \frac{1}{\lambda} \int_0^\lambda \mathrm{VaR}_\gamma(X) d\gamma.$$

This representation is given by the *average VaR* of level λ, denoted by AVaR_λ.

10.1.2 Representation of Coherent Risk Measures

First we consider the model-free case, i.e., assume $\mathcal{G} = L^\infty(\Omega, \mathcal{F})$. Let $\mathcal{M}_{1,f}$ and $\mathcal{M}_1$ denote the set of all finitely additive probability measures and the set of all (countably additive) probability measures on $(\Omega, \mathcal{F})$, respectively.

The next theorem is due to Artzner et al. (1999), the proof presented here simplifies its original proof. It is similar to the proof for Theorem 10.6 below.

Theorem 10.2 $\rho : \mathcal{G} \to \mathbb{R}$ *is a coherent risk measure if and only if there is a subset* $\mathcal{Q}$ *of* $\mathcal{M}_{1,f}$ *such that*

$$\rho(X) = \sup_{\mathbb{Q} \in \mathcal{Q}} \mathbb{E}_{\mathbb{Q}}(X), \ \forall X \in \mathcal{G}. \tag{10.3}$$

Moreover, in (10.3), $\mathcal{Q}$ *can be taken as a convex subset of* $\mathcal{M}_{1,f}$, *such that the supremum of the above formula can be attained on* $\mathcal{Q}$. *For example, one can let*

$$\mathcal{Q} = \{\lambda \in \mathcal{M}_{1,f} : \lambda(Y) \leqslant \rho(Y), \forall Y \in \mathcal{G}\}.$$

If ρ *is continuous from above on* $\mathcal{G}$, *i.e., for* $X_n, X \in \mathcal{G}$, *it holds that*

$$X_n \searrow X \implies \rho(X_n) \searrow \rho(X),$$

then $\mathcal{Q}$ *can further be taken as a convex subset of* $\mathcal{M}_1$.

Proof The sufficiency is easy to verify. We prove the necessity. To this end, we only need to prove that if ρ is a coherent risk measure, then for any $X \in \mathcal{G}$, there is $\mathbb{Q}_X \in \mathcal{M}_{1,f}$ such that $\mathbb{Q}_X(X) = \rho(X)$ and $\mathbb{Q}_X(Y) \leqslant \rho(Y), \forall Y \in \mathcal{G}$. This is because in this way, we have

$$\rho(X) = \sup_{\mathbb{Q} \in \mathcal{Q}_0} \mathbb{E}_{\mathbb{Q}}(X), \quad \forall \mathcal{G}, \tag{10.4}$$

where $\mathcal{Q}_0 = \{\mathbb{Q}_X : X \in \mathcal{G}\}$. Clearly, by replacing $\mathcal{Q}$ with the convex hull of $\mathcal{Q}_0$, (10.4) still holds.

Due to the translation invariance of ρ and $\mathbb{E}_{\mathbb{Q}}$, we only need to consider those X with $\rho(X) = 1$. Put

$$\mathcal{B} = \{Y \in \mathcal{G} : \rho(Y) < 1\}.$$

It is easy to see that $\mathcal{B}$ is a convex set, $B_1 := \{Y \in \mathcal{G} : ||Y||_\infty < 1\} \subset \mathcal{B}$ and $X \notin \mathcal{B}$.

Since the dual space of $L^\infty(\Omega, \mathcal{F})$ is the set $ba(\Omega, \mathcal{F})$ of all finitely additive set functions with finite total variation on $(\Omega, \mathcal{F})$ (see Theorem 1.12), by convex sets separation theorem, there is a non-zero element $\lambda \in ba(\Omega, \mathcal{F})$, such that

$$\sup_{Y \in \mathcal{B}} \lambda(Y) \leqslant \lambda(X).$$

Because of $B_1 \subset \mathcal{B}$, we have $\lambda(X) > 0$. Thus, one can select a λ such that $\lambda(X) = 1$. We will prove that λ has the following properties:

(1) For any $Y \geqslant 0$, $Y \in \mathcal{G}$, it holds that $\lambda(Y) \geqslant 0$;
(2) $\lambda(1) = 1$;
(3) For any $Y \in \mathcal{G}$, it holds that $\lambda(Y) \leqslant \rho(Y)$.

In fact, for any $Y \geqslant 0$, $Y \in \mathcal{G}$, $c > 0$, we have $-cY \in \mathcal{B}$. Thus, $\lambda(-cY) \leqslant \lambda(X) = 1$. Since c is arbitrary, (1) must hold. For any $0 < c < 1$, it holds that $c \in \mathcal{B}$, this implies $\lambda(1) \leqslant 1$. On the other hand, for any $c > 1$, we have $2 - c\lambda(1) = \lambda(2X - c) \leqslant \lambda(X) = 1$. This implies $\lambda(1) \geqslant 1/c$, so that we must have $\lambda(1) \geqslant 1$. Then (2) is proved. Finally, for any Y, let $Y_1 = Y - \rho(Y) + 1$, and then for any $c > 1$, it holds that $Y_1/c \in \mathcal{B}$. Thus, $\lambda(Y_1) \leqslant 1$, from which it follows $\lambda(Y) \leqslant \rho(Y)$, and (3) is proved. Hence, $\lambda \in \mathcal{M}_{1,f}$, it can be taken as $\mathbb{Q}_X$.

In addition, if further ρ is continuous from above, then because of $\lambda \leqslant \rho$, λ is continuous from above at 0, and consequently, $\lambda \in \mathcal{M}_1$. □

Now we consider the model-dependent case, i.e., assume $\mathcal{G}(\mathbb{P}) = L^\infty(\Omega, \mathcal{F}, \mathbb{P})$. We use $\mathcal{M}_1(\mathbb{P})$ to represent the set of all probability measures on $(\Omega, \mathcal{F})$, which are absolutely continuous w.r.t. $\mathbb{P}$.

The next theorem is due to Artzner et al. (1999) and Delbaen (2002). One can directly give its proof by using Theorem 10.2 (see the proof of Theorem 10.6). In the context of robust statistics, a special case of Theorem 10.3 appears in Huber (1981).

Theorem 10.3 *For a coherent risk measure ρ on $\mathcal{G}$, the following conditions are equivalent:*

(1) *ρ is continuous from above: $X_n \searrow X \implies \rho(X_n) \searrow \rho(X)$;*
(2) *There is a subset $\mathcal{Q}$ of $\mathcal{M}_1(\mathbb{P})$, such that*

$$\rho(X) = \sup_{\mathbb{Q} \in \mathcal{Q}} E_{\mathbb{Q}}(X), \quad \forall X \in \mathcal{G}(\mathbb{P});$$

(3) *ρ has the "Fatou property": for any bounded sequence (X_n) which $\mathbb{P}$-a.s. converges to certain X,*

$$\rho(X) \leqslant \liminf_{n \uparrow \infty} \rho(X_n).$$

10.2   Co-monotonic Subadditive Risk Measures

Subadditivity is a basic assumption of coherent risk measures. From the perspective of risk aversion point of view, this seems to be reasonable, because one always expect to be able to spread out some risks by putting together different risks. However, the subadditivity assumption of risk measures sometimes is too strong. For example, in insurance mathematics, there are many "premium principles" that have subadditivity only for co-monotonic claims.

Two functions $X, Y : \Omega \to \mathbb{R}$ are called *co-monotonic*, if there are no $\omega_1, \omega_2 \in \Omega$ such that $X(\omega_1) < X(\omega_2)$, and $Y(\omega_1) > Y(\omega_2)$.

Theorem 10.4 *Let $X, Y : \Omega \to \mathbb{R}$. The following two conditions are equivalent:*

(1) *X and Y are co-monotonic.*
(2) *There exist continuous non-decreasing functions u and v on $\mathbb{R}$, such that $u(z) + v(z) = z, z \in \mathbb{R}$, $X = u(X + Y)$, and $Y = v(X + Y)$.*

Proof We only need to prove (1)$\Rightarrow$(2). Assume that X and Y are co-monotonic. Set $Z = X + Y$. It is easy to know that any $z \in Z(\Omega)$ has a unique decomposition: there is a $\omega \in \Omega$, such that $z = Z(\omega)$, $x = X(\omega)$, $y = Y(\omega)$, and $z = x + y$. We denote $u(z)$ and $v(z)$ by x and y, respectively. Then by the co-monotonicity of X and Y, it is easy to see that u and v are non-decreasing functions on $Z(\Omega)$.

We will show that u and v are continuous on $Z(\Omega)$. First, noting that for $h > 0$ and $z, z + h \in Z(\Omega)$, we have

$$z + h = u(z + h) + v(z + h) \geqslant u(z + h) + v(z) = u(z + h) + z - u(z).$$

Thus, it holds that

$$u(z) \leqslant u(z + h) \leqslant u(z) + h.$$

Similarly, for $h > 0$ and $z, z - h \in Z(\Omega)$, we have

$$u(z) - h \leqslant u(z - h) \leqslant u(z).$$

The above two inequalities imply the continuity of u; thus v is also continuous.

Below it suffices to show u and v can be extended to a continuous non-decreasing functions on $\mathbb{R}$. First, we extend them on the closure $\overline{Z(\Omega)}$ of $Z(\Omega)$. If z is a unidirectional boundary point of $Z(\Omega)$, since u and v are non-decreasing functions, by taking the limit, the continuous extension can be performed. If z is a bidirectional limit point of $Z(\Omega)$, the above two inequalities imply the agreement of two-way limits, allowing the continuous extension. Finally, by performing the linear extension on each connected interval of open set $\mathbb{R} \setminus \overline{Z(\Omega)}$, one can extend u and v from $\overline{Z(\Omega)}$ to $\mathbb{R}$, and keep $u(z) + v(z) = z$ holds. The theorem is proved. $\square$

Song and Yan (2006) suggested to replace the subadditivity axiom by a weaker co-monotonic subadditivity axiom:

(3)′ *co-monotonic subadditivity*: for all co-monotonic $X_1, X_2 \in \mathcal{G}$, it holds that

$$\rho(X_1 + X_2) \leqslant \rho(X_1) + \rho(X_2).$$

A risk measure satisfying monotonicity, translation invariance, positive homogeneity, and co-monotonic subadditivity is called a *co-monotonic sub-additive risk measure*. It can be proved that the VaR actually has co-monotonic subadditivity.

10.2.1 Representation: The Model-Free Case

Now we consider the case of $\mathcal{G} = L^\infty(\Omega, \mathcal{F})$. We use $\mathcal{M}_{1,m}$ to represent the set of all monotonic set functions $\mu : \mathcal{F} \to [0, 1]$ satisfying $\mu(\Omega) = 1$.

Let $\mu \in \mathcal{M}_{1,m}$, $X \in \mathcal{G}$, the Choquet *integral* of X w.r.t. μ is defined by

$$\mu(X) := \int_{-\infty}^{0} [\mu(X \geqslant x) - 1]dx + \int_{0}^{\infty} \mu(X \geqslant x)dx. \tag{10.5}$$

This integral was first introduced by Choquet (1953–54) in the study of capacity theory.

Here we present some conclusions related to Choquet integral, and their proofs can be found in Denneberg (1994) or Yan (2010).

Let $\mu \in \mathcal{M}_{1,m}$, $X, Y \in \mathcal{G}$. Choquet integral has the following properties:

(1) positive homogeneity: if $c \geqslant 0$, then $\mu(cX) = c\mu(X)$;
(2) monotonicity: $X \leqslant Y$ implies $\mu(X) \leqslant \mu(Y)$;
(3) translation invariance: $\mu(X + c) = \mu(X) + c, \forall c \in \mathbb{R}$;
(4) co-monotonic additivity: if X and Y are co-monotonic, then $\mu(X + Y) = \mu(X) + \mu(Y)$.

The next theorem is known as Greco's *representation theorem*.

Theorem 10.5 *Let $\mathcal{H}$ be a family of bounded functions on Ω satisfying the following properties:*

(1) *If $X \in \mathcal{H}$, then for any $a \in R_+$, we have $aX, X \wedge a, X - X \wedge a \in \mathcal{H}$;*
(2) *If $X \in \mathcal{H}$, then $X + 1 \in \mathcal{H}$.*

If Γ is a real functional on $\mathcal{H}$, having the monotonicity and co-monotonic additivity, then there exists a monotonic set function γ on 2^Ω (the set of all subsets of Ω), such that $\forall X \in \mathcal{H}$, $\gamma(X) = \Gamma(X)$.

Let $\mathcal{Q}$ be a subset of $\mathcal{M}_{1,m}$. Define a risk measure ρ:

$$\rho(X) = \sup_{\mu \in \mathcal{Q}} \mu(X), \quad X \in \mathcal{G},$$

where $\mu(X)$ is the Choquet integral of X w.r.t. μ. It is easy to verify that ρ satisfies monotonicity, positive homogeneity, translation invariance, and co-monotonic subadditivity.

In fact, the next theorem (due to Song and Yan (2006)) shows that any risk measure possessing positive homogeneity and co-monotonic subadditivity has the above representation.

Theorem 10.6 *Let* $\rho : \mathcal{G} \to \mathbb{R}$ *be a risk measure possessing positive homogeneity and co-monotonic subadditivity, then*

$$\rho(X) = \max_{\mu \in \mathcal{M}} \mu(X), \quad X \in \mathcal{G},$$

where

$$\mathcal{M} = \{\mu \in \mathcal{M}_{1,m} : \mu(Y) \leqslant \rho(Y), \forall Y \in \mathcal{G}\},$$

$\mu(X)$ *is the Choquet integral of X w.r.t. μ.*

Proof Similar to the case of Theorem 10.2, it suffices to prove for any $X \in \mathcal{G}$, there exists $\mu_X \in \mathcal{M}_{1,m}$, such that $\mu_X(X) = \rho(X)$, and for any $Y \in \mathcal{G}$, it holds that $\mu_X(Y) \leqslant \rho(Y)$. Due to the translation invariance of ρ and μ, we need only to consider those X with $\rho(X) = 1$. For such a X, let $[X] := \{u(X) : u$ are continuous functions on R$\}$ and

$$\mathcal{B} = \{Y \in \mathcal{G} : \exists Z \in [X], \text{ such that } \rho(Z) < 1, \text{ and } Y \leqslant Z\}.$$

It is easy to prove that $\mathcal{B}$ is convex, $\mathcal{B}_1 := \{Y \in \mathcal{G} : \|Y\| < 1\} \subset \mathcal{B}$ and $X \notin \mathcal{B}$. Since the dual space of $L^\infty(\Omega, \mathcal{F})$ is the set $ba(\Omega, \mathcal{F})$ of all finitely additive set functions with finite total variation on $(\Omega, \mathcal{F})$ (see Theorem 1.12), by convex sets separation theorem, there is a non-travel $\lambda \in ba(\Omega, \mathcal{F})$, such that

$$\sup_{Y \in \mathcal{B}} \lambda(Y) \leqslant \lambda(X).$$

Because of $\mathcal{B}_1 \subset \mathcal{B}$, we have $\lambda(X) > 0$. Thus, we can select a λ with $\lambda(X) = 1$. We confirm that λ possesses the following properties:

1° for any $Y \geqslant 0, Y \in \mathcal{G}$, it holds that $\lambda(Y) \geqslant 0$;
2° $\lambda(1) = 1$;
3° for any $Y \in \mathcal{G}$, it holds that $\lambda(Y) \leqslant \rho(Y)$.

In fact, for any $Y \geqslant 0, Y \in \mathcal{G}, c > 0$, we have $-cY \in \mathcal{B}$. Thus,

$$\lambda(-cY) \leqslant \lambda(X) = 1.$$

Since c is arbitrary, we obtain 1°.

For any $0 < c < 1, c \in \mathcal{B}$. Since $c < 1$ is arbitrary, this implies $\lambda(1) \leqslant 1$. On the other hand, for any $c > 1$, we have

$$2 - c\lambda(1) = \lambda(2X - c) \leqslant \lambda(X) = 1,$$

it implies $\lambda(1) \geqslant 1/c$. Since $c > 1$ is arbitrary, we deduce that $\lambda(1) \geqslant 1$. Then $2°$ is proved.

Finally, for any $Y \in [X]$, let $Y_1 = Y - \rho[Y] + 1$. Then for $c > 1$, we have $Y_1/c \in \mathcal{B}$. Since $c > 1$ is arbitrary, this implies that $\lambda(Y_1) \leqslant 1$. Hence, we obtain $\lambda(Y) \leqslant \rho(Y)$. Thus $3°$ is proved.

Define $\pi^* : \mathcal{G} \to R$

$$\pi^*(Y) := \sup\{\lambda(Z) : Z \leqslant Y, Z \in [X]\}, \quad Y \in \mathcal{G}.$$

Obviously, π^* has monotonicity, positive homogeneity, and the following properties:

$$\pi^*(Y) \leqslant \rho(Y), \ \forall Y \in \mathcal{G}; \qquad \pi^*(Y) = \lambda(Y), \ \forall Y \in [X]. \tag{10.6}$$

We claim that π^* also has the co-monotonic additivity. In fact, let $Y_1, Y_2 \in \mathcal{G}$ be co-monotonic; then there are continuous non-decreasing functions u, v on $\mathbb{R}$, such that $u(z) + v(z) = z, z \in R$, and

$$Y_1 = u(Y_1 + Y_2), \ Y_2 = v(Y_1 + Y_2).$$

For any $Z \in [X]$ satisfying $Z \leqslant Y_1 + Y_2$, we have $u(Z), v(Z) \in [X]$, and

$$u(Z) \leqslant u(Y_1 + Y_2) = Y_1,$$
$$v(Z) \leqslant v(Y_1 + Y_2) = Y_2.$$

Hence,

$$\pi^*(Y_1) + \pi^*(Y_2) \geqslant \lambda(u(Z)) + \lambda(v(Z)) = \lambda(Z).$$

From the definition of π^* it follows

$$\pi^*(Y_1 + Y_2) \leqslant \pi^*(Y_1) + \pi^*(Y_2).$$

But the opposite inequality is obvious. Therefore, π^* has the co-monotonic additivity.

By Greco's representation theorem, there is a monotone set function μ_X on 2^Ω, such that $\mu_X(Y) = \pi^*(Y), \ \forall \, Y \in \mathcal{G}$. From (10.6) it follows $\mu_X \in \mathcal{M}$ and

$$\mu_X(X) = \pi^*(X) = \lambda(X) = \rho(X).$$

The theorem is proved.   □

10.2.2   Representation: The Model-Dependent Case

Now suppose that there is a probability model in the market, that is, a given probability measure $\mathbb{P}$. In this case, $\mathcal{G}(\mathbb{P}) = L^\infty(\Omega, \mathcal{F}, \mathbb{P})$.

Below we will extend the concept of co-monotonic function and Choquet integral to the random variable circumstances.

Definition 10.7 Two real random variables X, Y are called co-monotonic, if there is a $\Omega_0 \in \mathcal{F}$ such that $\mathbb{P}(\Omega_0) = 1$, and there do not exist $\omega_1, \omega_2 \in \Omega_0$, such that $X(\omega_1) < X(\omega_2)$ and $Y(\omega_1) > Y(\omega_2)$.

By Theorem 10.4 it is easy to prove: random variables X and Y are co-monotonic if and only if there exists continuous non-decreasing functions u, and v on $\mathbb{R}$ such that $u(z) + v(z) = z$, $z \in \mathbb{R}$, $X = u(X + Y)$, and $Y = v(X + Y)$ a.s.

Definition 10.8 For $\mu \in \mathcal{M}_{1,m}$, μ is called *absolutely continuous w.r.t.* $\mathbb{P}$, if for any A, $B \in \mathcal{F}$ satisfying $\mathbb{P}(A \triangle B) = 0$, it holds that $\mu(A) = \mu(B)$.

We denote by $\mathcal{M}_{1,m}(\mathbb{P})$ the set of all capacities in $\mathcal{M}_{1,m}$ which are absolutely continuous w.r.t. $\mathbb{P}$.

For $\mu \in \mathcal{M}_{1,m}$, $X \in \mathcal{G}(\mathbb{P})$, Choquet integral $\mu(X)$ may not exist. Thus, for $X \in \mathcal{G}(\mathbb{P})$, we only consider the Choquet integral of X w.r.t. $\mu \in \mathcal{M}_{1,m}(\mathbb{P})$. In this case, the Choquet integral has the following "a.s. monotonicity": $X \leqslant Y$ a.s. implies $\mu(X) \leqslant \mu(Y)$, and "a.s. co-monotonic additivity": If X, Y are a.s. co-monotonic, then $\mu(X + Y) = \mu(X) + \mu(Y)$. These properties are simple inference of the Choquet integral on spaces $L^\infty(\Omega, \mathcal{F})$. Therefore, in the future, when referring to these properties, we omit the prefix "a.s."

The next theorem is a simple inference of Greco's representation theorem.

Theorem 10.9 *If* Γ *is a real functional on* $\mathcal{G}(\mathbb{P})$ *satisfying* $\Gamma(1) = 1$, *monotonicity and co-monotonic additivity, then there is* $\gamma \in \mathcal{M}_{1,m}(\mathbb{P})$, *such that for all* $X \in \mathcal{G}(\mathbb{P})$, *it holds* $\gamma(X) = \Gamma(X)$.

Proof Let $\mathcal{H} = L^\infty(\Omega, \mathcal{F})$. Then Γ, restricted on $\mathcal{H}$, satisfies the condition of Greco's representation theorem. Therefore, there is a co-monotonic set function γ on 2^Ω such that for all $X \in \mathcal{H}$, $\gamma(X) = \Gamma(X)$. For A, $B \in \mathcal{F}$ with $P(A \triangle B) = 0$, we have $\Gamma(I_A) = \Gamma(I_B)$, $\gamma(A) = \gamma(B)$. Thus, $\gamma \in \mathcal{M}_{1,m}(\mathbb{P})$. From this it is easy to prove that for all $X \in \mathcal{G}(\mathbb{P})$, it holds $\gamma(X) = \Gamma(X)$. $\square$

Let $\mathcal{Q}$ be a subset of $\mathcal{M}_{1,m}(\mathbb{P})$. Define a risk measure ρ:

$$\rho(X) = \sup_{\mu \in \mathcal{Q}} \mu(X), \quad X \in \mathcal{G}(\mathbb{P}),$$

where $\mu(X)$ is the Choquet integral of X w.r.t. μ. It is easy to verify that ρ satisfies the monotonicity, positive homogeneity, translation invariance, and the co-monotonic subadditivity.

In fact, the next theorem (due to Yan and Song (2006)) shows that any risk measure on $\mathcal{G}(\mathbb{P})$ having the positive homogeneity and the co-monotonic subadditivity has the above representation.

Theorem 10.10 *Let* $\rho : \mathcal{G}(\mathbb{P}) \rightarrow \mathbb{R}$ *be a risk measure having the positive homogeneity and co-monotonic subadditivity. Then*

$$\rho(X) = \max_{\mu \in \mathcal{M}(\mathbb{P})} \mu(X), \quad X \in \mathcal{G}(\mathbb{P}),$$

where

$$\mathcal{M} = \{\mu \in \mathcal{M}_{1,m}(\mathbb{P}) : \mu(Y) \leqslant \rho(Y), \forall Y \in \mathcal{G}(\mathbb{P})\},$$

$\mu(X)$ *is the Choquet integral of X w.r.t.* μ.

Proof Let $ba(\Omega, \mathcal{F}, \mathbb{P})$ represents the space constructed by all finitely additive measures with finite variation which are absolutely continuous w.r.t. $\mathbb{P}$. Noting that $ba(\Omega, \mathcal{F}, \mathbb{P})$ is the dual space of $L^{\infty}(\Omega, \mathcal{F}, \mathbb{P})$ (see Theorem 1.12), similar to the proof of Theorem 10.6, we can prove this theorem.

Based on Theorem 10.6, now we give another proof of the theorem. We regard ρ as a functional on $L^{\infty}(\Omega, \mathcal{F})$. Then by Theorem 10.6, we have

$$\rho(X) = \max_{\mu \in \mathcal{M}} \mu(X), \quad \forall X \in L^{\infty}(\Omega, \mathcal{F}).$$

Therefore, for any given $X \in L^{\infty}(\Omega, \mathcal{F})$, there is a $\mu_X \in \mathcal{M}$, such that $\mu_X(X) = \rho(X)$. Set

$$[X] := \{u(X) : u \text{ is continuous non-decreasing on } \mathbb{R}\}.$$

We define $\Gamma : L^{\infty}(\Omega, \mathcal{F}) \rightarrow \mathbb{R}$ by

$$\Gamma(Y) := \sup\{\mu_X(Z) : Z \leqslant Y \text{ a.s. } Z \in [X]\}.$$

Similar to the proof of Theorem 10.6, Γ has monotonicity, positive homogeneity, co-monotonic additivity, and the following property:

$$\Gamma(Y) \leqslant \rho(Y), \forall Y \in L^{\infty}(\Omega, \mathcal{F}); \quad \Gamma(Y) \geqslant \mu_X(Y), \forall Y \in [X];$$

$$\Gamma(Y_1) = \Gamma(Y_2), \text{ if } Y_1, Y_2 \in L^{\infty}(\Omega, \mathcal{F}), Y_1 = Y_2 \text{a.s.}$$

By Theorem 10.6, there is a $\mu_X^* \in \mathcal{M}_{1,m}$, such that for all $Y \in L^{\infty}(\Omega, \mathcal{F})$, it holds $\mu_X^*(Y) = \Gamma(Y)$. Consequently, we get $\mu_X^* \in \mathcal{M}(\mathbb{P})$ and $\mu_X^*(X) = \Gamma(X) = \rho(X)$. The theorem is proved. $\square$

10.3 Convex Risk Measures

Positive homogeneity of a coherent risk measure implies that the risk's growth is linear with respect to positions. It is not quite realistic. For example, the growth of liquidity risk is nonlinear with respect to positions. Föllmer and Schied (2002) and Frittelli and Gianin (2002) independently suggested to use a weaker "convexity" hypothesis to replace "positive homogeneity" and "subadditivity" hypothesis:

(5) *Convexity*: Assume $X, Y \in \mathcal{G}$, then

$$\forall \lambda \in (0, 1), \ \rho(\lambda X + (1 - \lambda)Y) \leqslant \lambda \rho(X) + (1 - \lambda)\rho(Y).$$

A risk measure having the convexity is called a *convex risk measure*. Since a convex risk measure ρ needs not to have positive homogeneity, so it needs not have $\rho(0) = 0$.

This section describes the main results of the convex risk measure in Föllmer and Schied (2004). These results can be proved by using the reasoning concerning the representation of co-monotonic convex risk measure in Song and Yan (2006) (see the proof of Theorems 10.15 and 10.16). So here the proof is omitted.

10.3.1 Representation: The Model-Free Case

We consider the case $\mathcal{G} = L^{\infty}(\Omega, \mathcal{F})$. Let $\alpha : \mathcal{M}_{1,f} \to \mathbb{R} \cup \{+\infty\}$ be a function such that

$$\inf_{\mathbb{Q} \in \mathcal{M}_{1,f}} \alpha(\mathbb{Q}) \in \mathbb{R}.$$

For every $\mathbb{Q} \in \mathcal{M}_{1,f}$, the function $X \to \mathbb{E}_{\mathbb{Q}}[X] - \alpha(\mathbb{Q})$ is convex, monotonic, and translation invariant. In taking the supremum of $\mathbb{Q} \in \mathcal{M}_{1,f}$, these properties are maintained. Therefore,

$$\rho(X) = \sup_{\mathbb{Q} \in \mathcal{M}_{1,f}} (\mathbb{E}_{\mathbb{Q}}[X] - \alpha(\mathbb{Q}))$$

define a convex risk measure on $\mathcal{G}$, and

$$\rho(0) = - \inf_{\mathbb{Q} \in \mathcal{M}_{1,f}} \alpha(\mathbb{Q}).$$

We call such a functional α on $\mathcal{M}_{1,f}$ a *penalty function* for representing convex risk measures.

Theorem 10.11 *Any convex risk measure ρ on $\mathcal{G}$ possesses the following form of representation:*

$$\rho(X) = \sup_{\mathbb{Q} \in \mathcal{M}_{1,f}} (\mathbb{E}_{\mathbb{Q}}[X] - \alpha_\rho(\mathbb{Q})), \ X \in \mathcal{G},$$

where

$$\alpha_\rho(\mathbb{Q}) = \sup_{X \in \mathcal{G}} (\mathbb{E}_{\mathbb{Q}}[X] - \rho(X)),$$

is the Fenchel-Legendre transformation (or conjugate function) of ρ, called the penalty function representing ρ. In addition, α_ρ is the smallest penalty function representing ρ, namely, for any penalty function α representing ρ,

$$\alpha(\mathbb{Q}) \geqslant \alpha_\rho(\mathbb{Q}), \ \ \forall \mathbb{Q} \in \mathcal{M}_{1,f}.$$

Theorem 10.12 *Assume that ρ is a convex risk measure which is continuous from above, and α is a penalty function on $\mathcal{M}_{1,f}$ representing ρ with $\alpha(\mathbb{Q}) < \infty$. Then $\mathbb{Q}$ is a probability measure, and*

$$\rho(X) = \sup_{\mathbb{Q} \in \mathcal{M}_1} (\mathbb{E}_{\mathbb{Q}}[X] - \alpha_\rho(\mathbb{Q})), \ X \in \mathcal{G}.$$

10.3.2   Representation: The Model-Dependent Case

In this part we assume that there is a probability model, namely, a probability measure is given. In this case, $\mathcal{G}(\mathbb{P}) = L^\infty(\Omega, \mathcal{F}, \mathbb{P})$.

Let $\mathcal{M}_{1,f}(\mathbb{P})$ denote the set of all finitely additive probability measures in $\mathcal{M}_{1,f}$ which are absolutely continuous w.r.t. $\mathbb{P}$. For any convex risk measure ρ, we put

$$\alpha_\rho(\mathbb{Q}) = \sup_{X \in \mathcal{G}(\mathbb{P})} (\mathbb{E}_{\mathbb{Q}}[X] - \rho(X)).$$

Then α_ρ is a penalty function.

Theorem 10.13 *Any convex risk measure ρ on $\mathcal{G}(\mathbb{P})$ possesses the following form of representation:*

$$\rho(X) = \sup_{\mathbb{Q} \in \mathcal{M}_{1,f}(\mathbb{P})} (\mathbb{E}_{\mathbb{Q}}[X] - \alpha_\rho(\mathbb{Q})), \ X \in \mathcal{G},$$

and α_ρ is the smallest penalty function representing ρ, namely, for any penalty function α representing ρ,

$$\alpha(\mathbb{Q}) \geqslant \alpha_\rho(\mathbb{Q}), \ \ \forall \mathbb{Q} \in \mathcal{M}_{1,f}(\mathbb{P}).$$

Theorem 10.14 *For a convex risk measure, the following two conditions are equivalent:*

(1) ρ *is continuous from above:* $X_n \searrow X \implies \rho(X_n) \searrow \rho(X)$.
(2) *in the representation formula of ρ, the smallest penalty function $\alpha_{\min}$ can be restricted on $\mathcal{M}_1(\mathbb{P})$:*

$$\rho(X) = \sup_{\mathbb{Q} \in \mathcal{M}_1(\mathbb{P})} (E_{\mathbb{Q}}(X) - \alpha_{\min}(\mathbb{Q})).$$

(3) ρ *has the "Fatou property": for any bounded sequence (X_n) which converges $\mathbb{P}$-a.s. to certain X, it holds*

$$\rho(X) \leqslant \liminf_{n \uparrow \infty} \rho(X_n).$$

10.4   Co-monotonic Convex Risk Measures

Song and Yan (2006) suggested to replace the convexity axiom in convex risk measures by a weaker co-monotonic convexity axiom:

(5)′ *co-monotonic convexity*: For all co-monotonic $X_1, X_2 \in \mathcal{G}$ and all $\lambda \in (0, 1)$, it holds

$$\rho(\lambda X + (1 - \lambda)Y) \leqslant \lambda\rho(X) + (1 - \lambda)\rho(Y),$$

and call a risk measure satisfying the monotonicity, translation invariance, positive homogeneity, and the co-monotonic convexity, a *co-monotonic convex risk measure*.

10.4.1   The Model-Free Case

In this part, we consider the case $\mathcal{G} = L^\infty(\Omega, \mathcal{F})$. Let $\alpha : \mathcal{M}_{1,m} \to \mathbb{R} \cup \{+\infty\}$ be any functional such that $\inf_{\mu \in \mathcal{M}_{1,m}} \alpha(\mu)$ is finite. If we define $\rho : \mathcal{G} \to \mathbb{R}$ as

$$\rho(X) := \sup_{\mu \in \mathcal{M}_{1,m}} (\mu(X) - \alpha(\mu)), \quad \forall X \in \mathcal{G},$$

then ρ is a co-monotonic convex risk measure.

The next theorem shows that any co-monotonic convex risk measure has the above representation.

Theorem 10.15 *Any co-monotonic convex risk measure $\rho : \mathcal{G} \to \mathbb{R}$ possesses the following form:*

$$\rho(X) = \max_{\mu \in \mathcal{M}_{1,m}} (\mu(X) - \alpha(\mu)), \quad \forall X \in \mathcal{G},$$

where

$$\alpha(\mu) = \sup_{\rho(X) \leqslant 0} \mu(X), \quad \mu \in \mathcal{M}_{1,m}.$$

Proof Without loss of generality, we assume $\rho(0) = 0$. First, we prove

$$\rho(X) \geqslant \sup_{\mu \in \mathcal{M}_{1,m}} (\mu(X) - \alpha(\mu)), \quad \forall X \in \mathcal{G}.$$

In fact, for any $X \in \mathcal{G}$, $X_1 = X - \rho(X)$, then $\rho(X_1) = 0$. Hence,

$$\alpha(\mu) \geqslant \mu(X_1) = \mu(X) - \rho(X), \quad \forall \mu \in \mathcal{M}_{1,m},$$

thereby the above conclusion holds. Therefore, in order to prove this theorem, we only need to prove that for any $X \in \mathcal{G}$, there is $\mu_X \in \mathcal{M}_{1,m}$ such that

$$\rho(X) \leqslant \mu_X(X) - \alpha(\mu_X).$$

By the translation invariance, it suffices to prove this conclusion for those X which satisfy $\rho(X) = 0$. For such a $X \in \mathcal{G}$, set

$$[X] := \{u(X) : u \text{ is continuous non-decreasing on } \mathbb{R}\},$$

and let

$$\mathcal{B} = \{Y \in \mathcal{G} : \exists Z \in [X] \text{ such that } \rho(Z) < 0, \ Y \leqslant Z\}.$$

It is easy to prove that $\mathcal{B}$ is convex, $\mathcal{B}_1 := \{Y \in \mathcal{G} : \| Y + 1 \| < 1\} \subset \mathcal{B}$, and $X \notin \mathcal{B}$. By convex set separation theorem, there exists a non-trivial $\lambda \in ba(\Omega, \mathcal{F})$, such that

$$b := \sup_{Y \in \mathcal{B}} \lambda(Y) \leqslant \lambda(X).$$

We claim:

(1) For any $Y \geqslant 0$, $Y \in \mathcal{G}$, it holds $\lambda(Y) \geqslant 0$;
(2) $\lambda(1) > 0$.

In fact, for any $Y \geqslant 0$, $Y \in \mathcal{G}$, and $c > 0$, we have $-1 - cY \in \mathcal{B}$ and $\lambda(-1 - cY) \leqslant \lambda(X)$. Since c is arbitrary, we obtain $\lambda(Y) \geqslant 0$. As for (2), because λ is non-trivial, there is a certain $Y \in \mathcal{G}$, $\|Y\| < 1$, such that $\lambda(Y) > 0$. Therefore, $\lambda(Y^+) > 0$, $\lambda(1 - Y^+) \geqslant 0$, from which it follows $\lambda(1) > 0$. Thus, we can choose λ such that $\lambda(1) = 1$.

Define $\rho^* : \mathcal{G} \to \mathbb{R}$ as

$$\rho^*(Y) := \sup\{\lambda(Z) : Z \leqslant Y, Z \in [X]\}.$$

Similar to the proof of Theorem 10.6, we can prove that ρ^* has monotonicity, positive homogeneity, co-monotonic additivity, and the following property:

$$\rho^*(Y) = \lambda(Y), \ \forall Y \in [X].$$

Therefore, by Greco's representation theorem, there is a monotonic set function μ_X on 2^Ω which represents ρ^*:

$$\rho^*(Y) = \mu_X(Y), \ \forall Y \in \mathcal{G}.$$

So we have

$$\mu_X(1) = \rho^*(1) = \lambda(1) = 1.$$

If $\rho(Y) \leqslant 0$, then for any $\varepsilon > 0$, it holds $\rho(Y - \varepsilon) < 0$. As a result, we get

$$\begin{aligned}
\mu_X[Y] - \varepsilon &= \mu_X(Y - \varepsilon) = \rho^*(Y - \varepsilon) \\
&= \sup\{\lambda(Z) : Z \leqslant Y - \varepsilon, Z \in [X]\} \\
&\leqslant \sup_{Z \in B} \lambda(Z) = b.
\end{aligned}$$

Finally, we obtain

$$\alpha(\mu_X) = \sup_{\rho(Y) \leqslant 0} \mu_X(Y) \leqslant b,$$

$$\mu_X(X) - \alpha(\mu_X) \geqslant \mu_X(X) - b = \lambda(X) - b \geqslant 0 = \rho(X).$$

The theorem is proved. $\square$

10.4.2 The Model-Dependent Case

In this part we assume that there is a probability model, namely, a probability measure is given. In this case, $\mathcal{G}(\mathbb{P}) = L^\infty(\Omega, \mathcal{F}, \mathbb{P})$.

Theorem 10.16 *If $\rho : \mathcal{G}(\mathbb{P}) \to \mathbb{R}$ is a co-monotonic convex risk measure, then ρ has the following form:*

$$\rho(X) = \max_{\mu \in \mathcal{M}_{1,m}(\mathbb{P})} (\mu(X) - \alpha(\mu)), \ X \in \mathcal{G}(\mathbb{P}),$$

where

$$\alpha(\mu) = \sup_{\rho(X) \leqslant 0} \mu(X), \ \mu \in \mathcal{M}_{1,m}(\mathbb{P}).$$

Proof Similar to the proof of Theorem 10.15, according to the fact that $ba(\Omega, \mathcal{F}, \mathbb{P})$ is the dual space of $L^\infty(\Omega, \mathcal{F}, \mathbb{P})$, we can prove this theorem. Below we give a proof which is based on Theorem 10.15.

We can regard ρ as a convex risk measure defined on $\mathcal{G} = L^\infty(\Omega, \mathcal{F})$. Then by Theorem 10.15, ρ has the following form:

$$\rho(X) = \max_{\mu \in \mathcal{M}_{1,m}} (\mu(X) - \alpha(\mu)), \quad X \in \mathcal{G},$$

where

$$\alpha(\mu) = \sup_{\rho(X) \leqslant 0} \mu(X), \quad \mu \in \mathcal{M}_{1,m}.$$

Hence, for any given $X \in \mathcal{G}$, there is $\mu_X \in \mathcal{M}_{1,m}$, such that $\mu_X(X) - \alpha(\mu_X) = \rho(X)$. Set

$$[X] := \{u(X) : u \text{ are continuous non-decreasing functions on } \mathbb{R}\},$$

and define $\Gamma : \mathcal{G} \to \mathbb{R}$ as

$$\Gamma(Y) := \sup\{\mu_X(Z) : Z \leqslant Y \text{ a.s. } Z \in [X]\}.$$

Similar to the proof of Theorem 10.9, we can prove that Γ has monotonicity, positive homogeneity, co-monotonic additivity, and the following properties:

$$\Gamma(Y) \leqslant \alpha(\mu_X) + \rho(Y), \ \forall Y \in \mathcal{G}; \quad \Gamma(Y) \geqslant \mu_X(Y), \ \forall Y \in [X]. \tag{10.7}$$

$$\Gamma(Y_1) = \Gamma(Y_2), \quad \text{if } Y_1, Y_2 \in \mathcal{G}, Y_1 = Y_2 \text{ a.s..} \tag{10.8}$$

Now we prove $\Gamma(1) = 1$. From (10.7) we get $\Gamma(1) \geqslant 1$. Therefore, it suffices to prove $\Gamma(1) \leqslant 1$. We claim that if $Y \in \mathcal{G}$, $Y = 0$ a.s., then $\mu_X(Y) \leqslant 0$. In fact, if there is a $Y \in \mathcal{G}$, $Y = 0$ a.s., such that $\mu_X(Y) > 0$, then for any $n \in \mathbb{N}$, it holds $\rho(nY) = 0$. So we have

$$\alpha(\mu_X) \geqslant \mu_X(nY) = n\mu_X(Y), \quad \forall n \in \mathbb{N}.$$

Thereby we get $\alpha(\mu_X) = +\infty$, which contradicts $\mu_X(X) - \alpha(\mu_X) = \rho(X)$. Therefore, for any $Z \in [X]$ satisfying $Z \leqslant 1$ a.s., we have

$$Z \vee 1 - 1 = 0 \text{ a.s., and } \mu_X(Z \vee 1 - 1) \leqslant 0.$$

Thus,

$$\mu_X(Z) \leqslant \mu_X(Z \vee 1) = \mu_X(Z \vee 1 - 1) + 1 \leqslant 1.$$

Therefore, it holds $\Gamma(1) \leqslant 1$.

By Greco's representation theorem, there is a co-monotonic set function μ_X^* on 2^Ω, such that

$$\forall Y \in \mathcal{G}, \ \mu_X^*(Y) = \Gamma(Y).$$

From (10.7) and (10.8) we know that $\mu_X^* \in \mathcal{M}_{1,m}(\rho)$ and $\mu_X^*(X) = \mu_X(X)$. We claim that $\alpha(\mu_X^*) \leqslant \alpha(\mu_X)$. In fact, for any $Y \in \mathcal{G}$ with $\rho(Y) \leqslant 0$, from (10.4) we get

$$\mu_X^*(Y) \leqslant \rho(Y) + \alpha(\mu_X) \leqslant \alpha(\mu_X).$$

Thus, $\alpha(\mu_X^*) \leqslant \alpha(\mu_X)$. The theorem is proved. $\square$

10.5   Law-Invariant Risk Measures

We assume that there is a probability model, namely, a probability measure $\mathbb{P}$ is given. In this case, $\mathcal{G}(\mathbb{P}) = L^\infty(\Omega, \mathcal{F}, \mathbb{P})$.

Definition 10.17 A monetary risk measure ρ on $\mathcal{G}(\mathbb{P})$ is a *law-invariant risk measure* if for any X and Y that have the same law, one has $\rho(X) = \rho(Y)$.

In this section we describe law-invariant risk measures and their representations.

10.5.1   Law-Invariant Coherent Risk Measures

Clearly, $\mathrm{VaR}_\lambda(X)$ is law-invariant. For $\lambda \in (0, 1]$, let

$$A\mathrm{VaR}_\lambda(X) = \frac{1}{\lambda} \int_0^\lambda \mathrm{VaR}_\gamma(X) d\gamma.$$

For $\lambda = 0$, we define

$$A\mathrm{VaR}_0(X) = \lim_{\lambda \to 0} A\mathrm{VaR}_\lambda(X) = \mathrm{VaR}_0(X) = \mathrm{ess\,sup}(X).$$

Hence, $A\mathrm{VaR}_\lambda$ (called the average VaR) is also a law-invariant risk measure.

Now we define the weighted VaR.

Definition 10.18 Let μ be a probability measure on $[0, 1]$. Put

$$W\mathrm{VaR}_\mu(X) = \int_{[0,1]} A\mathrm{VaR}_\lambda(X)\mu(d\lambda), \ \ \forall X \in \mathcal{G}(\mathbb{P}),$$

We call $W\mathrm{VaR}_\mu$ a *weighted VaR* w.r.t. μ.

It is easy to see that there is one-to-one correspondence between weighted VaRs and probability measures on $[0, 1]$: Assume that μ_1 and μ_2 are two probability measures on $\big([0, 1], \mathcal{B}([0, 1])\big)$, then we have

$$W\mathrm{VaR}_{\mu_1} = W\mathrm{VaR}_{\mu_2} \Longrightarrow \mu_1 = \mu_2.$$

Definition 10.19 Let X be a random variable, $\lambda \in (0, 1)$. The λ-*quantile* of X is any real q satisfying the following condition:

$$\mathbb{P}(X \leqslant q) \geqslant \lambda, \quad \mathbb{P}(X < q) \leqslant \lambda.$$

Set

$$q_X(t) = \inf\{x \in \mathbb{R} : F_X(x) \geqslant t\}, \quad \forall t \in (0, 1),$$

$$q_X^+(t) = \inf\{x \in \mathbb{R} : F_X(x) > t\}, \quad \forall t \in (0, 1),$$

where F_X is the distribution function of X. We call q_X and q_X^+ the *quantile function* and the *upper quantile function* of X, respectively.

We have

$$\mathrm{VaR}_\lambda(X) = q_X(1 - \lambda).$$

The next theorem shows that a continuous from above average VaR risk measure is a coherent risk measure and so is a continuous from above weighted VaR.

Theorem 10.20 *Let $\lambda > 0$. Then $A\mathrm{VaR}_\lambda$ is a continuous from above coherent risk measure and has the following representation:*

$$A\mathrm{VaR}_\lambda(X) = \max_{\mathbb{Q} \in \mathcal{Q}_\lambda} \mathbb{E}_{\mathbb{Q}}[X],$$

where

$$\mathcal{Q}_\lambda := \{\mathbb{Q} \ll \mathbb{P} : d\mathbb{Q}/d\mathbb{P} \leqslant 1/\lambda\}.$$

The proof of this theorem is based on the following version of the classical *Neyman-Pearson lemma*.

Lemma 10.21 *For a given $\lambda \in [0, 1]$ and $\mathbb{Q} \ll \mathbb{P}$, consider the optimization problem Maximize $\mathbb{E}_{\mathbb{Q}}[\psi]$, subject to $0 \leqslant \psi \leqslant 1$, $\mathbb{E}[\psi] = \lambda$. This problem has the following optimal solution:*

$$\psi^0 = 1_{[\varphi > q]} + \kappa \, 1_{[\varphi = q]},$$

where q is a $1 - \lambda$ quantile of $\varphi := d\mathbb{Q}/d\mathbb{P}$ w.r.t. $\mathbb{P}$, and the choice of constant κ is such that $\mathbb{E}[\psi^0] = \lambda$. Besides, any optimal solution to this problem ψ is equal to ψ^0 on $[\varphi \neq q]$, $\mathbb{P}$-a.s.

Proof Let ψ be a measurable function satisfying the constraint condition. Then $(\psi^0 - \psi)(\varphi - q) \geqslant 0$. Thus,

$$\mathbb{E}_\mathbb{Q}[\psi^0 - \psi] = \mathbb{E}[(\psi^0 - \psi)\varphi] \geqslant q\mathbb{E}[\psi^0 - \psi] = 0,$$

and ψ^0 is an optimal solution.

On the other hand, for any optimal solution ψ,

$$\mathbb{E}[(\psi^0 - \psi)(\varphi - q)] = \mathbb{E}_\mathbb{Q}[\psi^0 - \psi] - q\mathbb{E}[\psi^0 - \psi] = 0.$$

Hence, $(\psi^0 - \psi)(\varphi - q) = 0$, $\mathbb{P}$-a.s. Therefore, we have $\psi^0 = \psi$, $\mathbb{P}$-a.s. on $[\varphi \neq q]$.
□

Proof of Theorem 10.20 Let $\rho_\lambda(X) := \sup_{\mathbb{Q} \in \mathcal{Q}_\lambda} \mathbb{E}_\mathbb{Q}[X]$. Due to the translation invariance and positive homogeneity of ρ_λ and $A\text{VaR}_\lambda$, we only need to prove that for $X > 0$ and random variable X with $\mathbb{E}[X] = 1$, it holds that $\rho_\lambda(X) = A\text{VaR}_\lambda X$, where

$$\rho_\lambda(X) = \sup_{\mathbb{Q} \in \mathcal{Q}_\lambda} \mathbb{E}_\mathbb{Q}[X] = \frac{1}{\lambda} \sup\{\mathbb{E}[X\psi] \mid 0 \leqslant \psi \leqslant 1, \mathbb{E}[\psi] = \lambda\}.$$

By Lemma 10.21, $\rho_\lambda(X) = (1/\lambda)\mathbb{E}[X\psi^0]$, where $\psi^0 = 1_{[X>q]} + \kappa 1_{[X=q]}$, q is a $1 - \lambda$ quantile of X w.r.t. $\mathbb{P}$, and constant κ is such that $\mathbb{E}[\psi^0] = \lambda$. Therefore,

$$\rho_\lambda(X) = 1/\lambda\mathbb{E}[(X-q)^+] + q = A\text{VaR}_\lambda(X).$$
□

If probability space $(\Omega, \mathcal{F}, \mathbb{P})$ has no atom, Kusuoka (2001), for the first time, gives the following representation of law-invariant coherent risk measures.

Theorem 10.22 *If probability space $(\Omega, \mathcal{F}, \mathbb{P})$ has no atom, the following conditions are equivalent:*

(1) *ρ is a law-invariant coherent risk measure;*
(2) *There is a family $\mathcal{M}_0^*$ of probability measures on $(0, 1]$, such that*

$$\rho(X) = \sup_{\mu \in \mathcal{M}_0^*} W\text{VaR}_\mu = \sup_{\mu \in \mathcal{M}_0^*} \int_{(0,1]} A\text{VaR}_\alpha(X)\mu(d\alpha).$$

Now we define the risk measure generated by a distorted probability.

Definition 10.23 For certain non-decreasing function $g : [0, 1] \to [0, 1]$ satisfying $g(0) = 0$, and $g(1) = 1$ (called *distortion function*), we let $g \circ \mathbb{P}$ denote the composition of g and $\mathbb{P}$ and call $g \circ \mathbb{P}$ the *distorted probability*. If distortion function g is concave, we call $g \circ \mathbb{P}$ a *concave distorted probability*.

For a distorted probability $(g \circ \mathbb{P})$, put

$$\rho_g(X) = g \circ \mathbb{P}(X), \quad \forall X \in \mathcal{G}(\mathbb{P}),$$

where $g \circ \mathbb{P}(X)$ is the Choquet integral of X w.r.t. $g \circ \mathbb{P}$. Then we have

$$(g \circ \mathbb{P})(X) = \int_0^1 q_X(1 - x)dg(x) = \int_0^1 q_X(t)d\gamma(t),$$

where $\gamma(t) = 1 - g(1 - t)$.

Now we investigate the relationship between risk measures produced by distorted probabilities and weighted VaRs. First we prove a lemma (from Föllmer and Schied 2004).

Lemma 10.24 *The density*

$$g'_+(t) = \int_{(t,1]} s^{-1}\mu(ds), \quad 0 < t < 1,$$

defines a one-to-one correspondence between probability measures μ on [0, 1] and distortion functions g. In addition, it holds $g(0+) = \mu(\{0\})$.

Proof Given a probability measure μ on [0, 1]. We define an increasing function g on $(0, 1)$ by using the density g'_+ and let $g(1) = 1$. Then we have

$$1 - g(0+) = \int_0^1 g'_+(t)dt = \int_{(0,1]} s^{-1} \int_0^1 I_{(t,1]}(s)\mu(ds) = \mu((0, 1]) \leqslant 1.$$

Hence, we can let $g(0) = 0$, then g is a distortion function, and $g(0+) = \mu(\{0\})$.

Conversely, given a concave distortion function g, since the right derivative $g'_+(t)$ of g is a non-increasing right-continuous function, there is a locally finite positive measure μ on $(0,1]$, such that $\nu((t, 1]) = g'_+(t)$. Define a measure $\mu : \mu(ds) = s\nu(ds)$ on $(0,1]$, then by Fubini's theorem, it holds

$$\mu((0, 1]) = \int_0^1 \int_{(0,1]} I_{[t<s]}\nu(ds)dt = 1 - g(0+) \leqslant 1.$$

Thus, letting $\mu(\{0\}) = g(0+)$, we obtain a probability measure μ on [0, 1]. □

The following theorem shows that if probability space $(\Omega, \mathcal{F}, \mathbb{P})$ has no atom, the weighted VaRs are totally identical to the concave distorted Choquet integrals with $g(0+) = 0$.

Theorem 10.25 *We assume that probability space $(\Omega, \mathcal{F}, \mathbb{P})$ has no atom. Let $\rho :$ $\mathcal{G}(\mathbb{P}) \to \mathbb{R}$; then the following conditions are equivalent:*

(1) *there is a probability measure μ on $(0, 1]$, such that $\rho = WVaR_\mu$;*
(2) *there is a concave distortion function g with $g(0+) = 0$, such that $\rho = \rho_g$.*

Proof First we prove $(1) \Rightarrow (2)$. Assume (1) holds, then

$$\rho_\mu(X) = \int_{(0,1]} A\text{VaR}_\lambda(X)\mu(d\lambda)$$

$$= \int_{(0,1]} \frac{1}{\lambda} \int_0^\lambda \text{VaR}_\gamma(X)d\gamma\,\mu(d\lambda)$$

$$= \int_{(0,1]} \text{VaR}_\gamma(X) \int_{(\gamma,1]} \frac{1}{\lambda}\mu(d\lambda)d\gamma.$$

Let $g(t) = \int_0^t \int_{(\gamma,1]} \frac{1}{\lambda}\mu(d\lambda)d\gamma$, then g is a nondecreasing concave function, and satisfies $g(0+) = 0$, $g(1) = 1$. The right derivative of g is $g\prime_+(t) = \int_t^1 s^{-1}\mu(ds)$. Thus, we have

$$\rho_\mu(X) = \int_{(0,1]} \text{VaR}_\gamma(X)g\prime_+(\gamma)d\gamma$$

$$= \int_{[0,1)} \left[\int_0^\infty 1_{[F_X(x) \leqslant \gamma]}dx + \int_{-\infty}^0 (1_{[F_X(x) \leqslant \gamma]} - 1)dx \right]g\prime_+(1 - \gamma)d\gamma$$

$$= \int_0^\infty \int_{[0,1)} 1_{[F_X(x) \leqslant \gamma]}g\prime_+(1 - \gamma)d\gamma dx$$

$$+ \int_{-\infty}^0 \int_{[0,1)} (1_{[F_X(x) \leqslant \gamma]} - 1)g\prime_+(1 - \gamma)d\gamma dx$$

$$= \int_0^\infty g(1 - F_X(x))dx + \int_{-\infty}^0 [g(1 - F_X(x)) - 1]dx$$

$$= g \circ P(X).$$

Now we prove $(2) \Rightarrow (1)$. For a concave distortion function g with $g(0+) = 0$, put

$$\mu((0, t]) := -\int_0^t sdg\prime_+(s).$$

Similar to the above derivation, we can get $(2) \Rightarrow (1)$.   □

Later, we will denote the set of all distortion functions on $[0, 1]$ by $\mathcal{D}$ and denote the set of all concave distortion function on $[0, 1]$ by $\mathcal{D}^{cc}$.

Since a probability measure μ on $[0, 1]$ corresponds one-to-one to a concave distortion function g, from Theorem 10.22 we know that a law-invariant coherent risk measure can be represented as the supremum of a family of concave distorted Choquet integrals with $g(0+) = 0$:

$$\rho(X) = \sup_{g \in \mathcal{D}^*} g \circ \mathbb{P}(X),$$

where $\mathcal{D}^*$ is the family of concave distortion functions with $g(0+) = 0$.

The next theorem (see Föllmer and Schied (2004) for the proof) characterizes a relationship between $A\text{VaR}_\lambda$ and VaR_λ, which reflects the importance of $A\text{VaR}_\lambda$ as a special law-invariant coherent risk measure.

Theorem 10.26 *$A\text{VaR}_\lambda$ is the smallest law-invariant coherent risk measure which dominates VaR_λ.*

10.5.2 Law-Invariant Convex Risk Measures

Föllmer and Schied (2004) first investigate the law-invariant convex risk measures (see also Dana 2005). Below we introduce their main results about the representation of the law-invariant convex risk measures. The proof is omitted.

The next theorem is from Föllmer and Schied (2004) and Dana (2005).

Theorem 10.27 *Let ρ be a convex risk measure and assume that ρ is continuous from above. Then ρ is law-invariant if and only if for $Q \in \mathcal{M}_1(\mathbb{P})$, the smallest penalty function $\alpha_{\min}(Q)$ depends only on the law of $\varphi_Q := dQ/d\mathbb{P}$ under $\mathbb{P}$. In this case, ρ has the following representation:*

$$\rho(X) = \sup_{Q \in \mathcal{M}_1(\mathbb{P})} \left(\int_0^1 q_X(t) q_{\varphi_Q}(t) dt - \alpha_{\min}(Q) \right),$$

where the smallest penalty function satisfies

$$\alpha_{\min}(Q) = \sup_{X \in \mathcal{A}_\rho} \int_0^1 q_X(t) q_{\varphi_Q}(t) dt$$

$$= \sup_{X \in \mathcal{G}(\mathbb{P})} \left(\int_0^1 q_X(t) q_{\varphi_Q}(t) dt - \rho(X) \right).$$

10.5.3   Some Results About Stochastic Orders and Quantiles

The definitions of the first- and second-order stochastic dominance are given in Chap. 2. We recall them for reader's convenience. Let X and Y be two random variables. We say that X dominates Y in the sense of first-order stochastic dominance, if for any $x \in \mathbb{R}$, we have $F_X(x) \leqslant F_Y(x)$, where F_X is the distribution function of X. We say that X dominates Y in the sense of second-order stochastic dominance, if for any $t \in \mathbb{R}$, we have

$$\int_{-\infty}^{t} F_X(s)ds \leqslant \int_{-\infty}^{t} F_Y(s)ds.$$

Now we define another stochastic dominance: the stop-loss order.

Definition 10.28 Let X and Y be two real random variables. We say Y is dominating X in the sense of *stop-loss order*, denoted as $X \leqslant_{sl} Y$, if for all $d \in \mathbb{R}$, we have $\mathbb{E}[(X - d)_+] \leqslant \mathbb{E}[(Y - d)_+]$.

From Theorem 2.13 we know that Y is better than X in the sense of stop-loss order, is equivalent to that $-X$ is better than $-Y$ in the sense of second-order stochastic dominance.

The following two theorems give the characterizations of the first-order stochastic dominance and the stop-loss order, respectively. For proofs see Föllmer and Schied (2004) and Dhaene et al. (2006), respectively.

Theorem 10.29 $X \leqslant_{st} Y$ *if and only if there are a probability space* $(\Omega', \mathcal{F}', \mathbb{P}')$ *and two random variables* X', Y' *on it, such that* X' *and* X *have the same law,* Y' *and* Y *have the same law, and*

$$X' \leqslant Y', \quad \mathbb{P}'\text{-}a.s.$$

$X \leqslant_{sl} Y$ *if and only if there are a probability space* $(\Omega', \mathcal{F}', \mathbb{P}')$ *and two random variables* X' *and* Y' *on it, such that* X' *and* X *have the same law,* Y' *and* Y *have the same law, and*

$$\mathbb{E}_{\mathbb{P}'}[Y'|X'] \geqslant X'.$$

Theorem 10.30 $X \leqslant_{st} Y$ *if and only if for all distortion functions* g,

$$g \circ \mathbb{P}(X) \leqslant g \circ \mathbb{P}(Y).$$

$X \leqslant_{sl} Y$ *if and only if for all concave distortion functions* g,

$$g \circ \mathbb{P}(X) \leqslant g \circ \mathbb{P}(Y).$$

Proof of the next lemma is left to the reader.

Lemma 10.31 *Let Y be a nonnegative random variable. If there is an increasing function f on $[0, \infty)$ with $X = f(Y)$, then*

$$q_X(t) = f(q_Y(t)), \text{ for a.e. } t \in (0, 1),$$

where $q_X(t)$ and $q_Y(t)$ are right-inverse functions of F_X and F_Y. If f is continuous, then

$$q_X(t) = f(q_Y(t)), \ \forall t \in (0, 1).$$

Lemma 10.32 *Let $X_1, X_2 \in \mathcal{G}(\mathbb{P})$. If X_1 and X_2 are co-monotonic, then*

$$q_{X_1+X_2}(t) = q_{X_1}(t) + q_{X_2}(t), \quad \forall t \in (0, 1).$$

Proof It suffices to prove the lemma for the case where X_1 and X_2 are nonnegative. Since X_1 and X_2 are co-monotonic, there is a non-decreasing continuous function u_1, u_2 on $\mathbb{R}$, such that $u_1(z) + u_2(z) = z$ and $X_i = u_i(X_1 + X_2), i = 1, 2$. By Lemma 10.31,

$$u_i(q_{X_1+X_2}(t)) = q_{X_i}(t), \quad \forall t \in (0, 1), i = 1, 2.$$

Thus, for all $t \in (0, 1)$, we have

$$q_{X_1+X_2}(t) = u_1(q_{X_1+X_2}(t)) + u_2(q_{X_1+X_2}(t)) = q_{X_1}(t) + q_{X_2}(t). \qquad \square$$

The next lemma is from Dhaene et al. (2006).

Lemma 10.33 *Assume that X, Y, X^c, and Y^c are random variables, defined on the same probability space $(\Omega, \mathcal{F}, \mathbb{P})$. If X^c and X have the same law, Y^c and Y have the same law, and X^c, Y^c are co-monotonic, then $X + Y \leqslant_{sl} X^c + Y^c$.*

Proof By Theorem 10.30, it suffices to show that for any concave distortion function g,

$$g \circ \mathbb{P}(X + Y) \leqslant g \circ \mathbb{P}(X^c + Y^c).$$

In fact, by the co-monotonic additivity of the Choquet integral we have

$$\begin{aligned} g \circ \mathbb{P}(X + Y) &\leqslant g \circ \mathbb{P}(X) + g \circ \mathbb{P}(Y) \\ &= g \circ \mathbb{P}(X^c) + g \circ \mathbb{P}(Y^c) \\ &= g \circ \mathbb{P}(X^c + Y^c). \end{aligned}$$

The theorem is proved. $\qquad \square$

The next lemma is from Song and Yan (2009a).

Lemma 10.34 *Let $X_1, X_2, Y_1, Y_2 \in \mathcal{G}(\mathbb{P})$. Assume X_1 and X_2 are co-monotonic, Y_1 and Y_2 are co-monotonic. If $X_1 \leqslant_{st} Y_1$ and $X_2 \leqslant_{st} Y_2$, then*

$$X_1 + X_2 \leqslant_{st} Y_1 + Y_2;$$

if $X_1 \leqslant_{sl} Y_1$, and $X_2 \leqslant_{sl} Y_2$, then

$$X_1 + X_2 \leqslant_{sl} Y_1 + Y_2.$$

Proof By definition, $X \leqslant_{st} Y$ if and only if for all $t \in (0, 1)$, $q_X(t) \leqslant q_Y(t)$. By Lemma 10.32, for all $t \in (0, 1)$,

$$q_{(X_1+X_2)}(t) = q_{X_1}(t) + q_{X_2}(t) \leqslant q_{Y_1}(t) + q_{Y_2}(t) = q_{(Y_1+Y_2)}(t).$$

Hence,

$$X_1 + X_2 \leqslant_{st} Y_1 + Y_2.$$

As for the case of stop-loss order, noting that $X \leqslant_{sl} Y$ if and only if

$$\int_t^1 q_X(s)ds \leqslant \int_t^1 q_Y(s)ds \quad \forall t \in (0, 1).$$

By Lemma 10.32, for all $t \in (0, 1)$,

$$
\begin{aligned}
\int_t^1 q_{(X_1+X_2)}(s)ds &= \int_t^1 q_{X_1}(s)ds + \int_t^1 q_{X_2}(s)ds \\
&\leqslant \int_t^1 q_{Y_1}(s)ds + \int_t^1 q_{Y_2}(s)ds \\
&= \int_t^1 q_{(Y_1+Y_2)}(s)ds.
\end{aligned}
$$

Hence, $X_1 + X_2 \leqslant_{sl} Y_1 + Y_2$. $\qquad\qquad\qquad\qquad\qquad\qquad\qquad\qquad\qquad\qquad\square$

10.5.4   Law-Invariant Co-monotonic Subadditive Risk Measures

Let $\rho : \mathcal{G}(\mathbb{P}) \to \mathbb{R}$. If $X \leqslant_{st} Y$ implies $\rho(X) \leqslant \rho(Y)$, then we say that ρ satisfies the first-order stochastic dominance. It is easy to prove: the fact that ρ satisfies the first-order stochastic dominance implies monotonicity and law-invariance of ρ. Conversely, if the probability space $(\Omega, \mathcal{F}, \mathbb{P})$ has no atom, then monotonicity and law-invariance of ρ imply that ρ satisfies the first-order stochastic dominance. In fact, in this case there is a random variable U on the probability space which is uniformly distributed on $(0, 1)$. If $X \leqslant_{st} Y$, then for all $x \in \mathbb{R}$, it holds $F_X(x) \geqslant F_Y(x)$. Thus, for all $t \in (0, 1)$, we have $q_X(t) \leqslant q_Y(t)$, and consequently, $q_X(U) \leqslant q_Y(U)$, $\mathbb{P}$-a.s. Noting that X and $q_X(U)$ have the same law and that Y and $q_Y(U)$ have the same law, we have

$$\rho(Y) = \rho(q_Y(U)) \leqslant \rho(q_X(U)) = \rho(X).$$

Therefore, ρ satisfies the first-order stochastic dominance.

A law-invariant coherent risk measure satisfies the first-order stochastic dominance. If a premium principle can represent a distorted Choquet integral, then it satisfies naturally the first-order stochastic dominance. Such a premium principle is very common in insurance, called *distorted premium principle*.

The results of this section are mainly from Song and Yan (2009a).

Theorem 10.35 *Let $\mathcal{M}^{st}(\mathbb{P})$ be the set of those $\mu \in \mathcal{M}_{1,m}(\mathbb{P})$, which satisfy the following properties:*

$$X \leqslant_{st} Y \Longrightarrow \mu(X) \leqslant \mu(Y).$$

Then $\mathcal{M}^{st}(\mathbb{P}) = \{g \circ \mathbb{P} : g \in \mathcal{D}\}$.

Proof If there is a $g \in \mathcal{D}$ with $\mu = g \circ \mathbb{P}$, then by Theorem 10.30, we know $\mu \in \mathcal{M}^{st}(\mathbb{P})$. Now assume $\mu \in \mathcal{M}^{st}(\mathbb{P})$. For $A, B \in \mathcal{F}$ with $\mathbb{P}(A) = \mathbb{P}(B)$, we have $\mu(A) = \mu(B)$. In fact, I_A and I_B have the same law, and $\mu \in \mathcal{M}^{st}(\mathbb{P})$ is law-invariant. Hence, for $A \in \mathcal{F}$, $\mu(A)$ is uniquely determined by $\mathbb{P}(A)$. Thus, let $E = \{\mathbb{P}(A) : A \in \mathcal{F}\}$, we can define a map $g : E \to \mathbb{R}$ as follows:

$$\forall x \in E, \ g(x) = \mu(A), \ \text{for certain } A \in \mathcal{F}, \ \text{such that } \mathbb{P}(A) = x.$$

For $x, y \in E$, $x \leqslant y$, by the definition of E, there are $A, B \in \mathcal{F}$, such that $\mathbb{P}(A) = x \leqslant y = \mathbb{P}(B)$. Thus, $1_A \leqslant_{st} 1_B$, and consequently, $g(x) = \mu(A) \leqslant \mu(B) = g(y)$, namely, g is an increasing function on E. Therefore, g can be extended to an increasing function on the closure of E and further extended to an increasing function on $[0, 1]$, such that $g(0) = 0$, $g(1) = 1$ and $\mu = g \circ \mathbb{P}$. $\square$

Remark By Theorem 10.35, a functional $\rho : \mathcal{G}(\mathbb{P}) \to \mathbb{R}$ satisfying the translation invariance, positive homogeneity, co-monotonic additivity, and the first-order stochastic dominance can be represented as a distorted Choquet integral.

Let $\mathcal{H}$ be any subset of $\mathcal{D}$. If we define $\rho : \mathcal{G} \to \mathbb{R}$ by

$$\rho(X) = \sup_{g \in \mathcal{H}} (g \circ \mathbb{P})(X),$$

then, obviously, ρ satisfies the translation invariance, positive homogeneity, co-monotonic sub-additivity, and the first-order stochastic dominance. The following theorem shows that the converse is also true.

Theorem 10.36 *If* $\rho : \mathcal{G}(\mathbb{P}) \to \mathbb{R}$ *satisfies the translation invariance, positive homogeneity, co-monotonic sub-additivity, and the first-order stochastic dominance, then* ρ *can be represented as*

$$\rho(X) = \max_{g \in \mathcal{D}^\rho} (g \circ \mathbb{P})(X),$$

where

$$\mathcal{D}^\rho = \{g \in \mathcal{D} : (g \circ \mathbb{P})(Y) \leqslant \rho(Y), \quad \forall Y \in \mathcal{G}(\mathbb{P})\}.$$

Proof By Theorem 10.35, in order to prove this theorem, it suffices to prove that ρ has the following representation:

$$\rho(X) = \max_{\mu \in \mathcal{M}^\rho(\mathbb{P})} \mu(X),$$

where

$$\mathcal{M}^\rho(\mathbb{P}) = \{\mu \in \mathcal{M}^{st}(\mathbb{P}) : \mu(Y) \leqslant \rho(Y), \forall Y \in \mathcal{G}(\mathbb{P})\}.$$

To this end, we only need to prove that for any $X \in \mathcal{G}(\mathbb{P})$, there exists $\mu_X \in \mathcal{M}^\rho(\mathbb{P})$ such that $\mu_X(X) = \rho(X)$. By the translation invariance of ρ and μ, we only need to consider those X with $\rho(X) = 1$. For such a X, let

$$[X] := \{u(X) : u \text{ are continuous increasing functions on } \mathbb{R}\},$$

and put

$$\mathcal{B} = \{Y \in \mathcal{G}(\mathbb{P}) : \exists Z \in [X] \text{ such that } \rho(Z) < 1, \text{ and } Y \leqslant Z \text{ a.s.}\}.$$

It is easy to verify that $\mathcal{B}$ is convex, $\mathcal{B}_1 := \{Y \in \mathcal{G}(\mathbb{P}) : \|Y\| < 1\} \subset \mathcal{B}$, and $X \notin \mathcal{B}$. By the convex separation theorem, there is a non-trivial $\lambda \in ba(\Omega, \mathcal{F}, \mathbb{P})$, such that

$$\sup_{Y \in \mathcal{B}} \lambda(Y) \leqslant \lambda(X).$$

Here $ba(\Omega, \mathcal{F}, \mathbb{P})$ represents the space constructed by all finitely additive measures with finite variation which are absolutely continuous w.r.t. $\mathbb{P}$. Since $\mathcal{B}_1 \subset \mathcal{B}$, we have $\lambda(X) > 0$. Hence, we can choose λ such that $\lambda(X) = 1$. We claim that λ has the following properties:

$1°$ for any $Y \geqslant 0$, $Y \in \mathcal{G}(\mathbb{P})$, $\lambda(Y) \geqslant 0$;
$2°$ $\lambda(1) = 1$;
$3°$ for any $Y \in [X]$, $\lambda(Y) \leqslant \rho(Y)$.

In fact, for any $Y \geqslant 0, Y \in \mathcal{G}(\mathbb{P}), c > 0$, we have $-cY \in \mathcal{B}$. Then, $\lambda(-cY) \leqslant \lambda(X) = 1$. Since c is arbitrary, we get $1°$.

For any $0 < c < 1$, clearly, $c \in \mathcal{B}$. Since $c < 1$ is arbitrary, this implies $\lambda(1) \leqslant 1$. On the other hand, for any $c > 1$, we have

$$2 - c\lambda(1) = \lambda(2X - c) \leqslant \lambda(X) = 1,$$

which implies $\lambda(1) \geqslant 1/c$. Since $c > 1$ is arbitrary, we obtain $\lambda(1) \geqslant 1$. Thus $2°$ is proved.

Finally, for any $Y \in [X]$, let $Y_1 = Y - \rho[Y] + 1$, then for $c > 1$, we have $Y_1/c \in \mathcal{B}$. Since $c > 1$ is arbitrary, this implies $\lambda(Y_1) \leqslant 1$. Therefore, we get $\lambda(Y) \leqslant \rho(Y)$, and $3°$ is proved.

We define $\rho_* : \mathcal{G}(\mathbb{P}) \to \mathbb{R}$ by

$$\rho_*(Y) := \sup\{\lambda(Z) : Z \leqslant_{st} Y, Z \in [X]\}, \quad Y \in \mathcal{G}(\mathbb{P}).$$

Obviously, ρ_* satisfies the first-order stochastic dominance, positive homogeneity, and the following property:

$$\rho_*(Y) \leqslant \rho(Y), \ \forall Y \in \mathcal{G}(\mathbb{P}); \quad \rho_*(Y) \geqslant \lambda(Y), \ \forall Y \in [X]. \tag{10.9}$$

We claim that ρ_* also has the co-monotonic additivity. In fact, let $Y_1, Y_2 \in \mathcal{G}(\mathbb{P})$ be co-monotonic, then there are continuous increasing functions u, v on $\mathbb{R}$ such that $u(z) + v(z) = z, z \in \mathbb{R}$, and

$$Y_1 = u(Y_1 + Y_2), \quad Y_2 = v(Y_1 + Y_2) \text{ a.s.}.$$

For any $Z \in [X]$ with $Z \leqslant_{st} Y_1 + Y_2$, we have $u(Z) \in [X]$, $v(Z) \in [X]$, and

$$u(Z) \leqslant_{st} u(Y_1 + Y_2) = Y_1, \quad v(Z) \leqslant_{st} v(Y_1 + Y_2) = Y_2.$$

Thus,

$$\rho_*(Y_1) + \rho_*(Y_2) \geqslant \lambda(u(Z)) + \lambda(v(Z)) = \lambda(Z).$$

By the definition of ρ_*, we obtain

$$\rho_*(Y_1 + Y_2) \leqslant \rho_*(Y_1) + \rho_*(Y_2).$$

On the other hand, for all $Z_1, Z_2 \in [X]$ satisfying $Z_1 \leqslant_{st} Y_1$, $Z_2 \leqslant_{st} Y_2$, by Lemma 10.34 we obtain $Z_1 + Z_2 \leqslant_{st} Y_1 + Y_2$. Hence,

$$\lambda(Z_1) + \lambda(Z_2) = \lambda(Z_1 + Z_2) \leqslant \rho_*(Y_1 + Y_2).$$

By the definition of ρ_*, we obtain

$$\rho_*(Y_1 + Y_2) \geqslant \rho_*(Y_1) + \rho_*(Y_2).$$

Therefore, ρ_* has the co-monotonic additivity.

By Theorem 10.9, there exists $\mu_X \in \mathcal{M}_{1,m}(\mathbb{P})$, such that for all $Y \in \mathcal{G}(\mathbb{P})$, we have $\mu_X(Y) = \rho_*(Y)$. By (10.9), $\mu_X \in \mathcal{M}(\mathbb{P})$, and

$$\mu_X(X) = \rho_*(X) = \lambda(X) = \rho(X).$$

The theorem is proved.   □

Remark Heyde et al. (2006) introduced the concept of so-called natural risk statistic. A natural risk statistic ρ is defined on the observed data $x = (x_1, x_2, \cdots, x_n) \in \mathbb{R}^n$ of random variables and satisfies the monotonicity, positive homogeneity, translation invariance, co-monotonic subadditivity, and permutation invariance for any permutation $(i_1, \cdots, i_n)$,

$$\rho((x_1, \cdots, x_n)) = \rho((x_{i_1}, \cdots, x_{i_n})).$$

They proved that any natural risk statistics ρ can be expressed as the supremum of a family of weighted averages.

With our notations we can describe their problem as follows. Let $\Omega = \{1, 2, \cdots, n\}$;

$$\mathcal{F} = 2^\Omega; \quad \mathbb{P}(i) = 1/n, i = 1, 2, \cdots, n; \quad \mathcal{G}(\mathbb{P}) = \{X | X : \Omega \to \mathbb{R}\}.$$

Here, $X, Y \in \mathcal{G}(\mathbb{P})$, $X \leqslant_{st} Y$ imply that there is a $\widetilde{Y} \in \mathcal{G}(\mathbb{P})$, $\widetilde{Y} =^d Y$, such that $X \leqslant \widetilde{Y}$. In fact, assume $X_1 \leqslant X_2 \leqslant \cdots \leqslant X_n$, $Y_{\sigma(1)} \leqslant Y_{\sigma(2)} \leqslant \cdots \leqslant Y_{\sigma(n)}$. Define $\widetilde{Y}_i = Y_{\sigma(i)}, i = 1, 2, \cdots, n$, then $\widetilde{Y}$ satisfies required conditions.

In this framework, ρ can be regarded as a functional on $\mathcal{G}(\mathbb{P})$ possessing the monotonicity, positive homogeneity, translation invariance, co-monotonic subadditivity, and the law-invariance. From the above discussion, in this framework, monotonicity and law-invariance are equivalent to first-order stochastic dominance. Therefore, ρ satisfies the first-order stochastic dominance, positive homogeneity, translation invariance, and the co-monotonic subadditivity. Accordingly, their result about the representation of the natural risk statistics is a special case of Theorem 10.36.

Below we establish a relationship between the concave distortion and the stop-loss order. Theorem 10.30 implies the following conclusion.

Theorem 10.37 *Let g be a concave distortion function. If $X \leqslant_{sl} Y$, then*

$$g \circ \mathbb{P}(X) \leqslant g \circ \mathbb{P}(Y).$$

The next theorem shows that if probability space $(\Omega, \mathcal{F}, \mathbb{P})$ has no atom, the converse of the above theorem is also true.

Theorem 10.38 *Assume that probability space $(\Omega, \mathcal{F}, \mathbb{P})$ has no atom. Let $\mathcal{M}^{sl}(\mathbb{P})$ denote the set of those $\mu \in \mathcal{M}_{1,m}(\mathbb{P})$ satisfying the following property:*

$$X \leqslant_{sl} Y \implies \mu(X) \leqslant \mu(Y).$$

Then

$$\mathcal{M}^{sl}(\mathbb{P}) = \{g \circ \mathbb{P} : g \in \mathcal{D}, \ g \text{ is concave}\}.$$

All elements of $\mathcal{M}^{sl}(\mathbb{P})$ are called concave distortions.

Proof Because $X \leqslant_{st} Y$ implies $X \leqslant_{sl} Y$, by Theorem 10.37, there is a $g \in \mathcal{D}$, such that $\mu = g \circ \mathbb{P}$. We only need to prove that g is concave. We will prove it by contradiction. Suppose that g is not concave. Then there are $0 \leqslant a < b < c \leqslant 1$, such that

$$(1 - \alpha)g(a) + \alpha g(c) - g(b) > 0,$$

where $\alpha = (b - a)/(c - a)$. Since probability space $(\Omega, \mathcal{F}, \mathbb{P})$ has no atom, we can take $A, B, C \in \mathcal{F}$ such that $A \subset B \subset C$ and

$$\mathbb{P}(A) = a, \quad \mathbb{P}(B) = b, \quad \mathbb{P}(C) = c.$$

Put

$$X = 3I_A + (1 + \alpha)I_{C \setminus A}, \quad Y = 3I_A + 2I_{B \setminus A} + I_{C \setminus B}.$$

We claim that $X \leqslant_{sl} Y$. In fact, we only need to verify that for any $d \in [1, 1 + \alpha]$, we have

$$\mathbb{E}[(X - d)_+] \leqslant \mathbb{E}[(Y - d)_+].$$

This is true, because for $d \in [1, 1 + \alpha]$, we have

$$\mathbb{E}[(Y - d)_+] - \mathbb{E}[(X - d)_+] = (d - 1)(c - b) \geqslant 0.$$

On the other hand, by the definition of Choquet integral, we have

$$(g \circ \mathbb{P})(X) - (g \circ \mathbb{P})(Y) = (2 - \alpha)g(a) + (1 + \alpha)g(c) - (g(a) + g(b) + g(c))$$

$$= (1 - \alpha)g(a) + \alpha g(c) - g(b) > 0,$$

which contradicts the assumption. Hence, g is concave. □

Remark By Theorem 10.38, a functional $\rho : \mathcal{G}(\mathbb{P}) \to \mathbb{R}$ satisfying the translation invariance, positive homogeneity, co-monotonic additivity, and stop-loss order can be represented as a concave distorted Choquet integral.

The Stop-loss order represents a common preference of risk adverse decision-makers. Thus, the translation invariance, homogeneity, co-monotonic sub-additivity, and the stop-loss order are very natural requirements for risk measures.

Now we demonstrate that a functional $\rho : \mathcal{G}(\mathbb{P}) \to \mathbb{R}$ satisfying translation invariance, positive homogeneity, co-monotonic sub-additivity, and stop-loss order can be represented as the supremum of a family of concave distortions. $\rho : \mathcal{G}(\mathbb{P}) \to \mathbb{R}$. To this end, we first prove a lemma.

Lemma 10.39 *Let g be an increasing function on $[0, 1]$ with $g(0) = 0$, $g(1) = 1$, and let $\bar{g}$ be the smallest concave function dominating g. Then $\bar{g}$ is also an increasing function on $[0, 1]$ with $\bar{g}(0) = 0$, $\bar{g}(1) = 1$, and satisfies the following property: $\forall b \in (0, 1)$, $\varepsilon > 0$, $\exists 0 \leqslant a < b < c \leqslant 1$, such that*

$$\bar{g}(b) - \varepsilon \leqslant (1 - \alpha)g(a) + \alpha g(c),$$

where $\alpha = (b - a)/(c - a)$.

Proof We just need to prove the last assertion. We prove it by contradiction. Assume the assertion does not hold. Then there is $b \in (0, 1)$, $\varepsilon > 0$, such that for any $0 \leqslant a < b < c \leqslant 1$,

$$A := \bar{g}(b) - \varepsilon > (1 - \alpha)g(a) + \alpha g(c),$$

where $\alpha = (b - a)/(c - a)$. Let

$$f_1(x) = Ax/b, \quad f_2(x) = (1 - A)(x - b)/(1 - b) + A.$$

Then $f_1 \wedge f_2$ is concave, and $g \leqslant f_1 \vee f_2$. Put

$$G = \{(x, y) : x \in [0, 1], f_1 \wedge f_2(x) < y \leqslant g(x)\},$$

and

$$k = \sup\{(y - A)/(x - b) : (x, y) \in G\} \vee (1 - A)/(1 - b).$$

Then $k \leqslant A/b$, and

$$f := k(x - b) + A \geqslant g(x).$$

Consequently, $f \geqslant \bar{g}$. However, we have $f(b) = A < \bar{g}(b)$. This leads to a contradiction. $\qquad\square$

Theorem 10.40 *We assume that probability space $(\Omega, \mathcal{F}, \mathbb{P})$ has no atom. If $\rho : \mathcal{G}(\mathbb{P}) \to \mathbb{R}$ satisfies the translation invariance, positive homogeneity, co-monotonic subadditivity, and the stop-loss order, then ρ has the following representation:*

$$\rho(X) = \max_{g \in \mathcal{D}^{\rho,cc}} (g \circ \mathbb{P})(X),$$

where

$$\mathcal{D}^{\rho,cc} = \{g \in \mathcal{D}^{cc} : (g \circ \mathbb{P})(Y) \leqslant \rho(Y), \forall Y \in \mathcal{G}(\mathbb{P})\},$$

$$\mathcal{D}^{cc} = \{g \in \mathcal{D} : \ g \text{ isconcave}\}.$$

Proof It suffices to prove that for any $X \in \mathcal{G}(\mathbb{P})$, there is $g \in \mathcal{D}^{\rho,cc}$, such that $g \circ \mathbb{P}(X) = \rho(X)$. By the translation invariance of ρ and $g \circ \mathbb{P}$, we only need to consider those X with $\rho(X) = 1$. For such X, set

$$[X] := \{u(X) : \ u \text{ is a continuous increasing function on } \mathbb{R}\},$$

and put

$$\mathcal{B} = \{Y \in \mathcal{G}(\mathbb{P}) : \ \exists Z \in [X] \text{ such that } \rho(Z) < 1, \text{ and } Y \leqslant Z$$

$$a.s.\}.$$

It is easy to verify that $\mathcal{B}$ is convex, $\mathcal{B}_1 := \{Y \in \mathcal{G}(\mathbb{P}) : \|Y\| < 1\} \subset \mathcal{B}, X \notin \mathcal{B}$. By the convex set separation theorem, there is a non-trivial $\lambda \in ba(\Omega, \mathcal{F}, \mathbb{P})$, such that

$$\sup_{Y \in \mathcal{B}} \lambda(Y) \leqslant \lambda(X).$$

Here $ba(\Omega, \mathcal{F}, \mathbb{P})$ represents the space constructed by all finitely additive measures with finite variations which are absolutely continuous w.r.t. $\mathbb{P}$. Due to $\mathcal{B}_1 \subset \mathcal{B}$, we have $\lambda(X) > 0$. Therefore, we can chose λ, such that $\lambda(X) = 1$. We claim that λ possesses the following properties:

1° for any $Y \in \mathcal{G}(\mathbb{P})$ with $Y \geqslant 0$, we have $\lambda(Y) \geqslant 0$
2° $\lambda(1) = 1$;
3° for any $Y \in [X], \lambda(Y) \leqslant \rho(Y)$.

In fact, for any $Y \in \mathcal{G}(\mathbb{P})$ with $Y \geqslant 0$ and $c > 0$, we have $-cY \in \mathcal{B}$. Thus, $\lambda(-cY) \leqslant \lambda(X) = 1$. Since c is arbitrary, we get 1°.

For any $0 < c < 1$, clearly, $c \in \mathcal{B}$. Since $c < 1$ is arbitrary, this implies $\lambda(1) \leqslant 1$. On the other hand, for any $c > 1$, we have

$$2 - c\lambda(1) = \lambda(2X - c) \leqslant \lambda(X) = 1,$$

which implies $\lambda(1) \geqslant 1/c$. Since $c > 1$ is arbitrary, we get $\lambda(1) \geqslant 1$. Thus, 2° is proved.

Finally, for any $Y \in [X]$, let $Y_1 = Y - \rho[Y] + 1$, then for $c > 1$, it holds $Y_1/c \in \mathcal{B}$. Since $c > 1$ in arbitrary, it implies $\lambda(Y_1) \leqslant 1$. Hence, we get $\lambda(Y) \leqslant \rho(Y)$. 3° is proved.

We define ρ_* and $\rho^*: \mathcal{G}(\mathbb{P}) \to \mathbb{R}$ by

$$\rho_*(Y) := \sup\{\lambda(Z) : Z \leqslant_{st} Y, \ Z \in [X]\}, \quad Y \in \mathcal{G}(\mathbb{P}).$$

$$\rho^*(Y) := \sup\{\lambda(Z) : Z \leqslant_{sl} Y, \ Z \in [X]\}, \quad Y \in \mathcal{G}(\mathbb{P}).$$

According the proof of Theorem 10.36, there is a distortion function $g \in \mathcal{D}$ such that $g \circ \mathbb{P} = \rho_* \leqslant \rho$ and $g \circ P(X) = \rho_*(X) = \rho(X)$.

Obviously, ρ^* satisfies the translation invariance, positive homogeneity, stop-loss order, and the following property:

$$\rho_*(Y) \leqslant \rho^*(Y) \leqslant \rho(Y), \quad \forall Y \in \mathcal{G}(\mathbb{P}); \quad \rho^*(Y) \geqslant \lambda(Y), \quad \forall Y \in [X].$$

We claim that ρ^* also has the co-monotonic upper additivity. In fact, let $Y_1, Y_2 \in \mathcal{G}(\mathbb{P})$ be co-monotonic. For any $Z_1, Z_2 \in [X]$ with $Z_1 \leqslant_{sl} Y_1$, $Z_2 \leqslant_{sl} Y_2$, by Lemma 10.34, we get

$$Z_1 + Z_2 \leqslant_{sl} Y_1 + Y_2.$$

Thus,

$$\lambda(Z_1) + \lambda(Z_2) = \lambda(Z_1 + Z_2) \leqslant \rho^*(Y_1 + Y_2).$$

By the definition of ρ^*, we obtain

$$\rho^*(Y_1 + Y_2) \geqslant \rho^*(Y_1) + \rho^*(Y_2).$$

Therefore, ρ^* has co-monotonic upper additivity.

Let $\overline{g} : [0, 1] \to [0, 1]$ be the smallest concave function which is larger than g. For $B \in \mathcal{F}$ with $b := \mathbb{P}(B) \in (0, 1)$ and $\varepsilon > 0$, there is $0 \leqslant a < b < c \leqslant 0$ such that

$$\overline{g}(b) - \varepsilon < (1 - \alpha)g(a) + \alpha g(c),$$

where $\alpha = (b - a)/(c - a)$. Since the probability space has no atom, we can choose $A \subset B \subset C$ such that $\mathbb{P}(A) = a$, $P(C) = c$. Then

$$\overline{g} \circ \mathbb{P}(B) - \varepsilon < (1 - \alpha)g \circ \mathbb{P}(A) + \alpha g \circ \mathbb{P}(C).$$

By (10.3),

$$(1 - \alpha)g \circ \mathbb{P}(A) + \alpha g \circ P(C) \leqslant (1 - \alpha)\rho^*(I_A) + \alpha \rho^*(I_C).$$

Since ρ^* has positive homogeneity and co-monotonic upper additivity, we get

$$(1 - \alpha)\rho^*(I_A) + \alpha \rho^*(I_C) \leqslant \rho^*((1 - \alpha)I_A + \alpha I_C).$$

Noting that $(1 - \alpha)I_A + \alpha I_C \leqslant_{sl} I_B$ and ρ^* satisfies the stop-loss order, we obtain

$$\rho^*((1 - \alpha)I_A + \alpha I_C) \leqslant \rho^*(I_B).$$

Thus, $\overline{g} \circ \mathbb{P}(B) \leqslant \rho^*(I_B)$.

For all $Y \in \mathcal{G}(\mathbb{P})$, define $\rho(Y) = -\rho^*(-Y)$. Since $X \leqslant_{st} Y$ implies $-Y \leqslant_{sl} -X$, we know that ρ satisfies the translation invariance, positive homogeneity, co-monotonic subadditivity, and stop-loss order. By Theorem 10.36, there is a subset $\mathcal{M}^\rho(\mathbb{P})$ of $\mathcal{M}^{st}(\mathbb{P})$, such that

$$\rho(Y) = \max_{\mu \in \mathcal{M}^\rho(\mathbb{P})} \mu(Y).$$

Thus,

$$\rho^*(Y) = \min_{\mu \in \mathcal{M}^{\rho^*}(\mathbb{P})} \mu(Y),$$

where $\mathcal{M}^{\rho^*}(\mathbb{P})$ is another subset of $\mathcal{M}^{st}(\mathbb{P})$. Therefore, for any $B \in \mathcal{F}$ and $\mu \in \mathcal{M}^{\rho^*}(P)$, $\overline{g} \circ \mathbb{P}(B) \leqslant \mu(B)$. Consequently, for any $Y \in \mathcal{G}(P)$ and $\mu \in \mathcal{M}^{\rho^*}(\mathbb{P})$, $\overline{g} \circ \mathbb{P}(Y) \leqslant \mu(Y)$. Thus, for all $Y \in \mathcal{G}(\mathbb{P})$, we have $\overline{g} \circ \mathbb{P}(Y) \leqslant \rho^*(Y) \leqslant \rho(Y)$, and $\overline{g} \circ \mathbb{P}(X) = \rho(X)$. Theorem is proved. $\square$

From Theorem 10.40 we see that a risk measure satisfying the translation invariance, positive homogeneity, co-monotonic subadditivity, and the stop-loss order has exactly the same representation as a coherent risk measure satisfying the law-invariance. So we have the following corollary.

Corollary 10.41 *Assume that probability space $(\Omega, \mathcal{F}, \mathbb{P})$ has no atom, and $\rho : \mathcal{G}(\mathbb{P}) \to \mathbb{R}$ has the translation invariance and positive homogeneity. Then ρ satisfies the co-monotonic sub-additivity and stop-loss order if and only if ρ satisfies the sub-additivity the first-order stochastic dominance.*

Proof As mentioned above, this is a direct consequence of Theorem 10.40. However, we give here a proof which is based on the proof of Lemma 10.33.

First, we prove the necessity. Suppose ρ satisfies the translation invariance, positive homogeneity, co-monotonic subadditivity, and the stop-loss order, we only need to prove that ρ satisfies sub-additivity. Since $(\Omega, \mathcal{F}, \mathbb{P})$ has no atom, on this probability space, there exists a random variable U, uniformly distributed in $[0,1]$. For any $X, Y \in \mathcal{G}(\mathbb{P})$, let $X^c = q_X(U)$, $Y^c = q_Y(U)$, where $q_X(t)$ and $q_Y(t)$ are right-inverse functions of F_X and F_Y, respectively. Then we have $X^c, Y^c \in \mathcal{G}(\mathbb{P})$ $X =^d X^c, Y =^d Y^c$, and X^c, Y^c are co-monotonic. By Lemma 10.33, it holds that

$$X + Y \leqslant_{sl} X^c + Y^c,$$

and consequently,

$$X^c + Y^c \leqslant_{sd} X + Y.$$

Since ρ satisfies the co-monotonic subadditivity and the stop-loss order,

$$\rho(X + Y) \leqslant \rho(X^c + Y^c) \leqslant \rho(X^c) + \rho(Y^c) = \rho(X) + \rho(Y).$$

Thus, ρ satisfies the sub-additivity.

Now we prove the sufficiency. Suppose that ρ satisfies the translation invariance, positive homogeneity, sub-additivity, and the first-order stochastic dominance, we only need to prove that ρ satisfies the stop-loss order. By Theorem 10.30 and the representation theorem of the coherent risk measure satisfying the law-invariance, the conclusion is obvious.   □

Remark Since a risk measure satisfying the stop-loss order reflects risk aversion, the above corollary gives a reasonable explanation for the requested subadditivity of coherent risk measures. In particular, since VaR does not have subadditivity, VaR does not reflect risk aversion.

Since any $g \in \mathcal{D}^{cc}$ is a limit of certain increasing sequence $\{g_n\} \subset \mathcal{D}^{cc}$ of continuous functions, by the definition of the Choquet integral, we deduce the following corollary from Theorem 10.40.

Corollary 10.42 *Assume that probability space $(\Omega, \mathcal{F}, \mathbb{P})$ has no atom. If $\rho : \mathcal{G}(\mathbb{P}) \to \mathbb{R}$ satisfies translation invariance, positive homogeneity, co-monotonic subadditivity, and the stop-loss order, then ρ has the following representation:*

$$\rho(X) = \sup_{g \in \mathcal{D}_0^{\rho,cc}} (g \circ \mathbb{P})(X),$$

$$\mathcal{D}_0^{\rho,cc} = \{g \in \mathcal{D}^{cc} : g(0+) = 0, \quad (g \circ \mathbb{P})(Y) \leqslant \rho(Y), \quad \forall Y \in \mathcal{G}(\mathbb{P})\}.$$

10.5.5 Law-Invariant Co-monotonic Convex Risk Measures

In this section, we will consider law-invariant co-monotonic convex risk measures as well as their special case: co-monotonic convex risk measures satisfying the stop-loss order. We will give a mathematical representation for these two kinds of risk measures, and from this representation, we can see that a co-monotonic convex risk measure satisfying the stop-loss order is actually a law-invariant convex risk measure. The main results of this section are from Song and Yan (2009a).

Let $\alpha : \mathcal{D} \to \mathbb{R} \cup \{+\infty\}$ be any functional such that $\inf_{g \in \mathcal{D}} \alpha(g)$ be finite. If we define $\rho : \mathcal{G}(\mathbb{P}) \to \mathbb{R}$ by

$$\rho(X) = \sup_{g \in \mathcal{D}} \big((g \circ \mathbb{P})(X) - \alpha(g)\big), \quad X \in \mathcal{G}(\mathbb{P}),$$

then ρ satisfies the monotonicity, translation invariance, and co-monotonic convexity.

Let us prove the converse.

Theorem 10.43 *If $\rho : \mathcal{G}(\mathbb{P}) \to \mathbb{R}$ satisfies the first-order stochastic dominance, translation invariance, and co-monotonic convexity, then ρ has the following representation:*

$$\rho(X) = \max_{g \in \mathcal{D}} \left((g \circ \mathbb{P})(X) - \alpha(g) \right), \quad X \in \mathcal{G}(\mathbb{P}),$$

where

$$\alpha(g) = \sup_{\rho(X) \leqslant 0} (g \circ \mathbb{P})(X), \quad g \in \mathcal{D}.$$

Proof By Theorem 10.35, it suffices to prove

$$\rho(X) = \max_{\mu \in \mathcal{M}^{st}} \left(\mu(X) - \alpha(\mu) \right), \quad X \in \mathcal{G},$$

where

$$\alpha(\mu) = \sup_{\rho(X) \leqslant 0} \mu(X), \quad \mu \in \mathcal{M}^{st}.$$

Without loss of generality, we assume $\rho(0) = 0$. First, we prove

$$\rho(X) \geqslant \sup_{\mu \in \mathcal{M}^{st}} (\mu(X) - \alpha(\mu)), \quad X \in \mathcal{G}(\mathbb{P}). \tag{10.10}$$

In fact, for any $X \in \mathcal{G}(\mathbb{P})$, let $X_1 = X - \rho(X)$, then $\rho(X_1) = 0$. Thus,

$$\alpha(\mu) \geqslant \mu(X_1) = \mu(X) - \rho(X), \quad \forall \mu \in \mathcal{M}^{st},$$

from which (10.10) follows. Therefore, in order to prove this theorem, we need only to prove that for any $X \in \mathcal{G}(\mathbb{P})$, there is a $\mu_X \in \mathcal{M}^{st}$ such that

$$\rho(X) \leqslant \mu_X(X) - \alpha(\mu_X).$$

By translation invariance, it suffices to consider those X satisfying $\rho(X) = 0$. For such a given $X \in \mathcal{G}(\mathbb{P})$, let

$$[X] := \{u(X) : u \text{ is continuous increasing on } \mathbb{R}\},$$

and put

$$\mathcal{B} = \{Y \in \mathcal{G}(P) : \text{there is } Z \in [X], \text{ such that } \rho(Z) < 0, \text{ and } Y \leqslant Z$$

$$\text{a.s.}\}.$$

It is easy to verify that $\mathcal{B}$ is convex, $B_1 := \{Y \in \mathcal{X}(\mathbb{P}) : \| Y + 1 \| < 1\} \subset \mathcal{B}$, and $X \notin \mathcal{B}$. By the convex set separation theorem, there is a non-trivial $\lambda \in ba(\Omega, \mathcal{F}, \mathbb{P})$, such that

$$b := \sup_{Y \in \mathcal{B}} \lambda(Y) \leqslant \lambda(X).$$

We claim that

(1) for any $Y \geqslant 0$ a.s., $Y \in \mathcal{G}(\mathbb{P})$, $\lambda(Y) \geqslant 0$;
(2) $\lambda(1) > 0$.

In fact, for any $Y \in \mathcal{G}(\mathbb{P})$ with $Y \geqslant 0$ a.s., and $c > 0$, we have $-1 - cY \in \mathcal{B}$, $\lambda(-1 - cY) \leqslant \lambda(X)$. Since c is arbitrary, we obtain $\lambda(Y) \geqslant 0$.

As for (2), since λ is non-trivial, there is a $Y \in \mathcal{G}(\mathbb{P})$, $\|Y\| < 1$, such that $\lambda(Y) > 0$. Thus, $\lambda(Y^+) > 0$, $\lambda(1 - Y^+) \geqslant 0$. From this we know $\lambda(1) > 0$. Therefore, we can choose λ such that $\lambda(1) = 1$. We define $\rho_* : \mathcal{G}(\mathbb{P}) \to \mathbb{R}$ by

$$\rho_*(Y) := \sup\{\lambda(Z) : Z \leqslant_{st} Y, Z \in [X]\}.$$

It is easy to prove that ρ_* satisfies the first-order stochastic dominance, positive homogeneity, and the following property:

$$\rho_*(Y) \geqslant \lambda(Y), \quad \forall Y \in [X].$$

We claim that ρ_* has also the co-monotonic additivity. In fact, let $Y_1, Y_2 \in \mathcal{G}(\mathbb{P})$ be co-monotonic. Then there are continuous increasing functions u, v on R, such that $u(z) + v(z) = z, z \in R$, and

$$Y_1 = u(Y_1 + Y_2), \quad Y_2 = v(Y_1 + Y_2) \text{ a.s.}.$$

For any $Z \in [X]$ satisfying $Z \leqslant_{st} (Y_1 + Y_2)$, we have $u(Z) \in [X]$, $v(Z) \in [X]$, and

$$u(Z) \leqslant_{st} u(Y_1 + Y_2) = Y_1, \quad v(Z) \leqslant_{st} v(Y_1 + Y_2) = Y_2.$$

Hence,

$$\rho_*(Y_1) + \rho_*(Y_2) \geqslant \lambda(u(Z)) + \lambda(v(Z)) = \lambda(Z).$$

By the definition of ρ_*, we obtain

$$\rho_*(Y_1 + Y_2) \leqslant \rho_*(Y_1) + \rho_*(Y_2).$$

On the other hand, for any $Z_1, Z_2 \in [X]$ satisfying $Z_1 \leqslant_{st} Y_1$, $Z_2 \leqslant_{st} Y_2$, by Lemma 10.34, we get $Z_1 + Z_2 \leqslant_{st} (Y_1 + Y_2)$. Consequently,

$$\lambda(Z_1) + \lambda(Z_2) = \lambda(Z_1 + Z_2) \leqslant \rho_*(Y_1 + Y_2).$$

By the definition of ρ_*, we obtain

$$\rho_*(Y_1 + Y_2) \geqslant \rho_*(Y_1) + \rho_*(Y_2).$$

Therefore, ρ_* has the co-monotonic additivity. Consequently, by Theorem 10.9, there exists a monotone finitely additive set function μ_X on 2^{Ω} which represents ρ_*. By the definition of ρ_*,

$$\mu_X(1) = \rho_*(1) = 1.$$

If $\rho(Y) \leqslant 0$, then for any $\varepsilon > 0$, we have $\rho(Y - \varepsilon) < 0$. Thus, we get

$$
\begin{aligned}
\mu_X[Y] - \varepsilon &= \mu_X(Y - \varepsilon) = \rho_*(Y - \varepsilon) \\
&= \sup\{\lambda(Z) : Z \leqslant_{st} Y - \varepsilon, Z \in [X]\} \\
&\leqslant \sup_{Z \in B} \lambda(Z) = b.
\end{aligned}
$$

Finally, we get

$$\alpha(\mu_X) = \sup_{\rho(Y) \leqslant 0} \mu_X(Y) \leqslant b,$$

$$\mu_X(X) - \alpha(\mu_X) \geqslant \mu_X(X) - b \geqslant \lambda(X) - b \geqslant 0 = \rho(X).$$

The theorem is proved. □

Theorem 10.44 *We assume that probability space $(\Omega, \mathcal{F}, \mathbb{P})$ has no atom. If $\rho : \mathcal{G}(\mathbb{P}) \to \mathbb{R}$ satisfies the stop-loss order, translation invariance, and co-monotonic convexity, then ρ has the following representation:*

$$\rho(X) = \max_{g \in \mathcal{D}^{cc}} \big((g \circ \mathbb{P})(X) - \alpha(g)\big), \quad X \in \mathcal{G}(\mathbb{P}),$$

$$\alpha(g) = \sup_{\rho(X) \leqslant 0} (g \circ P)(X), \quad g \in \mathcal{D}^{cc}.$$

Proof By Theorem 10.38, it suffices to prove

$$\rho(X) = \max_{\mu \in \mathcal{M}^{sl}} \big(\mu(X) - \alpha(\mu)\big), \quad X \in \mathcal{G}(\mathbb{P}),$$

where

$$\alpha(\mu) = \sup_{\rho(X) \leqslant 0} \mu(X), \quad \mu \in \mathcal{M}^{sl}.$$

Without loss of generality, we assume $\rho(0) = 0$. First, we prove

$$\rho(X) \geqslant \sup_{\mu \in \mathcal{M}^{sl}} (\mu(X) - \alpha(\mu)), \quad X \in \mathcal{G}(\mathbb{P}). \tag{10.11}$$

In fact, for any $X \in \mathcal{G}(\mathbb{P})$, let $X_1 = X - \rho(X)$, then $\rho(X_1) = 0$. Thus,

$$\alpha(\mu) \geqslant \mu(X_1) = \mu(X) - \rho(X), \ \forall \mu \in \mathcal{M}^{sl},$$

whence (10.11) holds. Therefore, in order to prove this theorem, we need only to prove that for any $X \in \mathcal{G}(\mathbb{P})$, there is a $\mu_X \in \mathcal{M}^{sl}$ such that

$$\rho(X) \leqslant \mu_X(X) - \alpha(\mu_X).$$

By translation invariance of ρ, it suffices to consider those X satisfying $\rho(X) = 0$. For such a given $X \in \mathcal{G}(\mathbb{P})$, let

$$[X] := \{u(X) : \ u \ \text{are continuous increasing functions on } \mathbb{R}\},$$

and put

$$\mathcal{B} = \{Y \in \mathcal{G}(\mathbb{P}) : \text{there is } Z \in [X], \ \text{such that } \rho(Z) < 0, \ \text{and } Y \leqslant Z \ \text{a.s.}\}.$$

It is easy to verify that $\mathcal{B}$ is convex, $\mathcal{B}_1 := \{Y \in \mathcal{X}(\mathbb{P}) : \| \ Y + 1 \ \| < 1\} \subset \mathcal{B}$, and $X \notin \mathcal{B}$. By the convex set separation theorem, there is a non-trivial $\lambda \in ba(\Omega, \mathcal{F}, \mathbb{P})$, such that

$$b := \sup_{Y \in \mathcal{B}} \lambda(Y) \leqslant \lambda(X).$$

We claim that

(1) for any $Y \geqslant 0$ a.s., $Y \in \mathcal{G}(\mathbb{P})$, $\lambda(Y) \geqslant 0$;
(2) $\lambda(1) > 0$.

In fact, for any $Y \in \mathcal{G}(\mathbb{P})$ with $Y \geqslant 0$ a.s., and $c > 0$, we have $-1 - cY \in \mathcal{B}$, $\lambda(-1 - cY) \leqslant \lambda(X)$. Since c is arbitrary, we obtain $\lambda(Y) \geqslant 0$.

As for (2), since λ is non-trivial, there is a $Y \in \mathcal{G}(\mathbb{P})$, $\|Y\| < 1$, such that $\lambda(Y) > 0$. Thus, $\lambda(Y^+) > 0$, $\lambda(1 - Y^+) \geqslant 0$. From this we know $\lambda(1) > 0$. Therefore, we can choose λ such that $\lambda(1) = 1$. We define ρ_* and $\rho^* : \mathcal{G}(\mathbb{P}) \to \mathbb{R}$ by

$$\rho_*(Y) := \sup\{\lambda(Z) : Z \leqslant_{st} Y, Z \in [X]\}, \ \ Y \in \mathcal{G}(\mathbb{P}),$$

$$\rho^*(Y) := \sup\{\lambda(Z) : Z \leqslant_{sl} Y, Z \in [X]\}, \ \ Y \in \mathcal{G}(\mathbb{P}).$$

By the proofs of Theorems 10.36 and 10.43, there is a distortion function $g \in \mathcal{D}$ such that $g \circ \mathbb{P} = \rho_* \leqslant \rho$ $g \circ \mathbb{P}(X) = \rho_*(X) = \rho(X)$. Clearly, ρ^* satisfies the stop-loss order, positive homogeneity, translation invariance, and the following property:

$$\rho_*(Y) \leqslant \rho^*(Y) \leqslant \rho(Y), \ \forall Y \in \mathcal{G}(\mathbb{P}); \ \ \ \rho^*(Y) \geqslant \lambda(Y), \ \forall Y \in [X]. \tag{10.12}$$

We claim that ρ_* has also the co-monotonic upper additivity. In fact, let $Y_1, Y_2 \in \mathcal{G}(\mathbb{P})$ be co-monotonic. For all $Z_1, Z_2 \in [X]$ with $Z_1 \leqslant_{sl} Y_1$, $Z_2 \leqslant_{sl} Y_2$, from Lemma 10.34, we get

$$Z_1 + Z_2 \leqslant_{sl} Y_1 + Y_2.$$

Thus,

$$\lambda(Z_1) + \lambda(Z_2) = \lambda(Z_1 + Z_2) \leqslant \rho^*(Y_1 + Y_2).$$

By the definition of ρ^*, we obtain

$$\rho^*(Y_1 + Y_2) \geqslant \rho^*(Y_1) + \rho^*(Y_2).$$

Therefore, ρ^* has the co-monotonic upper additivity.

Let $\overline{g} : [0, 1] \to [0, 1]$ be the smallest concave function which is bigger than g. For $B \in \mathcal{F}$ with $b := \mathbb{P}(B) \in (0, 1)$ and $\varepsilon > 0$, there are $0 \leqslant a < b < c \leqslant 0$ such that

$$\overline{g}(b) - \varepsilon < (1 - \alpha)g(a) + \alpha g(c),$$

where $\alpha = (b - a)/(c - a)$. Since the probability space has no atom, we can choose $A \subset B \subset C$ such that $\mathbb{P}(A) = a$, $\mathbb{P}(C) = c$. Then

$$\overline{g} \circ \mathbb{P}(B) - \varepsilon < (1 - \alpha)g \circ \mathbb{P}(A) + \alpha g \circ \mathbb{P}(C).$$

Thus, from (10.12) we get

$$(1 - \alpha)g \circ \mathbb{P}(A) + \alpha g \circ P(C) \leqslant (1 - \alpha)\rho^*(I_A) + \alpha \rho^*(I_C).$$

Since ρ^* has positive homogeneity and co-monotonic upper additivity, we have

$$(1 - \alpha)\rho^*(I_A) + \alpha \rho^*(I_C) \leqslant \rho^*((1 - \alpha)I_A + \alpha I_C).$$

Noting that $(1 - \alpha)I_A + \alpha I_C \leqslant_{sl} I_B$ and ρ^* satisfies the stop-loss order, we obtain

$$\rho^*((1 - \alpha)I_A + \alpha I_C) \leqslant \rho^*(I_B).$$

Therefore, $\overline{g} \circ \mathbb{P}(B) \leqslant \rho^*(I_B)$.

For any $Y \in \mathcal{G}(\mathbb{P})$, we define $\pi(Y) = -\rho^*(-Y)$. Since $X \leqslant_{st} Y$ implies $-Y \leqslant_{sl} - X$, we know that π satisfies the stop-loss order, positive homogeneity, translation invariance, and the co-monotonic subadditivity. By Theorem 10.36, there is a subset $\mathcal{M}^\rho(\mathbb{P})$ of $\mathcal{M}^{st}(\mathbb{P})$, such that

$$\pi(Y) = \max_{\mu \in \mathcal{M}^\rho(\mathbb{P})} \mu(Y).$$

Thus,

$$\rho^*(Y) = \min_{\mu \in \mathcal{M}^{\rho^*}(\mathbb{P})} \mu(Y),$$

where $\mathcal{M}^{\rho^*}(\mathbb{P})$ is another subset of $\mathcal{M}^{st}(\mathbb{P})$.

Therefore, for any $B \in \mathcal{F}$ and $\mu \in \mathcal{M}^{\rho^*}(\mathbb{P})$, we have $\bar{g} \circ \mathbb{P}(B) \leqslant \mu(B)$. Consequently, for any $Y \in \mathcal{G}(\mathbb{P})$ and $\mu \in \mathcal{M}^{\rho^*}(\mathbb{P})$, it holds that $\bar{g} \circ \mathbb{P}(Y) \leqslant \mu(Y)$. Thus, for all $Y \in \mathcal{G}(\mathbb{P})$, we have $\bar{g} \circ \mathbb{P}(Y) \leqslant \rho^*(Y) \leqslant \rho(Y)$, and $\bar{g} \circ \mathbb{P}(X) = \rho(X)$.

If $\rho(Y) \leqslant 0$, then for any $\varepsilon > 0$, it holds that $\rho(Y - \varepsilon) < 0$. Thus, we have

$$\begin{aligned}
\bar{g} \circ \mathbb{P}(Y) - \varepsilon &= \bar{g} \circ \mathbb{P}(Y - \varepsilon) \leqslant \rho^*(Y - \varepsilon) \\
&= \sup\{\lambda(Z) : Z \leqslant_{sl} Y - \varepsilon, Z \in [X]\} \\
&\leqslant \sup_{Z \in B} \lambda(Z) = b.
\end{aligned}$$

Finally, we get

$$\alpha(\bar{g}) = \sup_{\rho(Y) \leqslant 0} \bar{g} \circ P(Y) \leqslant b,$$

$$\bar{g} \circ \mathbb{P}(X) - \alpha(\bar{g}) \geqslant \bar{g} \circ \mathbb{P}(X) - b \geqslant \lambda(X) - b \geqslant 0 = \rho(X).$$

The theorem is proved. □

Corollary 10.45 *Assume that probability space* $(\Omega, \mathcal{F}, \mathbb{P})$ *has no atom. Let* $\rho : \mathcal{G}(\mathbb{P}) \to \mathbb{R}$. *Then* ρ *satisfies the stop-loss order, translation invariance, and the co-monotonic convexity if and only if* ρ *satisfies the first-order dominance, translation invariance, and the convexity.*

Proof This is a direct consequence of Theorem 10.44. However, we give here a proof which is based on the proof of Lemma 10.33.

First, we assume that ρ satisfies the stop-loss order, translation invariance, and the co-monotonic convexity. We only need to prove that ρ satisfies the convexity. Since $(\Omega, \mathcal{F}, \mathbb{P})$ has no atom, on this probability space, there exists a random variable U, uniformly distributed in $[0,1]$. For any $X, Y \in \mathcal{G}(\mathbb{P})$, let $-X^c = q_{(-X)}(U)$, $-Y^c = q_{(-Y)}(U)$, where $q_{(-X)}(t)$ and $q_{(-Y)}(t)$ are right-inverse functions of $F_{(-X)}$ and $F_{(-Y)}$, respectively. Then $-X^c, -Y^c \in \mathcal{G}(\mathbb{P})$, $-X =^d -X^c$, $-Y =^d -Y^c$, and $-X^c, -Y^c$ are co-monotonic. By Lemma 10.33, for any $\alpha \in (0, 1)$, we have

$$\alpha(-X) + (1 - \alpha)(-Y) \leqslant_{sl} \alpha(-X^c) + (1 - \alpha)(-Y^c),$$

and thus,

$$\alpha X^c + (1 - \alpha)Y^c \leqslant_{sd} \alpha X + (1 - \alpha)Y.$$

Since ρ satisfies the stop-loss order and the co-monotonic convexity, we have

$$\begin{aligned}
\rho(\alpha X + (1 - \alpha)Y) &\leqslant \rho(\alpha X^c + (1 - \alpha)Y^c) \\
&\leqslant \alpha\rho(X^c) + (1 - \alpha)\rho(Y^c) \\
&= \alpha\rho(X) + (1 - \alpha)\rho(Y).
\end{aligned}$$

Therefore, ρ satisfies the convexity.

Conversely, suppose that ρ satisfies the first-order dominance, translation invariance, and convexity, we need to prove that ρ satisfies the stop-loss order. By Theorem 10.30 and the representation theorem of law-invariant convex risk measures, this conclusion is obvious. □

Chapter 11
Stochastic Calculus and Semimartingale Model

K. Itô invented his famous stochastic calculus on Brownian motion in the 1940s. In the 1960s and 1970s, the "Strasbourg school," headed by P.A. Meyer, developed a modern theory of martingales, the general theory of stochastic processes, and stochastic calculus on semimartingales. It turned out soon that semimartingales constitute the largest class of right continuous adapted integrators with respect to which stochastic integrals of simple predictable integrands satisfy the theorem of dominated convergence in probability. Stochastic calculus on semimartingales not only became an important tool for modern probability theory and stochastic processes but also has broad applications to many branches of mathematics, physics, engineering, and mathematical finance.

Option pricing when the underlying stock returns are discontinuous was first studied by Merton (1976). Afterward, many people have used more general discontinuous processes (e.g., jump-diffusion processes or Lévy processes) to model the asset returns. These models have the feature of allowing random jumps in asset values. The most general model for asset returns is the semimartingale model.

In this chapter we first follow closely Yan (2002b) to give a short overview of semimartingales and stochastic calculus (see He et al. (1992) for details). Then in Sect. 11.2, following Xia and Yan (2001, 2003), we introduce some basic concepts and notations for the semimartingale model and establish a version of Kramkov's optional decomposition theorem in the setting of equivalent martingale measures. Based on this theorem, we obtain a numeraire-free expression for the cost of superhedging in Sect. 11.3. Finally, in Sect. 11.4 we introduce the notion of regular strategy and give numeraire-free characterizations of attainable contingent claims and the completeness of the market.

© Springer Nature Singapore Pte Ltd. and Science Press 2018
J.-A. Yan, *Introduction to Stochastic Finance*, Universitext,
https://doi.org/10.1007/978-981-13-1657-9_11

11.1   Semimartingales and Stochastic Calculus

Let $(\Omega, \mathcal{F}, (\mathcal{F}_t), \mathbb{P})$ be a stochastic basis satisfying the usual conditions. Throughout this chapter we assume that $\mathcal{F} = \mathcal{F}_\infty = \sigma(\cup_t \mathcal{F}_t)$.

Definition 11.1 The smallest σ-field on $\Omega \times \mathbb{R}_+$ such that all cadlag (resp. left-continuous) adapted processes are measurable is called the *optional* (resp. *predictable*) σ-field and denoted by $\mathcal{O}$ (resp. $\mathcal{P}$). A random set or stochastic process is called an *optional* (resp. *predictable*) *set* or *process* if it is $\mathcal{O}$ (resp. $\mathcal{P}$)-measurable.

11.1.1   Doob-Meyer's Decomposition of Supermartingales

A measurable process X is said to be of class (D) if $\{X_T I_{[T < \infty]} : T \in \mathcal{T}\}$ is uniformly integrable, where $\mathcal{T}$ is the set of all stopping times.

The following Doob-Meyer's decomposition theorem for supermartingales of class (D) is due to Meyer (1962).

Theorem 11.2 *Let X be a supermartingale of class (D). Then X can be decomposed uniquely as*

$$X = M - A, \tag{11.1}$$

where M is a uniformly integrable martingale and A is a predictable integrable increasing process with $A_0 = 0$. Equation (11.1) is called the Doob-Meyer's decomposition of X.

A martingale M is called a *uniformly square integrable martingale*, if $\sup_t \mathbb{E}[M_t^2] < \infty$. We denote by $\mathcal{M}^2$ the collection of all uniformly square integrable martingales and $\mathcal{M}^{2,c}$ the collection of all continuous uniformly square integrable martingales.

Let M be a uniformly integrable martingale. Then $M \in \mathcal{M}^2$ if and only if $\mathbb{E}[M_\infty^2] < \infty$. In fact, we have

$$\mathbf{E}[M_\infty^2] = \sup_t \mathbb{E}[M_t^2].$$

Moreover, $\mathcal{M}^2$ is a Hilbert space with inner product given by $(M, N) = \mathbb{E}[M_\infty N_\infty]$, and it is isomorphic to $L^2(\Omega, \mathcal{F}, \mathbf{P})$ through the mapping $M \mapsto M_\infty$. Let $\mathcal{M}^{2,d}$ denote the orthogonal complement of $\mathcal{M}^{2,c}$ in $\mathcal{M}^2$. We call elements of $\mathcal{M}^{2,d}$ *purely discontinuous uniformly square integrable martingales*.

If $M \in \mathcal{M}^{2,d}$, then we have $M_0 = 0$ a.s. Any $M \in \mathcal{M}^2$ admits the following unique decomposition:

$$M = M_0 + M^c + M^d,$$

where $M^c \in \mathcal{M}_0^{2,c}$ and $M^d \in \mathcal{M}^{2,d}$. We call M^c the *continuous martingale part* of M and M^d the *purely discontinuous martingale part* of M.

Let $M \in \mathcal{M}^2$. Then M^2 is a submartingale of class (D), because by Doob's inequality we have $M_\infty^* = \sup_t |M_t| \in L^2$. Thus, according to Doob-Meyer's decomposition theorem, there exists a unique predictable integrable increasing process, denoted by $\langle M \rangle$, such that $M^2 - \langle M \rangle$ is a uniformly integrable martingale with initial value 0.

Definition 11.3 Let $M \in \mathcal{M}^2$. We call $\langle M \rangle$ the *predictable quadratic variation* or the *oblique bracket process* of M. For $M, N \in \mathcal{M}^2$, put

$$\langle M, N \rangle = \frac{1}{2}[\langle M + N \rangle - \langle M \rangle - \langle N \rangle]. \tag{11.2}$$

We call $\langle M, N \rangle$ the *predictable quadratic covariation* or the *oblique bracket process* of M and N.

Definition 11.4 For $M, N \in \mathcal{M}^2$, put

$$[M, N]_t = M_0 N_0 + \langle M^c, N^c \rangle_t + \sum_{0 < s \leqslant t} \Delta M_s \Delta N_s, \quad t \geqslant 0.$$

$[M, N]$ is an adapted process of integrable variation, called *quadratic covariation* of M and N. The process $[M, M]$ (or simply, $[M]$) is an adapted integrable increasing process, called the *quadratic variation* or *bracket process* of M.

The following theorem is a basis for the definition of stochastic integrals.

Theorem 11.5 (Kunita-Watanabe inequality) *Let $M, N \in \mathcal{M}^2$. Then for any two measurable processes H and K,*

$$\int_{[0,\infty[} |H_s K_s| |d\langle M, N \rangle_s|$$
$$\leqslant \left(\int_{[0,\infty[} H_s^2 d\langle M \rangle_s \right)^{1/2} \left(\int_{[0,\infty[} K_s^2 d\langle N \rangle_s \right)^{1/2} \quad a.s.,$$
$$\mathbf{E}\left[\int_{[0,\infty[} |H_s K_s| |d\langle M, N \rangle_s| \right] \tag{11.3}$$
$$\leqslant \left\| \sqrt{\int_{[0,\infty[} H_s^2 d\langle M \rangle_s} \right\|_p \left\| \sqrt{\int_{[0,\infty[} K_s^2 d\langle N \rangle_s} \right\|_q,$$

where (p, q) is a pair of conjugate indices, i.e., $1 < p, q < \infty$, and $1/p + 1/q = 1$, and $\| \cdot \|_p$ is the L^p-norm.

A similar result holds for $[M, N]$, $[M]$ and $[N]$.

11.1.2 Local Martingales and Semimartingales

Definition 11.6 Let M be a cadlag adapted process. If there exist stopping times $T_n \uparrow +\infty$ such that each $M^{T_n} - M_0$ is uniformly integrable martingale, then M is called a *local martingale*.

We denote by $\mathcal{M}_{\mathrm{loc}}$ (resp. $\mathcal{W}_{\mathrm{loc}}$) the collection of all local martingales (resp. local martingales of locally integrable variation). We set $\mathcal{M}_{\mathrm{loc},0} = \{M \in \mathcal{M}_{\mathrm{loc}} : M_0 = 0\}$, and $\mathcal{W}_{\mathrm{loc},0} = \{M \in \mathcal{W}_{\mathrm{loc}} : M_0 = 0\}$.

The following is the fundamental theorem for local martingales, due to Yan (cf. Meyer (1997); Yan (1981)).

Theorem 11.7 *Let M be a local martingale. Then for any $\epsilon > 0$, M admits the following decomposition:*

$$M = M_0 + U + V,$$

where $U \in \mathcal{M}_{\mathrm{loc},0}$ with $|\Delta U| \leqslant \epsilon$ and $V \in \mathcal{W}_{\mathrm{loc},0}$.

If $M \in \mathcal{M}_{\mathrm{loc},0}$ has a decomposition $M = U + V$ with $U \in \mathcal{M}_{\mathrm{loc}}^{2,d}$ and $V \in \mathcal{W}_{\mathrm{loc}}$, we call M a *purely discontinuous local martingale*. We denote by $\mathcal{M}_{\mathrm{loc}}^{c}$ (resp. $\mathcal{M}_{\mathrm{loc}}^{d}$) the collection of all continuous (resp. purely discontinuous) local martingales.

Theorem 11.8 *Any local martingale M admits the following unique decomposition:*

$$M = M_0 + M^c + M^d, \tag{11.4}$$

where $M^c \in \mathcal{M}_{\mathrm{loc},0}^{c}$, and $M^d \in \mathcal{M}_{\mathrm{loc}}^{d}$. We call M^c the continuous martingale part of M and M^d the purely discontinuous martingale part of M.

Let M and N be two local martingales. Put

$$[M, N]_t = M_0 N_0 + \langle M^c, N^c \rangle_t + \sum_{0 < s \leqslant t} \Delta M_s \Delta N_s. \tag{11.5}$$

Then $[M, N]$ is an adapted process of finite variation, called the *quadratic covariation* of M and N; $[M, M]$ (or simply, $[M]$) is an adapted increasing process, called the *quadratic variation* or *bracket process* of M.

Definition 11.9 A right continuous adapted process X is called a *semimartingale* if X can be expressed as $X = X_0 + M + A$, where M is a local martingale with $M_0 = 0$ and A is a process of finite variation.

The continuous martingale part of M in the above decomposition is uniquely determined by X. We call it the *continuous martingale part* of X and denote it by X^c.

We denote by $\mathcal{S}$ the collection of all semimartingales.

Let X, Y be two semimartingales. Put

$$[X, Y]_t = X_0 Y_0 + \langle X^c, Y^c \rangle_t + \sum_{s \leqslant t} \Delta X_s \Delta Y_s, \ t \geqslant 0.$$

Then $[X, Y]$ is called the *quadratic covariation* of X and Y; $[X, X]$ (or simply, $[X]$) is an adapted increasing process, called the *quadratic variation* or *bracket process* of X.

11.1.3 Stochastic Integrals w.r.t. Local Martingales

We begin with one-dimensional case. Let M be a real local martingale with the decomposition $M = M_0 + M^c + M^d$ and H be a predictable process. We want to define the "stochastic integral" of H w.r.t. M, denoted by $H.M$. If $H = \xi I_{\rrbracket S, T \rrbracket}$, where $S \leqslant T$ are two stopping times and ξ is an $\mathcal{F}_S$-measurable random variable, $H.M$ should be naturally defined as $H.M = \xi(M^T - M^S)$. Then for any local martingale N, we have $[H.M, N] = H.[M, N]$. This property characterizes uniquely an element $H.M$ of $\mathcal{M}_{\text{loc}}$.

If we want that $H.M$ satisfies the above property for general integrands H, then a necessary condition for H is that $H^2 \in L_S([M])$ and $\sqrt{H^2.[M]} \in \mathcal{A}^+_{\text{loc}}$, where $\mathcal{A}^+_{\text{loc}}$ is the set of all locally integrable increasing adapted processes, and $L_S(A)$ is the set of all measurable processes which are Stieltjes integrable w.r.t. A. Fortunately, this condition is also sufficient for defining a local martingale $H.M$ to meet that property.

First, by using the Kunita-Watanabe inequality for square integrable martingales, we can define a continuous local martingale L' such that $[L', N] = H.[M^c, N]$ for any local martingale N. Second, by using the characterization for jump processes of local martingales, we can define uniquely an $L'' \in \mathcal{M}^d_{\text{loc}}$ such that $\Delta L'' = H \Delta M$. Finally, we put $H.M = L' + L''$. Then for any local martingale N, we have

$$[H.M, N] = H.[M, N].$$

We call $H.M$ the *stochastic integral* of H w.r.t. M. Sometimes we denote this integral by $H_{\dot{m}} M$ to emphasize that the obtained process is required to be a local martingale.

Let M be a local martingale. We denote by $L_m(M)$ the set of all predictable processes H such that $H^2 \in L_S([M])$ and $\sqrt{H^2.[M]} \in \mathcal{A}^+_{\text{loc}}$.

In the sequel, we also use the following notations to denote stochastic integrals. For $t \geqslant 0$,

$$\int_{[0,t]} H_s\, dM_s = (H.M)_t,$$

$$\int_0^t H_s\, dM_s = \int_{(0,t]} H_s\, dM_s = ((H I_{\rrbracket 0,\infty\llbracket}).M)_t.$$

The concept of stochastic integral will be generalized below. But we keep the same notations for stochastic integrals.

The following theorem characterizes the stochastic integrals.

Theorem 11.10 *Let M be a local martingale and $H \in L_m(M)$. Then $H.M$ is the unique local martingale such that $[H.M, N] = H.[M, N]$ holds for every local martingale N.*

The following theorem summarizes the fundamental properties of stochastic integrals.

Theorem 11.11 *Let M be a local martingale, $H, K \in L_m(M)$.*

(1) $L_m(M) = L_m(M^c) \cap L_m(M^d)$, $(H.M)^c = H.M^c$, $(H.M)^d = H.M^d$.
(2) $(H.M)_0 = H_0 M_0$, $\Delta(H.M) = H\Delta M$.
(3) $H + K \in L_m(M)$, and $(H + K).M = H.M + K.M$.
(4) *If H' is a predictable process, then $H' \in L_m(H.M)$ if and only if $HH' \in L_m(M)$. If it is the case, then*

$$H'.(H.M) = (H'H).M .$$

It is easy to show that the stochastic integrals coincide with the Stieltjes integral when the integrator is a local martingale of finite variation and both integrals exist.

Now we turn to the vector stochastic integrals (cf. Jacod 1979). Let $M = (M^i)_{i \leqslant n}$ be an $\mathbb{R}^n$-valued local martingale and $H = (H^i)_{i \leqslant n}$ be an $\mathbb{R}^n$-valued predictable process. If for each i, $H^i \in L_m(M^i)$, then we define naturally the *componentwise stochastic integral* of H against M as

$$H.M = \sum_{i=1}^n H^i.M^i .$$

In order for the stochastic integral to have good properties, such as representing a real local martingale as a stochastic integral w.r.t. a vector local martingale, we need to consider a larger class of integrands. To this end, we take an adapted increasing process Γ (e.g., $\Gamma = \sum_{i=1}^n [M^i, M^i]$) such that $d[M^i, M^j] \ll d\Gamma, \forall i, j \leqslant n$, and let

$$\gamma^{ij} = \frac{d[M^i, M^j]}{d\Gamma} .$$

We denote by $L_m(M)$ the set of all $\mathbb{R}^n$-valued predictable processes H such that

$$\sqrt{\Big(\sum_{i,j=1}^n H^i \gamma^{ij} H^j\Big).\Gamma} \in \mathcal{A}_{\text{loc}}^+.$$

It is easy to see that the space $L_m(M)$ does not depend on the choice of Γ. Similar to the real local martingale case, for $H \in L_m(M)$, we can define uniquely a real local martingale, denoted by $H.M$, such that for any real local martingale N,

$$[H.M, N] = \Big(\sum_{i=1}^n H^i \gamma^{iN}\Big).\Gamma,$$

where $\gamma^{iN} = d[M^i, N]/d\Gamma$. We call $H.M$ the *(vector) stochastic integral* of H w.r.t. M. Sometimes we denote also this integral by $H_{\dot{m}} M$. If $H, K \in L_m(M)$, then

$$[H.M, K.M] = \Big(\sum_{i,j=1}^n H^i \gamma^{ij} K^j\Big).\Gamma.$$

The properties of vector stochastic integrals are similar to that of the scalar case.

Theorem 11.12 *Let $M = (M^i)_{i \leqslant n}$ be a vector local martingale. If $[M^i, M^j] = 0, \forall i \neq j$, then $L_m(M) = \{H = (H^i)_{i \leqslant n} : H^i \in L_m(M^i), \forall i \leqslant n\}$, and the vector stochastic integral coincides with the componentwise stochastic integral.*

11.1.4 Stochastic Integrals w.r.t. Semimartingales

We only formulate results for the real-valued semimartingale case.

Lemma 11.13 *Let X be a semimartingale and H a predictable process. Let $X = M + A$ and $X = N + B$ be two decompositions of X, where $M, N \in \mathcal{M}_{\text{loc}}$ and $A, B \in \mathcal{V}_0$. Here we denote by $\mathcal{V}_0$ the set of all cadlag adapted processes with finite variation. If $H \in L_m(M) \cap L_S(A)$ and $H \in L_m(N) \cap L_S(B)$, then*

$$H_{\dot{m}} M + H_{\dot{s}} A = H_{\dot{m}} N + H_{\dot{s}} B.$$

Based on Lemma 11.13, we propose the following definition.

Definition 11.14 Let X be a semimartingale and H be a predictable process. If there exists a decomposition $X = M + A$, where $M \in \mathcal{M}_{\text{loc}}$ and $A \in \mathcal{V}_0$, such that $H \in L_m(M) \cap L_S(A)$, we say that H is integrable w.r.t. X (or simply H is *X-integrable*) and call $X = M + A$ an *H-decomposition* of X. In this case we put

$$H.X = H_{\dot{m}} M + H_{\dot{s}} A.$$

Then $H.X$ is independent of H-decompositions of X and is called the *stochastic integral* of H w.r.t. X. We denote by $L(X)$ the collection of all predictable processes which are integrable w.r.t. semimartingale X.

Remark Let M be a local martingale. Then $L_m(M) \subset L(M)$, and for $H \in L_m(M)$ two stochastic integrals $H_{\dot{m}}M$ and $H.M$ coincide. In general, $H \in L(M)$ does not imply that $H.M$ is a local martingale.

The next theorem summarizes the fundamental properties of stochastic integrals of predictable processes w.r.t. semimartingales.

Theorem 11.15 *Let X be a semimartingale.*

(1) *Let $H \in L(X)$. Then $(H.X)^c = H.X^c$, $\Delta(H.X) = H\Delta X$, $(H.X)_0 = H_0 X_0$.*
(2) *For any semimartingale Y, $[H.X, Y] = H.[X, Y]$.*
(3) *$H, K \in L(X) \Longrightarrow H + K \in L(X)$.*
(4) *Let $H \in L(X)$ and K be a predictable process. Then $K \in L(H.X)$ if and only if $K.H \in L(X)$. In this case, we have $K.(H.X) = (KH).X$.*

It turns out that semimartingales constitute the largest class of integrators, upon which stochastic integrals of predictable processes can be reasonably defined. Moreover, the semimartingale property and the stochastic integrals are invariant under an equivalent change of probability.

11.1.5 Itô's Formula and Doléans Exponential Formula

In this subsection we present the change of variables formula for semimartingales (Itô's formula). It is the most powerful tool in stochastic calculus.

To begin with, similar to Theorem 4.12, we can prove the following so-called *integration by parts formula*.

Theorem 11.16 *If X and Y are two semimartingales, then we have the following formula of integration by parts:*

$$X_t Y_t = \int_0^t X_{s-}dY_s + \int_0^t Y_{s-}dX_s + [X, Y]_t, \quad t \geqslant 0.$$

From this formula one can prove easily the following Itô's formula (cf. the proof of Theorem 4.11).

Theorem 11.17 (Itô's Formula) *Let $X^1, \cdots, X^d$ be semimartingales and F be a C^2-function on $\mathbf{R}^d$ (i.e., F has continuous partial derivatives of the first and the second order). Put $X_t = (X_t^1, \cdots, X_t^d)$ ((X_t) is also called a d-dimensional semimartingale). Then*

$$F(X_t) - F(X_0) = \sum_{j=1}^d \int_0^t D_j F(X_{s-})dX_s^j + \sum_{0 < s \leqslant t} \eta_s(F) + \frac{1}{2}A_t(F),$$

where

$$\eta_s(F) = F(X_s) - F(X_{s-}) - \sum_{j=1}^{d} D_j F(X_{s-})\Delta X_s^j,$$

$$A_t(F) = \sum_{i,j=1}^{d} \int_0^t D_{ij} F(X_{s-})d\langle (X^j)^c, (X^j)^c \rangle_s,$$

$D_j F = \frac{\partial F}{\partial x_j}$, $D_{ij} F = \frac{\partial^2 F}{\partial x_i \partial x_j}$, *and the series* $\sum_{0 < s \leqslant t} \eta_s(F)$ *is absolutely convergent.*

Theorem 11.18 *Put*

$$Z_t = \exp\left\{ X_t - X_0 - \frac{1}{2}\langle X^c \rangle_t \right\} \prod_{0 < s \leqslant t} (1 + \Delta X_s)e^{-\Delta X_s}. \tag{11.6}$$

Then $Z = (Z_t)$ *is the unique semimartingale satisfying the stochastic integral equation*

$$Z_t = 1 + \int_0^t Z_{s-}dX_s.$$

We call Z *the Doléans* (stochastic) *exponential of* X *and denote it by* $\mathcal{E}(X)$. *Equation* (11.6) *is called the Doléans* exponential formula, *due to* Doléans-Dade (1970).

11.2 Semimartingale Model

We fix a finite time horizon $[0, T]$ and consider a security market which consists of $m + 1$ assets whose price processes (S_t^i), $i = 0, \cdots, m$ are assumed to be strictly positive semimartingales, defined on a filtered probability space $(\Omega, \mathcal{F}, (\mathcal{F}_t), \mathbb{P})$ satisfying the usual conditions. Moreover, we assume that $\mathcal{F}_0$ is the trivial σ-algebra. We take arbitrarily an asset, say asset 0, as the numeraire asset. We set $\gamma_t \hat{=} (S_t^0)^{-1}$ and call γ_t a *deflator* at time t. We set $S_t = (S_t^1, \cdots, S_t^m)$ and $\widetilde{S}_t = (\widetilde{S}_t^1, \cdots, \widetilde{S}_t^m)$, where $\widetilde{S}_t^i = \gamma_t S_t^i$, $1 \leqslant i \leqslant m$. We call $(\widetilde{S}_t)$ the deflated price process of the assets. Note that the deflated price process of asset 0 is constant 1.

A widely adopted setting for "arbitrage-free" financial markets assumes that there exists an equivalent local martingale measure for the deflated price process of assets. According to the fundamental theorem of asset pricing (FTAP for short), due to Kreps (1981) and Delbaen and Schachermayer (1994), if the deflated price process is locally bounded, this assumption is equivalent to the condition of "no free lunch with vanishing risk" (NFLVR for short). For further results on the subject see Delbaen and Schachermayer (1995, 2006). However, the property of NFLVR is not invariant under a change of numeraire. Moreover, under this setting, the market is "arbitrage-free" only for admissible strategies; the market may allow arbitrage for static trading strategies with short selling; and a pricing system using an equivalent local martingale measure may not be consistent with the original prices of some primitive assets. In order to remedy these drawbacks, Yan (1998) introduced the numeraire-free notions of "allowable strategy" and fair market.

11.2.1 Basic Concepts and Notations

A *trading strategy* is an $\mathbb{R}^{m+1}$-valued $(\mathcal{F}_t)$-predictable process $\phi = \{\theta^0, \theta\}$ such that ϕ is integrable w.r.t semimartingale (S_t^0, S_t), where

$$\theta(t) = (\theta^1(t), \cdots, \theta^m(t)), \quad S_t = (S_t^1, \cdots, S_t^m),$$

and $\theta^i(t)$ represents the numbers of units of asset i held at time t. The wealth $V_t(\phi)$ at time t of a trading strategy $\phi = \{\theta^0, \theta\}$ is

$$V_t(\phi) = \theta^0(t)S_t^0 + \theta(t) \cdot S_t,$$

where $\theta(t) \cdot S_t = \sum_{i=1}^m \theta^i(t)S_t^i$. The deflated wealth at time t is $\tilde{V}_t(\phi) = V_t(\phi)\gamma_t$. A trading strategy $\{\theta^0, \theta\}$ is called *self-financing* if

$$V_t(\phi) = V_0(\phi) + \int_0^t \phi(u)d(S_u^0, S_u).$$

A strategy $\phi = \{\theta^0, \theta\}$ is self-financing if and only if its wealth process (V_t) satisfies

$$d\tilde{V}_t = \theta(t)d\tilde{S}_t,$$

where $\tilde{V}_t = V_t\gamma_t$.

It is easy to see that for any given $\mathbb{R}^m$-valued predictable process θ which is integrable w.r.t (S_t) and a real number x there exists a real-valued predictable process (θ_t^0) such that $\phi = \{\theta^0, \theta\}$ is a self-financing strategy with initial capital x.

Let $\mathbb{Q}$ be a probability measure which is equivalent to the historical probability measure $\mathbb{P}$. If the deflated price processes $(\tilde{S}_t)$ is a (vector-valued) $\mathbb{Q}$-martingale (resp. $\mathbb{Q}$-local martingale), then we call $\mathbb{Q}$ an *equivalent martingale measure* (resp. *equivalent local martingale measure*) for the market.

A security market is said to be *fair* if there exists an equivalent martingale measure for the market.

We denote by $\mathcal{P}^j$ the set of all equivalent martingale measures for the market, if asset j is taken as the numeraire asset. It is shown in Yan (1998) that the fairness of a market is invariant under the change of numeraire.

A self-financing strategy is called *admissible* if its wealth process is nonnegative. A self-financing strategy ϕ is called *tame* if there exists a positive constant c such that its wealth process $V_t(\phi)$ is bounded from below by $-cS_t^0$. A self-financing strategy is said to be *allowable* if there exists a positive constant c such that its wealth process V_t is bounded from below by $-c\sum_{i=0}^m S_t^i$. An admissible or tame strategy is allowable. The deflated wealth process of an allowable strategy is a $\mathbb{Q}$-local martingale and a $\mathbb{Q}$-supermartingale for any $\mathbb{Q} \in \mathcal{P}^0$.

Let $\mathcal{P}$ (resp. $\mathcal{P}_{loc}$) denote the set of all equivalent martingale measures (resp. equivalent local martingale measures) when asset 0 is taken as the numeraire. If $\mathcal{P}$ contains only one element, the market is called complete. Otherwise, the market is called incomplete. Stochastic volatility models, jump-diffusion models, and models driven by Lévy processes are incomplete models.

A *contingent claim* with maturity T is a nonnegative $\mathcal{F}_T$-measurable random variable ξ. Put

$$V_0^b(\xi) = \inf_{\mathbb{Q} \in \mathcal{P}} E_{\mathbb{Q}}[\gamma_T \xi], \quad V_0^a(\xi) = \sup_{\mathbb{Q} \in \mathcal{P}} E_{\mathbb{Q}}[\gamma_T \xi].$$

We call $V_0^b(\xi)$ and $V_0^a(\xi)$ the bid and ask prices of ξ, respectively.

Option pricing in a market of semimartingale modeles consists in finding a suitable martingale measure.

Now we give a general principle of finding martingale measures. Assume that there are two assets in the market whose price processes are $B_t = B_0\mathcal{E}_t(h)$ and $S_t = S_0\mathcal{E}_t(H)$, respectively, where h and H are semimartingales. We take B as the numeraire and want to find all martingale measures for the market.

Let $X_t = S_t/B_t$. Then

$$X_t = X_0\mathcal{E}(H)_t\mathcal{E}_t(h)^{-1}.$$

Since $\mathcal{E}(h)_t^{-1} = \mathcal{E}(-h^*)_t$, where

$$h_t^* = h_t - \langle h^c, h^c \rangle_t - \sum_{s \leqslant t} \frac{(\Delta h_s)^2}{1 + \Delta h_s},$$

by the formula $\mathcal{E}(Y)\mathcal{E}(Z) = \mathcal{E}(Y + Z + [Y, Z])$, we have $X_t = X_0\mathcal{E}_t(\Psi(h, H))$, where

$$\Psi_t(h, H) = H_t - h_t + \langle h^c, h^c + H^c \rangle_t + \sum_{s \leqslant t} \frac{\Delta h_s(\Delta h_s - \Delta H_s)}{1 + \Delta h_s}.$$

Let $\mathbb{P}^* \sim \mathbb{P}$ and $Z_t = d\mathbb{P}^*/d\mathbb{P}|_{\mathcal{F}_t}$. Then X is a $\mathbb{P}^*$-local martingale if and only if XZ is a $\mathbb{P}$-local martingale. Since $Z_t = \mathcal{E}_t(N)$ with some $\mathbb{P}$-local martingale N, we have

$$X_t\mathcal{E}_t(N) = X_0\mathcal{E}_t(\Psi(h, H) + N + [\Psi(h, H), N]).$$

The problem of finding martingale measures is reduced to determining when $\Psi(h, H) + N + [\Psi(h, H), N]$ is a local martingale. We refer the reader to Mel'nikov et al. (2002) for more details and examples.

11.2.2 Vector Stochastic Integrals w.r.t. Semimartingales

The following theorem is due to Xia and Yan (2001).

Theorem 11.18   *Let X be an $\mathbb{R}^n$-valued semimartingale and H an $\mathbb{R}^n$-valued predictable process. If $H \in L(X)$ and*

$$H_t \cdot X_t = H_0 \cdot X_0 + \int_0^t H_s dX_s, \tag{11.7}$$

where $\cdot$ denotes the inner product of two vectors, then for any real-valued semi-martingale y, we have $H \in L(yX)$ and

$$y_t(H \cdot X)_t = y_0(H \cdot X)_0 + \int_0^t H_s d(yX)_s. \tag{11.8}$$

Proof   Without loss of generality, we assume that $X_0 = 0$. By integration by parts formula and (11.7), we get

$$y_t(H \cdot X)_t = \int_0^t y_{s-} d(H_s \cdot X_s) + \int_0^t (H \cdot X)_{s-} dy_s + [y, H \cdot X]_t$$

$$= \int_0^t y_{s-} H_s dX_s + \int_0^t (H \cdot X)_{s-} dy_s + \int_0^t H_s d[y, X]_s. \tag{11.9}$$

Since $H \cdot \Delta X = \Delta(H.X)$, by (11.7), we have

$$(H \cdot X)_{t-} - H_t \cdot X_{t-} = (H.X)_{t-} - H_t \cdot X_{t-} = (H.X)_t - (H \cdot X)_t = 0.$$

Consequently, if we let $Z_t = \int_0^t X_{s-} dy_s$, then H is integrable w.r.t. Z, and we have

$$\int_0^t H_s dZ_s = \int_0^t (H \cdot X)_{s-} dy_s.$$

On the other hand, we have

$$(yX)_t = \int_0^t y_{s-} dX_s + Z_t + [y, X]_t.$$

Thus, H is integrable w.r.t. yX, and from (11.9) we have

$$\int_0^t H_s d(yX)_s = \int_0^t H_s y_{s-} dX_s + \int_0^t (H \cdot X)_{s-} dy_s + \int_0^t H_s d[y, X]_s = y_t(H \cdot X)_t.$$

The theorem is proved.   □

From Theorem 11.18 we deduce immediately the following lemma.

Lemma 11.19 *A strategy* $\phi = \{\theta^0, \theta\}$ *is self-financing if and only if its wealth process* $(V_t(\phi))$ *satisfies*

$$d\left(\frac{V_t(\phi)}{S_t^0}\right) = \theta_t d\left(\frac{S_t}{S_t^0}\right).$$

Lemma 11.19 leads easily to the following lemma, which will be used in the sequel.

Lemma 11.20 *For any given* $\mathbb{R}^m$*-valued predictable process* θ *which is integrable w.r.t. S and a real number x, there exists a real-valued predictable process* θ^0 *such that* $\{\theta^0, \theta\}$ *is a self-financing strategy with initial wealth x.*

The following theorem is due to Yan (1998).

Theorem 11.21 *The market is fair if and only if there is no sequence* (ϕ_n) *of allowable strategies with initial wealth 0 such that*

$$V_T(\phi_n) \geqslant -\frac{1}{n}\sum_{j=0}^m S_T^j \text{ a.s.,}$$

for all $n \geqslant 1$ *and* $V_T(\phi_n)$ *a.s., tends to a nonnegative random variable* ξ *with* $\mathbb{P}(\xi > 0) > 0$.

11.2.3 Optional Decomposition Theorem

The optional decomposition theorem (due to Kramkov 1996) is a very useful tool in mathematical finance. It was given in the equivalent local martingale measure setting (see also Föllmer and Yu 1998). In order to be able to apply this theorem to a fair market, we will follow Xia and Yan (2001, 2003) to establish its version in the equivalent martingale measure setting.

Theorem 11.22 *Assume that* $\mathcal{P}$ *is nonempty. If* Y *is a nonnegative* $\mathcal{P}$*-supermartingale, i.e.,* $\mathbb{Q}$*-supermartingale for all* $\mathbb{Q} \in \mathcal{P}$*, then there exist an adapted, right continuous, and increasing process* C *with* $C_0 = 0$ *and an S-integrable predictable process* φ *such that*

$$Y = Y_0 + \varphi.X - C.$$

Moreover, $\varphi.S$ *is a local martingale.*

Proof Let $X = (X^1, X^2, \cdots, X^m)$ and $X^* = 1 + X^1 + X^2 + \cdots + X^m$. Denote by $\mathcal{P}^*$ the set of all equivalent martingale measures for $\widetilde{X} = (X^1/X^*, \cdots, X^m/X^*)$. For each $\mathbb{Q} \in \mathcal{P}$, X^* is a $\mathbb{Q}$-martingale. We can define a probability measure $\mathbb{Q}^*$ by $d\mathbb{Q}^*/d\mathbb{Q}|_{\mathcal{F}_t} = X_t^*/X_0^*$. It is easy to see that $\mathbb{Q} \longrightarrow \mathbb{Q}^*$ is a bijection from $\mathcal{P}$ onto $\mathcal{P}^*$.

Since Y is a $\mathcal{P}$-supermartingale, by Bayes' rule of conditional expectation, for all $\mathbb{Q} \in \mathcal{P}$ and $t > s \geqslant 0$, we have

$$\mathbb{E}_{\mathbb{Q}^*}\left[\frac{Y_t}{X_t^*}\bigg|\mathcal{F}_s\right] = \frac{\mathbb{E}_{\mathbb{Q}}[Y_t|\mathcal{F}_s]}{X_s^*} \leqslant \frac{Y_s}{X_s^*}.$$

Thus, Y/X^* is a $\mathbb{Q}^*$-supermartingale and hence an $\mathcal{P}^*$-supermartingale.

Since $(X^1/X^*, \cdots, X^m/X^*)$ is uniformly bounded, $\mathcal{P}^*$ is also the set of all equivalent local martingale measures for it. From Kramkov's optional decomposition theorem, we know that there exists an $\widetilde{X}$-integrable predictable process θ and an adapted, right continuous and increasing process D with $D_0 = 0$ such that

$$\frac{Y}{X^*} = \frac{Y_0}{X_0^*} + \theta.\widetilde{X} - D. \tag{11.10}$$

Let $\theta_t^0 = Y_0 / X_0^* + (\theta.\widetilde{X})_{t-} - \theta_t \cdot \widetilde{X}_{t-}$, and put $\phi = \{\theta^0, \theta\}$. Then we have

$$\phi_t \cdot (1, \widetilde{X}_t) = \frac{Y_0}{X_0^*} + (\theta.\widetilde{X})_{t-} - \theta_t \cdot \widetilde{X}_{t-} + \theta_t \cdot \widetilde{X}_t$$

$$= \frac{X_0}{X_0^*} + (\theta.\widetilde{X})_t = \frac{Y_0}{X_0^*} + \int_0^t \phi_u d(1, \widetilde{X}_u). \tag{11.11}$$

It follows from Theorem 11.18 that ϕ is integrable w.r.t. (X^*, X) and we have

$$X^*\phi \cdot (1, \widetilde{X}) = Y_0 + \phi.(X^*, X). \tag{11.12}$$

Since $X^*\phi \cdot (1, \widetilde{X}) = Y + X^*D$ is nonnegative, from (11.12), $\phi.(X^*, X)$ must be a $\mathcal{P}_{\text{loc}}$-local martingale. Thus, from a result of Jacka (1992) or Ansel and Stricker (1994), which is also a corollary to the optional decomposition theorem, there exists an X-integrable predictable process σ such that $\phi.(X^*, X) = \sigma.X$. On the other hand,

$$X^i D = X_-^i.D + D_-.X^i + [D, X^i] = X_-^i.D + D_-.X^i + \Sigma \Delta X^i \Delta D$$

$$= X_-^i.D + D_-.X^i + \Delta X^i.D = D_-.X^i + X^i.D.$$

Thus

$$X^*D = (D_-e_m).(X^1, X^2, \cdots, X^m) + X^*.D, \tag{11.13}$$

where e_m is the m-dimensional vector $(1, 1, \cdots, 1)$. From (11.10), (11.11), (11.12), and (11.13), we get

$$Y = Y_0 + \varphi.X - C,$$

where $\varphi = \sigma - D_-e_m$ and $C = X^*.D$. This completes the proof.   □

11.3 Superhedging

Let ξ be a contingent claim at time T. A self-financing strategy ϕ is called a *superhedging strategy* for ξ if

$$V_t(\phi) + \int_t^T \phi d(S^0, S^1, \cdots, S^m) \geqslant \xi \text{ a.s.}$$

and

$$V_t(\phi) + \int_t^u \phi d(S^0, S^1, \cdots, S^m) \geqslant 0 \text{ a.s., for all } u \in [t, T].$$

In general, one cannot find a self-financing strategy ϕ to replicate ξ (i.e., such that the equality holds in the first inequality), but one can always find superhedging strategies for ξ. We define essinf $V_t(\phi)$ as the cost at time t of superhedging for ξ, where ϕ runs over the class of all superhedging strategies for ξ. Note that the above definition of the cost of superhedging does not involve the numeraire.

In the literature, for example, in Kramkov (1996) and Föllmer and Yu (1998), the superhedging problem was solved by using the optional decomposition theorem in the equivalent local martingale measure setting. The result can be stated as follows. We take S^0 as the numeraire, and let $\mathcal{P}^0_{\text{loc}}$ denote the set of all equivalent local martingale measures for $(S^1/S^0, \cdots, S^m/S^0)$. Assume that $\mathcal{P}^0_{\text{loc}}$ is nonempty. Then the cost at time t of a superhedging for the claim ξ is given by

$$U_t = \operatorname{esssup}_{Q \in \mathcal{P}^0_{\text{loc}}} S^0_t \mathbb{E}_Q \left[\left. \frac{\xi}{S^0_T} \right| \mathcal{F}_t \right] \tag{11.14}$$

if

$$\sup_{Q \in \mathcal{P}^0_{\text{loc}}} S^0_0 \mathbb{E}_Q \left[\frac{\xi}{S^0_T} \right] < \infty.$$

However, in such a market, for some numeraire the corresponding local martingale measure may not exist. The model $S = (1, R)$ in Delbaen and Schachermayer (1995) is such an example. In this case, one should use a so-called "strict martingale density" instead of a local martingale measure (see Stricker and Yan 1998). More precisely, for any $0 \leqslant j \leqslant m$ if we take the asset j as the numeraire, we call Z^j a strict martingale density for the deflated assets price process $(S^j)^{-1}(S^0, S^1, \cdots, S^m)$ if Z^j is a strictly positive local martingale such that $Z^j_0 = 1$ and $Z^j(S^j)^{-1}(S^0, S^1, \cdots, S^m)$ is a local martingale. The existence of a strict martingale density is a numeraire-free property. In fact, we denote by $\mathcal{Z}(j)$ the set of all strict martingale densities for the deflated price process $(S^j)^{-1}(S^0, S^1, \cdots, S^m)$. For any $0 \leqslant i, j \leqslant m$, it is easy to see that $Z^i \longrightarrow \frac{S^j S^i_0}{S^i S^j_0} Z^i$ is a bijection from $\mathcal{Z}(i)$ onto $\mathcal{Z}(j)$. If we choose asset i as the numeraire, then by an extended optional decomposition theorem in Stricker and Yan (1998), we know that the cost at time t of superhedging for ξ is

$$\operatorname{esssup}_{Z^i \in \mathcal{Z}(i)} S_t^i \mathbb{E}\left[Z_T^i \cdot \frac{\xi}{S_T^i} \middle| \mathcal{F}_t \right] (Z_t^i)^{-1}. \tag{11.15}$$

Since

$$S_t^j \mathbb{E}\left[\frac{S_T^j S_0^i}{S_T^i S_0^j} Z_T^i \cdot \frac{\xi}{S_T^j} \middle| \mathcal{F}_t \right] \left(\frac{S_t^j S_0^i}{S_t^i S_0^j} Z_t^i \right)^{-1} = S_t^i \mathbb{E}\left[Z_T^i \cdot \frac{\xi}{S_T^i} \middle| \mathcal{F}_t \right] (Z_t^i)^{-1},$$

formula (11.15) expressing the cost of superhedging is independent of the choice of numeraire.

Now we show how to express the cost of superhedging in a fair market. First of all, we augment the market by a new asset with price process $S^{m+1} = \sum_{j=0}^{m} S^j$. It is obvious that any self-financing strategy in the augmented market $(S^0, S^1, \cdots, S^{m+1})$ can be expressed as the one in the original market $(S^0, S^1, \cdots, S^m)$. Therefore, the costs of the superhedging in these two markets are the same. Denote by $\mathcal{P}^{m+1}$ the set of all equivalent martingale measures corresponding to the numeraire S^{m+1}. Since $\left(\frac{S^0}{S^{m+1}}, \frac{S^1}{S^{m+1}}, \cdots, \frac{S^m}{S^{m+1}} \right)$ is uniformly bounded, $\mathcal{P}^{m+1}$ is just the set of all equivalent local martingale measures for it. Thus, from (11.14), the cost at time t of superhedging claim ξ is given by

$$U_t = \operatorname{esssup}_{\mathbb{Q} \in \mathcal{P}^{m+1}} S_t^{m+1} \mathbb{E}_{\mathbb{Q}}\left[\frac{\xi}{S_T^{m+1}} \middle| \mathcal{F}_t \right].$$

Now let $0 \leqslant j \leqslant m$. By Bayes' rule, we obtain

$$U_t = \operatorname{esssup}_{\mathbb{Q} \in \mathcal{P}^j} S_t^j \mathbb{E}_{\mathbb{Q}}\left[\frac{\xi}{S_T^j} \middle| \mathcal{F}_t \right], \tag{11.16}$$

which does not depend on the choice of numeraire.

11.4 Fair Prices and Attainable Claims

First of all, we show that even in the Black-Scholes economy, the principle of NFLVR w.r.t. admissible strategies cannot determine uniquely the price of a contingent claim. In fact, let W be a standard Brownian motion defined on the probability space $(\Omega, \mathcal{F}, \mathbb{P})$, $(\mathcal{F}_t)$ be the usual augmentation of the natural filtration of W, and $\mathcal{F} = \mathcal{F}_T$. We set $S^0 \equiv 1$ and

$$dS_t^1 = S_t^1(\mu dt + \sigma dW_t),$$

where μ and σ are constants. Define $Z = \mathcal{E}(-(\mu/\sigma)W)$ and $d\mathbb{Q} = Z_T d\mathbb{P}$, where $\mathcal{E}(X)$ denotes the Doléans' exponential of X, and then

$$\mathcal{P}^0_{\text{loc}} = \mathcal{P}^0 = \{\mathbb{Q}\}.$$

In this market, the NFLVR property w.r.t. admissible strategies holds whenever either of the two assets is chosen as numeraire. We can find a nonnegative $\mathbb{Q}$-local martingale X such that X is not a $\mathbb{Q}$-martingale. It is clear that X is a $\mathbb{Q}$-supermartingale and $\mathbb{E}_\mathbb{Q}[X_T] < X_0$. Let $Y_t = \mathbb{E}_\mathbb{Q}[X_T|\mathcal{F}_t]$. It is clear that $X \neq Y$. Both (S^1, X) and (S^1, Y) are $\mathbb{Q}$-local martingales. Thus both X and Y can be candidates for the price process of the contingent claim X_T, under the principle of NFLVR w.r.t. admissible strategies. Which one should we choose? The principle of NFLVR w.r.t. admissible strategies cannot give an answer, but the principle of NFLVR w.r.t. allowable strategies can. Under the principle of NFLVR w.r.t. allowable strategies, the price process of the claim X_T should be a $\mathbb{Q}$-martingale. So it must be Y.

This example suggests that we should study the pricing of European contingent claims under the principle of NFLVR w.r.t. allowable strategies. That means we should work in a fair market. Let ξ be a contingent claim. Assume that $(S_T^j)^{-1}\xi$ is $\mathbb{Q}$-integrable for some $0 \leqslant j \leqslant m$ and some $\mathbb{Q} \in \mathcal{P}^j$. We put

$$V_t = S_t^j \mathbb{E}_\mathbb{Q}\left[(S_T^j)^{-1}\xi \mid \mathcal{F}_t\right]. \tag{11.17}$$

If we consider (V_t) as the price process of an asset, then the market augmented with this derivative asset is still fair, because when we take asset j as the numeraire, the deflated price process of this derivative asset is a $\mathbb{Q}$-martingale. So it seems that (V_t) can be considered as a candidate for a "fair" price process of ξ. However, generally speaking, if the martingale measure is not unique, we cannot define uniquely the "fair" price of a contingent claim.

It is clear that a "fair price" of ξ needs not to be the "cost of replicating" ξ by an allowable strategy and the "cost of replicating" ξ by an allowable strategy needs not to be a "fair price" of ξ, either. In general, the fair prices of ξ are not unique. Even if ξ is replicable by an allowable strategy, the costs of different allowable strategies replicating ξ are usually different. In order to eliminate this uncertainty, we propose to introduce the following notions of trading strategies:

Definition 11.23 A trading strategy ϕ is called *regular* (resp. *strongly regular*), if it is allowable and there is a j, $0 \leqslant j \leqslant m$, such that $\frac{V(\phi)}{S^j}$ is a $\mathbb{Q}$-martingale (or,

equivalently, $\mathbb{E}_\mathbb{Q}\left[\dfrac{V_T(\phi)}{S_T^j}\right] = \dfrac{V_0(\phi)}{S_0^j}$) for some $\mathbb{Q} \in \mathcal{P}^j$ (resp. for all $\mathbb{Q} \in \mathcal{P}^j$).

From Bayes' rule for conditional expectations, it is easy to see that if ϕ is regular (resp. strongly regular), then for each j, $0 \leqslant j \leqslant m$, $\frac{V(\phi)}{S^j}$ is a $\mathbb{Q}$-martingale (or,

equivalently, $\mathbb{E}_\mathbb{Q}\left[\dfrac{V_T(\phi)}{S_T^j}\right] = \dfrac{V_0(\phi)}{S_0^j}$) for some $\mathbb{Q} \in \mathcal{P}^j$ (resp. for all $\mathbb{Q} \in \mathcal{P}^j$).

Lemma 11.24 *Let ϕ and ϕ' be regular strategies such that $V_T(\phi) = V_T(\phi')$. Then $V_t(\phi) = V_t(\phi')$ for all $t \in [0, T]$.*

Proof Since ϕ and ϕ' are regular, there exist $\mathbb{Q} \in \mathcal{P}^0$ and $\mathbb{Q}' \in \mathcal{P}^0$ such that $\frac{V(\phi)}{S^0}$ (resp. $\frac{V(\phi')}{S^0}$) is a $\mathbb{Q}$-martingale (resp. $\mathbb{Q}'$-martingale). Since $\frac{V(\phi')}{S^0}$ is a $\mathbb{Q}$-supermartingale, for all $t \in [0, T]$ we have

$$\frac{V_t(\phi)}{S_t^0} = \mathbb{E}_{\mathbb{Q}}\left[\frac{V_T(\phi)}{S_T^0} \Big| \mathcal{F}_t \right] = \mathbb{E}_{\mathbb{Q}}\left[\frac{V_T(\phi')}{S_T^0} \Big| \mathcal{F}_t \right] \leqslant \frac{V_t(\phi')}{S_t^0}.$$

Similarly, we can get $\frac{V_t(\phi')}{S_t^0} \leqslant \frac{V_t(\phi)}{S_t^0}$. Thus the conclusion of the lemma follows. $\square$

A contingent claim ξ is called *attainable* if ξ can be replicated by a regular strategy. From Lemma 11.24 we see that the costs of different regular strategies replicating an attainable claim ξ are the same and it is a fair price of ξ. From Theorem 11.22 we can easily deduce the following characterization for attainable claims, which is a numeraire-free version of the well-known one in the literature.

Theorem 11.25 *Let ξ be a contingent claim such that $\sup_{\mathbb{Q}\in\mathcal{P}^0} \mathbb{E}_{\mathbb{Q}}\left[\frac{\xi}{S_T^0} \right] < \infty$ (or, equivalently, for all j, $0 \leqslant j \leqslant m$, $\sup_{\mathbb{Q}\in\mathcal{P}^j} \mathbb{E}_{\mathbb{Q}}\left[\frac{\xi}{S_T^j} \right] < \infty$). Then the following conditions are equivalent:*

(1) there is a j, $0 \leqslant j \leqslant m$, and some $\mathbb{Q}^j \in \mathcal{P}^j$ such that

$$\sup_{\mathbb{Q}\in\mathcal{P}^j} \mathbb{E}_{\mathbb{Q}}\left[\frac{\xi}{S_T^j} \right] = \mathbb{E}_{\mathbb{Q}^j}\left[\frac{\xi}{S_T^j} \right]; \tag{11.18}$$

(2) for all j, $0 \leqslant j \leqslant m$, there exists some $\mathbb{Q}^j \in \mathcal{P}^j$ such that (11.18) holds;
(3) ξ is attainable.

As a consequence of the above theorem, we obtain a numeraire-free version of a result of Jacka (1992) as follows.

Theorem 11.26 *Let ξ be a contingent claim. Then the following conditions are equivalent:*

(1) there is a j, $0 \leqslant j \leqslant m$ such that for all $\mathbb{Q} \in \mathcal{P}^j$, $\mathbb{E}_{\mathbb{Q}}\left[\frac{\xi}{S_T^j} \right]$ are the same constant;

(2) for all j, $0 \leqslant j \leqslant m$ and all $\mathbb{Q} \in \mathcal{P}^j$, $\mathbb{E}_{\mathbb{Q}}\left[\frac{\xi}{S_T^j} \right]$ are the same constant;
(3) ξ can be replicated by a strongly regular strategy.

Now we turn to study the completeness of a market. The following theorem is the key for this study.

Theorem 11.27 *Let ϕ be a regular strategy such that $0 \leqslant V_T(\phi) \leqslant c \sum_{j=0}^{m} S_T^j$ a.s. for some constant c, then ϕ is strongly regular.*

Proof Since ϕ is a regular strategy ϕ, there exists some $\mathbb{Q}^j \in \mathcal{P}^j$ such that $\frac{V(\phi)}{S^j}$ is a $\mathbb{Q}^j$-martingale. Thus we have

$$0 \leqslant \frac{V_t(\phi)}{S_t^j} = \mathbb{E}_{\mathbb{Q}^j}\left[\left.\frac{V_T(\phi)}{S_T^j}\right|\mathcal{F}_t\right] \leqslant c \sum_{i=0}^{m} \frac{S_t^i}{S_t^j}, \quad 0 \leqslant t \leqslant T.$$

It yields that $\frac{V(\phi)}{S^j}$ is a $\mathbb{Q}$-martingale for all $\mathbb{Q} \in \mathcal{P}^j$. $\qquad\square$

We propose the following definition of complete markets which does not depend on the choice of numeraire.

Definition 11.28 The market is called *complete* if any contingent claim ξ satisfying $0 \leqslant \xi \leqslant \sum_{j=0}^{m} S_T^j$ is attainable.

The following theorem gives a characterization for the completeness of a market.

Theorem 11.29 *The following conditions are equivalent:*

(1) *the market is complete;*
(2) *$\mathcal{P}^0$ is a singleton;*
(3) *any contingent claim ξ satisfying $\sup_{\mathbb{Q} \in \mathcal{P}^0} \mathbb{E}_{\mathbb{Q}}\left[\frac{\xi}{S_T^0}\right] < \infty$ (or, equivalently, for all j, $0 \leqslant j \leqslant m$, $\sup_{\mathbb{Q} \in \mathcal{P}^j} \mathbb{E}_{\mathbb{Q}}\left[\frac{\xi}{S_T^j}\right] < \infty$) is attainable;*
(4) *there is a j, $0 \leqslant j \leqslant m$, such that any contingent claim ξ satisfying $0 \leqslant \xi \leqslant S_T^j$ is attainable.*

Proof (2)$\Rightarrow$ (3), (3)$\Rightarrow$ (1), and (1)$\Rightarrow$ (4) are trivial. We only need to prove (4)$\Rightarrow$ (2). Assume that (4) holds. From Theorems 11.26 and 11.27, we know that any bounded nonnegative $\mathcal{F}_T$-measurable random variable η, $\mathbb{E}_{\mathbb{Q}}[\eta]$ is the same constant for all $\mathbb{Q} \in \mathcal{P}^j$. This is equivalent to that $\mathcal{P}^j$ is a singleton, i.e., $\mathcal{P}^0$ is a singleton.

$\qquad\square$

Remark For continuous-time models and under the setting of local martingale measures, the characterizations for attainable contingent claims and complete markets are mainly due to Jacka (1992), who used a martingale representation result. El Karoui and Quenez (1995) and Kramkov (1996) applied the optional decomposition theorem to simplify the proof of these results, but still under the setting of local martingale measures. All the results in this section are established under the setting of martingale measures and hence have an advantage that they do not depend on the choice of numeraire.

Chapter 12
Optimal Investment in Incomplete Markets

It is an important topic in mathematical finance to maximize the expected utility from the (discounted) terminal wealth of an agent's trading strategies. In a continuous-time model, the problem was studied for the first time by Merton (1969, 1971), who derived the Bellman equation for the value function of the optimization problem with the method of stochastic control. This method, however, requires the state processes to be Markovian. The martingale and duality method, as another approach to the problem, allows us to work in non-Markovian settings. In the case of complete markets, this methodology was developed by Pliska (1986), Cox and Huang (1989, 1991), and Karatzas et al. (1987). In the much more complicated case of incomplete markets, the problem was studied by He and Pearson (1991a,b) and Karatzas et al. (1991) for some specified models. In the general semimartingale model, for utility functions U with effective domains $\mathcal{D}(U) = \mathbb{R}_+$, the portfolio optimization problem was completely solved by Kramkov and Schachermayer (1999, 2003), and Cvitanić et al. (2001). Bellini and Frittelli (2002), Biagini and Frittelli (2005), and Schachermayer (2002) studied the problem for utility functions U with $\mathcal{D}(U) = \mathbb{R}$. The relationship between portfolio optimization and contingent claim pricing was studied in Frittelli (2000a) and Goll and Rüschendorf (2001) among others.

In Sect. 12.1 we follow Xia and Yan (2008) to give a survey on convex duality theory for optimal investment, developed by Kramkov and Schachermayer (1999, 2003) under the setting of equivalent local martingale measures and for deflated wealthes. In Sect. 12.2 we present a numeraire-free and original probability-based framework for financial markets (due to Yan 2002a) and give a reformation of the results of Kramkov and Schachermayer (1999, 2003) under this framework and for non-deflated wealths. In Sect. 12.3, we introduce two utility-based approaches to option pricing, one closely related to the duality method for the expected utility maximization and the other being the marginal utility-based approach of Hugonnier et al. (2005).

© Springer Nature Singapore Pte Ltd. and Science Press 2018   327
J.-A. Yan, *Introduction to Stochastic Finance*, Universitext,
https://doi.org/10.1007/978-981-13-1657-9_12

12.1   Convex Duality on Utility Maximization

In this section we work on the market of semimartingale models in Chap. 11.

12.1.1   The Problem

We take asset 0 as the numeraire asset and deal with deflated wealth processes. Thus, we may assume $S^0 \equiv 1$ without loss of generality. For notational convenience, we set $S = (S^1, \cdots, S^d)$. The family of equivalent local martingale measures (resp. martingale measures) is denoted by $\mathcal{P}_{\text{loc}}$ (resp. $\mathcal{P}$). We assume that $\mathcal{P}_{\text{loc}} \neq \emptyset$.

A self-financing trading strategy with initial capital x is a pair (x, H), where $H = (H^1, \cdots, H^d)$ is a predictable S-integrable process. The deflated wealth process $X = (X_t, 0 \leqslant t \leqslant T)$ of such a self-financing strategy (x, H) is given by

$$X_t = x + \int_0^t H_u dS_u, \quad 0 \leqslant t \leqslant T. \tag{12.1}$$

We denote by $\mathcal{X}(x)$ the set of the wealth processes of all admissible strategies and denote $\mathcal{X}(1)$ by $\mathcal{X}$. Clearly, $\mathcal{X}(x) = x\mathcal{X} = \{xX : X \in \mathcal{X}\}$, $x \geqslant 0$.

Assume that an agent has an initial capital $x > 0$ and that his utility function $U : (0, \infty) \longrightarrow (-\infty, \infty)$ is a continuously differentiable, strictly increasing, and strictly concave, and satisfies the Inada conditions:

$$U'(0) := \lim_{x \downarrow 0} U'(x) = \infty, \quad U'(\infty) := \lim_{x \to \infty} U'(x) = 0.$$

The objective of the agent is to maximize the expected utility of the deflated terminal wealth at time T over family $\mathcal{X}(x)$:

$$\text{Maximize } \mathbb{E}[U(X_T)] \text{ over } X \in \mathcal{X}(x). \tag{12.2}$$

For any $x > 0$, let

$$\mathcal{C}(x) = \{g \in L^0(\Omega, \mathcal{F}, \mathbb{P}) : 0 \leqslant g \leqslant X_T \text{ a.s., for some } X \in \mathcal{X}(x)\}.$$

By monotonicity of utility function U, problem (12.2) is equivalent to

$$\text{Maximize } \mathbb{E}[U(g)] \text{ subject to } g \in \mathcal{C}(x). \tag{12.3}$$

By the related results on superhedging (see, e.g., El Karoui and Quenez (1995); Kramkov (1996); Föllmer and Kabanov (1998)), it is well-known that

$$g \in \mathcal{C}(x) \Longleftrightarrow g \geqslant 0 \text{ a.s., and } \mathbb{E}_{\mathbb{Q}}[g] \leqslant x \text{ for all } \mathbb{Q} \in \mathcal{P}_{\text{loc}}. \tag{12.4}$$

So problem (12.3) is equivalent to

$$
\begin{cases}
\text{Maximize } \mathbb{E}[U(g)] \\
\text{subject to } g \geq 0 \text{ a.s., and } \sup_{\mathbb{Q} \in \mathcal{P}_{\text{loc}}} \mathbb{E}_{\mathbb{Q}}[g] \leq x.
\end{cases}
\tag{12.5}
$$

12.1.2 Complete Market Case

For a complete market, there exists a unique equivalent martingale measure $\mathbb{Q}$, i.e., $\mathcal{P}_{\text{loc}} = \{\mathbb{Q}\}$ is a singleton. In this case, problem (12.5) is reduced to

$$
\begin{cases}
\text{Maximize } \mathbb{E}[U(g)] \\
\text{subject to } g \geq 0 \text{ a.s., and } \mathbb{E}_{\mathbb{Q}}[g] \leq x.
\end{cases}
\tag{12.6}
$$

By the method of Lagrange multiplier, we should consider the following problem without the budget constraint:

$$
\begin{cases}
\text{Maximize } L(\lambda, g) := \mathbb{E}[U(g)] - \lambda(\mathbb{E}_{\mathbb{Q}}[g] - x) \\
\text{subject to } g \geq 0 \text{ a.s.}
\end{cases}
\tag{12.7}
$$

If $\widehat{\lambda} > 0$, $\widehat{g} \geq 0$ a.s., and $(\widehat{\lambda}, \widehat{g})$ is a saddle point of L, that is, it holds for all $\lambda > 0$ and $g \geq 0$ a.s. that

$$
\mathbb{E}[U(g)] - \widehat{\lambda}(\mathbb{E}_{\mathbb{Q}}[g]-x) \leq \mathbb{E}[U(\widehat{g})]-\widehat{\lambda}(\mathbb{E}_{\mathbb{Q}}[\widehat{g}] - x) \leq \mathbb{E}[U(\widehat{g})] - \lambda(\mathbb{E}_{\mathbb{Q}}[\widehat{g}]-x),
$$

then we have $\mathbb{E}_{\mathbb{Q}}[\widehat{g}] = x$ and $\mathbb{E}[U(g)] \leq \mathbb{E}[U(\widehat{g})]$, if $g \geq 0$ a.s. and $\mathbb{E}_{\mathbb{Q}}[g] \leq x$. Thus, $\widehat{g}$ solves problem (12.6).

The remaining questions are the following. Does the saddle point exists? If so, what is the saddle point? From the above arguments, we can see that $(\widehat{\lambda}, \widehat{g})$ is a saddle point if and only if $\mathbb{E}_{\mathbb{Q}}[\widehat{g}] = x$ and

$$
\mathbb{E}[U(g)] - \widehat{\lambda}(\mathbb{E}_{\mathbb{Q}}[g] - x) \leq \mathbb{E}[U(\widehat{g})] - \widehat{\lambda}(\mathbb{E}_{\mathbb{Q}}[\widehat{g}] - x) \text{ for all } g \geq 0 \text{ a.s.}
$$

or, equivalently,

$$
\mathbb{E}\left[U(g) - \widehat{\lambda}\frac{d\mathbb{Q}}{d\mathbb{P}}g\right] \leq \mathbb{E}\left[U(\widehat{g}) - \widehat{\lambda}\frac{d\mathbb{Q}}{d\mathbb{P}}\widehat{g}\right] \text{ for all } g \geq 0 \text{ a.s...}
\tag{12.8}
$$

The conjugate function V of U is defined as

$$V(y) := \sup_{x>0}[U(x) - xy], \quad y > 0. \tag{12.9}$$

It is well known (see, e.g., Rockafellar 1970) that V is a continuously differentiable, strictly decreasing, and strictly convex function satisfying $V'(0) = -\infty$, $V'(\infty) = 0$, $V(0) = U(\infty)$, and $V(\infty) = U(0)$ and that the following relation holds:

$$U(x) = \inf_{y>0}[V(y) + xy], \quad x > 0.$$

Denote by I the inverse function of U', then $I = -V'$, $I(0) = \infty$, $I(\infty) = 0$, and the supremum in (12.9) is attained at $x = I(y)$, i.e.,

$$V(y) = U(I(y)) - yI(y).$$

For any given $\widehat{\lambda} > 0$, let

$$\widehat{g} = I\left(\widehat{\lambda}\frac{d\mathbb{Q}}{d\mathbb{P}}\right), \tag{12.10}$$

then we have (12.8). Thus, if $\widehat{\lambda} > 0$ satisfies

$$\mathbb{E}\left[\frac{d\mathbb{Q}}{d\mathbb{P}}I\left(\widehat{\lambda}\frac{d\mathbb{Q}}{d\mathbb{P}}\right)\right] = x, \tag{12.11}$$

i.e., $\mathbb{E}_{\mathbb{Q}}[\widehat{g}] = x$, then $(\widehat{\lambda}, \widehat{g})$ is a saddle point for L, and $\widehat{g}$ solves problem (12.6).

12.1.3 Incomplete Market Case

For an incomplete market, there exist various equivalent martingale measures $\mathbb{Q} \in \mathcal{P}_{\text{loc}}$. In this case, corresponding to problem (12.7), we should consider the following problem without the budget constraint:

$$\begin{cases} \text{Maximize } L(\lambda, g) := \mathbb{E}[U(g)] - \lambda\left(\sup_{\mathbb{Q}\in\mathcal{P}_{\text{loc}}} \mathbb{E}_{\mathbb{Q}}[g] - x\right) \\ \\ \text{subject to } g \geqslant 0 \text{ a.s.} \end{cases} \tag{12.12}$$

If $\widehat{\lambda} > 0$, $\widehat{g} \geqslant 0$ a.s., and $(\widehat{\lambda}, \widehat{g})$ is a saddle point of L, i.e., it holds that for all $\lambda > 0$ and $g \geqslant 0$ a.s.,

$$\mathbb{E}[U(g)] - \widehat{\lambda}\left(\sup_{Q \in \mathcal{P}_{\mathrm{loc}}} \mathbb{E}_Q[g] - x\right)$$

$$\leqslant \mathbb{E}[U(\widehat{g})] - \widehat{\lambda}\left(\sup_{Q \in \mathcal{P}_{\mathrm{loc}}} \mathbb{E}_Q[\widehat{g}] - x\right)$$

$$\leqslant \mathbb{E}[U(\widehat{g})] - \lambda\left(\sup_{Q \in \mathcal{P}_{\mathrm{loc}}} \mathbb{E}_Q[\widehat{g}] - x\right).$$

Then we have

$$\begin{cases} \sup_{Q \in \mathcal{P}_{\mathrm{loc}}} \mathbb{E}_Q[\widehat{g}] = x \\ \mathbb{E}[U(g)] \leqslant \mathbb{E}[U(\widehat{g})], \quad \text{if } g \geqslant 0 \text{ a.s., and } \sup_{Q \in \mathcal{P}_{\mathrm{loc}}} \mathbb{E}_Q[g] \leqslant x, \end{cases}$$

which implies that $\widehat{g}$ solves problem (12.5).

It is easy to see that $(\widehat{\lambda}, \widehat{g})$ is a saddle point of L if and only if

$$\begin{cases} \sup_{Q \in \mathcal{P}_{\mathrm{loc}}} \mathbb{E}_Q[\widehat{g}] = x, \\ \mathbb{E}[U(g)] - \widehat{\lambda}\left(\sup_{Q \in \mathcal{P}_{\mathrm{loc}}} \mathbb{E}_Q[g] - x\right) \leqslant \mathbb{E}[U(\widehat{g})] - \widehat{\lambda}\left(\sup_{Q \in \mathcal{P}_{\mathrm{loc}}} \mathbb{E}_Q[\widehat{g}] - x\right) \text{ for all } g \geqslant 0, \end{cases}$$

or, equivalently,

$$\begin{cases} \sup_{Q \in \mathcal{P}_{\mathrm{loc}}} \mathbb{E}_Q[\widehat{g}] = x, \\ \inf_{Q \in \mathcal{P}_{\mathrm{loc}}} \mathbb{E}\left[U(g) - \widehat{\lambda}\frac{dQ}{d\mathbb{P}}g\right] \leqslant \inf_{Q \in \mathcal{P}_{\mathrm{loc}}} \mathbb{E}\left[U(\widehat{g}) - \widehat{\lambda}\frac{dQ}{d\mathbb{P}}\widehat{g}\right] \text{ for all } g \geqslant 0 \text{ a.s.} \end{cases}$$

$$(12.13)$$

Motivated by (12.13), we should consider the following maxmin problem:

$$\text{Maximize } \inf_{Q \in \mathcal{Q}_{\mathrm{loc}}} \mathbb{E}\left[U(g) - \widehat{\lambda}\frac{dQ}{d\mathbb{P}}g\right], \quad \text{subject to } g \geqslant 0 \text{ a.s.} \tag{12.14}$$

For a given $\widehat{\lambda} > 0$, if $(\widehat{g}, \widehat{Q})$ is the saddle point for problem (12.14), then we have

$$\sup_{g \geqslant 0 \text{ a.s.}} \inf_{Q \in \mathcal{Q}_{\mathrm{loc}}} \mathbb{E}\left[U(g) - \widehat{\lambda}\frac{dQ}{d\mathbb{P}}g\right] = \inf_{Q \in \mathcal{P}_{\mathrm{loc}}} \sup_{g \geqslant 0 \text{ a.s.}} \mathbb{E}\left[U(g) - \widehat{\lambda}\frac{dQ}{d\mathbb{P}}g\right]$$

$$= \sup_{g \geqslant 0 \text{ a.s.}} \mathbb{E}\left[U(g) - \widehat{\lambda}\frac{d\widehat{Q}}{d\mathbb{P}}g\right]$$

$$= \mathbb{E}\left[U(\widehat{g}) - \widehat{\lambda}\frac{d\widehat{Q}}{d\mathbb{P}}\widehat{g}\right],$$

and $\widehat{g}$ satisfies the inequality in (12.13). By the last equality and the arguments in Sect. 12.1.2, we know

$$\widehat{g} = I\left(\widehat{\lambda}\frac{d\widehat{\mathbb{Q}}}{d\mathbb{P}}\right).$$

Furthermore, if

$$\sup_{\mathbb{Q}\in\mathcal{P}_{\text{loc}}} \mathbb{E}_{\mathbb{Q}}[\widehat{g}] = \sup_{\mathbb{Q}\in\mathcal{P}_{\text{loc}}} \mathbb{E}\left[\frac{d\mathbb{Q}}{d\mathbb{P}}I\left(\widehat{\lambda}\frac{d\widehat{\mathbb{Q}}}{d\mathbb{P}}\right)\right] = x,$$

then (12.13) follows, and $(\widehat{\lambda}, \widehat{g})$ is a saddle point of L, which implies that $\widehat{g}$ solves problem (12.5).

12.1.4   Results of Kramkov and Schachermayer

Unfortunately, the saddle point $(\widehat{g}, \widehat{\mathbb{Q}})$ for problem (12.14), generally speaking, does not exist. The reason is that the family $\{\frac{d\mathbb{Q}}{d\mathbb{P}} : \mathbb{Q} \in \mathcal{P}_{\text{loc}}\}$ is not necessarily closed in the topology of convergence in probability. In order to overcome this drawback of $\mathcal{P}_{\text{loc}}$, Kramkov and Schachermayer (1999) introduced the following family of non-negative semimartingales:

$$\mathcal{Y} = \{Y \geqslant 0 : Y_0 = 1 \text{ and } XY \text{ is a supermartingale for all } X \in \mathcal{X}(1)\}.$$

Note that, as $1 \in \mathcal{X}$, any $Y \in \mathcal{Y}$ is a supermartingale. Note also that the set $\mathcal{Y}$ contains the density processes of all $\mathbb{Q} \in \mathcal{P}_{\text{loc}}$. Let

$$\mathcal{D} = \{g \in L^0(\Omega, \mathcal{F}, \mathbb{P}) : 0 \leqslant g \leqslant Y_T \text{ a.s. for some } Y \in \mathcal{Y}\}.$$

Then by Proposition 3.1 in Kramkov and Schachermayer (1999), both $\mathcal{C}(x)$ and $\mathcal{D}$ are convex and closed in the topology of convergence in probability. Furthermore, $\mathcal{C}(x)$ and $\mathcal{D}$ have the bipolar relation as follows:

$$g \in \mathcal{C}(x) \Longleftrightarrow g \geqslant 0 \text{ a.s. and } \mathbb{E}[gh] \leqslant x \quad \text{for all } h \in \mathcal{D}, \tag{12.15}$$

$$h \in \mathcal{D} \Longleftrightarrow h \geqslant 0 \text{ a.s. and } \mathbb{E}[gh] \leqslant x \quad \text{for all } g \in \mathcal{C}(x). \tag{12.16}$$

By (12.15), we can replace $\mathcal{P}_{\text{loc}}$ by $\mathcal{D}$ and the budget constraint is equivalent to $\sup_{h\in\mathcal{D}} \mathbb{E}[gh] \leqslant x$. More precisely, problem (12.5) can be written as

$$\begin{cases} \text{Maximize } \mathbb{E}[U(g)] \\ \text{subject to } g \geqslant 0 \text{ a.s. and } \sup_{h \in \mathcal{D}} \mathbb{E}[gh] \leqslant x. \end{cases} \qquad (12.17)$$

By the method of Lagrange multiplier, we should consider the following problem without the budget constraint:

$$\begin{cases} \text{Maximize } l(\lambda, g) := \mathbb{E}[U(g)] - \lambda \Big(\sup_{h \in \mathcal{D}} \mathbb{E}[gh] - x \Big) \\ \text{subject to } g \geqslant 0 \text{ a.s.} \end{cases} \qquad (12.18)$$

Corresponding to problem (12.14), we should consider the following maximization problem:

$$\text{Maximize } \inf_{h \in \mathcal{D}} \mathbb{E}\Big[U(g) - \widehat{\lambda} gh \Big] \text{ subject to } g \geqslant 0 \text{ a.s.} \qquad (12.19)$$

It turns out that problem (12.19) admits a saddle point $(\widehat{g}, \widehat{h})$, that is,

$$\sup_{g \geqslant 0 \text{ a.s.}} \inf_{h \in \mathcal{D}} \mathbb{E}\Big[U(g) - \widehat{\lambda} gh \Big] = \inf_{h \in \mathcal{D}} \sup_{g \geqslant 0 \text{ a.s.}} \mathbb{E}\Big[U(g) - \widehat{\lambda} gh \Big], \qquad (12.20)$$

$$\sup_{g \geqslant 0 \text{ a.s.}} \mathbb{E}\Big[U(g) - \widehat{\lambda} g\widehat{h} \Big] = \mathbb{E}\Big[U(\widehat{g}) - \widehat{\lambda} \widehat{g}\widehat{h} \Big]. \qquad (12.21)$$

By (12.21), we have

$$\widehat{g} = I(\widehat{\lambda h}). \qquad (12.22)$$

Furthermore, if $\sup_{h \in \mathcal{D}} \mathbb{E}[\widehat{g}h] = x$, i.e., $\sup_{h \in \mathcal{D}} \mathbb{E}[I(\widehat{\lambda h})h] = x$, then $(\widehat{\lambda}, \widehat{g})$ is the saddle point of l, and $\widehat{g}$ solves problem (12.5).

On the other hand, since $(\widehat{g}, \widehat{h})$ is the saddle point of problem (12.19), the infimum in (12.20) is attained at $\widehat{h}$. Thus we have

$$\inf_{h \in \mathcal{D}} \sup_{g \geqslant 0 \text{ a.s.}} \mathbb{E}\Big[U(g) - \widehat{\lambda} gh \Big] = \inf_{h \in \mathcal{D}} \mathbb{E}[V(\widehat{\lambda}h)] = \mathbb{E}[V(\widehat{\lambda h})],$$

which implies that $\widehat{h}$ solves the dual problem to

$$\text{Minimize } \mathbb{E}[V(\widehat{\lambda}h)] \text{ over } h \in \mathcal{D}.$$

The solutions of the original problem (12.3) and the previous dual problem, $\widehat{g}$ and $\widehat{h}$, are related by (12.22).

For all $x > 0$ and $y > 0$, denote

$$u(x) = \sup_{X \in \mathcal{X}(x)} \mathbb{E}[U(X_T)], \qquad (12.23)$$

$$v(y) = \inf_{Y \in \mathcal{Y}} \mathbb{E}[V(yY_T)]. \qquad (12.24)$$

Now we can state the main result of the convex duality theory on the utility maximization which is due to Kramkov and Schachermayer (1999, 2003).

Theorem 12.1 (Kramkov and Schachermayer) *If $v(y) < \infty$ for all $y > 0$, then we have the following conclusions:*

(1) *Function u is finitely valued, strictly increasing, strictly concave, and continuously differentiable on $(0, \infty)$. Furthermore, for all $x > 0$, there exists a unique $\widehat{X} \in \mathcal{X}(x)$ which attains the supremum in (12.23).*

(2) *Function v is finitely valued, strictly decreasing, strictly convex, and continuously differentiable on $(0, \infty)$. Furthermore, for all $y > 0$, there exists a unique $\widehat{Y} \in \mathcal{Y}$ which attains the infimum in (12.24).*

(3) *Functions u and v are mutually conjugate:*

$$v(y) = \sup_{x>0}[u(x) - xy], \quad u(x) = \inf_{y>0}[v(y) + xy],$$

and with $y = u'(x)$ for some $x > 0$, we have that the corresponding optimizers are related by $\widehat{X}_T = I(y\widehat{Y}_T)$ (or, equivalently, $y\widehat{Y}_T = U'(\widehat{X}_T)$) and $\widehat{X}\widehat{Y}$ is a martingale.

(4) *Function v can be represented by*

$$v(y) = \inf_{\mathbb{Q} \in \mathcal{P}_{loc}} \mathbb{E}\left[V\left(y\frac{d\mathbb{Q}}{d\mathbb{P}}\right)\right].$$

From the previous theorem, the problem of the expected utility maximization for a single agent is well understood in a fairly general setting.

12.2   A Numeraire-Free Framework

In this section we present a numeraire-free and original probability-based framework for financial markets, introduced in Yan (2002a).

The market consists of d (primitive) assets whose price processes $(S_t^i), i = 1, \cdots, d$ are assumed to be non-negative semimartingales with initial values nonzero. We further assume that the process $\sum_{i=1}^d S_t^i$ is strictly positive and that each S_t^i vanishes on $[T^i, \infty)$, where $T^i(\omega) = \inf\{t > 0 : S_t^i(\omega) = 0 \text{ or } S_{t-}^i(\omega) = 0\}$ stands for the ruin time of the company issuing asset i. We will see later that this latter assumption is automatically satisfied for a fair market, since any non-negative supermartingale satisfies this property. In the literature, it was assumed that all primitive assets have strictly positive prices.

Let $S_t = (S_t^1, \cdots, S_t^d)$. In the sequel, we will define

$$S_t^* = \Big(\sum_{i=1}^{d} S_0^i \Big)^{-1} \sum_{i=1}^{d} S_t^i.$$

By assumption, S_t^* is a strictly positive semimartingale. In the literature on mathematical finance, one often takes a primitive asset, whose price never vanishes, as numeraire. In our model, such a primitive numeraire asset may not exist. However, by our assumption on the model, we can always take S_t^* as numeraire. By using this numeraire, we define the allowable strategy, the fair market, and the equivalent martingale measure as before.

12.2.1 Martingale Deflators and Superhedging

In principle, we can take any strictly positive semimartingale as a numeraire and its reciprocal as a deflator. A strictly positive semimartingale M_t with $M_0 = 1$ is called a *martingale deflator* for the market if the deflated price processes $(S_t^i M_t), i = 1, \cdots, d$ are martingales under the original probability measure P. Obviously, a market is fair if and only if there exists a martingale deflator for the market.

Assume that the market is fair. We denote by $\mathcal{M}$ the set of all martingale deflators and denote by $\mathcal{P}$ the set of all equivalent martingale measures, when S_t^* is taken as numeraire. Note that there exists a one-to-one correspondence between $\mathcal{M}$ and $\mathcal{P}$. If $M \in \mathcal{M}$, then $\frac{dQ}{dP} = M_T S_T^*$ defines an element Q of $\mathcal{P}$. If $Q \in \mathcal{P}$, then we can define an element M of $\mathcal{M}$ with $M_T = \frac{dQ}{dP}(S_T^*)^{-1}$. If $\mathcal{P}$ contains only one element, the market is called complete. Otherwise, the market is called incomplete.

Using the main result in Delbaen and Schachermayer (1994), Yan (1998) obtained an intrinsic characterization of fair markets. The same result is valid for our more general model.

Theorem 12.2 *The market is fair if and only if there is no sequence (θ_n) of allowable strategies with initial wealth 0 such that $W_T(\theta_n) \geqslant -\frac{1}{n} S_T^*$ a.s., $\forall n \geqslant 1$, and $W_T(\theta_n)$ a.s. tends to a non-negative random variable ξ satisfying $P(\xi > 0) > 0$.*

Remark In deflated terms, the condition in Theorem 12.2 is just the NFLVR condition introduced in Delbaen and Schachermayer (1994).

The following theorem is a reformulation of Theorem 11.22 in terms of martingale deflators.

Theorem 12.3 *Assume that the market is fair. Let X be a semimartingale. If XM is a local supermartingale for all $M \in \mathcal{M}$, then there exists an adapted, right continuous, and increasing process C with $C_0 = 0$ and an S-integrable predictable process φ such that*

$$X = X_0 + \varphi.S - C.$$

Moreover, if X is non-negative, then $(\varphi.S)M$ is a local martingale for all $M \in \mathcal{M}$.

Proof Let $\widetilde{X}_t = X_t(S_t^*)^{-1}$. Then $\widetilde{X}$ is a local $\mathcal{P}$-supermartingale. By Theorem 12.3, we have

$$\widetilde{X} = X_0 + \psi.\widetilde{S} - D,$$

where D is an adapted, right continuous, and increasing process with $D_0 = 0$. Let $\theta^*(t)$ be a real-valued predictable process such that $\{\theta^*(t)1_d + \psi(t)\}$ is a self-financing strategy with initial wealth X_0. Since $\sum_{i=1}^d \widetilde{S}_t^i = \sum_{i=1}^d S_0^i$, we have $\theta^*(t)1_d.\widetilde{S} = 0$. Consequently,

$$X_0 + ((\theta^* 1_d + \psi).S)_t = S_t^*(X_0 + ((\theta^* 1_d + \psi).\widetilde{S})_t) = S_t^*(X_0 + (\psi.\widetilde{S})_t) = X_t + S_t^* D_t.$$

Putting $\varphi = (\theta^* - D_-)1_d + \psi$ and $C = S^*.D$, we get the desired decomposition. $\qquad\square$

In a fair market setting, if $\sup_{Q \in \mathcal{P}} E_Q\left[(S_T^*)^{-1}\xi\right] < \infty$, then by (11.15) the cost at time t of superhedging claim ξ is given by

$$U_t = \operatorname{esssup}_{Q \in \mathcal{Q}} S_t^* E_Q\left[(S_T^*)^{-1}\xi \middle| \mathcal{F}_t\right]. \tag{12.25}$$

Note that U is the smallest non-negative $\mathcal{Q}$-supermartingale with $U_T \geqslant \xi$. In terms of martingale deflators, we can rewrite (12.25) as

$$U_t = \operatorname{esssup}_{M \in \mathcal{M}} M_t^{-1} E\left[M_T \xi \middle| \mathcal{F}_t\right]. \tag{12.26}$$

Using the optional decomposition theorem Föllmer and Leukert (2000) showed that the optional decomposition of a suitably modified claim gives a more realistic hedging (called *efficient hedging*) of a contingent claim. This result can also be reformulated in terms of martingale deflators.

Now we reformulate the notions of regular and strongly regular strategies in terms of martingale deflators.

Definition 12.4 A self-financing strategy ψ is said to be *regular* (resp. *strongly regular*) if for some (resp. for all) $M \in \mathcal{M}$, $W_t(\psi)M_t$ is a martingale. A contingent claim is called *attainable* if it can be replicated by a regular strategy.

By Theorem 12.3, one can easily deduce the following characterizations for attainable claims and complete markets.

Theorem 12.5 *Let ξ be a contingent claim such that* $\sup_{M \in \mathcal{M}} E\left[\xi M_T\right] < \infty$. *Then ξ is attainable (resp. replicatable by a strongly regular strategy) if and only if the above supremum is attained by an $M^* \in \mathcal{M}$ (resp. $E[M_T \xi]$ doesn't depend on $M \in \mathcal{M}$).*

Theorem 12.6 *The market is complete if and only if any contingent claim ξ dominated by S_T^* is attainable or, equivalently, $E[M_T \xi]$ doesn't depend on $M \in \mathcal{M}$.*

12.2.2 Reformulation of Theorem 12.1

For $x > 0$, we denote by $\mathcal{A}(x)$ the set of all admissible strategies θ with initial wealth x. For $x > 0$, $y > 0$, we put

$$\mathcal{X}(x) = \{W_T(\psi) : \psi \in \mathcal{A}(x)\}, \qquad \mathcal{X} = \mathcal{X}(1),$$

$$\mathcal{Y} = \{Y \geqslant 0 : Y_0 = 1, YX \text{ is a supermartingale } \forall X \in \mathcal{X}\}, \qquad \mathcal{Y}(y) = y\mathcal{Y},$$

$$\mathcal{C}(x) = \{g \in L^0(\Omega, \mathcal{F}_T, P), 0 \leqslant g \leqslant X_T, \text{ for some } X \in \mathcal{X}(x)\}, \qquad \mathcal{C} \hat{=} \mathcal{C}(1),$$

$$\mathcal{D}(y) = \{h \in L^0(\Omega, \mathcal{F}_T, P), 0 \leqslant h \leqslant Y_T, \text{ for some } Y \in \mathcal{Y}(y)\}, \qquad \mathcal{D} \hat{=} \mathcal{D}(1).$$

The agent's optimization problem is

$$\widehat{\psi}(x) = \arg \max_{\psi \in \mathcal{A}(x)} E\left[U(W_T(\psi))\right].$$

To solve this problem, we consider two optimization problems (I) and (II):

$$\widehat{X}(x) = \arg \max_{X \in \mathcal{X}(x)} E\left[U(X_T)\right]; \qquad \widehat{Y}(y) = \arg \min_{Y \in \mathcal{Y}(y)} E\left[V(Y_T)\right].$$

Problem (II) is the dual of problem (I). Their value functions are

$$u(x) = \sup_{X \in \mathcal{X}(x)} E\left[U(X_T)\right], \qquad v(y) = \inf_{Y \in \mathcal{Y}(y)} E\left[V(Y_T)\right].$$

The following theorem is a reformulation of Theorem 12.1, i.e., results of Kramkov and Schachermayer (henceforth, K-S), under our framework.

Theorem 12.7 *Assume that there is an $X \in \mathcal{X}(1)$ such that $X_T \geqslant K$ for a positive constant K (e.g., $S_T^* \geqslant K$). If $v(y) < \infty$, $\forall y > 0$, then the value functions $u(x)$ and $v(y)$ are conjugate in the sense that*

$$v(y) = \sup_{x>0}[u(x) - xy], \quad u(x) = \inf_{y>0}[v(y) + xy],$$

and we have

(1) *For any $x > 0$ and $y > 0$, both optimization problems (I) and (II) have unique solutions $\widehat{X}(x)$ and $\widehat{Y}(y)$.*

(2) *If $y = u'(x)$, then $\widehat{X}_T(x) = I(\widehat{Y}_T(y))$, and the process $\widehat{X}(x)\widehat{Y}(y)$ is a martingale.*

(3) $v(y) = \inf_{M \in \mathcal{P}} E\left[V(yM_T)\right].$

Proof The proof is almost the same as that in K-S (1999, 2001). The main differences are indicated as follows. Obviously, $\mathcal{C}$ and $\mathcal{D}$ are convex sets. By Proposition 3.1 and a slight modification of Lemma 4.2 in K-S (1999), one can show that $\mathcal{C}$ and $\mathcal{D}$ are closed under the convergence in probability. For (1) and (2), as in Lemma 3.2 of K-S (1999) and Lemma 1 of K-S (2003), in order to prove the families $(V^-(h))_{h \in \mathcal{D}(y)}$ and $(U^+(g))_{g \in \mathcal{C}(x)}$ are uniformly integrable, we need to use a fact that $\mathcal{C}$ contains a positive constant. In our case, we have indeed $K \in \mathcal{C}$, since by assumption $K \leqslant W_T(\psi)$ for some $\psi \in \mathcal{A}(1)$. As for (3), according to Proposition 1 in K-S (2003), we only need to show $\widehat{\mathcal{D}} = \{M_T : M \in \mathcal{P}\}$ satisfies the following conditions:

- For any $g \in \mathcal{C}$, $\sup_{h \in \widehat{\mathcal{D}}} E[gh] = \sup_{h \in \mathcal{D}} E[gh]$;
- $\widehat{\mathcal{D}} \subset \mathcal{D}$, and $\widehat{\mathcal{D}}$ is convex and closed under countable convex combinations.

The first condition can be easily proved, and the second one is trivial.   □

12.3   Utility-Based Approaches to Option Pricing

12.3.1   *Minimax Martingale Deflator Approach*

In this subsection, under the framework of Sect. 12.2 and in terms of martingale deflator, we study the utility-based approach to contingent claim pricing, which are closely related to the duality method for the expected utility maximization.

Assume that the market is fair. Let ξ be a contingent claim such that $M_T\xi$ is integrable for some $M \in \mathcal{M}$. We put

$$V_t = (M_t)^{-1} E\left[M_T\xi \mid \mathcal{F}_t\right]. \tag{12.27}$$

We can define (V_t) as a "fair price process" of ξ. Thus, in deflated terms, pricing of contingent claims in an incomplete market consists in choosing a reasonable martingale measure. There are several approaches to make such a choice. A well-known one is the so-called utility-based approach. The basic idea of this approach is as follows. Assume that the representative agent in the market has preference represented by a utility function. In certain cases, the dual optimization problem (II) may produce the so-called minimax martingale measure (MMM for short).

In the sequel, we assume that $v(y) < \infty$ for all $y > 0$. First of all we consider the case where for the dual problem

$$v(y) = \inf_{\mathbb{Q} \in \mathcal{P}} \mathbb{E}\left[V\left(y \frac{d\mathbb{Q}}{d\mathbb{P}}\right) \right],$$

the minimizer $\widehat{\mathbb{Q}}(y)$ exists. In this case, $\widehat{\mathbb{Q}}(y)$ is called a *minimax martingale measure*. Bellini and Frittelli (2002), for utility functions $U : (-\infty, \infty) \to (-\infty, \infty)$, give some sufficient conditions for the existence of minimax martingale measures.

We show that the expected utility maximization problem is linked by duality to a martingale deflator. Assume that the solution $\widehat{Y}(y)$ of the dual optimization problem (II) lies in $y\mathcal{P}$. We put $\widehat{M}(y) = y^{-1}\widehat{Y}(y)$. Then $\widehat{M}(y) \in \mathcal{M}$, and we have

$$\widehat{M}(y) = \arg\min_{M \in \mathcal{P}} E\left[V(yM_T) \right].$$

We call $\widehat{M}(y)$ the *minimax martingale deflator*.

The following theorem gives a necessary and sufficient condition for the existence of the minimax martingale deflator.

Theorem 12.8 *Assume that there is a $\psi \in \mathcal{A}(1)$ such that $W_T(\psi) \geqslant K$ for a positive constant K (e.g., $S_T^* \geqslant K$) and that $v(y) < \infty$ for all $y > 0$. Let $x > 0$ be the agent's initial wealth and $M^* \in \mathcal{M}$. Furthermore, M^* is the minimax martingale deflator corresponding to the utility function U if and only if there exist $y > 0$ and $X^* \in \mathcal{X}(x)$ such that $X_T^* = I(yM_T^*)$ and $E[M_T^* X_T^*] = x$. If it is the case, then X^* solves the optimization problem (I).*

Proof We only need to prove the sufficiency of the condition. We have the following inequality

$$U(I(z)) \geqslant U(w) + z[I(z) - w], \quad \forall w > 0, z > 0.$$

If we replace z and w by yM_T^* and $X_T \in \mathcal{X}(x)$ and take the expectation w.r.t. P, we get immediately that $E[U(X_T^*)] \geqslant E[U(X_T)]$ for all $X \in \mathcal{X}(x)$. This shows that X^* solves optimization problem (I). On the other hand, since $X_T^* = I(yM_T^*)$ and assumption $E[M_T^* X_T^*] = x$ implies that $M^* X^*$ is a martingale, by Theorem 12.19, yM^* must solve optimization problem (II). In particular, M^* is the minimax martingale deflator. $\square$

Now assume that a minimax martingale deflator $\widehat{M}(y)$ exists. Let ξ be a contingent claim. If we use $\widehat{M}(y)$ to compute a fair price of ξ by (12.27), then it coincides with the fair price of Davis (1997), which is derived through the so-called "marginal rate of substitution" argument (see Chap. 14, Sect. 14.3). In fact, Davis' fair price of ξ is defined by

$$\hat{\pi}(\xi) = \frac{E[U'(\widehat{X}_T(x))\xi]}{u'(x)}.$$

Since $y = u'(x)$ and $U'(\widehat{X}_T(x)) = \widehat{Y}(y)$, we have $\hat{\pi}(\xi) = E[\widehat{M}_T(y)\xi]$.

Now we explain the economic meaning of Davis' fair price of a contingent claim. Let ξ be a contingent claim with $E[\widehat{M}_T(y)\xi] < \infty$. Put $\xi_t = (\widehat{M}_t(y))^{-1}E[\widehat{M}_T(y)\xi|\mathcal{F}_t]$. We augment the market with derivative asset ξ and consider the portfolio maximization problem in the new market. Then it is easy to see that $\widehat{Y}(y)$ is still the solution to the dual optimization problem (II) in the new market. Consequently, the value function v and its conjugate function u remain unchanged. By Theorem 4.1, $\widehat{X}_T(x)$ solves again the optimization problem (I) in the new market. This shows that if the price of a contingent claim is defined by Davis' fair price, no trade on this contingent claim increases the maximal expected utility in comparison to an optimal trading strategy. This fact was observed in Goll and Rüschendorf (2001).

Note that in general the MMM (i.e., minimax martingale measure) depends on the agent's initial wealth x. This is a disadvantage of the utility-based approach to contingent claim pricing. However, for utility functions $\ln x$ and x^p/p, where $p \in (-\infty, 1) \setminus \{0\}$, the MMM is independent of the agent's initial wealth x.

For $U(x) = -e^{-x}$, the corresponding MMM is called the *minimal entropy martingale measure*. We refer the reader to Frittelli (2000b), Miyahara (1999, 2001), Fujiwara and Miyahara (2003), and Xia and Yan (2000b) for studies on the subject. If $U(x) = \ln x$, the minimax martingale deflator $\widehat{M}$, if it exists, is nothing but the reciprocal of the wealth process $\widehat{X}(1)$ of the growth optimal portfolio (see Chap. 14).

12.3.2 Marginal Utility-Based Approach

Hugonnier et al. (2005) developed another approach to utility-based pricing, which permits us to deal with the case where the minimizer for the dual problem

$$v(y) = \inf_{\mathbb{Q} \in \mathcal{P}_{\mathrm{loc}}} \mathbb{E}\left[V(y\frac{d\mathbb{Q}}{d\mathbb{P}})\right]$$

does not exist. We now present their main result.

A process $X \in \mathcal{X}(x)$ is called *maximal* if $X' \in \mathcal{X}(x)$ and that $X_T \leqslant X'_T$ implies $X' = X$. A process X is called *acceptable* if it admits a decomposition of the form $X = X' - X''$ where X' is admissible and X'' is maximal. Let $B \in L^0(\Omega \mathcal{F}, \mathbb{P})$ be a contingent claim. For a pair $(x, q) \in \mathbb{R}^2$, we denote by $\mathcal{X}(x, q|B)$ the set of acceptable processes with initial value x and whose terminal value dominates $-qB$. A real number p is called a *marginal utility-based price* of B given the initial capital x if

$$\mathbb{E}[U(X_T + qB)] \leqslant u(x), \quad q \in \mathbb{R}, \ X \in \mathcal{X}(x - qp, q|B).$$

The economic meaning of a marginal utility-based price p is that, given the possibility to buy and hold the claim at price p, the agent's optimal demand for the contingent claim is equal to zero.

The main result of Hugonnier et al. (2005) is the following theorem which provides sufficient conditions for the uniqueness of the marginal utility-based price.

Theorem 12.9 *Assume $v(y) < \infty$ for all $y > 0$. For $x > 0$, define $y = u'(x)$, and let $\widehat{Y}(y)$ denote the corresponding solution to (12.24). Then for any maximal admissible process $\widetilde{X} \in \mathcal{X}$ we have:*

(1) *If the product $\widehat{Y}(y)\widetilde{X}$ is a uniformly integrable martingale, then every contingent claim with the property $|B| \leqslant c\widetilde{X}_T$ for some $c > 0$ admits a unique marginal utility-based price w.r.t. the initial capital x, which is given by*

$$p(B|x) := \mathbb{E}[\widehat{Y}(y)B].$$

(2) *If the product $\widehat{Y}(y)\widetilde{X}$ fails to be a uniformly integrable martingale, then there exists a contingent claim $0 \leqslant B \leqslant \widetilde{X}_T$ and a constant $\alpha \geqslant 0$ such that every $\alpha \leqslant \pi \leqslant \alpha + \delta$ is a marginal utility-based price for B.*

As a consequence, the dual optimizer $\widehat{Y}(y)$ is the density process of a $\widehat{\mathbb{Q}}(y) \in \mathcal{P}_{\text{loc}}$ w.r.t. $\mathbb{P}$ (or equivalently, the minimizer of $v(y) = \inf_{\mathbb{Q} \in \mathcal{P}_{\text{loc}}} \mathbb{E}\left[V(y\frac{d\mathbb{Q}}{d\mathbb{P}})\right]$ exists) if and only if every bounded contingent claim admits a unique marginal utility-based price given the initial capital x. If it is the case, the marginal utility-based price $p(B|x)$ is given by $\mathbb{E}_{\widehat{\mathbb{Q}}(y)}[B]$.

Chapter 13
Martingale Method for Utility Maximization

The problem of the expected utility maximization for diffusion models in complete markets has been studied by many authors; see the review article of Karatzas (1989). The same problem in incomplete markets has been extensively studied by Karatzas et al. (1991). They judiciously augmented the stocks with fictitious ones to create a complete market, such that the fictitious stocks are superfluous in the optimal portfolio for the completed market. In this case, their solution is also optimal for the original incomplete market.

In this chapter we follow closely Xia and Yan (2000a,b) to present the expected utility maximization and valuation problems in a general semimartingale setting. In this case, the method of "fictitious completion" is no longer applicable, because this model is an infinite-dimensional one. A martingale measure method is proposed to solve this problem. We refer the reader to Kallsen (2000) for more discussions on the subject. Our market model is the same as those described in Chap. 12. We denote by $\mathcal{P}$ the set of all equivalent martingale measures.

In Sect. 13.1, for any $x > 0$ and $\mathbb{Q} \in \mathcal{P}$, we construct a contingent claim $\xi_{\mathbb{Q}}^x(T)$ at time T in a way similar to Karatzas et al. (1991). We show that if for some $\tilde{\mathbb{Q}} \in \mathcal{P}$, $\xi_{\tilde{\mathbb{Q}}}^x(T)$ can be replicated by a self-financing strategy with initial wealth x, then this strategy is optimal, and the contingent claim pricing using martingale measure $\mathbb{Q}$ is related to Davis' option pricing rule. In Sect. 13.2, we consider three particular types of utility function: HARA utility function $\log x$ (resp. $\frac{1}{\gamma}(x^\gamma - 1)(\gamma < 0)$), utility function $W_\gamma(x) = -(1 - \gamma x)^{\frac{1}{\gamma}} (\gamma < 0)$, and utility function $W_0(x) = -e^{-x}$. In Sect. 13.3, for a market driven by a Lévy process, the optimal portfolio and the related martingale measure are explicitly worked out for each of the three utility functions.

© Springer Nature Singapore Pte Ltd. and Science Press 2018
J.-A. Yan, *Introduction to Stochastic Finance*, Universitext,
https://doi.org/10.1007/978-981-13-1657-9_13

13.1   Expected Utility Maximization and Valuation

We assume that asset 0 is a risk-free bond, whose price at time t is

$$S_t^0 = \exp\left\{ \int_0^t r_u du \right\}, \quad 0 \leqslant t \leqslant T,$$

where $r = (r_t)_{0 \leqslant t \leqslant T}$ is a nonnegative adapted process. We take asset 0 as the numeraire asset. The discount process is $\beta_t = \exp\{-\int_0^t r_u du\}$, $0 \leqslant t \leqslant T$.

13.1.1   Expected Utility Maximization

The agent in our model has a utility function $U : (D_U, \infty) \longrightarrow \mathbb{R}$ for wealth, $-\infty < D_U < \infty$. We assume throughout that U is strictly increasing, strictly concave, continuous, and continuously differentiable and satisfies

$$U'(D_U) \hat{=} \lim_{x \downarrow D_U} U'(x) = \infty, \quad U'(\infty) \hat{=} \lim_{x \to \infty} U'(x) = 0.$$

The (continuous, strictly decreasing) inverse of U' is denoted by I : $(0, \infty) \longrightarrow (D_U, \infty)$. It is obvious that

$$I(0) \hat{=} \lim_{y \downarrow 0} I(y) = \infty, \quad I(\infty) \hat{=} \lim_{y \to \infty} I(y) = D_U.$$

The following inequality is very useful:

$$U(I(y)) \geqslant U(x) + y[I(y) - x], \quad \text{for all } x > D_U, y > 0. \tag{13.1}$$

For $K \in (-\infty, \infty)$, we denote by $\mathcal{A}_K(x)$ the family of all self-financing strategies ψ with initial wealth x such that $\beta_t V_t(\psi) > K$ for $t \in [0, T]$. The discounted wealth process of a strategy in $\mathcal{A}_K(x)$ is a $\mathbb{Q}$-supermartingale for any $\mathbb{Q} \in \mathcal{P}$.

For a given utility function U and a given initial capital $x > 0$, we consider the problem to maximize the expected utility from terminal wealth $\mathbb{E}\,[U(V_T(\psi))]$, over the class $\mathcal{A}_{D_U}(x)$. A strategy $\psi \in \mathcal{A}_{D_U}(x)$ which maximizes the expected utility is called optimal. By the strict concavity of U, it is clear that the terminal wealth of optimal strategies is unique.

In the sequel, we assume that r_t is deterministic. Then so is β_t.

Denote $Z_t^{\mathbb{Q}} = \mathbb{E}\left[\frac{d\mathbb{Q}}{d\mathbb{P}} \Big| \mathcal{F}_t \right]$ for all $\mathbb{Q} \in \mathcal{P}$, and set

$$\mathcal{P}_t = \{\, \mathbb{Q} \in \mathcal{P} : \mathbb{E}[\beta_t Z_t^{\mathbb{Q}} I(y \beta_t Z_t^{\mathbb{Q}})] \text{ is finite for all } y \in (0, \infty)\}, \quad t \in [0, T].$$

Following Karatzas et al. (1991), for every $t \in [0, T]$ and $\mathbb{Q} \in \mathcal{P}_t$, we define a function $\mathcal{X}_t^{\mathbb{Q}}$ by

$$\mathcal{X}_t^{\mathbb{Q}}(y) \hat{=} \mathbb{E}\left[\beta_t Z_t^{\mathbb{Q}} I(y\beta_t Z_t^{\mathbb{Q}})\right], \quad 0 < y < \infty,$$

which inherits from I the property of being a continuous, strictly decreasing mapping from $(0, \infty)$ onto $(D_U \beta_t, \infty)$. And so $\mathcal{X}_t^{\mathbb{Q}}$ has an (continuous, strictly decreasing) inverse $\mathcal{Y}_t^{\mathbb{Q}}$ from $(D_U \beta_t, \infty)$ onto $(0, \infty)$. We define

$$\xi_{\mathbb{Q}}^x(t) \hat{=} I(\mathcal{Y}_t^{\mathbb{Q}}(x)\beta_t Z_t^{\mathbb{Q}}), \quad \text{for all } t \in [0, T], \ \mathbb{Q} \in \mathcal{P}_t, \ x \in (D_U \beta_t, \infty).$$

Then by definitions

$$
\begin{aligned}
\mathbb{E}_{\mathbb{Q}}\left[\beta_T \xi_{\mathbb{Q}}^x(T)\right] &= \mathbb{E}\left[\beta_T Z_T^{\mathbb{Q}} I(\mathcal{Y}_T^{\mathbb{Q}}(x)\beta_T Z_T^{\mathbb{Q}})\right] \\
&= \mathcal{X}_T^{\mathbb{Q}}(\mathcal{Y}_T^{\mathbb{Q}}(x)) = x, \quad x \in (D_U \beta_t, \infty).
\end{aligned}
\tag{13.2}
$$

Note that for every strategy $\psi \in \mathcal{A}_{D_U}(x)$, $\beta V(\psi)$ is a $\mathbb{Q}$-supermartingale, and hence

$$\mathbb{E}\left[\beta_T Z_T^{\mathbb{Q}} V_T(\psi)\right] \leqslant x, \quad \mathbb{Q} \in \mathcal{P}_T. \tag{13.3}$$

By (13.1), (13.2), and (13.3), we have

$$
\begin{aligned}
\mathbb{E}\left[U(\xi_{\mathbb{Q}}^x(T))\right] &\geqslant \mathbb{E}\left[U(V_T(\psi)) + \mathcal{Y}_T^{\mathbb{Q}}(x)\beta_T Z_T^{\mathbb{Q}}(\xi_{\mathbb{Q}}^x(T) - V_T(\psi))\right] \\
&= \mathbb{E}[U(V_T(\psi))] + \mathcal{Y}_T^{\mathbb{Q}}(x)\left(\mathbb{E}\left[\beta_T Z_T^{\mathbb{Q}}\xi_{\mathbb{Q}}^x(T)\right] - \mathbb{E}\left[\beta_T Z_T^{\mathbb{Q}} V_T(\psi)\right]\right) \\
&\geqslant \mathbb{E}[U(V_T(\psi))].
\end{aligned}
$$

$$\tag{13.4}$$

Consequently, for a given $x > 0$, if there exists a probability measure $\mathbb{Q} \in \mathcal{P}_T$ such that $x \in (D_U \beta_T, \infty)$ and $\xi_{\mathbb{Q}}^x(T) = V_T(\widehat{\psi})$ for a trading strategy $\widehat{\psi} \in \mathcal{A}_{D_U}(x)$. Then $\widehat{\psi}$ is optimal. Since $Z_T^{\mathbb{Q}}$ is uniquely determined by $\xi_{\mathbb{Q}}^x(T)$, such a $\mathbb{Q}$ is unique, and by (13.4) $\mathbb{Q}$ is optimal within $\mathcal{P}_T$ in the sense that

$$\mathbb{E}\left[U(\xi_{\mathbb{Q}}^x(T))\right] \leqslant \mathbb{E}\left[U(\xi_{\mathbb{R}}^x(T))\right], \quad \forall \mathbb{R} \in \mathcal{P}_T.$$

13.1.2 Utility-Based Valuation

In Sect. 12.3, we already introduced two utility-based approaches to option pricing. One is closely related to the duality method for the expected utility maximization, while the other is the marginal utility-based approach of Hugonnier et al. (2005). In this subsection we take another viewpoint to consider the utility-based valuation problem. It seems that this viewpoint has clear economic meaning. It will turn out that this utility-based valuation still coincides with that in Davis (1997).

Assume that for a given $x > 0$, there exists a probability measure $\mathbb{Q} \in \mathcal{P}_T$ such that $x \in (D_U \beta_T, \infty)$ and $\xi_{\mathbb{Q}}^x(T) = V_T(\widehat{\psi})$ for a trading strategy $\widehat{\psi} \in \mathcal{A}_{D_U}(x)$. Let B be a non-replicatable contingent claim at time T with $\mathbb{E}_{\mathbb{Q}}[\beta_T B] < \infty$. We will show that for an investor with initial wealth x and utility function U, $\mathbb{E}_{\mathbb{Q}}[\beta_T B]$ is a fair price of B.

Firstly, assume that the price of B is p and $p \geqslant \mathbb{E}_{\mathbb{Q}}[\beta_T B]$. We will show that the investor should buy no shares of B at the price p. Assume that the investor with initial wealth x buys δ shares of the contingent claim B with price p, where δp $(0 < \delta < \frac{x}{p})$, and invest the remaining wealth $x - \delta p$ in the market using a strategy $\psi \in \mathcal{A}_{D_U}(x - \delta p)$. Then we have

$$\mathbb{E}[U(V_T(\widehat{\psi}))] - \mathbb{E}[U(V_T(\psi) + \delta B)]$$

$$= \mathbb{E}[U(\xi_{\mathbb{Q}}^x(T)) - U(V_T(\psi) + \delta B)]$$

$$> \mathbb{E}[U'(\xi_{\mathbb{Q}}^x(T))(\xi_{\mathbb{Q}}^x(T) - V_T(\psi) - \delta B)]$$

$$= \mathbb{E}[\mathcal{Y}_T^{\mathbb{Q}}(x)\beta_T Z_T^{\mathbb{Q}}(I(\mathcal{Y}_T^{\mathbb{Q}}(x)\beta_T Z_T^{\mathbb{Q}}) - V_T(\psi) - \delta B)]$$

$$= \mathcal{Y}_T^{\mathbb{Q}}(x)(x - \mathbb{E}_{\mathbb{Q}}[\beta_T V_T(\psi)] - \delta \mathbb{E}_{\mathbb{Q}}[\beta_T B])$$

$$\geqslant \mathcal{Y}_T^{\mathbb{Q}}(x)(x - (x - \delta p) - \delta \mathbb{E}_{\mathbb{Q}}[\beta_T B])$$

$$= \mathcal{Y}_T^{\mathbb{Q}}(x)\delta(p - \mathbb{E}_{\mathbb{Q}}[\beta_T B]) \geqslant 0.$$

Here the strict inequality at the third line follows from the strict concavity of U and the fact that B is non-replicatable.

Secondly, assume that the price of B is $p < \mathbb{E}_{\mathbb{Q}}[\beta_T B]$. We will show that the investor should buy certain shares of contingent claim B at the price p. To this end we need the following additional assumption on the utility function U: For any $K_1 \in (0, \infty)$, there exists a $K_2 \in (0, \infty)$ such that $U'(K_1 x) \leqslant K_2 U'(x)$ for all $x \in (0, \infty)$. Note that the HARA utility functions $U_0(x) = \log x$ and $U_\gamma(x) = \frac{1}{\gamma}(x^\gamma - 1)(\gamma < 0)$ satisfy this condition.

For any $\varepsilon \in (0, \frac{x}{p}]$, we have

$$\frac{1}{\varepsilon} \left(U \left(\frac{x-\varepsilon p}{x} V_T(\widehat{\psi}) + \varepsilon B \right) - U(V_T(\widehat{\psi})) \right)^-$$

$$\leqslant U' \left(\frac{x-\varepsilon p}{x} V_T(\widehat{\psi}) + \varepsilon B \right) \left(B - \frac{p}{x} V_T(\widehat{\psi}) \right)^-$$

$$\leqslant U' \left(\frac{x-\varepsilon p}{x} V_T(\widehat{\psi}) \right) \left(\frac{p}{x} V_T(\widehat{\psi}) - B \right) I_{[\frac{p}{x} V_T(\widehat{\psi}) > B]}$$

$$\leqslant K_2 U'(V_T(\widehat{\psi})) \left(\frac{p}{x} V_T(\widehat{\psi}) - B \right) I_{[\frac{p}{x} V_T(\widehat{\psi}) > B]}$$

$$= K_2 \mathcal{Y}_T^{\mathbb{Q}}(x) \beta_T Z_T^{\mathbb{Q}} \left(\frac{p}{x} V_T(\widehat{\psi}) - B \right) I_{[\frac{p}{x} V_T(\widehat{\psi}) > B]} \in L^1(\mathbb{P}).$$

By Fatou's lemma we have

$$\liminf_{\varepsilon \downarrow 0} \frac{1}{\varepsilon} \mathbb{E} \left[U \left(\frac{x-\varepsilon p}{x} V_T(\widehat{\psi}) + \varepsilon B \right) - U(V_T(\widehat{\psi})) \right]$$

$$\geqslant \mathbb{E} \left[U'(V_T(\widehat{\psi})) \left(B - \frac{p}{x} V_T(\widehat{\psi}) \right) \right]$$

$$= \mathbb{E} \left[\mathcal{Y}_T^{\mathbb{Q}}(x) \beta_T Z_T^{\mathbb{Q}} \left(B - \frac{p}{x} V_T(\widehat{\psi}) \right) \right]$$

$$= \mathcal{Y}_T^{\mathbb{Q}}(x) \left(\mathbb{E}_{\mathbb{Q}}[\beta_T B] - \frac{p}{x} \mathbb{E}_{\mathbb{Q}}[\beta_T V_T(\widehat{\psi})] \right)$$

$$= \mathcal{Y}_T^{\mathbb{Q}}(x)(\mathbb{E}_{\mathbb{Q}}[\beta_T B] - p) > 0.$$

Thus, there exists an $\varepsilon_0 \in (0, \frac{x}{p}]$ such that for all $\varepsilon \in (0, \varepsilon_0]$ we have

$$\mathbb{E} \left[U \left(\frac{x - \varepsilon p}{x} V_T(\widehat{\psi}) + \varepsilon B \right) \right] > \mathbb{E}[U(V_T(\widehat{\psi}))].$$

This shows that if the investor buys certain shares of B with price p, then he will get more benefit than using the optimal strategy.

Now we are going to show that the price defined by martingale measure $\mathbb{Q}$ coincides with the fair price defined in Davis (1997). We fix $x > 0$ and assume that there exist an $\varepsilon > 0$ with $\varepsilon < x$ and a $\mathbb{Q} \in \mathcal{P}_T$ such that for each $y \in (x - \varepsilon, x + \varepsilon)$, $\xi_{\mathbb{Q}}^y(T)$ can be replicated. Define $V(y) = \sup_{\psi \in A_{D_U}(y)} \mathbb{E}[U(V_T(\psi))]$ and then

$$V(y) = \mathbb{E}[U(V_T(\widehat{\psi}))] = \mathbb{E}[U(\xi_{\mathbb{Q}}^y(T))] = \mathbb{E}[U(I(\mathcal{Y}_T^{\mathbb{Q}}(y) \beta_T Z_T^{\mathbb{Q}}))]. \tag{13.5}$$

Since I is the inverse of U', by a formal computation, it shows that $V'(x) = \mathcal{Y}_T^{\mathbb{Q}}(x)$. By the definition of $\xi_{\mathbb{Q}}^x(T)$, we have $U'(\xi_{\mathbb{Q}}^x(T)) = \mathcal{Y}_T^{\mathbb{Q}}(x) \beta_T Z_T^{\mathbb{Q}}$. Thus

$$Z_T^Q = \frac{U'(\xi_Q^x(T))}{y_T^Q(x)\beta_T} = \frac{U'(V_T(\widehat{\psi}))}{V'(x)\beta_T}.$$

Then the arbitrage price of an option B under the martingale measure Q is given by

$$\mathbb{E}_Q[\beta_T B] = \frac{\mathbb{E}[U'(V_T(\widehat{\psi}))B]}{V'(x)}$$

which coincides with the fair price $\widehat{p}(x)$ given by the "marginal rate of substitution" approach of Davis (1997).

13.2   Minimum Relative Entropy and Maximum Hellinger Integral

In this section, we consider three particular types of utility function. It is shown that for HARA utility function $\log x$ (resp. $\frac{1}{\gamma}(x^\gamma - 1)(\gamma < 0)$), the historical probability measure has the *minimum relative entropy* (resp. *maximizes the Hellinger integral* of order $\frac{\gamma}{\gamma-1}$) w.r.t. Q^* within $\mathcal{P}$; for utility function $-(1 - \gamma x)^{\frac{1}{\gamma}}(\gamma < 0)$, Q^*, within $\mathcal{P}$, maximizes the Hellinger integral of order $\frac{\gamma}{\gamma-1}$ w.r.t. the historical probability measure; for utility function, $-e^{-x}$ Q^* has also the minimum relative entropy w.r.t. the historical probability measure.

13.2.1   HARA Utility Functions

In this subsection, we consider a widely used class of HARA utility functions, which are given by

$$U_\gamma(x) = \begin{cases} \frac{1}{\gamma}(x^\gamma - 1), & \gamma < 0, \\ \log x, & \gamma = 0. \end{cases}$$

Here HARA stands for hyperbolic absolute risk aversion. For HARA utility functions $U_\gamma(\gamma \leqslant 0)$, we have

$$D_{U_\gamma} = 0, \quad U_\gamma'(x) = x^{\gamma-1}, \quad I(x) = x^{\frac{1}{\gamma-1}}, \quad \delta \widehat{=} \frac{\gamma}{\gamma - 1} \in [0, 1),$$

where δ satisfies $1/\delta + 1/\gamma = 1$ for $\gamma < 0$. Thus, for all $Q \in \mathcal{P}$, $t \in [0, T]$, and $y \in (0, \infty)$, it follows that

$$\mathcal{X}_t^{Q}(y) = \mathbb{E}\left[\beta_t Z_t^{Q}(y\beta_t Z_t^{Q})^{\frac{1}{\gamma-1}}\right] = y^{\frac{1}{\gamma-1}}\mathbb{E}\left[(\beta_t Z_t^{Q})^{\delta}\right] \leqslant y^{\frac{1}{\gamma-1}} < \infty, \qquad (13.6)$$

which shows that $\mathcal{P}_t = \mathcal{P}$ for all $t \in [0, T]$.

By (13.6),

$$x = \mathcal{X}_t^{Q}(\mathcal{Y}_t^{Q}(x)) = (\mathcal{Y}_t^{Q}(x))^{\frac{1}{\gamma-1}}\mathbb{E}[(\beta_t Z_t^{Q})^{\delta}],$$

so that

$$(\mathcal{Y}_t^{Q}(x))^{\frac{1}{1-\gamma}} = \frac{\beta_t^{\delta}}{x}\mathbb{E}\left[(Z_t^{Q})^{\delta}\right], \quad t \in [0, T]. \qquad (13.7)$$

On the other hand,

$$\xi_Q^x(t) = I(\mathcal{Y}_t^{Q}(x)\beta_t Z_t^{Q}) = (\mathcal{Y}_t^{Q}(x)\beta_t Z_t^{Q})^{\frac{1}{\gamma-1}} = \frac{x(Z_t^{Q})^{\frac{1}{\gamma-1}}}{\beta_t\mathbb{E}[(Z_t^{Q})^{\delta}]}, \quad t \in [0, T]. \qquad (13.8)$$

Thus, for $\gamma < 0$,

$$\begin{aligned}
\mathbb{E}\left[U_\gamma(\xi_Q^x(T))\right] &= \mathbb{E}\left[\frac{1}{\gamma}(\mathcal{Y}_T^{Q}(x)\beta_T Z_T^{Q})^{\delta}\right] - \frac{1}{\gamma} \\
&= \frac{1}{\gamma}(\mathcal{Y}_T^{Q}(x))^{\delta}\mathbb{E}\left[(\beta_T Z_T^{Q})^{\delta}\right] - \frac{1}{\gamma} \qquad (13.9) \\
&= \frac{x}{\gamma}(\mathcal{Y}_T^{Q}(x))^{\delta-\frac{1}{\gamma-1}} - \frac{1}{\gamma} = \frac{x}{\gamma}\mathcal{Y}_T^{Q}(x) - \frac{1}{\gamma}.
\end{aligned}$$

For $\gamma = 0$, by (13.7) we have $\mathcal{Y}_t^{Q}(x) = 1/x$ and

$$U_0(\xi_Q^x(T)) = \log x - \log \beta_T - \log Z_T^{Q}. \qquad (13.10)$$

From (13.7) and (13.9), we get easily the following theorem.

Theorem 13.1 *Let $\gamma < 0$, $x > 0$ and $Q \in \mathcal{P}$. For HARA utility function U_γ, the following statements are equivalent:*

(1) $\mathbb{E}\left[U(\xi_Q^x(T))\right] \leqslant \mathbb{E}\left[U(\xi_R^x(T))\right]$, *for all $R \in \mathcal{P}_T$.*
(2) $\mathcal{Y}_T^{Q}(x) \geqslant \mathcal{Y}_T^{R}(x)$, *for all $R \in \mathcal{P}$.*
(3) $\mathbb{E}\left[(Z_T^{Q})^{\delta}\right] \geqslant \mathbb{E}\left[(Z_T^{R})^{\delta}\right]$, *for all $R \in \mathcal{P}$.*

The following theorem is a consequence of (13.10).

Theorem 13.2 *For HARA utility function $U_0(x) = \log x$, the statement (1) in Theorem 13.1 is equivalent to the following one:*

(4) $\mathbb{E}[\log Z_T^{Q}] \geqslant \mathbb{E}[\log Z_T^{R}]$, *for all $R \in \mathcal{P}$.*

Assume that a probability measure $\mathbb{P}$ is absolutely continuous with respect to a probability measure $\mathbb{Q}$ (denoted by $\mathbb{P} \ll \mathbb{Q}$). For $\delta \in (0, 1)$, the *Hellinger integral* of order δ of $\mathbb{P}$ with respect to $\mathbb{Q}$ is defined by

$$H_\delta(\mathbb{Q}, \mathbb{P}) = \mathbb{E}_\mathbb{P}\left[\left(\frac{d\mathbb{Q}}{d\mathbb{P}}\right)^\delta\right] = \mathbb{E}_\mathbb{P}\left[(Z_T^\mathbb{Q})^\delta\right].$$

The *relative entropy* of $\mathbb{P}$ with respect to $\mathbb{Q}$ is defined by

$$I_\mathbb{Q}(\mathbb{P}) = \mathbb{E}_\mathbb{Q}\left[\frac{d\mathbb{P}}{d\mathbb{Q}} \log \frac{d\mathbb{P}}{d\mathbb{Q}}\right] = \mathbb{E}_\mathbb{P}\left[\log \frac{d\mathbb{P}}{d\mathbb{Q}}\right].$$

Both Hellinger integral and relative entropy are quantitative measures of the difference between $\mathbb{Q}$ and $\mathbb{P}$. One should be aware that in general, $H_\delta(\mathbb{Q}, \mathbb{P})$ (resp. $I_\mathbb{Q}(\mathbb{P})$) is not equal to $H_\delta(\mathbb{P}, \mathbb{Q})$ (resp. $I_\mathbb{P}(\mathbb{Q})$), if $\mathbb{P}$ and $\mathbb{Q}$ are equivalent.

For the utility function $U_0(x) = \log x$, by Theorem 13.2 and results in Sect. 13.1, we know that if there exists a probability measure $\mathbb{Q} \in \mathcal{P}$ such that $\xi_\mathbb{Q}^x(T)$ can be replicated by some strategy $\psi \in \mathcal{A}_0(x)$, then ψ is optimal, and the historical measure $\mathbb{P}$ has the minimum relative entropy w.r.t. $\mathbb{Q}$ over $\mathcal{P}$.

For the utility function $U_\gamma (\gamma < 0)$, by Theorem 13.1 and results in Sect. 13.1, we know that if there exists a probability measure $\mathbb{Q} \in \mathcal{P}$ such that $\xi_\mathbb{Q}^x(T)$ can be replicated by some strategy $\psi \in \mathcal{A}_0(x)$, then ψ is optimal, and the measure $\mathbb{Q}$ maximizes the Hellinger integral $H_\delta(\mathbb{Q}, \mathbb{P})$ over $\mathcal{P}$, where δ satisfies $1/\delta + 1/\gamma = 1$. For a market driven by a Lévy process, we will also give explicitly this maximum Hellinger integral.

In the HARA utility function case, by (13.7), (13.9), (13.10), and (13.5), it is easy to see that $V'(x) = \mathcal{Y}_T^\mathbb{Q}(x)$. In this case, it is more transparent that Davis' fair price of an option is the same as the one computed using the martingale measure $\mathbb{Q}$.

13.2.2 Another Type of Utility Function

In this subsection, we consider another type of utility function, given by

$$W_\gamma(x) = -(1 - \gamma x)^{\frac{1}{\gamma}}, \quad \gamma < 0.$$

It is easy to see that $U_\gamma(-W_\gamma(x)) = -x$, $\gamma < 0$. For utility function W_γ, we have

$$D_{W_\gamma} = \frac{1}{\gamma}, \quad W_\gamma'(x) = (1 - \gamma x)^{\frac{1-\gamma}{\gamma}}, \quad I(x) = \frac{1 - x^{\frac{\gamma}{1-\gamma}}}{\gamma}.$$

For all $\mathbb{Q} \in \mathcal{P}$, $t \in [0, T]$, and $y \in (0, \infty)$, we see that

$$\mathcal{X}_t^{\mathbb{Q}}(y) = \mathbb{E}[\beta_t Z_t^{\mathbb{Q}} I(y\beta_t Z_t^{\mathbb{Q}})]$$

$$= \mathbb{E}\left[\beta_t Z_t^{\mathbb{Q}} \frac{1 - (y\beta_t Z_t^{\mathbb{Q}})^{\frac{\gamma}{1-\gamma}}}{\gamma}\right]$$

$$= \frac{\beta_t}{\gamma} - \frac{1}{\gamma} y^{\frac{\gamma}{1-\gamma}} \beta_t^{\frac{1}{1-\gamma}} \mathbb{E}\left[(Z_t^{\mathbb{Q}})^{\frac{1}{1-\gamma}}\right]$$

is finite. Thus, $\mathcal{P}_t = \mathcal{P}$ for all $t \in [0, T]$. Let $\mathcal{Y}_t^{\mathbb{Q}}(x)$ be such that $\mathcal{X}_t^{\mathbb{Q}}(\mathcal{Y}_t^{\mathbb{Q}}(x)) = x$, then we have for $t \in [0, T]$,

$$\beta_t^{\frac{1}{1-\gamma}} \mathbb{E}\left[(Z_t^{\mathbb{Q}})^{\frac{1}{1-\gamma}}\right] = (\beta_t - \gamma x)(\mathcal{Y}_t^{\mathbb{Q}}(x))^{\frac{\gamma}{\gamma-1}}, \quad \gamma < 0.$$

In the sequel, we replace the notion $\xi_{\mathbb{Q}}^x(t)$ by $\zeta_{\mathbb{Q}}^x(t)$ for utility functions $W_\gamma (\gamma < 0)$. Namely, we put $\zeta_{\mathbb{Q}}^x(t) = I(\mathcal{Y}_t^{\mathbb{Q}}(x)\beta_t Z_t^{\mathbb{Q}})$. Then in the same way as in Sect. 13.2.1, we can obtain that

$$\zeta_{\mathbb{Q}}^x(t) = \frac{1}{\gamma}\left\{1 - \frac{\beta_t - \gamma x}{\beta_t} \frac{(Z_t^{\mathbb{Q}})^{\frac{\gamma}{1-\gamma}}}{\mathbb{E}\left[(Z_t^{\mathbb{Q}})^{\frac{1}{1-\gamma}}\right]}\right\}, \gamma < 0, \qquad (13.11)$$

$$\mathbb{E}\left[W_\gamma(\zeta_{\mathbb{Q}}^x(T))\right] = -(\beta_T - \gamma x)\mathcal{Y}_T^{\mathbb{Q}}(x), \quad \gamma < 0.$$

Thus, we can get the following result.

Theorem 13.3   *For utility functions $U = W_\gamma (\gamma < 0)$, the statement (1) in Theorem 13.1 is equivalent to the following statement:*

(5) $\mathbb{E}\left[(Z_T^{\mathbb{Q}})^{\frac{1}{1-\gamma}}\right] \geqslant \mathbb{E}\left[(Z_T^{\mathbb{R}})^{\frac{1}{1-\gamma}}\right]$, *for all* $\mathbb{R} \in \mathcal{P}$.

For utility function $W_\gamma(\gamma < 0)$, by Theorem 13.1 and results in Sect. 13.1, we know that if there exists a probability measure $\mathbb{Q} \in \mathcal{P}$ such that $\zeta_{\mathbb{Q}}^x(T)$ can be replicated by some strategy $\psi \in \mathcal{A}_{\frac{1}{\gamma}}(x)$, then ψ is optimal, and the measure $\mathbb{Q}$ maximizes the Hellinger integral $H_{\frac{1}{1-\gamma}}(\mathbb{Q}, \mathbb{P}) = H_{1-\frac{1}{1-\gamma}}(\mathbb{P}, \mathbb{Q}) = H_\delta(\mathbb{P}, \mathbb{Q})$ over $\mathcal{P}$, where δ satisfies $1/\delta + 1/\gamma = 1$.

As in the previous section, one can show that $V'(x) = \mathcal{Y}_T^{\mathbb{Q}}(x)$.

13.2.3   Utility Function $W_0(x) = -e^{-x}$

For utility function $W_0(x) = -e^{-x}$, we have $W_0'(x) = e^{-x}$ and $I(x) = -\log x$. For all $\mathbb{Q} \in \mathcal{P}$, $t \in [0, T]$ and $y \in (0, \infty)$, we have

$$\beta_t Z_t^{\mathbb{Q}} I(y\beta_t Z_t^{\mathbb{Q}}) = -\beta_t Z_t^{\mathbb{Q}} \log(y\beta_t) - \beta_t Z_t^{\mathbb{Q}} \log Z_t^{\mathbb{Q}}.$$

Thus

$$\mathcal{P}_t = \{ \mathbb{Q} \in \mathcal{P} : \mathbb{E}[Z_t^{\mathbb{Q}} \log Z_t^{\mathbb{Q}}] \text{ is finite } \}.$$

In particular, $\mathcal{P}_T = \{ \mathbb{Q} \in \mathcal{P} : I_{\mathbb{P}}(\mathbb{Q}) \text{ is finite } \}$, and

$$-\log \mathcal{Y}_T^{\mathbb{Q}}(x) = \log \beta_T + \frac{x}{\beta_T} + \mathbb{E}[Z_T^{\mathbb{Q}} \log Z_T^{\mathbb{Q}}].$$

$$\xi_{\mathbb{Q}}^x(t) = \frac{x}{\beta_t} + \mathbb{E}[Z_t^{\mathbb{Q}} \log Z_t^{\mathbb{Q}}] - \log Z_t^{\mathbb{Q}}, \tag{13.12}$$

$$\mathbb{E}\left[W_0(\xi_{\mathbb{Q}}^x(T)) \right] = -\beta_T \mathcal{Y}_T^{\mathbb{Q}}(x).$$

Thus, in the present case, the statement (1) is reduced to the following statement (2):
   (2) $\mathbb{E}[Z_T^{\mathbb{Q}} \log Z_T^{\mathbb{Q}}] \leqslant \mathbb{E}[Z_T^{\mathbb{Q}'} \log Z_T^{\mathbb{Q}'}]$, for all $\mathbb{Q}' \in \mathcal{P}_T$.
Consequently, we get the following:

Theorem 13.4   *Let $x > 0$. If there exists a probability measure $\mathbb{Q} \in \mathcal{P}_T$ such that $\xi_{\mathbb{Q}}^x(T) = V_T(\widehat{\psi})$ for a trading strategy $\widehat{\psi} \in \mathcal{A}(x)$, then $\widehat{\psi}$ is optimal, and within $\mathcal{P}_T$, $\mathbb{Q}$ has the minimum relative entropy w.r.t. historical measure $\mathbb{P}$.*

13.3   Market Driven by a Lévy Process

13.3.1   The Market Model

Let (X_t) be a cadlag version of a Lévy process with jump measure μ. For a Borel set Λ in $\mathbb{R}\backslash\{0\}$, we put

$$\nu(\Lambda) = \mathbb{E}[\mu([0, 1] \times \Lambda)], \quad \widetilde{\mu}(dt, dx) = dt\nu(dx).$$

Note that $\widetilde{\mu}$ is the predictable projection of μ.
   It is well known that (X_t) has the following Lévy decomposition (see Protter 2004 or He et al. 1992)

$$X_t = \alpha t + c B_t + M_t + A_t,$$

where α and c are constants, (B_t) is a standard Brownian motion, and

$$M_t = \int_{[0,t]\times[|x|<1]} x(\mu(ds, dx) - \widetilde{\mu}(ds, dx)),$$

$$A_t = \int_{[0,t]\times[|x|\geqslant 1]} x\mu(ds, dx) = \sum_{0<s\leqslant t} \Delta X_s I_{[|\Delta X_s|\geqslant 1]}.$$

The definition of stochastic integral with respect to $(\mu - \widetilde{\mu})$ can be seen in He et al. (1992) or Jacod and Shiryaev (1987). We recall the fact that $\int_{\mathbb{R}\setminus\{0\}} (x^2 \wedge 1)\nu(dx) < \infty$.

Now we consider a financial market in which there are two assets: a stock and a risk-free bond. We assume that the price process S_t of the stock satisfies the following equation:

$$dS_t = S_{t-}(\sigma_t dX_t + b_t dt),$$

where σ_t, b_t are deterministic functions of t. We assume that $\sigma_t \neq 0$. The bond price at time t is $S_t^0 = \exp\{\int_0^t r_u du\}$, where r_t is a nonnegative deterministic function of t. The discount process is $\beta_t = \exp\{-\int_0^t r_u du\}$. We assume that there exist $c_1 \in [0, 1)$ and $0 < c_2 \leqslant \infty$ such that $-c_1 \leqslant \Delta X \leqslant c_2$. Consequently, the Lévy measure ν is supported by $[-c_1, c_2]$. We assume that $\sigma_t \in (-c_2^{-1}, c_1^{-1})$ to ensure the strict positivity of S_t.

In the sequel, we always assume $\alpha = 0$, as otherwise we can replace b_t by $b_t + \alpha \sigma_t$.

The wealth $V_t(\psi)$ of a trading strategy $\psi = \{\varphi^0, \varphi\}$ at time t is $V_t(\psi) = \varphi_t^0 S_t^0 + \varphi_t S_t$. For a self-financing strategy ψ, let

$$\pi_t = \varphi_t S_{t-}/V_{t-}(\psi), \qquad (13.13)$$

then we have

$$d(\beta_t S_t) = \beta_t S_{t-} d\widetilde{X}_t, \qquad (13.14)$$

$$d(\beta_t V_t(\psi)) = \varphi_t d(\beta_t S_t) = \beta_t \varphi_t S_{t-} d\widetilde{X}_t = \beta_t V_{t-}(\psi)\pi_t d\widetilde{X}_t, \qquad (13.15)$$

where

$$d\widetilde{X}_t = (b_t - r_t)dt + \sigma_t(cdB_t + dM_t + dA_t). \qquad (13.16)$$

If $\pi_t \sigma_t$ takes value in $(-c_2^{-1}, c_1^{-1})$, then $(V_t(\psi))$ is strictly positive. The predictable process $\pi = (\pi_t)$ represents the proportion of the agent's wealth invested in the stock. We denote by $(V_t^{x,\pi})$ the wealth process corresponding to the portfolio π with initial wealth x. On the other hand, for any given predictable process π such that $\pi_t \sigma_t \in (-c_2^{-1}, c_1^{-1})$, let $\varphi_t = \pi_t V_{t-}^{x,\pi}/S_{t-}$, then there exists a unique self-financing strategy $\psi = (\varphi^0, \varphi)$ with initial capital x such that $V_t(\psi) = V_t^{x,\pi}$ or equivalently (13.15) holds.

Denote by $L(B)$ (resp. $\mathcal{G}(\mu)$) the set of all predictable processes (resp. predictable functions) which are integrable w.r.t. B (resp. $(\mu - \widetilde{\mu})$), i.e.,

$$L(B) = \left\{ \theta_1 : \theta_1 \text{ is a predictable process and } \int_0^T \theta_1^2(s)ds < \infty \text{ a.s.} \right\},$$

$$\mathcal{G}(\mu) = \left\{ W \in \widetilde{\mathcal{P}} : \sqrt{\sum_{0 < s \leqslant t} W^2(s, \Delta X_s) I_{[\Delta X_s \neq 0]}} \text{ is a locally integrable increasing} \right.$$

$$\left. \text{process and } \forall t \in [0, T], \int_{\mathbb{R} \setminus \{0\}} |W(t, x)| \nu(dx) < \infty \right\},$$

where $\widetilde{\mathcal{P}} = \mathcal{P} \times \mathcal{B}(\mathbb{R} \setminus \{0\})$.

For $\theta_1 \in L(B)$ and $1 - \theta_2 \in \mathcal{G}(\mu)$, denote $\theta = (\theta_1, \theta_2)$ and

$$Z_\theta = \mathcal{E} \left(-\int \theta_1 dB - \int (1 - \theta_2) d(\mu - \widetilde{\mu}) \right),$$

where $\mathcal{E}(Y)$ denotes the Doléans-Dade exponential of Y. By the Doléans-Dade formula, we have

$$Z_\theta(t) = \exp \left\{ -\int_0^t \theta_1(s) dB_s - \frac{1}{2} \int_0^t \theta_1^2(s) ds - \int_0^t \int_{\mathbb{R} \setminus \{0\}} (1 - \theta_2) d(\mu - \widetilde{\mu}) \right.$$

$$\left. + \int_0^t \int_{\mathbb{R} \setminus \{0\}} (\log \theta_2 + 1 - \theta_2) d\mu \right\}.$$

(13.17)

Denote

$$\Theta = \{\theta : \theta_1 \in L(B), \ 1 - \theta_2 \in \mathcal{G}(\mu) \text{ and } Z_\theta \text{ is a strictly positive martingale} \}.$$

For $\theta \in \Theta$, define the probability measure $\mathbb{P}_\theta$ on $(\Omega, \mathcal{F}_T)$ by $d\mathbb{P}_\theta = Z_\theta(T) d\mathbb{P}$, then $\mathbb{E} \left[\frac{d\mathbb{P}_\theta}{d\mathbb{P}} \Big| \mathcal{F}_t \right] = Z_\theta(t)$. It is well known that the predictable projection of μ under the probability measure P_θ is $\widetilde{\mu}_\theta(dt, dx) = \theta_2(t, x) \widetilde{\mu}(dt, dx) = \theta_2(t, x) dt \nu(dx)$.

Recall that $\mathcal{P}$ is the collection of all equivalent local martingale measures for $(\beta_t S_t)$. The following theorem can be proved by the usual argument.

Theorem 13.5 $\mathbb{Q} \in \mathcal{P}$ if and only if $\mathbb{Q} = \mathbb{P}_\theta$ for some $\theta \in \Theta$ such that

$$\frac{b_t - r_t}{\sigma_t} - c\theta_1(t) + \int_{[|x| \geqslant 1]} x\theta_2(t, x)\nu(dx), - \int_{[|x| < 1]} x(1 - \theta_2(t, x))\nu(dx) = 0$$

for all $t \in [0, T]$.

13.3.2   Results for HARA Utility Functions

In this subsection, we will explicitly work out the optimal strategies for HARA utility functions $U_\gamma (\gamma \leqslant 0)$. The minimum relative entropy martingale measure and maximum Hellinger integral martingale measure are also obtained.

Lemma 13.6 *If the following condition holds*

$$\int_0^T \int_{\mathbb{R}\setminus\{0\}} |g(s,x)|\nu(dx)ds < \infty, \tag{13.18}$$

then

$$g * (\mu - \widetilde{\mu}) = g * \mu - g * \widetilde{\mu}.$$

Proof By (13.18) and Theorem 11.21 in He et al. (1992), we see that $g * (\mu - \widetilde{\mu})$ is well defined and of locally integrable variation. Since $\sqrt{|g|} * (\mu - \widetilde{\mu})$ is also well defined, it follows that $|g| * \mu = \left[\sqrt{|g|} * (\mu - \widetilde{\mu})\right]$ and $g * \mu$ is well defined. Note that $g * (\mu - \widetilde{\mu}) - g * \mu$ is continuous, it is easy to see that $g * \mu$ is of integrable variation. Thus the conclusion follows. $\qquad\square$

For any $\theta \in \Theta$, $t \in [0, T]$, and the HARA utility function $U_\gamma (\gamma \leqslant 0)$, we put

$$\delta = \frac{\gamma}{\gamma - 1}, \quad \xi_\theta^x(t) = \frac{x}{\beta_t \mathbb{E}[Z_\theta^\delta(t)]} \cdot (Z_\theta(t))^{\frac{1}{\gamma-1}}.$$

It is easy to see that

$$\beta_t \xi_\theta^x(t) = \frac{x}{\mathbb{E}[Z_\theta^\delta(t)]} \cdot \exp\left\{ -\int_0^t \frac{1}{\gamma-1}\theta_1(s)dB_s - \frac{1}{2}\int_0^t \frac{1}{\gamma-1}\theta_1^2(s)ds \right.$$

$$\left. -\int_0^t \int_{\mathbb{R}\setminus\{0\}} \frac{1}{\gamma-1}(1 - \theta_2)d(\mu - \widetilde{\mu}) + \int_0^t \int_{\mathbb{R}\setminus\{0\}} \frac{1}{\gamma-1}(\log\theta_2 + 1 - \theta_2)d\mu \right\}.$$
$$\tag{13.19}$$

We have

$$Z_\theta^\delta(t) = \exp\left\{ -\int_0^t \delta\theta_1(s)dB_s - \frac{1}{2}\int_0^t \delta\theta_1^2(s)ds - \int_0^t \int_{\mathbb{R}\setminus\{0\}} \delta(1 - \theta_2)d(\mu - \widetilde{\mu}) \right.$$

$$\left. + \int_0^t \int_{\mathbb{R}\setminus\{0\}} \left(\log\theta_2^\delta + \delta(1 - \theta_2)\right) d\mu \right\}.$$

Under the following two conditions:

(6) $(1 - \theta_2^\delta) * (\mu - \widetilde{\mu})$ is well defined,

(7) $\int_0^T \int_{\mathbb{R}\setminus\{0\}} |\theta_2^\delta - 1 + \delta(1 - \theta_2)|(s, x)\nu(dx)ds < \infty$,

we have by Lemma 13.6

$$Z_\theta^\delta(t) = \mathcal{E}\left(-\int \delta\theta_1 dB - (1 - \theta_2^\delta) * (\mu - \widetilde{\mu}) \right)_t$$

$$\times \exp\left\{ \frac{1}{2}\int_0^t (\delta^2 - \delta)\theta_1^2(s)ds + \int_0^t \int_{\mathbb{R}\setminus\{0\}} [(\theta_2^\delta - 1) + \delta(1 - \theta_2)](s, x)\nu(dx)ds \right\}.$$

Under condition

(8) $\mathcal{E}\left(-\int \delta\theta_1 dB - (1 - \theta_2^\delta) * (\mu - \widetilde{\mu})\right)$ is a martingale and (θ_1, θ_2) is deterministic,

we have

$$\mathbb{E}[Z_\theta^\delta(t)] = \exp\left\{\frac{1}{2}\int_0^t (\delta^2 - \delta)\theta_1^2(s)ds + \int_0^t \int_{\mathbb{R}\setminus\{0\}} [(\theta_2^\delta - 1) + \delta(1 - \theta_2)](s, x)\nu(dx)ds\right\}.$$

(13.20)

It is also easy to see that

$$\beta_t V_t^{x,\pi} = x \exp\left\{\int_0^t \pi_s(b_s - r_s)ds + \int_0^t c\pi_s\sigma_s dB_s - \frac{1}{2}\int_0^t c^2\pi_s^2\sigma_s^2 ds\right.$$

$$+ \int_0^t \int_{\mathbb{R}\setminus\{0\}} \pi_s\sigma_s x I_{[|x|<1]}d(\mu - \widetilde{\mu})$$ (13.21)

$$\left. + \int_0^t \int_{\mathbb{R}\setminus\{0\}} (\log(1 + \pi_s\sigma_s x) - \pi_s\sigma_s x I_{[|x|<1]})d\mu\right\}.$$

By comparison between (13.19) and (13.21) and taking (13.20) into account, we see that in order to have $\xi_\theta^x(T) = V_T^{x,\pi}$, we should take

$$\theta_1(s) = (1 - \gamma)ca_s, \quad \theta_2(s, x) = (1 + a_s x)^{\gamma - 1},$$ (13.22)

where $a_s = \pi_s\sigma_s$. By Theorem 13.5, a_s solves the equation

$$\frac{b_s - r_s}{\sigma_s} + (\gamma - 1)c^2 a_s + \int_{\mathbb{R}\setminus\{0\}} \left(x(1 + a_s x)^{\gamma - 1} - x I_{[|x|<1]}\right)\nu(dx) = 0.$$ (13.23)

Conversely, assume that (a_s) is a function defined on $(-c_2^{-1}, c_1^{-1})$ and solves Eq. (13.23). We put $\pi_s = a_s/\sigma_s$ and define θ_1 and θ_2 by (13.22). If the above conditions (6), (7), and (8) and the following two conditions

(9) $\int_0^T \int_{\mathbb{R}\setminus\{0\}} |\log(1 + a_s x) - a_s x I_{[|x|<1]}|\nu(dx)ds < \infty$,

(10) $\int_0^T \int_{\mathbb{R}\setminus\{0\}} |\log\theta_2 + 1 - \theta_2|(s, x)\nu(dx)ds < \infty$,

are satisfied, then by Lemma 13.6, it is easy to verify from (13.19), (13.21), and (13.20) that we have $\xi_\theta^x(t) = V_t^{x,\pi}$, for $t \in [0, T]$.

Lemma 13.7 *For given $b, r, \sigma \neq 0$, and $\gamma \leqslant 0$, we put*

$$h(a) = \frac{b - r}{\sigma} + (\gamma - 1)c^2 a + \int_{\mathbb{R}\setminus\{0\}} \left(x(1 + ax)^{\gamma - 1} - x I_{[|x|<1]}\right)\nu(dx), \quad a \in (-c_2^{-1}, c_1^{-1}).$$

Then $h(a) = 0$ has a unique solution a^* in $(-c_2^{-1}, c_1^{-1})$ if and only if

$$\lim_{a\downarrow -c_2^{-1}} h(a) > 0, \quad \lim_{a\uparrow c_1^{-1}} h(a) < 0.$$ (13.24)

Proof For a fixed $a \in (-c_2^{-1}, c_1^{-1})$, both $x(1+ax)^{\gamma-1}I_{[x \geqslant 1]}$ and $(1+ax)^{\gamma-1}I_{[x \geqslant 1]}$ are bounded over $x \in [-c_1, c_2]$ by some constant $K_1 > 0$. Thus, we have

$$\left| x(1+ax)^{\gamma-1}I_{[x \geqslant 1]} \right| \leqslant K_1(x^2 \wedge 1) \quad \text{for } x \in [-c_1, c_2].$$

On the other hand, by the differential mean value theorem, for a fixed $x \in (-c_1, c_2)$, we have

$$x\left((1+ax)^{\gamma-1} - 1\right) = (\gamma - 1)ax^2(1+\xi x)^{\gamma-2}$$

for some $\xi \in (0, a)$ (resp. $\xi \in (a, 0)$), if $a > 0$ (resp. $a < 0$). Obviously, for $a \in (0, c_1^{-1})$ (resp. $a \in (-c_2^{-1}, 0)$), $(1+\xi x)^{\gamma-2} < (1 - ac_1)^{\gamma-2}$ (resp. $< (1 + ac_2)^{\gamma-2}$) for $x \in [-c_1, c_2]$. Thus, for some constant $K_2 > 0$, we have

$$\left| x\left((1+ax)^{\gamma-1} - 1\right) I_{[|x|<1]} \right| \leqslant K_2(x^2 \wedge 1) \quad \text{for } x \in [-c_1, c_2].$$

Since $0 \leqslant c_1 < 1$, ν is supported by $[-c_1, c_2] \subset (-1, \infty)$ and $\int_{\mathbb{R} \setminus \{0\}}(x^2 \wedge 1)\nu(dx) < \infty$, we see that $h(a)$ is well defined.

Let $a \in (-c_2^{-1}, c_1^{-1})$ and $\varepsilon > 0$ such that $a + \varepsilon \in (-c_2^{-1}, c_1^{-1})$. Since

$$x\left((1 + ax + \varepsilon x)^{\gamma-1} - (1 + ax)^{\gamma-1}\right) < 0 \quad \text{for } x \in [-c_1, c_2] \setminus \{0\},$$

h decreases strictly in $(-c_2^{-1}, c_1^{-1})$. By the monotone convergence theorem, h is continuous in $(-c_2^{-1}, c_1^{-1})$. Thus, the conclusion of the lemma follows. $\square$

In the sequel, for each $t \in [0, T]$, we put

$$h_t(a) = \frac{b_t - r_t}{\sigma_t} + (\gamma - 1)c^2 a + \int_{\mathbb{R} \setminus \{0\}} \left(x(1+ax)^{\gamma-1} - xI_{[|x|<1]} \right) \nu(dx),$$

$$a \in (-c_2^{-1}, c_1^{-1})$$

and assume that

$$\lim_{a \downarrow -c_2^{-1}} h_t(a) > 0 \quad \text{and} \quad \lim_{a \uparrow c_1^{-1}} h_t(a) < 0.$$

Then by Lemma 13.7 $h_t(a) = 0$ has a unique solution a_t^* in $(-c_2^{-1}, c_1^{-1})$. Define $\pi_t^* = \sigma_t^{-1}a_t^*$, and then obviously (V_t^{x,π^*}) is strictly positive.

Theorem 13.8 *Let $\widehat{\theta}_1(s) = (1 - \gamma)ca_s^*$, $\widehat{\theta}_2(s, x) = (1 + a_s^*x)^{\gamma-1}$, and $\widehat{\theta} = (\widehat{\theta}_1, \widehat{\theta}_2)$. If*

$$\int_0^T (a_s^*)^2 ds < \infty, \quad \int_0^T \int_{\mathbb{R}\backslash\{0\}} \left(1 - (1 + a_s^* x)^{\gamma-1}\right)^2 \nu(dx) ds < \infty, \quad (13.25)$$

then $\widehat{\theta} \in \Theta$ and $\mathbb{P}_{\widehat{\theta}} \in \mathcal{P}$. If furthermore

$$\int_0^T \int_{\mathbb{R}\backslash\{0\}} |\widehat{\theta}_2^\delta - 1 + \delta(1 - \widehat{\theta}_2)|(s, x)\nu(dx) ds < \infty, \quad (13.26)$$

$$\int_0^T \int_{\mathbb{R}\backslash\{0\}} (1 - \widehat{\theta}_2^\delta(s, x))^2 \nu(dx) ds < \infty, \quad (13.27)$$

$$\int_0^T \int_{\mathbb{R}\backslash\{0\}} |\log(1 + a_s^* x) - a_s^* x I_{[|x|<1]}| \nu(dx) ds < \infty, \quad (13.28)$$

$$\int_0^T \int_{\mathbb{R}\backslash\{0\}} |\log \widehat{\theta}_2 + 1 - \widehat{\theta}_2|(s, x)\nu(dx) ds < \infty, \quad (13.29)$$

then $\xi_{\widehat{\theta}}^x(T) = V_T^{x,\pi^*}$ for utility function $U(x) = U_\gamma(x)$, $\gamma \leqslant 0$.

Proof By (13.25) and Theorem 11.21 in He et al. (1992), we have $\widehat{\theta}_1 \in L(B)$ and $1 - \widehat{\theta}_2 \in \mathcal{G}(\mu)$. Thus, $Z_{\widehat{\theta}}$ is well defined. It is obvious that $-(1 - \widehat{\theta}_2(s, x)) > -1$ for $x \in [-c_1, c_2]$. Therefore, $Z_{\widehat{\theta}}$ is strictly positive. By (13.25),

$$\langle -\widehat{\theta}_1.B - (1 - \widehat{\theta}_2) * (\mu - \widetilde{\mu}) \rangle_T$$
$$= (1 - \gamma)^2 c^2 \int_0^T (a_s^*)^2 ds + \int_0^T \int_{\mathbb{R}\backslash\{0\}} (1 - (1 + a_s^* x)^{\gamma-1})^2 \nu(dx) ds$$

is a finite constant. Then by a result of Lépingle and Mémin (1978), $Z_{\widehat{\theta}}$ is a square integrable martingale. It follows that $\widehat{\theta} \in \Theta$. On the other hand, since $h_t(a_t^*)=0$, $\widehat{\theta}$ satisfies the condition of Theorem 13.5, we have $\mathbb{P}_{\widehat{\theta}} \in \mathcal{P}$.

By (13.27) and Theorem 11.21 in He et al. (1992), $(1 - \widehat{\theta}_2^\delta) * (\mu - \widetilde{\mu})$ is well defined. By (13.25), (13.27), and once again by a result of Lépingle and Mémin (1978), $\mathcal{E}\left(-\delta\widehat{\theta}_1.B - (1 - \widehat{\theta}_2^\delta) * (\mu - \widetilde{\mu})\right)$ is a square integrable martingale. Hence, according to the discussion at the beginning of this subsection, we complete the proof. □

By Theorem 13.8 and the results in Sect. 13.2, we know that the strategy corresponding to π^* is optimal for the HARA utility. Moreover, $\mathbb{P}_{\widehat{\theta}}$ defined in Theorem 13.8 is just the associated optimal equivalent local martingale measure. For $\gamma = 0$, $\mathbb{P}_{\widehat{\theta}}$ minimizes the relative entropy $I_{\mathbb{Q}}(\mathbb{P})$ over $\mathcal{P}$. Yan et al. (2000) also got this minimum relative entropy by the numeraire portfolio approach. For $\gamma < 0$, $\mathbb{P}_{\widehat{\theta}}$ maximizes the Hellinger integral $H_\delta(\mathbb{Q}, \mathbb{P})$ over $\mathcal{P}$, where δ satisfies $\frac{1}{\delta} + \frac{1}{\gamma} = 1$.

13.3.3 Results for Utility Functions of the Form W_γ ($\gamma < 0$)

In this subsection, we deal with utility functions of the form W_γ ($\gamma < 0$). In this case, we have $D_{W_\gamma} = 1/\gamma$, and

$$\zeta_\theta^x(t) = \frac{1}{\gamma} \left\{ 1 - \frac{\beta_t - \gamma x}{\beta_t} \cdot \frac{(Z_\theta(t))^{\frac{\gamma}{1-\gamma}}}{\mathbb{E}[(Z_\theta(t))^{\frac{1}{1-\gamma}}]} \right\}.$$

Let $1/\delta + 1/\gamma = 1$, i.e., $\delta = \gamma/(\gamma - 1)$. Let $\gamma' = 1/\gamma$, $1/\delta' + 1/\gamma' = 1$, and then $\delta' = \gamma'/(\gamma' - 1) = 1 - \delta$. Under the conditions and notations in Theorem 13.8, we have

$$\frac{(Z_{\widehat{\theta}}(t))^{\frac{\gamma}{1-\gamma}}}{\mathbb{E}\left[(Z_{\widehat{\theta}}(t))^{\frac{1}{1-\gamma}}\right]} = \frac{(Z_{\widehat{\theta}}(t))^{\frac{1}{\gamma'-1}}}{\mathbb{E}\left[(Z_{\widehat{\theta}}(t))^{\delta'}\right]} = \frac{\beta_t \xi_{\widehat{\theta}}^x(t)}{x} = \frac{\beta_t V_t^{x,\pi^*}}{x}.$$

Thus,

$$\zeta_{\widehat{\theta}}^x(t) = \frac{1}{\gamma} + \left(1 - \frac{\beta_t}{\gamma x}\right) V_t^{x,\pi^*}.$$

Now we construct a self-financing strategy $\widehat{\psi} = (\widehat{\varphi}^0, \widehat{\varphi})$ as follows. Let $\psi = (\varphi^0, \varphi)$ be the strategy corresponding to the portfolio π^*. We put $\widehat{\varphi}_t^0 = \left(1 - \frac{\beta_T}{\gamma x}\right)\varphi_t^0 + \frac{\beta_T}{\gamma}$, $\widehat{\varphi}_t = (1 - \frac{\beta_T}{\gamma x})\varphi_t$. Then

$$\beta_t V_t(\widehat{\psi}) = \frac{\beta_T}{\gamma} + \left(1 - \frac{\beta_T}{\gamma x}\right)\beta_t V_t^{x,\pi^*} > \frac{1}{\gamma}, \quad t \in [0, T],$$

which implies $\widehat{\psi} \in \mathcal{A}_{\frac{1}{\gamma}}(x)$. Moreover, we have $V_T(\widehat{\psi}) = \frac{1}{\gamma} + \left(1 - \frac{\beta_T}{\gamma x}\right) V_T^{x,\pi^*} = \zeta_{\widehat{\theta}}^x(T)$. Thus, by a result of Sect. 13.2, $\widehat{\psi}$ is optimal for utility function $W_\gamma(x) = -(1 - \gamma x)^{\frac{1}{\gamma}}$ ($\gamma < 0$), and the martingale measure $\mathbb{P}_{\widehat{\theta}}$ maximizes the Hellinger integral $H_{\delta'}(\mathbb{P}, \mathbb{Q})$ over $\mathcal{P}$.

13.3.4 Results for Utility Function $W_0(x) = -e^{-x}$

For $\theta \in \Theta$ and $t \in [0, T]$, we have by (13.12) and (13.17)

$$\beta_t \xi_\theta^x(t) = x + \beta_t \mathbb{E}[Z_\theta(t) \log Z_\theta(t)] + \beta_t \left[\int_0^t \theta_1(s)dB_s + \frac{1}{2}\int_0^t \theta_1^2(s)ds \right.$$
$$\left. + \int_0^t \int_{\mathbb{R}\backslash\{0\}} (1 - \theta_2)d(\mu - \widetilde{\mu}) - \int_0^t \int_{\mathbb{R}\backslash\{0\}} (\log \theta_2 + 1 - \theta_2)d\mu \right].$$

On the other hand, it is easy to see that

$$\beta_t V_t(\psi) = x + \int_0^t \beta_s \varphi_s S_{s-}[(b_s - r_s)ds + c\sigma_s dB_s + \sigma_s dM_s + \sigma_s dA_s].$$

Comparing these two expressions leads us to guess that in order to have $\xi_\theta^x(T) = V_T(\psi)$, we should take

$$\begin{cases} \beta_T \theta_1(s) = c\beta_s \varphi_s S_{s-}\sigma_s \\ \log \theta_2(s, x) = -\frac{\beta_s \varphi_s S_{s-}\sigma_s x}{\beta_T}. \end{cases}$$

Denoting $a_s = \beta_s \varphi_s S_{s-}\sigma_s / \beta_T$, then we have

$$\theta_1(s) = ca_s, \quad \theta_2(s, x) = e^{-a_s x}.$$

By Theorem 13.5, a_s should solve the equation

$$\frac{b_s - r_s}{\sigma_s} - c^2 a_s + \int_{\mathbb{R}\setminus\{0\}} (xe^{-a_s x} - xI_{[|x|<1]})\nu(dx) = 0.$$

Lemma 13.9 *For given b, r, and $\sigma \neq 0$, we put*

$$f(a) = \frac{b - r}{\sigma} - c^2 a + \int_{\mathbb{R}\setminus\{0\}} (xe^{-ax} - xI_{[|x|<1]})\nu(dx).$$

Then there exists some $k \in [-\infty, 0]$ such that $f(a)$ is $+\infty$ for $a < k$ and finite for $a > k$. Furthermore, $f(a) = 0$ has a unique solution $\tilde{a}$ in (k, ∞) if and only if $\lim_{a\downarrow k} f(a) > 0$.

Proof For any $a \in (-\infty, \infty)$, we have

$$|x(e^{-ax} - 1)I_{[|x|<1]}| \leqslant (|a|e^{|a|}x^2) \wedge (1 + e^{|a|}).$$

Thus, $\int_{\mathbb{R}\setminus\{0\}} |x(e^{-ax} - 1)I_{[|x|<1]}|\nu(dx) < \infty$. It is clear that there exists some $k(-\infty \leqslant k \leqslant 0)$ such that $\int_{\mathbb{R}\setminus\{0\}} xe^{-ax}I_{[x\geqslant 1]}\nu(dx)$ is $+\infty$ for $a < k$ and finite for $a > k$. Since $0 \leqslant c_1 < 1$ and ν is supported by $[-c_1, c_2] \subset (-1, \infty)$, the first property of $f(a)$ follows. For any $a \in (k, \infty)$ and $\varepsilon > 0$, it is obvious that $xe^{-(a+\varepsilon)x} - xe^{-ax} < 0$ for $x \in \mathbb{R} \setminus \{0\}$. It follows that

$$f(a + \varepsilon) - f(a) = -c^2\varepsilon + \int_{\mathbb{R}\setminus\{0\}} (xe^{-(a+\varepsilon)x} - xe^{-ax})\nu(dx) < 0.$$

Thus f decreases strictly in (k, ∞). By the monotone convergence theorem, f is also continuous in (k, ∞). The fact that $\lim_{a \to \infty} f(a) = -\infty$ is clear. Thus the conclusion follows.   □

For each $t \in [0, T]$, we put

$$f_t(a) = \frac{b_t - r_t}{\sigma_t} - c^2 a + \int_{\mathbb{R} \setminus \{0\}} (x e^{-ax} - x I_{[|x| < 1]}) \nu(dx), \ a \in (k, \infty)$$

and assume that $\lim_{a \downarrow k} f_t(a) > 0$. Then by Lemma 13.9, $f_t(a) = 0$ has a unique solution $\widetilde{a}_t$ in (k, ∞). For $t \in [0, T]$, we define $\varphi_s^t = \frac{\beta_t \widetilde{a}_s}{\beta_s \sigma_s S_{s-}}$, $s \in [0, t]$ and let ψ^t be the self-financing strategy corresponding to φ^t with initial capital x.

Theorem 13.10 Let $\widetilde{\theta}_1(s) = c\widetilde{a}_s$, $\widetilde{\theta}_2(s, x) = e^{-\widetilde{a}_s x}$, $\widetilde{\theta} = (\widetilde{\theta}_1, \widetilde{\theta}_2)$. If

$$\int_0^T (\widetilde{a}_s)^2 ds < \infty, \qquad \int_0^T \int_{\mathbb{R} \setminus \{0\}} \left(1 - e^{-\widetilde{a}_s x}\right)^2 \nu(dx) ds < \infty, \tag{13.30}$$

then $\widetilde{\theta} \in \Theta$ and $\mathbb{P}_{\widetilde{\theta}} \in \mathcal{P}$. Moreover, if

$$\int_0^T \int_{\mathbb{R} \setminus \{0\}} \left|1 - e^{-\widetilde{a}_s x} - \widetilde{a}_s x\right| e^{-\widetilde{a}_s x} \nu(dx) ds < \infty, \tag{13.31}$$

$$\int_0^T \int_{\mathbb{R} \setminus \{0\}} \left|\widetilde{a}_s x e^{-\widetilde{a}_s x}\right| \nu(dx) ds < \infty, \tag{13.32}$$

then for utility function $W_0(x) = -e^{-x}$, $\xi_{\widetilde{\theta}}^x(T) = V_T(\psi^T)$. Furthermore, if $\int_{\mathbb{R} \setminus \{0\}} x^2 \nu(dx) < \infty$, then $\psi^T \in \mathcal{A}(x)$.

Proof Under condition (13.30), one can easily prove that $\widetilde{\theta} \in \Theta$, $\mathbb{P}_{\widetilde{\theta}} \in \mathcal{P}$ and $Z_{\widetilde{\theta}}$ is a $\mathbb{P}$-square integrable martingale.

Now let $\widetilde{B}_t = B_t + \int_0^t \widetilde{\theta}_1(s) ds$. Then $\widetilde{B}$ is a standard Brownian motion under probability measure $\mathbb{P}_{\widetilde{\theta}}$. In the following proof, for any integrable predictable function W, the integral $W * (\mu - \widetilde{\mu})$ (resp. $W * (\mu - \widetilde{\mu}_{\widetilde{\theta}})$) will be taken under probability measure $\mathbb{P}$ (resp. $\mathbb{P}_{\widetilde{\theta}}$). Since $1 - \widetilde{\theta}_2 \in \mathcal{G}(\mu)$, then $(1 - \widetilde{\theta}_2)^2 * \mu$ is well defined (in fact, $(1 - \widetilde{\theta}_2)^2 * \mu = \left[(1 - \widetilde{\theta}_2) * (\mu - \widetilde{\mu})\right]$), where $[M]$ is the quadratic variation process of local martingale M. By (13.30), we have

$$\mathbb{E}[(1 - \widetilde{\theta}_2)^2 * \mu_T] = \int_0^T \int_{\mathbb{R} \setminus \{0\}} \left(1 - e^{-\widetilde{a}_s x}\right)^2 \nu(dx) ds < \infty. \tag{13.33}$$

Thus, $(1 - \widetilde{\theta}_2)^2 * \mu$ is an integrable increasing process. It is clear that

$$-\left[(1 - \widetilde{\theta}_2) * (\mu - \widetilde{\mu}), Z_{\widetilde{\theta}}\right] = \int Z_{\widetilde{\theta}}(-) d((1 - \widetilde{\theta}_2)^2 * \mu)$$

is a locally integrable increasing process, where $Z_{\widetilde{\theta}}(-)$ is the left limit of $Z_{\widetilde{\theta}}$. Thus,

$$\left(\frac{1}{Z_{\widetilde{\theta}}(-)}\right) \cdot \langle (1-\widetilde{\theta}_2)*(\mu-\widetilde{\mu}), Z_{\widetilde{\theta}} \rangle = -\langle (1-\widetilde{\theta}_2)*(\mu-\widetilde{\mu}) \rangle = -(1-\widetilde{\theta}_2)^2 * \widetilde{\mu}.$$

By Theorem 12.28 in He et al. (1992), $(1-\widetilde{\theta}_2)*(\mu-\widetilde{\mu}_{\widetilde{\theta}})$ is well defined and

$$(1-\widetilde{\theta}_2)*(\mu-\widetilde{\mu}) = (1-\widetilde{\theta}_2)*(\mu-\widetilde{\mu}_{\widetilde{\theta}}) - (1-\widetilde{\theta}_2)^2 * \widetilde{\mu}. \tag{13.34}$$

It follows from (13.32) that under $\mathbb{P}_{\widetilde{\theta}}$, $(\log\widetilde{\theta}_2 + 1 - \widetilde{\theta}_2)*\mu$ is well defined and of integrable variation. Therefore,

$$(\log\widetilde{\theta}_2 + 1 - \widetilde{\theta}_2)*(\mu-\widetilde{\mu}_{\widetilde{\theta}}) = (\log\widetilde{\theta}_2 + 1 - \widetilde{\theta}_2)*\mu - (\log\widetilde{\theta}_2 + 1 - \widetilde{\theta}_2)*\widetilde{\mu}_{\widetilde{\theta}} \tag{13.35}$$

is a martingale. By (13.33), (13.34), and (13.35),

$$\log Z_{\widetilde{\theta}}(t) = -\int_0^t \widetilde{\theta}_1(s)d\widetilde{B}_s - (1-\widetilde{\theta}_2)*(\mu-\widetilde{\mu}_{\widetilde{\theta}})_t + (\log\widetilde{\theta}_2 + 1 - \widetilde{\theta}_2)*(\mu-\widetilde{\mu}_{\widetilde{\theta}})_t$$

$$+ \frac{1}{2}\int_0^t \widetilde{\theta}_1^2(s)ds + (1-\widetilde{\theta}_2)^2 * \widetilde{\mu}_t + \left((\log\widetilde{\theta}_2 + 1 - \widetilde{\theta}_2)*\widetilde{\mu}_{\widetilde{\theta}}\right)_t$$

$$= -\int_0^t \widetilde{\theta}_1(s)d\widetilde{B}_s - (1-\widetilde{\theta}_2)*(\mu-\widetilde{\mu}_{\widetilde{\theta}})_t + (\log\widetilde{\theta}_2 + 1 - \widetilde{\theta}_2)*(\mu-\widetilde{\mu}_{\widetilde{\theta}})_t$$

$$+ \frac{1}{2}\int_0^t \widetilde{\theta}_1^2(s)ds + (\widetilde{\theta}_2\log\widetilde{\theta}_2 + 1 - \widetilde{\theta}_2)*\widetilde{\mu}_t.$$

By (13.30), we know that $\int \widetilde{\theta}_1 d\widetilde{B}$ is a $\mathbb{P}_{\widetilde{\theta}}$-martingale. Since $Z_{\widetilde{\theta}}$ is a $\mathbb{P}$-square integrable martingale, then by (13.33),

$$\mathbb{E}_{\widetilde{\theta}}\left[\sqrt{[(1-\widetilde{\theta}_2)*(\mu-\widetilde{\mu}_{\widetilde{\theta}})]_T}\right] = \mathbb{E}_{\widetilde{\theta}}\left[\sqrt{(1-\widetilde{\theta}_2)^2 * \mu_T}\right]$$

$$= \mathbb{E}\left[Z_{\widetilde{\theta}}(T)\sqrt{(1-\widetilde{\theta}_2)^2 * \mu_T}\right] \leqslant \left(\mathbb{E}\left[Z_{\widetilde{\theta}}^2(T)\right] \cdot \mathbb{E}\left[(1-\widetilde{\theta}_2)^2 * \mu_T\right]\right)^{1/2} < \infty.$$

Thus, $(1-\widetilde{\theta}_2)*(\mu-\widetilde{\mu}_{\widetilde{\theta}})$ is an $\mathcal{H}^1$-martingale under $\mathbb{P}_{\widetilde{\theta}}$. So we have

$$\mathbb{E}[Z_{\widetilde{\theta}}(t)\log Z_{\widetilde{\theta}}(t)] = \mathbb{E}_{\widetilde{\theta}}[\log Z_{\widetilde{\theta}}(t)] = \frac{1}{2}\int_0^t \widetilde{\theta}_1^2(s)ds + (\widetilde{\theta}_2\log\widetilde{\theta}_2 + 1 - \widetilde{\theta}_2)*\widetilde{\mu}_t,$$

and

$$\xi_{\widetilde{\theta}}^x(t) = \frac{x}{\beta_t} + \mathbb{E}[Z_{\widetilde{\theta}}(t) \log Z_{\widetilde{\theta}}(t)] - \log Z_{\widetilde{\theta}}(t)$$

$$= \frac{x}{\beta_t} + \int_0^t \widetilde{\theta}_1(s) d\widetilde{B}_s - (\log \widetilde{\theta}_2) * (\mu - \widetilde{\mu}_{\widetilde{\theta}})_t$$

$$= \frac{x}{\beta_t} + \int_0^t c\widetilde{a}_s d\widetilde{B}_s + (\widetilde{a}_s x) * (\mu - \widetilde{\mu}_{\widetilde{\theta}})_t.$$

For each $t \in [0, T]$, $\varphi_s^t = \frac{\beta_t \widetilde{a}_s}{\beta_s \sigma_s S_{s-}}$, $s \in [0, t]$. Then by (13.15), at time t, the wealth of self-financing strategy ψ^t corresponding to φ^t with initial capital x satisfies

$$\beta_t V_t(\psi^t) = x + \int_0^t \beta_s \varphi_s^t S_{s-} ((b_s - r_s)ds + c\sigma_s dB_s + \sigma_s dM_s + \sigma_s dA_s)$$

$$= x + \beta_t \left(\int_0^t \widetilde{a}_s \frac{b_s - r_s}{\sigma_s} ds + \int_0^t c\widetilde{a}_s dB_s + \int_0^t \widetilde{a}_s dM_s + \int_0^t \widetilde{a}_s dA_s \right).$$

Hence,

$$V_t(\psi^t) = \frac{x}{\beta_t} + \int_0^t \widetilde{a}_s \frac{b_s - r_s}{\sigma_s} ds + \int_0^t c\widetilde{a}_s dB_s$$

$$+ (\widetilde{a}_s x I_{[|x|<1]}) * (\mu - \widetilde{\mu}) + (\widetilde{a}_s x I_{[|x|\geqslant 1]}) * \mu. \tag{13.36}$$

Clearly,

$$\left[(\widetilde{a}_s x I_{[|x|<1]}) * (\mu - \widetilde{\mu}), Z_{\widetilde{\theta}} \right] = - \int Z_{\widetilde{\theta}}(-) d \left((\widetilde{a}_s x I_{[|x|<1]} (1 - e^{-\widetilde{a}_s x})) * \mu \right).$$

However, $\widetilde{a}_s x I_{[|x|<1]} (1 - e^{-\widetilde{a}_s x}) \leqslant \frac{\widetilde{a}_s^2}{2}(x^2 \wedge 1) + \frac{1}{2}(1 - e^{-\widetilde{a}_s x})^2$. By (13.30), we see that process $\left[(\widetilde{a}_s x I_{[|x|<1]}) * (\mu - \widetilde{\mu}), Z_{\widetilde{\theta}} \right]$ is of locally integrable variation. Similar to (13.34), we have

$$(\widetilde{a}_s x I_{[|x|<1]}) * (\mu - \widetilde{\mu}) = (\widetilde{a}_s x I_{[|x|<1]}) * (\mu - \widetilde{\mu}_{\widetilde{\theta}}) - (\widetilde{a}_s x I_{[|x|<1]} (1 - e^{-\widetilde{a}_s x})) * \widetilde{\mu}.$$

By (13.33), similar to (13.35), we have

$$(\widetilde{a}_s x I_{[|x|\geqslant 1]}) * (\mu - \widetilde{\mu}_{\widetilde{\theta}}) = (\widetilde{a}_s x I_{[|x|\geqslant 1]}) * \mu - (\widetilde{a}_s x I_{[|x|\geqslant 1]}) * \widetilde{\mu}_{\widetilde{\theta}}.$$

Then it holds that

$$V_t(\psi^t) = \frac{x}{\beta_t} + \int_0^t c\tilde{a}_s d\tilde{B}_s + (\tilde{a}_s x) * (\mu - \tilde{\mu}_{\tilde{\theta}})_t + \int_0^t \tilde{a}_s \frac{b_s - r_s}{\sigma_s} ds$$

$$- \int_0^t c^2 \tilde{a}_s^2 ds - (\tilde{a}_s x I_{[|x|<1]}(1 - e^{-\tilde{a}_s x})) * \tilde{\mu}_t + ((\tilde{a}_s x I_{[|x|\geq 1]}) * \tilde{\mu}_{\tilde{\theta}})_t$$

$$= \xi_{\tilde{\theta}}^x(t) + \int_0^t \tilde{a}_s \left(\frac{b_s - r_s}{\sigma_s} - c^2 \tilde{a}_s + \int_{\mathbb{R}\setminus\{0\}} (x e^{-\tilde{a}_s x} - x I_{[|x|<1]}) \nu(dx) \right) ds$$

$$= \xi_{\tilde{\theta}}^x(t) + \int_0^t \tilde{a}_s f_s(\tilde{a}_s) ds = \xi_{\tilde{\theta}}^x(t), \quad t \in [0, T].$$

$$(13.37)$$

In particular, we have $V_T(\psi^T) = \xi_{\tilde{\theta}}^x(T)$. For $t \in [0, T]$, note that $\varphi_s^T = \frac{\beta_T}{\beta_t}\varphi_s^t$, $s \leq t$. Thus, we have

$$\beta_t V_t(\psi^T) = x + \int_0^t \beta_s \varphi_s^T S_{s-} d\tilde{X}_s = x + \int_0^t \frac{\beta_T}{\beta_t} \beta_s \varphi_s^t S_{s-} d\tilde{X}_s$$

$$= x + \frac{\beta_T}{\beta_t} \int_0^t \beta_s \varphi_s^t S_{s-} d\tilde{X}_s = x + \frac{\beta_T}{\beta_t}(\beta_t V_t(\psi^t) - x) \qquad (13.38)$$

$$= x + \beta_T V_t(\psi^t) - \frac{\beta_T}{\beta_t} x = \left(1 - \frac{\beta_T}{\beta_t}\right) x + \beta_T \xi_{\tilde{\theta}}^x(t).$$

By (13.36), (13.37), and (13.38), it is clear that $\psi^T \in \mathcal{A}(x)$ if $\int_{\mathbb{R}\setminus\{0\}} x^2 \nu(dx) < \infty$.

$$\square$$

Chapter 14
Optimal Growth Portfolios and Option Pricing

In this chapter we introduce the "optimal growth strategy" and the associated "optimal growth portfolios" in markets of semimartingale models. We work out expressions of "optimal growth portfolios" in a geometric Lévy process model and a jump-diffusion-like process model. In Sect. 14.2, we present the "numeraire portfolio approach" to contingent claim pricing in a geometric Lévy process model. In Sect. 14.3 we give an overview of other martingale measure approaches to contingent claim pricing.

We use the same concepts and notations as those employed in Chap. 11.

14.1 Optimal Growth Portfolio

The optimal growth portfolio (also called the numeraire portfolio) approach to option pricing was proposed by Long (1990) and further developed by Bajeux-Besnainou and Portrait (1997), known as the benchmark approach. The starting point of this approach is to search for a suitable derivative asset as a numeraire asset such that the deflated price processes of primitive assets are martingales under the historical probability measure. It turned out that this numeraire must be the wealth process of the optimal growth portfolio. For a market with asset returns being diffusion processes, the optimal growth portfolio is well known (see Karatzas and Shreve 1998).

In this section, we first introduce a general result about the optimal growth strategy in markets of semimartingale model. Then we follow Yan et al. (2000) to work out the optimal growth portfolio in a market with asset returns being a Lévy process or a jump-diffusion-like process.

© Springer Nature Singapore Pte Ltd. and Science Press 2018
J.-A. Yan, *Introduction to Stochastic Finance*, Universitext,
https://doi.org/10.1007/978-981-13-1657-9_14

14.1.1 Optimal Growth Strategy

Definition 14.1 An admissible self-financing strategy $\{\theta^0, \theta\}$ is called an *optimal growth strategy* if its wealth process V_t satisfies the following condition: for the wealth process (X_t) of any other admissible self-financing strategy with the same initial wealth V_0, we have

$$\mathbb{E}\left[\log X_t\right] \leqslant \mathbb{E}\left[\log V_t\right], \quad t \geqslant 0.$$

The following well-known theorem can be traced to Samuelson (1963). For reader's convenience we include its proof.

Theorem 14.2 *Let (V_t) be the wealth process of an admissible self-financing strategy with $V_0 = 1$. Assume that for every $0 \leqslant j \leqslant m$, $\left(V_t^{-1} S_t^j\right)$ is a $\mathbb{P}$-martingale. Then (V_t) is the unique optimal growth wealth process with initial value 1.*

Proof We fix arbitrarily a j. By the assumptions, we can define a probability measure $\mathbb{P}^j$ such that

$$M_t^j \triangleq \frac{d\mathbb{P}^j}{d\mathbb{P}}\bigg|_{\mathcal{F}_t} = V_t^{-1} S_t^j (S_0^j)^{-1}.$$

Since $\frac{d\mathbb{P}}{d\mathbb{P}^j}|_{\mathcal{F}_t} = (M_t^j)^{-1}$, we know that for each $0 \leqslant i \leqslant m$, $S_t^i (S_t^j)^{-1} S_0^j = (M_t^j)^{-1} S_t^i V_t^{-1}$ is a $\mathbb{P}^j$-martingale, and thus $\mathbb{P}^j \in \mathcal{P}^j$.

Now we assume that (X_t) is the wealth process of an admissible self-financing strategy with $X_0 = 1$. Then $\left(X_t (S_t^j)^{-1}\right)$ is a nonnegative $\mathbb{P}^j$-local martingale, hence a $\mathbb{P}^j$-supermartingale. Thus we have

$$\mathbb{E}\left[\frac{X_t}{V_t}\right] = \mathbb{E}^{(j)}\left[\frac{X_t}{V_t}\frac{d\mathbb{P}}{d\mathbb{P}^j}\bigg|_{\mathcal{F}_t}\right] = \mathbb{E}^{(j)}\left[X_t (S_t^j)^{-1}\right] S_0^j \leqslant 1. \tag{14.1}$$

By Jensen's inequality, this implies

$$\mathbb{E}\left[\log \frac{X_t}{V_t}\right] \leqslant \log \mathbb{E}\left[\frac{X_t}{V_t}\right] \leqslant 0. \tag{14.2}$$

Consequently, we have $\mathbb{E}[\log X_t] \leqslant \mathbb{E}[\log V_t]$. This means that V_t is an optimal growth wealth process. Its uniqueness can be proved as follows. Assume that (X_t) is an optimal growth wealth process with initial value 1. Then we have $\mathbb{E}\log X_t \geqslant \mathbb{E}\log V_t$, i.e., $\log \mathbb{E}\frac{X_t}{V_t} \geqslant 0$, which together with (14.2) implies

$$\mathbb{E}\left[\log \frac{X_t}{V_t}\right] = \log \mathbb{E}\left[\frac{X_t}{V_t}\right] = 0.$$

Consequently, both X_t/V_t and $\log(X_t/V_t)$ are $\mathbb{P}$-martingales, because they are already known to be $\mathbb{P}$-supermartingales. Therefore, $X_t = V_t$ a.s. $\qquad\square$

Definition 14.3 For a self-financing strategy $\phi = \{\theta^0, \theta\}$, we put

$$\pi_t = \theta_t \cdot S_{t-}/V_{t-}(\phi),$$

where $V_t(\phi) = \theta^0(t)S_t^0 + \theta(t) \cdot S_t$. We call (π_t) the *portfolio* associated to trading strategy (ϕ_t). The portfolio associated to the optimal growth strategy is called the *optimal growth portfolio*. The wealth process (V_t) of the optimal growth portfolio is called the *growth optimal wealth process*.

Recall that the relative entropy $I_\mathbb{Q}(\mathbb{P})$ of a probability measure $\mathbb{P}$ with respect to $\mathbb{Q}$ is defined by

$$I_\mathbb{Q}(\mathbb{P}) = \mathbb{E}_\mathbb{Q}\left[\frac{d\mathbb{P}}{d\mathbb{Q}}\log\frac{d\mathbb{P}}{d\mathbb{Q}}\right] = -\mathbb{E}_\mathbb{P}\left[\log\frac{d\mathbb{Q}}{d\mathbb{P}}\right]. \tag{14.3}$$

$I_\mathbb{Q}(\mathbb{P})$ is a quantitative measure of the difference between $\mathbb{Q}$ and $\mathbb{P}$.

The following theorem shows that the objective measure $\mathbb{P}$ has minimum relative entropy with respect to the measure $\mathbb{P}^j$ within $\mathcal{P}^j$. We refer the reader to Chan (1999) for a similar result (in a "dual form") about the Esscher transform.

Theorem 14.4 *We have*

$$\mathbb{E}\left[\log\left(\frac{d\mathbb{P}^j}{d\mathbb{P}}\Big|_{\mathcal{F}_t}\right)\right] \geqslant \mathbb{E}\left[\log\left(\frac{d\mathbb{Q}}{d\mathbb{P}}\Big|_{\mathcal{F}_t}\right)\right], \quad \forall \mathbb{Q} \in \mathcal{P}^j. \tag{14.4}$$

Proof Let $\mathbb{Q} \in \mathcal{P}^j$. We put $N_t = \frac{d\mathbb{Q}}{d\mathbb{P}}\Big|_{\mathcal{F}_t}$ and $L_t = \frac{d\mathbb{Q}}{d\mathbb{P}^j}\Big|_{\mathcal{F}_t}$. Since V_t is the wealth process of an admissible self-financing strategy, $V_t(S_t^j)^{-1}$ must be a nonnegative $\mathbb{Q}$-local martingale, hence a $\mathbb{Q}$-supermartingale. Therefore, $L_t V_t^{-1} S_t^j$ is a $\mathbb{P}^j$-supermartingale. Consequently, L_t is a $\mathbb{P}$-supermartingale, and $\mathbb{E}[L_t] \leqslant \mathbb{E}[L_0] = 1$. Thus, By Jensen's inequality, we have $\mathbb{E}[\log L_t] \leqslant 0$. But we have

$$\mathbb{E}[\log N_t] = \mathbb{E}[\log L_t] + \mathbb{E}[\log(V_t^{-1}S_t^j)].$$

So (14.4) is proved. $\qquad\square$

14.1.2 A Geometric Lévy Process Model

Chan (1999) introduced a market model in which the stock price (S_t) is driven by a process:

$$dS_t = \sigma_t S_{t-}dX_t + b_t S_{t-}dt,$$

where σ_t and b_t are deterministic functions of t and $X_t = cB_t + N_t + \alpha t$ with (B_t) being a Brownian motion and (N_t) being a purely discontinuous martingale.

Our model is a little more general. Let (X_t) be a Lévy *process*, i.e., a process with stationary and independent increments. Put

$$\mu(\omega, dt, dx) = \sum_{s>0} I_{[\Delta X_s(\omega) \neq 0]}(s) \delta_{(s, \Delta X_s(\omega))}.$$

We call μ the *jump measure* of X. For a Borel set Λ in $\mathbb{R} \backslash \{0\}$, we put

$$N_t(\omega, \Lambda) = \mu(\omega, [0, t] \times \Lambda) = \sum_{0 < s \leqslant t} I_\Lambda(\Delta X_s(\omega)), \quad \nu(\Lambda) = E[N_1(\cdot, \Lambda)].$$

Then for each $t \in \mathbb{R}_+$ and $\omega \in \Omega$, $N_t(\omega, \cdot)$ and ν are σ-finite measures on $\mathbb{R} \backslash \{0\}$. We call ν the Lévy *measure*. For every $n \geqslant 2$, $N_t(\cdot, [\frac{1}{n}, 1))$ is an integrable increasing process, $M_t^{(n)} = \int_{[\frac{1}{n} \leqslant |x| < 1]} x N_t(\cdot, dx) - t \int_{[\frac{1}{n} \leqslant |x| < 1]} x \nu(dx)$ is a square-integrable martingale, and the sequence $(M_t^{(n)})$ tends to a square-integrable martingale (M_t). We denote the limit by

$$M_t = \int_{[|x| < 1]} x(N_t(\cdot, dx) - t\nu(dx)).$$

One should beware that $\int_{[|x| < 1]} x N_t(\cdot, dx)$ and $\int_{[|x| < 1]} x \nu(dx)$ individually may make no sense or equal infinity. What we know is only the fact that $\int_{\mathbb{R} \backslash \{0\}} (x^2 \wedge 1)\nu(dx) < \infty$. It is well known that (X_t) has the following Lévy *decomposition* (see Protter 2004):

$$X_t = \alpha t + cB_t + M_t + A_t,$$

where α and c are constants, (B_t) is a standard Brownian motion, and

$$A_t = \int_{[|x| \geqslant 1]} x N_t(\cdot, dx) = \sum_{0 < s \leqslant t} \Delta X_s I_{[|\Delta X_s| \geqslant 1]}.$$

In particular, (X_t) is a semimartingale.

Now we consider a security market in which there are two assets: a risky asset and a savings account. We assume that the price process S_t of the risky asset is a *geometric* Lévy *process*, which satisfies the following equation:

$$dS_t = S_{t-}[\sigma_t dX_t + b_t dt], \tag{14.5}$$

where σ_t and b_t are deterministic functions of t. We assume that σ_t is strictly positive. The value of the savings account at time t is $\beta_t = \int_0^t e^{rs} ds$, where r_t is assumed to be a deterministic function of t. We assume that there exist $c_1 \in [0, 1)$

and $0 < c_2 \leqslant \infty$ such that $-c_1 \leqslant \Delta X \leqslant c_2$. Consequently, the Lévy measure ν is supported by $[-c_1, c_2]$. We assume that $c_1 \sigma_t < 1$ to ensure the strict positivity of (S_t).

In what follows, we always assume $\alpha = 0$; otherwise we can replace b_t by $b_t + \alpha \sigma_t$.

Let (V_t) be the wealth process of an admissible self-financing strategy (θ_t^0, θ_t), i.e., $V_t = \theta_t S_t + \theta_t^0 \beta_t$, and then

$$dV_t = \theta_t dS_t + \theta_t^0 r_t \beta_t dt. \tag{14.6}$$

Let $\widetilde{V}_t = \beta_t^{-1} V_t$, $\widetilde{S}_t = \beta_t^{-1} S_t$, and $\pi_t = \theta_t S_{t-} / V_{t-}$. Then

$$d\widetilde{S}_t = \widetilde{S}_{t-}[(b_t - r_t)dt + \sigma_t(cdB_t + dM_t + dA_t)], \tag{14.7}$$

$$d\widetilde{V}_t = \theta_t d\widetilde{S}_t = \widetilde{V}_{t-}\pi_t[(b_t - r_t)dt + \sigma_t(cdB_t + dM_t + dA_t)]. \tag{14.8}$$

If $\pi_t \sigma_t$ takes values in $(-c_2^{-1}, c_1^{-1})$, then (V_t) is strictly positive.

Lemma 14.5 *We have*

$$\int_{\mathbb{R}\setminus\{0\}} \frac{x^2}{(1+ax)^2}\nu(dx) < \infty, \quad a \in (-c_2^{-1}, c_1^{-1}). \tag{14.9}$$

For given b, r, and $\sigma > 0$, we put

$$f(a) = b - r - a\sigma c^2 + \sigma \int_{\mathbb{R}\setminus\{0\}} \left(\frac{x}{1+ax} - x I_{[|x|<1]} \right) \nu(dx), \quad a \in (-c_2^{-1}, c_1^{-1}). \tag{14.10}$$

Then $f(a) = 0$ has a unique solution a^* in $(-c_2^{-1}, c_1^{-1})$ if and only if

$$\lim_{a \to -c_2^{-1}} f(a) > 0 \text{ and } \lim_{a \to c_1^{-1}} f(a) < 0. \tag{14.11}$$

Proof For $a \in (-c_2^{-1}, c_1^{-1})$, $x \in [-c_1, c_2]$, we have

$$\frac{x}{1+ax} - x I_{[|x|<1]} = \frac{x}{1+ax} I_{[|x|\geqslant 1]} - \frac{ax^2}{1+ax} I_{[|x|<1]},$$

$$\frac{|ax^2|}{1+ax} I_{[|x|<1]} \leqslant \frac{|a|x^2}{1-ac_1} \wedge \frac{|a|}{1-ac_1},$$

$$\frac{|x|}{1+ax} \leqslant \frac{|x|}{1-ac_1} \wedge \frac{c_2}{1+ac_2}, \quad \frac{|x|}{1+ax} I_{[|x|\geqslant 1]} \leqslant \frac{x^2}{1-ac_1} \wedge \frac{c_2}{1+ac_2}.$$

Since $\int_{\mathbb{R}\setminus\{0\}}(x^2 \wedge 1)v(dx) < \infty$ we see that $f(a)$ is well defined and (14.9) holds. Here $c_2/(1 + ac_2) = 1/a$ if $c_2 = \infty$. Thus by the dominated convergence theorem, it is easy to see that for each $a \in (-c_2^{-1}, c_1^{-1})$, $f'(a)$ exists and

$$f'(a) = -\sigma c^2 - \sigma \int_{\mathbb{R}\setminus\{0\}} \frac{x^2}{(1 + ax)^2} v(dx) < 0.$$

This implies that $f(a) = 0$ has a unique solution $a^* \in (-c_2^{-1}, c_1^{-1})$, if and only if (14.11) holds.   □

In the sequel, for each $t \in [0, T]$, we put

$$f_t(a) = b_t - r_t - a\sigma_t c^2 + \sigma_t \int_{\mathbb{R}\setminus\{0\}} \left(\frac{x}{1 + ax} - xI_{[|x|<1]}\right)v(dx), \quad a \in (-c_2^{-1}, c_1^{-1}),$$
$$(14.12)$$

and assume

$$\lim_{a \to -c_2^{-1}} f_t(a) > 0 \text{ and } \lim_{a \to c_1^{-1}} f_t(a) < 0, \quad t \in [0, T]. \tag{14.13}$$

By Lemma 14.5, there exists a unique solution a_t^* of $f_t(a) = 0$ in $(-c_2^{-1}, c_1^{-1})$.

Theorem 14.6 *Assume (14.13) holds. We denote by (V_t) the wealth process of portfolio $\pi_t^* = \sigma_t^{-1}a_t^*$ with $V_0 = 1$. If*

$$\int_0^T (a_s^*)^2 ds < \infty, \quad \int_0^T (1 - \pi_s^*)^2 \sigma_s^2 ds < \infty, \tag{14.14}$$

$$\int_0^t \int_{\mathbb{R}\setminus\{0\}} \frac{(a_s^*)^2 x^2}{(1 + a_s^* x)^2} v(dx) ds < \infty, \tag{14.15}$$

and

$$\int_0^t \int_{\mathbb{R}\setminus\{0\}} \frac{(1 - \pi_s^*)^2 \sigma_s^2 x^2}{(1 + a_s^* x)^2} v(dx) ds < \infty, \tag{14.16}$$

then $(V_t^{-1}\beta_t)$ and $(V_t^{-1}S_t)$ are strictly positive square-integrable martingales.

Proof Let (V_t) be the wealth process of an admissible self-financing strategy π such that for each $t \in [0, T]$, $\sigma_t \pi_t$ takes values in $(-c_2^{-1}, c_1^{-1})$. We shall use the notation $dY_t \sim dZ_t$ to stand for the fact that $Y_t - Z_t$ is a local martingale. By Itô's formula, we have

$$d\widetilde{V}_t^{-1} = -\widetilde{V}_{t-}^{-2}d\widetilde{V}_t + \widetilde{V}_{t-}^{-3}d\langle V^c, V^c\rangle_t + d\sum_{0<s\leqslant t}\left(\widetilde{V}_s^{-1} - \widetilde{V}_{s-}^{-1} + \widetilde{V}_{s-}^{-2}\Delta\widetilde{V}_s\right)$$

$$\sim -\widetilde{V}_{t-}^{-1}[\pi_t(b_t - r_t - \pi_t\sigma_t^2 c^2)dt + \pi_t\sigma_t dA_t] + d\sum_{0<s\leqslant t}\widetilde{V}_{s-}^{-1}\frac{\pi_s^2\sigma_s^2\Delta X_s^2}{1 + \pi_s\sigma_s\Delta X_s}$$

$$\sim -\widetilde{V}_{t-}^{-1}\left[\pi_t(b_t - r_t - \pi_t\sigma_t^2 c^2) - \int_{\mathbb{R}\backslash\{0\}}\left(\frac{\pi_t^2\sigma_t^2 x^2}{1 + \pi_t\sigma_t x} - \pi_t\sigma_t x I_{[|x|\geqslant 1]}\right)\nu(dx)\right]dt$$

$$\sim -\widetilde{V}_{t-}^{-1}\pi_t\left[b_t - r_t - \pi_t\sigma_t^2 c^2 + \sigma_t\int_{\mathbb{R}\backslash\{0\}}\left(\frac{x}{1 + \pi_t\sigma_t x} - x I_{[|x|<1]}\right)\nu(dx)\right]dt.$$

Similarly, we have

$$d(V_t^{-1}S_t) = d(\widetilde{V}_t^{-1}\widetilde{S}_t) = \widetilde{V}_{t-}^{-1}d\widetilde{S}_t + d[\widetilde{V}^{-1}, \widetilde{S}]_t + \widetilde{S}_{t-}d\widetilde{V}_t^{-1}$$

$$\sim \widetilde{V}_{t-}^{-1}\widetilde{S}_{t-}[(b_t - r_t)dt + \sigma_t dA_t] - \widetilde{V}_{t-}^{-2}d[\widetilde{V}, \widetilde{S}]_t$$

$$+ d\sum_{0<s\leqslant t}\left(\widetilde{V}_s^{-1} - \widetilde{V}_{s-}^{-1} + \widetilde{V}_{s-}^{-2}\Delta\widetilde{V}_s\right)\Delta\widetilde{S}_s + \widetilde{S}_{t-}d\widetilde{V}_t^{-1}$$

$$\sim \widetilde{V}_{t-}^{-1}\widetilde{S}_{t-}(1 - \pi_t)\left[b_t - r_t - \pi_t\sigma_t^2 c^2 + \sigma_t\int_{\mathbb{R}\backslash\{0\}}\left(\frac{x}{1 + \pi_t\sigma_t x} - x I_{[|x|<1]}\right)\nu(dx)\right]dt.$$

Thus, if (V_t) is the wealth process corresponding to $\pi_t^* = \sigma_t^{-1}a_t^*$, then $(V_t^{-1}\beta_t)$ and $(V_t^{-1}S_t)$ are local martingales.

Now we are going to prove that $(V_t^{-1}\beta_t)$ and $(V_t^{-1}S_t)$ are square-integrable martingales. In the following for any semimartingale Y with $Y_0 = 0$, we denote by $\mathcal{E}(Y)$ the Doléans' exponential of Y, i.e.:

$$\mathcal{E}(Y)_t = \exp\left\{Y_t - \frac{1}{2}\langle Y^c, Y^c\rangle_t\right\}\prod_{0<s\leqslant t}(1 + \Delta Y_s)e^{-\Delta Y_s},$$

where Y^c is the continuous martingale part of Y. Note that $\mathcal{E}(Y)$ is the unique solution of the equation $dW_t = W_{t-}dY_t$ with $W_0 = 1$.

Since $V_t^{-1}\beta_t = \widetilde{V}_t$, by Doléans exponential formula, we have

$$V_t^{-1}\beta_t = \exp\left\{-\int_0^t \pi_s^*\left[(b_s - r_s)ds + c\sigma_s dB_s - \frac{c^2\sigma_s^2\pi_s^*}{2}ds + \sigma_s(dM_s + dA_s)\right]\right\}$$

$$\times \prod_{0<s\leqslant t}(1 + a_s^*\Delta X_s)^{-1}e^{a_s^*\Delta X_s}$$

$$= \mathcal{E}(-c(a^*.B)_t)\exp\left\{-\int_0^t[\pi_s^*(b_s - r_s - \pi^*\sigma_s^2 c^2)ds + a_s^*(dM_s + dA_s)]\right\}$$

$$\times \prod_{0<s\leqslant t}\left(1 - \frac{a_s^*\Delta X_s}{1+a_s^*\Delta X_s}\right)e^{\frac{a_s^*\Delta X_s}{1+a_s^*\Delta X_s} + \frac{a_s^{*2}(\Delta X_s)^2}{1+a_s^*\Delta X_s}}.$$

Put

$$Z_t = \int_0^t \int_{\mathbb{R}\backslash\{0\}} \frac{\sigma_s x}{1 + a_s^* x}[\mu(\cdot, ds, dx) - \nu(dx)ds].$$

Then Z is a well-defined local martingale. By (14.7), the stochastic integral $\pi^*.Z$ is well defined and its oblique bracket process is a deterministic function:

$$\langle \pi^*.Z, \pi^*.Z\rangle_t = \int_0^t \int_{\mathbb{R}\backslash\{0\}} \frac{(\sigma_s\pi_s^*)^2 x^2}{(1 + a_s^* x)^2}\nu(dx)ds < \infty.$$

Hence, $\pi^*.Z$ is a square-integrable martingale.

Since we have

$$\mathcal{E}(-\pi^*.Z)_t = e^{-(\pi^*.Z)_t}\prod_{0<s\leqslant t}\left(1 - \frac{a_s^*\Delta X_s}{1+a_s^*\Delta X_s}\right)e^{\frac{a_s^*\Delta X_s}{1+a_s^*\Delta X_s}},$$

from the fact that $f_t(a_t^*) = 0$, we obtain

$$V_t^{-1}\beta_t = \mathcal{E}(-\pi^*.(c\sigma.B + Z))_t.$$

The oblique bracket process at time T of the martingale $\pi^*.(c\sigma.B + Z)$

$$\langle \pi^*.(c\sigma.B + Z), \pi^*.(c\sigma.B + Z)\rangle_T = c^2\int_0^T (a_s^*)^2 ds + \langle \pi^*.Z, \pi^*.Z\rangle_T$$

is a finite constant. Thus, by a result of Lépingle and Mémin (1978), $\mathcal{E}(-\pi^*.(c\sigma.B + Z))$ is a square-integrable martingale.

Similarly, we have

$$V_t^{-1}S_t = \widetilde{V}_t^{-1}\widetilde{S}_t$$

$$= \exp\left\{-\int_0^t [\pi_s^*(b_s - r_s) - \frac{(c\sigma_s\pi_s^*)^2}{2}]ds - (a^*.(cB + M + A))_t\right\}$$

$$\times \prod_{0<s\leqslant t} (1 + a_s^*\Delta X_s)^{-1}e^{a_s^*\Delta X_s}$$

$$\times \exp\left\{\int_0^t \left[b_s - r_s - \frac{(c\sigma_s)^2}{2}\right]ds + (\sigma.(cB + M + A))_t\right\}$$

$$\times \prod_{0<s\leqslant t} (1 + \sigma_s\Delta X_s)e^{-\sigma_s\Delta X_s},$$

from which and the fact that $f_t(a^*) = 0$, it is easy to prove that

$$V_t^{-1}S_t = \mathcal{E}((1 - \pi^*).(c\sigma.B + Z))_t.$$

Therefore, by (14.14), (14.16), and a result of Lépingle and Mémin (1978), we know that $(V_t^{-1}S_t)$ is also a square-integrable martingale. □

Remark The condition (14.12) is also necessary for (V_t) to be strictly positive, if $[-c_1, c_2]$ is exactly the support of the Lévy measure ν. This condition is easily checked for some concrete models (e.g., the compound Poisson process case).

14.1.3 A Jump-Diffusion-Like Process Model

Consider a financial market which consists of a risk-free asset (savings account) and a risky asset whose prices S_t^0 and S_t satisfy the following jump-diffusion-like process:

$$d S_t^0 = S_t^0 r(t)\, dt, \quad S_0^0 = 1,$$

$$d S_t = S_{t-}\{b(t)dt + \sigma(t)d B_t + \phi(t)d N_t\}, \tag{14.17}$$

where (B_t) is a $(\mathcal{F}_t, \mathbb{P})$-standard Brownian motion, $N(t)$ is $\mathcal{F}_t$-adapted counting process with intensity $\lambda(t) > 0$, and $r(t)$, $b(t)$, $\sigma(t)$, $\phi(t)$, and $\lambda(t)$ are assumed to be bounded $\mathcal{F}_t$-predictable processes. Furthermore, we assume that $\phi(t) > -1$; and that $|\sigma(t)|$, $|\phi(t)|$, and $\lambda(t)$ are uniformly bounded from below by a positive constant. If $b(t)$, $\sigma(t)$, and $\phi(t)$ are of the forms $b(t, S_t)$, $\sigma(t, S_t)$, and $\phi(t, S_t)$, respectively, with b, σ, and ϕ being deterministic functions and $\lambda(t)$ is a constant, then S_t is a jump-diffusion process.

Let $\{\theta^0(t), \theta(t)\}$ be an admissible self-financing strategy. Its wealth process V_t satisfies

$$d V_t = \theta(t)dS_t + \theta^0(t)S_t^0 r(t)dt.$$

Set $\widetilde{V}_t = e^{-\int_0^t r(s)ds} V_t$ and $\widetilde{S}_t = e^{-\int_0^t r(s)ds} S_t$. We have

$$d\widetilde{V}_t = \theta(t)d\widetilde{S}_t = \theta(t)\widetilde{S}_{t-}\left[\bar{b}(t)dt + \sigma(t)dB_t + \phi(t)dN_t\right]$$
$$= \widetilde{V}_{t-}\pi(t)\left[\bar{b}(t)dt + \sigma(t)dB_t + \phi(t)dN_t\right], \tag{14.18}$$

where

$$\pi(t) = \theta(t)S_{t-}/V_{t-}, \quad \bar{b}(t) = b(t) - r(t).$$

Namely, $(\pi(t))$ is the portfolio associated to strategy $\{\theta^0(t), \theta(t)\}$.

Obviously, $(\widetilde{V}_t)$ is strictly positive if and only if $\pi(t)\phi(t) > -1$ for each t. In the following we assume that (V_t) is strictly positive.

Lemma 14.7 *Assume that $\pi(t)\phi(t) > -1$ for each t. Then*

$$\widetilde{V}_t^{-1} - \widetilde{V}_{t-}^{-1} + \widetilde{V}_{t-}^{-2}\Delta\widetilde{V}_t = \widetilde{V}_{t-}^{-1}\frac{\pi(t)^2\phi(t)^2}{1+\pi(t)\phi(t)}\Delta N_t, \tag{14.19}$$

where we use notation $\Delta X = X - X_-$.

Proof By (14.18), $\Delta\widetilde{V}_t = \widetilde{V}_{t-}\pi(t)\phi(t)\Delta N_t$. Thus, we have

$$\widetilde{V}_t^{-1} - \widetilde{V}_{t-}^{-1} + \widetilde{V}_{t-}^{-2}\Delta\widetilde{V}_t = -\frac{\Delta\widetilde{V}_t}{\widetilde{V}_t\widetilde{V}_{t-}} + \widetilde{V}_{t-}^{-1}\pi(t)\phi(t)\Delta N_t$$
$$= -\frac{\pi(t)\phi(t)\Delta N_t}{\widetilde{V}_{t-}(1+\pi(t)\phi(t)\Delta N_t)} + \widetilde{V}_{t-}^{-1}\pi(t)\phi(t)\Delta N_t$$
$$= \widetilde{V}_{t-}^{-1}\frac{\pi(t)^2\phi(t)^2}{1+\pi(t)\phi(t)}\Delta N_t.$$
$\square$

Theorem 14.8 *Assume that $\pi(t)\phi(t) > -1$ and $\sigma^2(t) > 0$ for each t. Let*

$$\pi(t) = \frac{-(\sigma^2(t) - \phi(t)\bar{b}(t)) + \sqrt{(\sigma^2(t) + \phi(t)\bar{b}(t))^2 + 4\sigma^2(t)\phi^2(t)\lambda(t)}}{2\sigma^2(t)\phi(t)}, \tag{14.20}$$

and (V_t) be its wealth process. Then $(V_t^{-1}S_t^0)$ (i.e., $(\widetilde{V}_t^{-1})$) and $(V_t^{-1}S_t)$ are $\mathbb{P}$-martingales, i.e., $(\pi(t))$ is the optimal growth portfolio. Moreover, we have

$$d\widetilde{V}_t^{-1} = \widetilde{V}_{t-}^{-1}\left[-\pi(t)\sigma(t)dB_t + \left((1+\pi(t)\phi(t))^{-1} - 1\right)dM_t\right], \tag{14.21}$$

where (M_t) is the following $(\mathcal{F}_t, \mathbb{P})$-martingale:

$$M_t = N_t - \int_0^t \lambda(s)ds.$$

Proof In the following, we write $dX_t \sim dY_t$ to stand for the fact that $(X_t - Y_t)$ is a $\mathbb{P}$-local martingale. By Itô's formula and (14.19), we have

$$d\widetilde{V}_t^{-1} = -\widetilde{V}_{t-}^{-2} d\widetilde{V}_t + \widetilde{V}_{t-}^{-3} d\langle \widetilde{V}^c, \widetilde{V}^c \rangle_t + d \sum_{0 < s \leqslant t} \left(\widetilde{V}_s^{-1} - \widetilde{V}_{s-}^{-1} + \widetilde{V}_{s-}^{-2} \Delta \widetilde{V}_s \right)$$

$$\sim -\widetilde{V}_{t-}^{-1}[\pi(t)\bar{b}(t)dt + \pi(t)\phi(t)\lambda(t)dt] + \widetilde{V}_{t-}^{-1}\pi(t)^2\sigma(t)^2 dt + \widetilde{V}_{t-}^{-1}\frac{\pi(t)^2\phi(t)^2\lambda(t)}{1 + \pi(t)\phi(t)}dt$$

$$\sim \widetilde{V}_{t-}^{-1}\pi(t)\left(\bar{b}(t) + \phi(t)\lambda(t) - \pi(t)\sigma(t)^2 - \frac{\pi(t)\phi(t)^2\lambda(t)}{1 + \pi(t)\phi(t)} \right)dt,$$

and

$$d\left(V_t^{-1} S_t \right) = d\left(\widetilde{V}_t^{-1}\widetilde{S}_t \right) = \widetilde{S}_{t-}d\widetilde{V}^{-1} + \widetilde{V}_{t-}^{-1}d\widetilde{S}_t + d[\widetilde{V}^{-1}, \widetilde{S}]_t$$

$$\sim \widetilde{S}_{t-}d\widetilde{V}^{-1} + \widetilde{V}_{t-}^{-1}\widetilde{S}_{t-}[\bar{b}(t)dt + \phi(t)\lambda(t)dt] - \widetilde{V}_{t-}^{-2}d[\widetilde{V}^{-1}, \widetilde{S}]_t$$

$$+ d\sum_{0 < s \leqslant t} \left(\widetilde{V}_s^{-1} - \widetilde{V}_{s-}^{-1} + \widetilde{V}_{s-}^{-2}\Delta\widetilde{V}_s \right)\Delta\widetilde{S}_s$$

$$= \widetilde{S}_{t-}d\widetilde{V}^{-1} + \widetilde{V}_{t-}^{-1}\widetilde{S}_{t-}[\bar{b}(t)dt + \phi(t)\lambda(t)dt]$$

$$- \widetilde{V}_{t-}^{-1}\widetilde{S}_{t-}[\pi(t)\sigma(t)^2 dt + \pi(t)\phi(t)^2 dN_t]$$

$$+ d\sum_{0 < s \leqslant t} \left(\widetilde{V}_s^{-1} - \widetilde{V}_{s-}^{-1}\frac{\pi(s)^2\phi(s)^2}{1 + \pi(s)\phi(s)}\phi(t)\Delta N_s \right)$$

$$\sim \widetilde{S}_{t-}d\widetilde{V}^{-1}$$

$$+ \widetilde{V}_{t-}^{-1}\widetilde{S}_{t-}\left[\bar{b}(t) + \phi(t)\lambda(t) - \pi(t)\sigma(t)^2 - \pi(t)\phi(t)^2\lambda(t) + \frac{\pi(t)^2\phi(t)^3\lambda(t)}{1 + \pi(t)\phi(t)} \right]dt$$

$$= \widetilde{S}_{t-}d\widetilde{V}^{-1} + \widetilde{V}_{t-}^{-1}\widetilde{S}_{t-}\left[\bar{b}(t) + \phi(t)\lambda(t) - \pi(t)\sigma(t)^2 - \frac{\pi(t)\phi(t)^2\lambda(t)}{1 + \pi(t)\phi(t)} \right]dt.$$

Thus, in order that $(\widetilde{V}_t^{-1})$ and $(V_t^{-1}S_t)$ are $\mathbb{P}$-local martingales, the above "dt" terms must vanish, i.e., π must satisfy the following equation:

$$\bar{b} - \pi\sigma^2 + \frac{\phi\lambda}{1 + \pi\phi} = 0. \tag{14.22}$$

Equation (14.22) has two solutions. Only the solution given by (14.20) satisfies the condition $\pi\phi > -1$. From the above proof, it is easy to see that for such choice of π, we have

$$d\widetilde{V}_t^{-1} = -\widetilde{V}_{t-}^{-1}[\pi(t)\sigma(t)dB_t + \pi(t)\phi(t)dM_t] + \widetilde{V}_{t-}^{-1}\frac{\pi(t)^2\phi(t)^2}{1 + \pi(t)\phi(t)}dM_t,$$

from which (14.22) follows.

Now we are going to prove that $(\widetilde{V}_t^{-1})$ and $(V_t^{-1}S_t)$ are actually $\mathbb{P}$-martingales. In fact, it is easy to see that

$$\widetilde{V}_t^{-1} = V_0^{-1} \exp\left\{-\int_0^t \pi(s)\bar{b}(s)ds - \int_0^t \pi(s)\sigma(s)dB_s\right.$$
$$\left. - \int_0^t \log(1+\pi(s)\phi(s))dN_s - \frac{1}{2}\int_0^t \pi(s)^2\sigma(s)^2ds\right\},$$

and

$$V_t^{-1}S_t = \widetilde{V}_t^{-1}\widetilde{S}_t$$
$$= V_0^{-1}S_0 \exp\left\{\int_0^t (1-\pi(s))\bar{b}(s)ds + \int_0^t (1-\pi(s))\sigma(s)dB_s\right.$$
$$\left. + \int_0^t \log\frac{1+\phi(s)}{1+\pi(s)\phi(s)}dN_s - \frac{1}{2}\int_0^t (1-\pi(s)^2)\sigma(s)^2ds\right\}.$$

Since functions π, $\bar{b}$, σ, and ϕ are all bounded, $(\widetilde{V}_t^{-1})$ and $(V_t^{-1}S_t)$ are uniformly integrable on $[0, T]$, and $(\widetilde{V}_t^{-1})$ and $(V_t^{-1}S_t)$ are $\mathbb{P}$-martingales.   □

Remark   From the above proof, it is easy to see that $(\widetilde{V}_t^{-1})$ and $(V_t^{-1}S_t)$ are $\mathbb{P}$-martingales if and only if π is given by (14.20). For those $t \in [0, T]$ with $\bar{b}(t) + \phi(t)\lambda(t) = 0$, we have $\pi(t) = 0$.

Theorem 14.9   *Let $\pi(t)$ be given by (14.20) and V_t be its wealth process with $V_0 = 1$. Put*

$$\frac{d\widehat{\mathbb{P}}}{d\mathbb{P}}\bigg|_{\mathcal{F}_t} = \widetilde{V}_t^{-1}, \ 0 \leqslant t \leqslant T.$$

Then under $\widehat{\mathbb{P}}$, $\widehat{B}_t = B_t + \int_0^t \pi(s)\sigma(s)ds$ is an $(\mathcal{F}_t)$-standard Brownian motion, and N_t is an $(\mathcal{F}_t)$-adapted counting process with intensity $(1 + \pi(t)\phi(t))^{-1}\lambda(t)$. Moreover,

$$d\widetilde{S}_t = \widetilde{S}_{t-}\left[\sigma(t)d\widehat{B}_t + \phi(t)[dN_t - ((1+\pi(t)\phi(t))^{-1}\lambda(t))dt]\right].$$

Proof We have

$$d(\widehat{B}_t\widetilde{V}_t^{-1}) \sim \widetilde{V}_t^{-1}d\widehat{B}_t + d[\widehat{B}, \widetilde{V}^{-1}]_t$$
$$= \widetilde{V}_{t-}^{-1}(dB_t + \pi(t)\sigma(t)dt) - \widetilde{V}_{t-}^{-1}\pi(t)\sigma(t)dt$$
$$= \widetilde{V}_{t-}^{-1}dB_t.$$

This means that $(\widehat{B}_t\widetilde{V}_t^{-1})$ is a $\mathbb{P}$-local martingale, i.e., $(\widehat{B}_t)$ is a $\widehat{\mathbb{P}}$-local martingale. On the other hand, under $\mathbb{P}$, $[\widehat{B}, \widehat{B}]_t = [B, B]_t = t$, and $[\widehat{B}, \widehat{B}]$ is invariant under the equivalent change of probability. Therefore, by the well-known Lévy's theorem,

$(\widehat{B}_t)$ is an $(\mathcal{F}_t, \widehat{\mathbb{P}})$-standard Brownian motion. Similarly, we can prove that (N_t) is an $(\mathcal{F}_t, \widehat{\mathbb{P}})$-counting process with intensity $(1 + \pi(t)\phi(t))^{-1}\lambda(t)$. □

14.2 Pricing in a Geometric Lévy Process Model

In this section, we will study the problem of pricing contingent claims on a stock driven by a Lévy process. We will take the optimal growth wealth process as a numeraire, to derive an integro-differential equation for pricing a contingent claim. In this case, by Theorem 14.2, the historical probability measure is a martingale measure for the market. Thus, for a contingent claim paying ξ at time T, an arbitrage-free price process $U_t(\xi)$ of ξ can be defined by

$$U_t(\xi) = V_t \mathbb{E}[\xi V_T^{-1} | \mathcal{F}_t]. \tag{14.23}$$

It is more convenient to compute the price process $U_t(\xi)$ by a change of numeraire (see Geman et al. 1995). Since $(\widetilde{V}_t^{-1})$ is a $\mathbb{P}$-martingale, we define a new probability measure $\widehat{\mathbb{P}}$ by $\frac{d\widehat{\mathbb{P}}}{d\mathbb{P}}\Big|_{\mathcal{F}_t} = \widetilde{V}_t^{-1}$. Then by (14.17) and Beyes' rule, we obtain

$$U_t(\xi) = \beta_t \widetilde{V}_t \mathbb{E}[\beta_T^{-1} \xi \widetilde{V}_T^{-1} | \mathcal{F}_t] = \beta_t \widehat{\mathbb{E}}[\beta_T^{-1} \xi | \mathcal{F}_t]. \tag{14.24}$$

This means that $\widehat{\mathbb{P}}$ is a martingale measure for the market if the savings account is taken as a numeraire. Such a martingale measure is called a *risk-neutral measure*.

Theorem 14.10 *Let ξ be a contingent claim of the form $f(S_T)$. If f is continuous and there exists an $l \geqslant 1$ such that $f(x) \leqslant c(1 + x^l)$, $x \geqslant 0$, then we have*

$$U_t(\xi) = \beta_t \widehat{\mathbb{E}} \left[\beta_T^{-1} f(S_T) \Big| \mathcal{F}_t \right] = F(t, S_t),$$

where $F(t, x)$ satisfies the following integro-differential equation:

$$F_t' + rxF_x' + \frac{c\sigma_t^2 x^2}{2} F_{xx}'' - r_t F$$
$$+ \int_{-1}^{\infty} [F(t, x + \sigma_t xy) - F(t, x) - F_x'(t, x)\sigma_t xy]\hat{v}_t(dy) = 0,$$
$$F(T, x) = f(x),$$

where

$$v_t(dx) = (1 + \pi_t^* \sigma_t x)^{-1} v(dx),$$

and $v(dt, dx) = v_t(dx)dt$ is the compensator of the jump measure μ of Y under $\widehat{\mathbb{P}}$.

Proof Since $f_t(a^*) = 0$, it is easy to check that

$$d\widetilde{S}_t = \widetilde{S}_{t-}[\sigma_t(cd\widehat{B}_t + d\widehat{M}_t)],$$

where by Girsanov's theorem

$$\widehat{B}_t = B_t + c\int_0^t \pi_s^*\sigma_s ds$$

is a standard Brownian motion under $\widehat{\mathbb{P}}$ and

$$\widehat{M}_t = \int_0^t \int_{\mathbb{R}\backslash\{0\}} x(\mu(\cdot, ds, dx) - \nu_s(dx)ds).$$

Thus

$$dS_t = S_{t-}[rdt + \sigma_t(cd\widehat{B}_t + d\widehat{M}_t)].$$

It is easy to prove that for any $p > 1$, S_T is in $L^p(\widehat{\mathbb{P}})$. Thus by assumption on f, $f(S_T)$ is $\widehat{\mathbb{P}}$-integrable. Let $W_t = \beta_t^{-1}F(t, S_t)$. Then (W_t) is a $\widehat{\mathbb{P}}$-martingale. By Itô's formula,

$$dW_t = \beta_t^{-1}d F(t, S_t) - r_t\beta_t^{-1}F(t, S_{t-})dt$$

$$= \beta_t^{-1}\left[F_t'(t, S_{t-})+F_x'(t, S_{t-})S_{t-}r_t+\frac{1}{2}F_{xx}''(t, S_{t-})S_{t-}^2c^2\sigma_t^2 - r_tF(t, S_{t-})\right]dt$$

$$+\beta_t^{-1}d M_t^{(1)} \quad (M^{(1)} \text{ is a } \widehat{\mathbb{P}}\text{-local martingale})$$

$$+\beta_t^{-1}d\left[\sum_{0<s\leqslant t} (F(s, S_s) - F(s, S_{s-}) - F_x'(s, S_{s-})\Delta S_s)\right]$$

$$= \beta_t^{-1}\left[F_t'(t, S_{t-})+F_x'(t, S_{t-})S_{t-}r_t+\frac{1}{2}F_{xx}''(t, S_{t-})S_{t-}^2c^2\sigma_t^2 - r_tF(t, S_{t-})\right]dt$$

$$+\beta_t^{-1}d (M_t^{(1)} + d M_t^{(2)}) \quad (M^{(2)} \text{ is a } \widehat{\mathbb{P}}\text{-local martingale})$$

$$+\beta_t^{-1}\int_{\mathbb{R}\backslash\{0\}} \left(F(t, S_{t-}(1 + y)) - F(t, S_{t-}) - F_x'(t, S_{t-})S_{t-}y\right)\hat{\nu}_t(dy)\, dt.$$

Thus, $W_t - \int_0^t \beta_s^{-1}d(M_s^{(1)}+M_s^{(2)})$ is a continuous $\widehat{\mathbb{P}}$-local martingale. Consequently, "dt" term must vanish. The theorem is proved.   □

In the following, we apply the above results to a market model in which all parameters are constants. Specifically, we assume that the interest rate r_t is a constant r, $\sigma_t \equiv 1$, and b_t is a constant b. In this case, equation (14.5) is reduced to

$$dS_t = S_{t-}dX_t,$$

where (X_t) is a Lévy process with jump measure μ and Lévy measure ν. Namely, we have

$$X_t = bt + cB_t + M_t + A_t, \tag{14.25}$$

where b and c are constants, (B_t) is a standard Brownian motion,

$$M_t = \int_{[|x|<1]} x(N_t(\cdot, dx) - t\nu(dx)), \tag{14.26}$$

and

$$A_t = \int_{[|x|\geq 1]} x N_t(\cdot, dx) = \sum_{0<s\leq t} \Delta X_s I_{[|\Delta X_s|\geq 1]}. \tag{14.27}$$

Let $f(a)$ be defined by (14.10). In the following, we assume that (14.13) holds and denote by (V_t) the wealth process corresponding to $\pi_t^* \equiv a^*$. By Theorem 14.3, $(V_t^{-1}e^{rt})$ and $(V_t^{-1}S_t)$ are strictly positive square-integrable $\mathbb{P}$-martingales. We define a new probability measure $\widehat{\mathbb{P}}$ by $\frac{d\widehat{\mathbb{P}}}{d\mathbb{P}}\big|_{\mathcal{F}_t} = \widetilde{V}_t^{-1}$.

From Theorem 14.10, we obtain the following theorem.

Theorem 14.11   *Let ξ be a contingent claim of the form $f(S_T)$. If f is continuous and there exists an $l \geq 1$ such that $f(x) \leq c(1 + x^l)$, $x \geq 0$, then we have*

$$U_t(\xi) = \widehat{\mathbb{E}}\left[e^{-r(T-t)}f(S_T)\Big| \mathcal{F}_t\right] = F(t, S_t), \tag{14.28}$$

where $F(t, x)$ satisfies the following equations:

$$F_t' + rxF_x' + \frac{\sigma^2 x^2}{2}F_{xx}'' - rF$$
$$+ \int_{-1}^{\infty}(F(t, x+xy) - F(t, x) - F_x'(t, x)xy)\hat{\nu}(dy) = 0,$$
$$F(T, x) = f(x).$$

Now we are going to derive an analytical expression of the function F. To this end, we first consider a Lévy process with a finite Lévy measure and then we approximate a σ-finite Lévy measure by a sequence of finite Lévy measures. If the Lévy measure is finite, then the Lévy process has the following form:

$$X_t = bt + cB_t + \sum_{j=1}^{\Phi_t} U_j, \tag{14.29}$$

where (Φ_t) is a Poisson process with parameter λ, $(U_j)_{j\geqslant 1}$ is a sequence of square-integrable i.i.d. random variables taking values in $[-c_1, c_2]$, (Φ_t) is independent of $(U_j)_{j\geqslant 1}$, and (B_t) is a standard Brownian motion independent of $\sum_{j=1}^{\Phi_t} U_j$. We denote by $F(x)$ the distribution function of U_1. The process $\sum_{j=1}^{\Phi_t} U_j$ is usually called the *compound Poisson process* with arrival rate λ and jump law F. Rewriting (14.29) as

$$X_t = \left[b + \lambda \int_{[|x|<1]} x F(dx)\right] t + c B_t + \int_{[|x|<1]} x[N_t(\cdot, dx) - \lambda t F(dx)]$$

$$+ \int_{[|x|\geqslant 1]} x N_t(\cdot, dx),$$

we see that (14.10) reduces to

$$f(a) = b - r - ac^2 + \lambda \mathbb{E}\left[\frac{U_1}{1 + aU_1}\right]. \tag{14.30}$$

We have

$$\widehat{\mathbb{E}}\left[\exp\left\{iu(B_t + a^*ct) + iv \sum_{j=1}^{\Phi_t} U_j\right\}\right]$$

$$= \mathbb{E}\left[\widetilde{V}_t^{-1} \exp\left\{iu(B_t + a^*ct) + iv \sum_{j=1}^{\Phi_t} U_j\right\}\right]$$

$$= \mathbb{E}\left[\exp\left\{-a^*(\alpha - r)t - (a^*c - iu)B_t + \frac{a^{*2}c^2 t}{2}\right.\right.$$

$$= \exp\left\{-a^*(\alpha - r)t + a^{*2}c^2 t - \frac{u^2}{2}t\right\} \mathbb{E}\left[\exp\left\{\sum_{j=1}^{\Phi_t}[ivU_j - \log(1 + a^*U_j)]\right\}\right]$$

$$= \exp\left\{-a^*(\alpha - r)t + a^{*2}c^2 t - \frac{u^2}{2}t + \lambda t \mathbb{E}\left[\frac{e^{ivU_1}}{1 + a^*U_1} - 1\right]\right\}$$

$$= \exp\left\{\lambda t a^* \mathbb{E}\left[\frac{U_1}{1 + a^*U_1}\right] - \frac{u^2}{2}t + \lambda t \mathbb{E}\left[\frac{e^{ivU_1}}{1 + a^*U_1} - 1\right]\right\} \text{ (since f(a}^*) = 0)$$

$$= \exp\left\{-\frac{u^2}{2}t + \lambda t \int_{\mathbb{R}\setminus\{0\}} (e^{ivx} - 1)(1 + a^*x)^{-1} F(dx)\right\}.$$

Therefore, under $\widehat{\mathbb{P}}$, (Φ_t) is a Poisson process with parameter $\hat{\lambda} = \lambda \mathbb{E}[(1 + a^*U_1)^{-1}]$, $(U_j)_{j\geqslant 1}$ is a sequence of i.i.d. random variables with distribution

$$\widehat{F}(dx) = \left(1/\{(1+a^*x)\mathbb{E}[(1+a^*U_1)^{-1}]\}\right) F(dx),$$

$\sum_{j=1}^{\Phi_t} U_j$ is a compound Poisson process, and $\widehat{B}_t = B_t + a^*ct$ is a standard Brownian motion. By Theorem 11.43 in Heath et al. (1992), $(\widehat{B}_t)$ is independent of $\sum_{j=1}^{\Phi_t} U_j$. We have

$$d\widetilde{S}_t = \widetilde{S}_{t-}\left[cd\,\widehat{B}_t + d\left(\sum_{j=1}^{\Phi_t} U_j - \hat{\lambda}\mathbb{E}[U_1]t\right)\right]. \tag{14.31}$$

By a result of Lamberton and Lapeyre (1996), we get immediately the following theorem.

Theorem 14.12 *Let ξ be a contingent claim of the form $f(S_T)$. The price process of ξ is given by*

$$U_t(\xi) = \widehat{\mathbb{E}}\left[e^{-r(T-t)}f(S_T)\Big|\mathcal{F}_t\right] F(t, S_t),$$

where

$$F(t, x) = \sum_{n=0}^{\infty} \widehat{\mathbb{E}}\left[F_0\left(t, xe^{-\hat{\lambda}(T-t)\widehat{\mathbb{E}}[U_1]}\prod_{j=1}^{n}(1+U_j)\right)\right]\frac{\hat{\lambda}^n(T-t)^n}{n!}e^{-\hat{\lambda}(T-t)},$$

$$F_0(t, x) = \widehat{\mathbb{E}}\left[e^{-r(T-t)}f(xe^{(r-c^2/2)(T-t)} + c\widehat{B}_{T-t})\right].$$

Theorem 14.13 *Under the assumption of Theorem 14.8, we have*

$$F(t, x) = \lim_{m\to\infty}\sum_{n=0}^{\infty}\int_{\mathbb{R}^n\setminus(-1/m, 1/m)^n} F_0\left(t, xe^{-(T-t)\int_{[|y|\geqslant 1/m]}\frac{y}{1+a^*y}v(dy)}\prod_{i=1}^{n}(1+y_i)\right)$$
$$\times \prod_{i=1}^{n}(1+a^*y_i)^{-1}v(dy_1)\cdots v(dy_n)\frac{(T-t)^n}{n!}e^{-(T-t)\int_{[|y|\geqslant 1/m]}\frac{1}{1+a^*y}v(dy)},$$

$$\tag{14.32}$$

where $F_0(t, x) = \widehat{\mathbb{E}}\left[e^{-r(T-t)}f(xe^{(r-c^2/2)(T-t)} + c\widehat{B}_{T-t})\right].$

Proof By Lemma 14.39 in He et al. (1992), we have

$$\widehat{\mathbb{E}}\left[\exp\left\{iu\int_{[|x|\geqslant\frac{1}{m}]}xN_t(\cdot, dx)\right\}\right] = \exp\left\{t\int_{[|x|\geqslant\frac{1}{m}]}(e^{iux}-1)\hat{v}(dx)\right\}.$$

Thus, $Y_t^{(m)} = \int_{[|x| \geqslant \frac{1}{m}]} x N_t(\cdot, dx)$ is a compound Poisson process with arrival rate $\lambda_m = \hat{v}(\mathbb{R} \setminus (-\frac{1}{m}, \frac{1}{m}))$ and jump law $F_m(dx) = \lambda_m^{-1} I_{[|x| \geqslant \frac{1}{m}]} \hat{v}(dx)$. Put $\widetilde{S}_t^{(m)} = \mathcal{E}(Z^{(m)})_t$, where

$$Z_t^{(m)} = c\widetilde{B}_t + \int_{[|x| \geqslant \frac{1}{m}]} x(N_t(\cdot, dx) - \hat{v}(dx)).$$

Then

$$S_t^{(m)} = e^{Z_t^{(m)} - \frac{1}{2}c^2 t} \prod_{s \leqslant t} (1 + \Delta Z_s^{(m)}) e^{-\Delta Z_s^{(m)}} \leqslant e^{Z_t^{(m)} - \frac{1}{2}c^2 t}.$$

Similarly, we have $\widetilde{S}_t \leqslant e^{Z_t - \frac{1}{2}c^2 t}$, where

$$Z_t = c\widetilde{B}_t + \int_{\mathbb{R} \setminus \{0\}} x(N_t(\cdot, dx) - \hat{v}(dx)).$$

Once again by Lemma 14.39 in He et al. (1992), we have for any $p \geqslant 1$,

$$\widehat{\mathbb{E}}[e^{pZ_t^{(m)}}] = e^{t \int_{[|x| \geqslant \frac{1}{m}]} (e^{px} - 1 - px) \hat{v}(dx)} \longrightarrow e^{t \int_{\mathbb{R} \setminus \{0\}} (e^{px} - 1 - px) \hat{v}(dx)} = \widehat{\mathbb{E}}[e^{pZ_t}].$$

Consequently, as $m \to \infty$, $e^{Z_t^{(m)}}$ converges to e^{Z_t} in $L^p(\widehat{\mathbb{P}})$. Thus, by the dominated convergence theorem, we know that, as $m \to \infty$, $\mathcal{E}(Z^{(m)})_t$ converges to $\mathcal{E}(Z)_t$ in $L^p(\widehat{\mathbb{P}})$ for any $p \geqslant 1$. However, we have

$$\widehat{\mathbb{E}}\left[e^{-r(T-t)} f(S_T^{(m)}) \middle| \mathcal{F}_t\right] = F_m(t, S_t^{(m)}),$$

where

$$F_m(t, x) = \sum_{n=0}^{\infty} \int_{\mathbb{R}^n} F_0\left(t, xe^{-(T-t)\lambda_m \int_{\mathbb{R}} F_m(dx)} \prod_{i=1}^{n} (1 + y_i)\right) \tag{14.33}$$

$$\times F_m(dy_1) \cdots F_m(dy_n) \frac{\lambda_m^n (T-t)^n}{n!} e^{-\lambda_m(T-t)},$$

and

$$F_m(dx) = \lambda^{-1} I_{[|x| \geqslant \frac{1}{m}]} \frac{1}{1 + a^* x} v(dx).$$

Thus, we obtain (14.32) from (14.33).   □

14.3 Other Approaches to Option Pricing

14.3.1 The Föllmer-Schwarzer Approach

Assume that the discounted price process of a risky asset is a continuous semimartingale with decomposition $X_t = M_t + A_t$, where M is a square-integrable martingale and A is a process with square-integrable variation.

If there exists a martingale measure $\mathbb{P}^*$ such that for any square-integrable $\mathbb{P}$-martingale which is orthogonal to M under $\mathbb{P}$ remains a martingale under $\mathbb{P}^*$, then we call $\mathbb{P}^*$ the minimal martingale measure.

Let ξ be a contingent claim at time T. We denote by $\widetilde{\xi}$ the deflated value of ξ: $\widetilde{\xi} = \gamma_T \xi$. Define

$$\widetilde{U}_t(\xi) = \mathbb{E}^*[\widetilde{\xi}|\mathcal{F}_t].$$

Then

$$\widetilde{U}_t(\xi) = x^* + \int_0^t \varphi_s^* d\widetilde{X}_s + L_t, \quad 0 \leqslant t \leqslant T,$$

where L is a square-integrable $\mathbb{P}$-martingale which is orthogonal to M, and (x^*, φ^*) attains the following minimum:

$$\inf_{x,\varphi} \mathbb{E}\left[\left(\widetilde{\xi} - x - \int_0^T \varphi_s d\widetilde{X}_s\right)^2\right], \quad x \geqslant 0, \quad \varphi \in \mathcal{A}(x),$$

where $\mathcal{A}(x)$ is the set of all admissible trading strategies with initial wealth x.

For models with continuous price processes, the minimal martingale measure $\mathbb{Q}^*$ minimizes the *reverse relative entropy* $H(\mathbb{Q}, \mathbb{P})$ over all martingale measures $\mathbb{Q}$, where $H(\mathbb{Q}, \mathbb{P}) = \mathbb{E}\left(\frac{d\mathbb{Q}}{d\mathbb{P}} log \frac{d\mathbb{Q}}{d\mathbb{P}}\right)$, if $\mathbb{Q} \ll \mathbb{P}$; $= +\infty$, otherwise.

See Föllmer and Schweizer (1991), Schweizer (1994), and Chan (1999) for more details.

14.3.2 The Davis' Approach

Let $\mathcal{A}(x)$ denote the set of all admissible trading strategies with initial wealth x. For $\varphi \in \mathcal{A}(x)$, we denote by X_T^φ the wealth at time T of the trading strategy φ and set

$$V(x) = \sup_{\varphi \in \mathcal{A}(x)} \mathbb{E}[U(X_T^\varphi)],$$

where U is a utility function. For δ, $p > 0$, put

$$W(\delta, p, x) = \sup_{\varphi \in \mathcal{A}(x-\delta)} \mathbb{E}\left[U\left(X_T^\varphi + \frac{\delta}{p}\xi\right)\right].$$

Definition 14.14 If $\hat{\pi}(x)$ is a unique solution to the equation

$$\frac{\partial W}{\partial \delta}(0, p, x) = 0,$$

then $\hat{\pi}(x)$ is called the fair price of ξ at time 0.

The following theorem is due to Davis (1997).

Theorem 14.15 *Suppose that $V'(x) > 0$ on R_+. Then*

$$\hat{\pi}(x) = \frac{\mathbb{E}[U'(X_T^{\varphi^*}\xi)]}{V'(x)},$$

where φ^* is the optimal strategy which attains $V(x)$.

Under certain condition, one can prove that $\hat{\pi}(x) = D_+ p(x)$.

14.3.3 Esscher Transform Approach

This approach to option pricing was first introduced by Gerber and Shiu (1994). Assume that the stock price at time t is $S_t = S_0 e^{X_t}$, where X_t is a Lévy process (i.e., a process with stationary and independent increments). Then for any $h \in R$,

$$E[e^{hX_t}] = E[e^{hX_1}]^t \hat{=} e^{t\psi(h)},$$

where

$$\psi(h) = \exp\left\{hb + \frac{h^2}{2}c + \int_{\mathbb{R}}(e^{hx} - 1 - hxI_{[|x|\leqslant 1]})\nu(dx)\right\},$$

and $M_t^h = \frac{e^{hX_t}}{e^{t\psi(h)}}$ is a martingale with $M_0^h = 1$. So we can define an equivalent probability measure P^h by

$$\frac{dP^h}{dP}\Big|_{\mathcal{F}_t} = M_t^h.$$

The discounted price of the stock is

$$\widetilde{S}_t = e^{-rt} S_t = S_0 e^{X_t - rt}.$$

Thus, P^h is an equivalent martingale measure for $\widetilde{S}_t$ if and only if

$$\widetilde{S}_t M_t^h = S_0 \frac{e^{(1+h)X_t - rt}}{e^{t\psi(h)}}$$

is a P-martingale, i.e., h satisfies the following equation:

$$\psi(h) = \psi(1+h) - r.$$

For further results, see Bühlmann et al. (1996).

References

Ait-Sahalia, Y.: Testing continuous-time models of the spot interest rates. Rev. Financ. Stud. **9**, 385–426 (1996)

Ansel, J.-P., Stricker, C.: Quelques remarques sur un théorème de Yan. In: Séminaire de Probabilités XXIV. Lecturer Notes in Mathematics, vol. 1426, pp. 266–274. Springer, Berlin (1990)

Ansel, J.-P., Stricker, C.: Couverture des actifs contingents. Ann. Inst. H. Poincaré Probab. Stat. **30**, 303–315 (1994)

Arrow, K.: Aspects of the Theory of Risk-Bearing. Yrjö Hahnsson Foundation, Helsinki (1965)

Artzner, P., Delbaen, F., Eber, J.M., Heath, D.: Thinking coherently. Risk **10**, 68–71 (1997)

Artzner, P., Delbaen, F., Eber, J.M., Heath, D.: Coherent measures of risk. Math. Financ. **9**(3), 203–228 (1999)

Bachelier, L.: Théorie de la speculation. Ann. Sci. Ecole Norm. Sup. **17**, 21–86 (1900)

Bajeux-Besnainou, I., Portrait, R.: The numeraire portfolio: a new perspective on financial theory. Eur. J. Financ. **3**, 291–309 (1997)

Bass, R.F.: Probabilistic Techniques in Analysis. Springer, New York (1995)

Bass, R.F.: Doob-Meyer decomposition revisited. Can. Math. Bull. **39**(2), 138–150 (1996)

Baxter, M.: General interest-rate models and the universality of HJM. In: Dempster, A.H., Pliska, S.R. (eds.) Mathematics of Derivative Securities. Publications of the Newton Institute, pp. 315–335. Cambridge University Press, Cambridge (1997)

Bellini, F., Frittelli, M.: On the existence of minimax martingale measures. Math. Financ. **12**, 1–21 (2002)

Biagini, S., Frittelli, M.: Utility maximization in incomplete markets for unbounded processes. Financ. Stoch. **9**, 493–517 (2005)

Björk, T.: Interest rate theory. In: Runggaldier, W. (ed.) Financial Mathematics, pp. 53–122. Springer, Berlin (1997)

Björk, T.: Abitrage Theory in Continous Time. Oxford University Press, Oxford (1998)

Black, F.: Capital market equilibrium with restricted borrowing. J. Bus. **45**, 444–454 (1972)

Black, F., Karasinski, P.: Bond and option pricing when short rates are lognormal. Financ. Anal. J. **47**, 52–59 (1991)

Black, F., Scholes, M.: The pricing of options and corporate liabilities. J. Polit. Econ. **81**, 635–654 (1973)

Black, F., Derman, E., Toy, W.: A one factor model of interest rates and its application to treasury bond options. Financ. Anal. J. **46**, 33–39 (1990)

Bollerslev, T.: Generalized autoregressive conditional heteroscedasticity. J. Econ. **31**, 307–327 (1986)

© Springer Nature Singapore Pte Ltd. and Science Press 2018

J.-A. Yan, *Introduction to Stochastic Finance*, Universitext,

https://doi.org/10.1007/978-981-13-1657-9

Bollerslev, T., Engle, R.F., Nelson, D.B.: ARCH models. In: Engle, R.F., McFadden, D.L. (eds.) The Handbook of Econometrics, vol. 4. Elsevier, Amsterdam (1994)

Brace, A., Gatarek, D., Musiela, M.: The market model of interest rate dynamics. Math. Financ. **7**, 127–154 (1997)

Breeden, D., Litzenberger, R.: Prices of state-contingent claims implicit in options prices. J. Bus. **51**, 621–651 (1978)

Brown, R.G., Schaefer, S.M.: Interest rate volatility and the shape of the term structure. Philos. Trans. R. Soc. Lond. A **347**, 563–576 (1994)

Bühlmann, H., Delbaen, F., Embrechts, P., Shiryaeve, A.N.: No-arbitrage, change of measure and conditional Esscher transforms. CWI Q. Amst. **9**(4), 291–317 (1996)

Chamberlain, G.: A characterization of the distributions that imply mean-variance utility functions. J. Econ. Theory **29**, 185–201 (1983)

Chan, T.: Pricing contingent claims on stocks driven by Lévy processes. Ann. Appl. Probab. **9**(2), 504–528 (1999)

Chen, L.: Interest Rate Dynamics, Derivatives Pricing and Risk Management. Lecture Notes in Economics and Mathematical Systems, vol. 435, Springer, Berlin (1996)

Choquet, G.: Theory of capacities. Ann. Inst. Fourier Grenoble. **5**, 131–295 (1953–54)

Constantinides, G.M., Ingersoll, J.E.: Optimal bond trading with personal taxes. J. Financ. Econ. **13**, 299–335 (1984)

Conze, A., Viswanathan, R.: Path dependent options: the case of lookback options. J. Financ. **46**(5), 1893–1907 (1991)

Corcuera, J.M., Guerra, J., Nualart, D., Schoutens, W.: Optimal investment in a Lévy market. Preprint (2006)

Cox, J.C.: Notes on option pricing I: constant elasticity of variance diffusions. Working paper, Stanford University (1975)

Cox, J.C., Huang, C.F.: Optimal consumption and portfolio policies when asset prices follow a diffusion process. J. Econ. Theory **49**, 33–83 (1989)

Cox, J.C., Huang, C.F.: A variational problem arising in financial economics. J. Math. Econ. **20**, 465–487 (1991)

Cox, J.C., Ross, S.A.: The valuation of options for alternative stochastic processes. J. Financ. Econ. **3**, 145–166 (1976)

Cox, J.C., Ross, S.A., Rubinstein, M.: Option pricing: a simplified approach. J. Financ. Econ. **7**, 229–263 (1979)

Cox, J.C., Ingersoll, J.E., Ross, S.A.: A theory of term structure of interest rates. Econometrica **53**, 385–407 (1985)

Cui, X., Li, D., Yan, J.A.: Classical mean-variance model revisited: pseudo efficiency. J. Oper. Res. Soc. **66**, 1646–1655 (2015)

Cvitanić, J., Schachermayer, W., Wang, H.: Utility maximization in incomplete markets with random endowment. Financ. Stoch. **5**(2), 259–272 (2001)

Dalang, R.C., Morton, A., Willinger, W.: Equivalent martingale measures and no-arbitrage in stochastic securities market models. Stoch. Stoch. Rep. **29**(2), 185–202 (1990)

Dana, R.-A.: A representation result for concave Schur concave functions. Math. Financ. **15**, 615–634 (2005)

Dana, R.A., Jeanblanc, M.: Financial Markets in Continuous Time. Springer, New York (2003)

Davis, M.H.A.: On Margrabe's Formula. Mitsubishi Capital, London (1994)

Davis, M.H.A.: Option pricing in incomplete markets. In: Dempster, A.H., Pliska, S.R. (eds.) Mathematics of Derivative Securities, pp. 216–226. Publications of the Newton Institute, Cambridge University Press, Cambridge (1997)

Delbaen, F.: Coherent measures of risk on general probability spaces. In: Sandmann, K., Schönbucher, P.J. (eds.) Advances in Finance and Stochastics, Essays in Honour of Dieter Sondermann, pp. 1–37. Springer, Berlin (2002)

Delbaen, F., Schachermayer, W.: A general version of the fundamental theorem of asset pricing. Math. Ann. **300**, 463–520 (1994)

Delbaen, F., Schachermayer, W. (1995): The no-arbitrage property under a change of numeraire. Stoch. Stoch. Rep. **53**, 213–226

Delbaen, F., Schachermayer, W.: The Mathematics of Arbitrage. Springer, Berlin/New York (2006)

Delbaen, F., Grandits, P., T. Rheinländer, Samperi, D., Schweizer, M., Stricker, C.: Exponential hedging and entropic penalties. Math. Financ. **12**, 99–123 (2002)

Dellacherie, C., Meyer, P.A.: Probabilities and Potential B. North-Holland, Amsterdam/New York (1982)

Denneberg, D.: Non-additive Measure and Integral. Kluwer Academic, Boston (1994)

Dhaene, J., Vanduffel, S., Goovaerts, M.J., Kaas, R., Tang, Q., Vyncke, D.: Risk measures and comotononicity: a review. Stoch. Model. **22**, 573–606 (2006)

Dobin, J.: Liquidity preference as behavior towards risk. Rev. Econ. Stud. **25**, 65–86 (1958)

Doléans-Dade, C.: Quelques applications de la formule de changement de variables pour les semimartingales. Z. W. **16**, 181–194 (1970)

Dothan U.: On the term structure of interest rates. J. Financ. Econ. **6**, 59–69 (1978)

Duan, J.-C.: The GARCH option pricing model. Math. Financ. **5**, 13–32 (2005)

Dudley, R.M.: Wiener functionals as Itô integrals. Ann. Probab. **5**, 140–141 (1977)

Dudler, R.M.: Real Analysis and Probability. Wadsworth & Brooks/Cole, Pacific Grove (1989)

Duffie, D.: Dynamic Asset Pricing Theory, 2nd edn. Princeton University Press, Princeton (1996)

Duffie, D., Kan, R.: A yield-factor model of interest rates. Math. Financ. **6**, 379–406 (1996)

Dupire, B. (1997): Pricing and hedging with smiles. In: Dempster, A.H., Pliska, S.R. (eds.) Mathematics of Derivative Securities, pp. 103–111. Publications of the Newton Institute, Cambridge University Press, Cambridge

Einstein, A.: On the motion - required by the molecular kinetic theory of heat - of small particles suspended in a stationary liquid. Ann. Phys. **17**(8), 549–560 (1905)

El Karoui, N., Quenez, M.C.: Dynamic programming and pricing of contingent claims in an incomplete market. S. I. S. M. J. Control Optim. **33**, 29–66 (1995)

Embrechts, P., McNeil, A.J., Straumann, D.: Correlation and dependence in risk management: properties and pitfalls. In: Dempster, M., Moffatt, H.K. (eds.) Risk Management: Value at Risk and Beyond. Cambridge University Press, Cambridge (2000)

Engle, R.F.: Autoregressive conditional heteroscedasticity with estimates of the variance of United Kingdom inflation. Econometrica **50**, 987–1007 (1982)

Filipović, D.: A general characterization of one factor affine term structure models. Finance Stoch. **5**, 389–412 (2001)

Fishburn, P.: Utility Theory for Decision Making. Publications in Operations Research, vol. 18. Wiley, New York (1970)

Flesaker, B., Hughston, L.: Positive interest. Risk Mag. **9**(1), 46–49 (1996)

Föllmer, H., Leukert, P.: Efficient hedging: cost versus shortfall risk. Financ. Stoch. **4**, 117–146 (2000)

Föllmer, H., Schied, A.: Convex measures of risk and trading constraints. Financ. Stoch. **6**(4), 429–447 (2002)

Föllmer, H., Schied, A.: Stochastic Finance, an Introduction in Discrete Time, 2nd revised and extended edition. Welter de Gruyter, Berlin, New York (2004)

Föllmer, H., Schweizer, M.: Hedging of contingent claims under incomplete information. In: Davis, M.H.A., Elliott, R.J. (eds.) Applied Stochastic Analysis. Stochastics Monographs, vol. 5, pp. 389–414. Gordon & Breach Science Publishers, New York (1991)

Föllmer, H., Kabanov, Yu.M.: Optional decomposition theorem and Lagrange multipliers. Financ. Stoch. **2**, 69–81 (1998)

Fong, H.G., Vasicek, O.A.: Fixed income volatility management. J. Portf. Manag. Summer **17**, 41–46 (1991)

Freedman, D.: Brownian Motion and Diffusion. Springer, New York (1983)

Frey, R.: Derivative asset analysis in models with level-dependent and stochastic volatility. CWI Quart. (Amsterdam) **10**, 1–34 (1997)

Frittelli, M.: Semimartingales and asset pricing under constraints. In: Dempster, A.H., Pliska, S.R. (eds.) Mathematics of Derivative Securities. Publications of The Newton Institute, pp. 216–226. Cambridge University Press, Cambridge (2000a)

Frittelli, M.: The minimal entropy martingale measure and the valuation problem in incomplete markets. Math. Financ. **10**, 39–52 (2000b)

Frittelli, M., Gianin, E.R.: Putting order in risk measures. J. Bank. Financ. **26**(7), 1473–1486 (2002)

Fujisaki, M., Kallianpur, G., Kunita, H.: Stochastic differential equations for the non-linear filtering problem. Osaka J. Math. **9**, 19–40 (1972)

Fujiwara, T., Miyahara, Y.: The minimal entropy martingale measures for geometric Lévy processes. Financ. Stoch. **7**, 509–531 (2003)

Garman, M., Kohlhagen, S.W.: Foreign currency option values. J. Int. Money Financ. **2**, 231–237 (1983)

Geman, H., Yor, M.: Bessel processes, Asian options and perpetuities. Math. Financ. **3**, 349–375 (1993)

Geman, H., Yor, M.: Pricing and hedging double-barrier options: a probabilistic approach. Math. Financ. **6**(4), 365–378 (1996)

Geman, H., El Karoui, N., Rochet, J.-C.: Changes of numeraire, changes of probability measure and option pricing. J. Appl. Probab. **32**, 443–458 (1995)

Gerber H.U., Shiu, E.S.W.: Option pricing by Esscher transforms. Trans. Soc. Actuaries **46**, 99–191 (1994)

Geske, R.: The valuation of compound options. J. Financ. Econ. **7**, 63–81 (1979)

Glasserman, P., Wu, Q.: Forward and future implied volatility. Int. J. Theor. Appl. Financ. **14**(3), 407–432 (2011)

Goldenberg, D.H.: A unified method for pricing options on diffusion processes. J. Financ. Econ. **29**, 3–34 (1991)

Goldman, M.B., Sosin, H.B., Gatto, M.A.: Path dependent options: by at low, sell at the high. J. Financ. **34**, 1111–1128 (1979)

Goll, T., Ruschendorf, L.: Minimax and minmal distance martingale measures and their relationship to portfolio optimization. Finance Stochast. **5**, 557–581 (2001)

Gray, S, Whaley, R.: Reset put options: valuation, risk characteristics and an application. Aust. J. Manag. **24**, 1–20 (1999)

Hagan, P.S., Kumar, D., Lesniewski, A.S., Woodward, D.E.: Managing smile risk. Wilmott Magazine, pp. 84–108 (2002)

Harrison, M.J., Kreps, D.M.: Martingales and arbitrage in multiperiod securities markets. J. Econ. Theory **29**, 381–408 (1979)

Harrison, M.J., Pliska, S.R.: Martingales and stochastic integrals in the theory of continuous trading. Stoch. Pro. Appl. **11**, 215–260 (1981)

He, H., Pearson, N.D.: Consumption and portfolio policies with incomplete markets and short-sale constraints: the finite-dimensional case. Math. Financ. **1**, 1–10 (1991a)

He, H., Pearson, N.D.: Consumption and portfolio policies with incomplete markets and short-sale constraints: the infinite-dimensional case. J. Econ. Theory **54**, 259–304 (1991b)

Heath, D., Jarrow, A., Morton, A.: Bond pricing and the term structure of interest rates. preprint (1987)

Heath, D., Jarrow, A., Morton, A.: Bond pricing and the term structure of the interest rates; a new methodology. Econometrica **60**(1), 77–105 (1992)

He, S.W., Wang, J.G., Yan, J.A.: Semimartingale theory and stochastic calculus. Science Press/CRC Press, Beijing/Boca Raton (1992)

Heston, S.L.: A closed-form solution for options with stochastic volatility with applications to bond and currency options. Rev. Financ. Stud. **6**(2), 327–343 (1993)

Heyde, C.C., Kou, S.G., Peng, X.H.: What is a good risk measure: bridging the gaps between data, coherent risk measure, and insurance risk measure. Preprint (2006)

Ho, T.S, Lee, S.B.: Term structure movements and pricing interest rate contingent claims. J. Financ. **41**, 1011–1029 (1986)

Hofmann, N., Platen, E., Schweizer, M.: Option pricing under incompleteness and stochastic volatility. Math. Financ. **2**, 153–187 (1992)

Hoogland, J., Neumann, D.: Local scale invariance and contingent claim pricing. Int. J. Theor. Appl. Financ. **4**(1), 1–21 (2001)

Huang, C.F., Litzenberger, R.H.: Foundations for Financial Economics. North-Holland, New York (1988)

Huber, P.J.: Robust Statistics. Wiley, New York (1981)

Huberman, G.: A simplified approach to arbitrage pricing theory. J. Econ. Theory **28**, 1983–1991 (1983)

Hugonnier, J., Kramkov, D., Schachermayer, W.: On utility-based pricing of contingent claims in incomplete markets. Math. Financ. **15**, 203–212 (2005)

Hui, C.H.: One-touch double barrier binary option values. Appl. Financ. Econ. **6**, 343–346 (1996)

Hull, J., White, A.: The pricing of options on assets with stochastic volatilities. J. Financ. **42**, 281–300 (1987)

Hull, J., White, A.: Pricing interest rate derivatives securities. Rev. Financ. Stud. **3**, 573–592 (1990)

Hull, J., White, A.: Bond option pricing on a model for the evolution of bond prices. Adv. Futur. Opt. Res. **6**, 1–13 (1993a)

Hull, J., White, A.: One-factor interest rate models and the valuation of interest rate derivative securities. J. Financ. Quant. Anal. **28**, 235–254 (1993b)

Hull, J., White, A.: Numerical procedures for implementing term structure models II: two factor models. J. Deriv. **2**, 37–47 (1994)

Hull, J., White, A.: Hull-White on Derivatives. Risk Publications, London (1996)

Jacka, S.D.: A martingale representation result and an application to incomplete financial markets. Math. Financ. **2**(4), 239–250 (1992)

Jacod, J.: Calcul stochastique et problémes de martingales. Lecturer Notes in Mathematics, vol. 714, Springer, Berlin/Heidelberg/New York (1979)

Jacod, J., Shiryaev, A.N.: Limit Theorems for Stochastic Process. Springer, Berlin/Heidelberg (1987)

Jacod, J., Yor, M.: Etude des solutions extrémales et representation integrable de solutions pour certains problèmes de martingales. Z.W. **38**, 83–125 (1977)

Jamshidian, F.: Libor and swap market model and measures. Financ. Stoch. **1**, 293–330 (1997)

Jiang, L.: Mathematical Models and Methods of Option Pricing. High Education Press, Beijing (2003, in Chinese)

Jin, H., Yan, J.A., Zhou, X.Y.: Continuous-time mean-risk portfolio selection. Ann. I. H. Poincar – PR **41**, 559–580 (2005)

Jin, H., Markowitz, H.M., Zhou, X.Y.: A note on semivariance. Math. Financ. **16**, 53–61 (2006)

Kabanov, Yu.M., Kramkov, D.O.: No-arbitrage and equivalent martingale measures: an elementary proof of the Harrisson-Pliska theorem. Theory Probab. Appl. **39**, 523–527 (1994)

Kabanov, Yu.M., Stricker, C.: A teachers' note on no-arbitrage criteria. In: Séminaire de Probabilités XXXV. Lecturer Notes in Mathematics, vol. 1755, pp. 149–152. Springer, Berlin/London (2001)

Kallsen, J.: Optimal portfolios for exponential Lévy process. Math. Meth. Oper. Res. **51**, 357–374 (2000)

Karatzas, I.: On the pricing of American options. Appl. Math. Optim. **17**, 37–60 (1988)

Karatzas, I.: Optimization problems in the theory of continuous trading. SIAM J. Control Optim. **27**(6), 1221–1259 (1989)

Karatzas, I.: Lectures on the Mathematics of Finance. CRM Monograph Series, vol. 8. American Mathematical Society, Providence (1997)

Karatzas, I., Shreve, S.E.: Brownian Motion and Stochastic Calculus, 2nd edn. Springer, New York (1991)

Karatzas, I., Shreve, S.E.: Methods of Mathematical Finance. Springer, Berlin/Heidelberg/New York (1998)

Karatzas, I., Lehoczky, J.P., Shreve, S.E.: Optimal portfolio and consumption decisions for a "small investor" on a finite horizon. SIAM J. Control Optim. **27**, 1157–1186 (1987)

Karatzas, I., Lehoczky, J.P., Shreve, S.E., Xu, G.L.: Martingale and duality methods for utility miximization in an incomplete market. SIAM J. Control Optim. **29**, 702–730 (1991)

Kramkov, D.: Optional decomposition of supermartingales and hedging contingent claims in incomplete security markets. Probab. Theory Relat. Fields **105**, 459–479 (1996)

Kramkov, D., Schachermayer, W.: The asymptotic elasticity of utility functions and optimal investment in incomplete markets. Ann. Appl. Probab. **9**, 904–950 (1999)

Kramkov, D., Schachermayer, W.: Necessary and sufficient conditions in the problem of optimal investment in incomplete markets. Ann. Appl. Prob. **13**, 1504–1516 (2003)

Kreps, D.M.: Arbitrage and equilibrium in economies with infinitely many commodities. J. Math. Econ. **8**, 15–35 (1981)

Kunita, H.: Stochastic differential equations and stochastic flows of diffeomorphisms. Lecturer Notes in Mathematics, vol. 1097, pp. 143–303. Springer, Berlin (1984)

Kusuoka, S.: On law invariant coherent risk measures. Adv. Math. Econ. **3**, 83–95 (2001)

Kwok, Y.-K.: Mathematical Models of Financial Derivatives. Springer, Singapore/New York (1998)

Lamberton, D., Lapeyre, B.: Introduction to Stochastic Calculus Applied to Finance. Chapman & Hall, London (1996)

Lépingle, D., Mémin, J.: Sur l'intergrabilité uniforme des martingales exponentielles. Z.W. **42**, 175–203 (1978)

Li, D., Ng, W.-L.: Optimal dynamic portfolio selection: multiperiod mean-variance formulation. Math. Financ. **10**(3), 387–406 (2000)

Liao, S.-L, Wang, C.-W.: The valuation of reset options with multiple strike resets and reset dates. J. Futur. Mark. **23**(1), 87–107 (2003)

Li, P., Xia, J.M., Yan, J.A.: Martingale measure method for expected utility maximization in discrete-time incomplete markets. Ann. Econ. Financ. **2**(2), 445–465 (2001)

Lintner, J.: The valuation of risky assets and the selection of risky investments in stock portfolios and capital budgets. Rev. Econ. Stat. **47**, 13–37 (1965)

Long, J.B.: The numeraire portfolio. J. Financ. Econ. **26**, 29–69 (1990)

Longstaff, F., Schwartz, E.: Interest rate volatility and the term structure: a two-factor general equilibrium model. J. Financ. **47**, 1259–1282 (1992a)

Longstaff, F., Schwartz, E.: A two-factor interest rate model and contingent claim valuation. J. Fixed Income **3**, 16–23 (1992b)

Luo, S.L., Yan, J.A., Zhang, Q.: Arbitrage Pricing Systems in a Market Driven by an Itô Process. In: Yong, J. (eds.) Recent Developments in Mathematical Finance, pp. 263–271. World Scientific Publishing, New Jersey (2002)

Luo, S.L., Yan, J.A., Zhang, Q.: A functional transformation approach to interest rate modeling. In: Cohen, S. N. et al. (eds.) Stochastic Processes, Finance and Control, pp. 303–316. World Scientific Publishing, Singapore/Hackensack (2012)

Madan, D.B.: Purely discontinuous asset price process. In: Jouini, E. et al. (eds.) Option Pricing, Interest Rates and Risk Management, pp. 105–153. Cambridge University Press, Cambridge (2001)

Madan, D.B., Seneta, E.: The variance gamma (V. G.) model for share market returns. J. Bus. **63**, 511–524 (1990)

Madan, D.B., Carr, P., Chang, E.: The variance gamma process and option pricing. Eur. Financ. Rev. **2**, 79–105 (1998)

Maghsoodi, Y.: Solution to the extended CIR term structure and bond option valuation. Math. Financ **6**, 89–109 (1996)

Margrabe, W.: The value of an option to exchange one asset for another. J. Financ. **33**, 177–186 (1978)

Markowitz, H.M.: Portfolio selection. J. Financ. **7**, 77–91 (1952)

Markowitz, H.M.: Portfolio Selection: Efficient Diversification of Investments. Wiley, New York (1959)

Marsh, T., Rosenfeld, E.R.: Stochastic processes for interest rates and equilibrium bond prices. J. Financ. **38**, 635–646 (1983)

McKean, H.P.: Stochastic Integrals. Academic, New York (1969)

Mehra, R., Prescott, E.C.: The equity premium: a puzzle. J. Monet. Econ. **15**(2), 145–161 (1985)

Mel'nikov, A.V., Volkov, S.N., Nechaev, M.L.: Mathematics of Financial Obligations. Translations of Mathematical Monographs, vol. 212. American Mathematical Society, Providence (2002)

Merton, R.C.: Lifetime portfolio selection under uncertainty: the continuous-time case. Rev. Econ. Stat. **51**, 247–257 (1969)

Merton, R.C.: Optimum consumption and portfolio rules in continuous time model. J. Econ. Theory **3**, 373–413 (1971)

Merton, R.C.: An intertemporal capital asset pricing model. Econometrica **41**, 867–887 (1973a)

Merton, R.C.: Theory of rational option pricing. Bell J. Econ. Manag. Sci. **4**, 141–183 (1973b)

Merton, R.C.: Option pricing when underlying stock returns are discontinuous. J. Financ. Econ. **3**, 125–144 (1976)

Meyer, P.A.: A decomposition theorem for supermartingales. Ill. J. Math. **6**, 193–205 (1962)

Meyer, P.A.: Decomposition of supermartingales: the uniqueness theorem. Ill. J. Math. **7**, 1–17 (1963)

Meyer, P.A.: Martingales and Stochastic Integrals I. Lecturer Notes in Mathematics, vol. 284. Springer, Berlin (1972)

Meyer, P.A.: Notes sur les intégrales stochastiques. I-VI. In: Séminaire de Probabilités XIV. Lecturer Notes in Mathematics, vol. 581, pp. 446–481. Springer, Berlin (1977)

Miltersen, K., Sandmann, K., Sondermann, D.: Closed form solutions for term structure derivatives with log-normal interest rates. J. Financ. **52**, 409–430 (1997)

Miyahara, Y.: Minimal entropy martingale measures of jump type processes in incomplete assets markets. Asia-Pacific Financ. Mark. **6**, 97–113 (1999)

Miyahara, Y.: [Geometric Lévy processes & MEMM] pricing model and related estimation problem. Asia-Pacific Financ. Mark. **8**, 45–60 (2001)

Mossin, J.: Equilibrium in capital assets markets. Econometrica **34**, 261–276 (1966)

Musiela, M., Rutkowski, M.: Martingale Methods in Financial Modeling. Springer, Berlin/Heidelberg/New York (1997)

Nelson, D.B.: ARCH as diffusion appriximations, J. of Econ. **45**, 7-38 (1990)

Novikov, A.A.: On a identity for stochastic integrals. Theor. Probab. Appl. **17**, 717–720 (1972)

Oksendal, B.: Stochastic differential equations, 5th edn. Springer, Berlin (1998)

Pearson, N.D., Sun, T.-S.: Exploiting the conditional density in estimating the term structure: an application to the Cox, Ingersoll, and Ross model. J. Financ. **49**, 1279–1304 (1994)

Pliska, S.R.: A stochastic calculus model of continuous trading: optimal portfolios. Math. Oper. Res. **11**, 371–382 (1986)

Pratt, J.: Risk aversion in the small and in the large. Econometrica **32**, 122–36 (1964)

Protter, P.: Stochastic Integration and Differential Equations, 2nd edn. Springer, Berlin (2004)

Rao, K.M.: On decomposition theorem of Meyer. Math. Scand. **24**, 66–78 (1969)

Revuz, D., Yor, M.: Continuous Martingales and Brownian Motion, 3rd edn. Springer, Berlin (1999)

Rockafellar, R.T.: Convex Analysis. Princeton University Press, Princeton (1970)

Rogers, L.C.G.: Equivalent martingale measures and no-arbitrage. Stoch. Stoch. Rep. **51**(1+2), 41–49 (1994)

Rogers, L.C.G.: Which model for the term-structure of interest rates should one use? In: Davis, M.H.A., et al. (eds.) Mathematical Finance, IMA vol. 65, pp. 93–116. Springer, New York (1995)

Rogers, L.C.G., Shi, Z.: The value of an Asian option. J. Appl. Probab. **32**, 1077–1088 (1995)

Ross, S.A.: The arbitrage theory of capital asset pricing. J. Econ. Theory **13**(3), 341–60 (1976)

Rubinstein, M.: Exotic options. In: FORC Conference, Warwick (1992)

Rutkowski, M.: A note on the Flesaker-Hughston model of the term structure of interest rates. Appl. Math. Financ. **4**, 151–163 (1997)

Samuelson, P.: Risk and ambiguity: a fallacy of large numbers. Scientia **57**, 108 (1963)

Schachermayer, W.: A Hilbert space proof of the fundamental theorem of asset pricing in finite discrete time. Insur. Math. Econ. **11**, 249–257 (1992)

Schachermayer, W.: Martingale measures for discrete-time processes with infinite horizon. Math. Financ. **4**, 25–56 (1994)

Schachermayer, W.: Optimal investment in incomplete financial markets. In: Mathematical Finance – Bachelier Congress, 2000, Paris. Springer Finance. Springer, Berlin (2002)

Schmidt, W.M.: On a general class of one-factor models for the term structure of interest rates. Finance Stoch. **1**, 3–24 (1997)

Schweizer, M.: On the minimal martingale measure and the Föllmer- Schweizer decomposition. Stoch. Anal. Appl. **13**, 573–579 (1994)

Sharpe, W.F.: Capital asset prices: a theory of market equilibrium under conditions of risk. J. Financ. **19**, 425–442 (1964)

Song, Y., Yan, J.A.: The representations of two types of functionals on $L^\infty(\Omega, \mathcal{F})$ and $L^\infty(\Omega, \mathcal{F}, P)$. Sci. China, Ser. A Math. **49**(10), 1376–1382 (2006)

Song, Y., Yan, J.A.: Risk measures with co-monotonic subadditivity or convexity and respecting stochastic orders. Insur. Math. Econ. **45**, 459–465 (2009a)

Song, Y., Yan, J.A.: An overview of representation theorems for static risk measures. Sci. China Ser. A Math. **52**(7), 1412–1422 (2009b)

Stricker, C.: Arbitrage et lois de martingale. Ann. Inst. Henri Poincaré **26**, 451–460 (1990)

Stricker, C., Yan, J.A.: Some remarks on the optional decomposition theorem. In: Séminaire de Probabilités XXXII. Lecturer Notes in Mathematics, vol. 1686, pp. 56–66 (1998). Springer, Berlin

Taqqu, M.S., Willinger, W.: The analysis of finite security markets using martingales. Adv. Appl. Probab. **19**, 1–25 (1987)

Tobin, J.: The theory of portfolio selection. In: Hahn, F.H., Brechling, F.P.R. (eds.) The Theory of Interest Rates, pp. 3–51. Macmillan, London (1965)

Vasicek, O.A.: An equilibrium characterization of the term structure. J. Financ. Econ. **5**, 177–188 (1997)

Vecer, J.: A new PDE approach for pricing arithmetic average Asian options. J. Comput. Financ. **4**(4), 105–113 (2001)

Vecer, J.: Unified pricing of Asian options. Risk **15**(6), 113–116 (2002)

von Neumann, J., Morgenstern, O.: Theory of Games and Economic Behavior. Princeton University Press, Princeton (1944)

Wang, J.: Financial Economics. Chinese Renming University Press, Beijing (2006, in Chinese)

Willinger, W., Taqqu, M.S.: Pathwise approximations of processes based on the structure of their filtrations. In: Séminaire de Probabilités XXII. Lecturer Notes in Mathematics, vol. 1321, pp. 542–599. Springer, Berlin (1988)

Wilmott, P., Dewynne, J., and Howison, S.: Option Pricing: Mathematical Models and Computations. Oxford Financial Press, and Oxford (1993)

Xia J.M., Yan J.A. (2000a) Martingale measure method for expected utility maximization and valuation in incomplete markets, unpublished

Xia J.M., Yan J.A. (2000b) The utility maximization approach to a martingale measure constructed via Esscher transform, unpublished

Xia, J.M., Yan, J.A.: Some remarks on arbitrage pricing theory. In: Yong, J. (eds.) Recent Developments. In: Mathematical Finance. World Scientific Publishing, New Jersey (2001)

Xia, J.M., Yan, J.A.: A new look at some basic concepts in arbitrage pricing theory. Sci. China Ser. A **46**(6), 764–774 (2003)

Xia, J.M., Yan, J.A.: Convex duality for optimal investment. AMS/IP Stud. Adv. Math. **42**, 663–678 (2008)

Yamada, T., Watanabe, S.: On the uniqueness of solutions of stochastic differential equations. J. Math. Kyoto Univ. **11**, 155–167 (1971)

Yan, J.A.: Caractérization d'une class d'ensembles convexes de L^1 ou H^1. In: Séminaire de Probabilités XIV. Lecturer Notes in Mathematics, vol. 784, pp. 220–222. Springer, Berlin (1980a)

Yan, J.A.: Propriété de représentation prévisible pour les semimartingales spéciales. Sientia Sinica **23**(7), 803–813 (1980b)

Yan, J.A.: Critères d'intégrabilité uniforme des martingales exponentielles. Acta. Math. Sinica **23**, 311–318 (1980c)

Yan, J.A.: Introduction to Martingales and Stochastic Integrals. Shanghai Science and Technology Press, Shanghai (1981, in Chinese)

Yan, J.A.: A new look at the fundamental theorem of asset pricing. J. Korean Math. Soc. **35**(3), 659–673 (1998)

Yan, J.A.: A numeraire-free and original probability based framework for financial markets. In: Proceedings of the ICM 2002, Beijing, vol. III, pp. 861–871 (2002a)

Yan, J.A.: Semimartingale theory and stochastic calculus. In: Kannan, D., Lakshmikantham, V. (eds.) Handbook of Stochastic Analysis and Applications. Marcel Dekker, Inc., New York, pp. 47–106 (2002b)

Yan, J.A.: Lectures on Measure Theory, 2nd edn. Science Press, Beijing (2009, in Chinese)

Yan, J.A.: A short presentation of Choquet integral. In: Duan, J. et al., (eds.) Recent Development in Stochastic Dynamics and Stochastic Analysis. Interdisciplinary Mathematical Science, vol. 8, pp. 269–291. Wold Scientific, New Jersey (2010)

Yan, J.A., Peng, S.G., Wu, L.M., Fang, S.Z.: Selected Topics in Stochastic Analysis. Science Press, Beijing (1997, in Chinese)

Yan, J.A., Zhang, Q., Zhang, S.: Growth optimal portfolio in a market driven by a jump-diffusion-like process or a Lévy process. Ann. Econ. Financ. **1**, 101–116 (2000)

Yang, Z., Huang, L., Ma, C.: Explicit expressions for the valuation and hedging of the arithmetic Asian option. J. Syst. Sci. Complex. **16**(4), 557–561 (2003)

Yor, M.: On some exponential functionals of Brownian motion. Adv. Appl. Probab. **24**, 509–531 (1992)

Zhang, P.G.: Exotic Options, 2nd edn. World Scientific, New Jersey (2006)

Zhou, X.Y.: Continuous-time asset allocation. In: Melnick, E.L., Everitt, B.S. (eds.) Encyclopedia of Quantitative Risk Analysis and Assessment. Wiley, Hoboken, US (2008)

Index

Printed in the United States
By Bookmasters